PURE AND APPLIED MATHEMATICS

NAYFEH—Perturbation Methods
NAYFEH and MOOK—Nonlinear Oscillations
PETRICH—Inverse Semigroups
PIER—Amenable Locally Compact Groups
***PRENTER—Splines and Variational Methods**
RAO—Measure Theory and Integration
RENELT—Elliptic Systems and Quasiconformal Mappings
RICHTMYER and MORTON—Difference Methods for Initial-Value Problems, 2nd Edition
ROCKAFELLAR—Network Flows and Monotropic Optimization
ROITMAN—Introduction to Modern Set Theory
SCHUMAKER—Spline Functions: Basic Theory
SENDOV and POPOV—The Averaged Moduli of Smoothness
SHAPIRO—Introduction to the Theory of Numbers
***SIEGEL—Topics in Complex Function Theory**
Volume 1—Eliptic Functions and Uniformization Theory
Volume 2—Automorphic Functions and Abelian Integrals
Volume 3—Abelian Functions and Modular Functions of Several Variables
STAKGOLD—Green's Functions and Boundary Value Problems
***STOKER—Differential Geometry**
STOKER—Nonlinear Vibrations in Mechanical and Electrical Systems
TURÁN-
WHITH
ZAUDEI **2nd edition**

LINEAR OPERATORS IN SPACES WITH AN INDEFINITE METRIC

LINEAR OPERATORS IN SPACES WITH AN INDEFINITE METRIC

T. Ya. Azizov

I. S. Iokhvidov
Voronezh State University, USSR

Translated by

E. R. Dawson
University of Dundee

A Wiley-Interscience Publication

JOHN WILEY & SONS
Chichester · New York · Brisbane · Toronto · Singapore

Originally published under the title
Osnovy Teorii Lineynykh Operatorov v Prostranstvakh s Indefinitnoy Metrikoy,
by T. Ya. Asizov and I. S. Iokhvidov, Nauka Publishing House, Moscow

Library of Congress Cataloging-in-Publication Data:

Azizov, T. IA. (Tomas IAkovlevich)
  [Osnovy teorii lineĭnykh operatorov v prostranstvakh s
indefinitnoĭ metrikoĭ. English]
  Linear operators in spaces with an indefinite metric / T.Ya.
Azizov, I.S. Iokhvidov ; translated by E. R. Dawson.
    p.   cm. — (Pure and applied mathematics)
  Translation of: Osnovy teorii lineĭnykh operatorov v
prostranstvakh s indefinitnoĭ metrikoĭ.
  Bibliography:
  Includes index.
  ISBN 0 471 92129 7
  1. Linear operators. I. Iokhvidov, I. S. (Iosif Semenovich)
II. Title. III. Series: Pure and applied mathematics (John Wiley &
Sons)
QA329.2.A9913   1989
515′.7246—dc20

89-14655
CIP

British Library Cataloguing in Publication Data

Azizov, Ia
  Linear operators in spaces with an
  indefinite metric.
  1. Mathematics. Linear operations
  I. Title II. Iukvidov, I. S.
  III. Osnovy teorii lineinykh operatorov
  v prostranstvakh s indefinitnoi metrikoi.
  English
  515.7′246

ISBN 0 471 92129 7

Phototypesetting by MCS Ltd. Salisbury, Great Britain
Printed and bound in Great Britain by Courier International, Tiptree, Essex

CONTENTS

Note: The symbols □ and ■ are used to mark the beginning and end of some section complete in itself.
A reference in the text such as 2.§6.3 refers the reader to §6.3 of Chapter 2.

PREFACE

L. S. Pontryagin's article 'Hermitian operators in spaces with an indefinite metric' appeared in *Izvestiya Acad. Nauk, U.S.S.R.,* more than 40 years ago. The hard war years followed, and probably because of this, the author learnt of the significance of such operators for the solution of certain mechanical problems only from a small footnote to an article by S. L. Sobolev. We mention that Sobolev's article [I] itself appeared only in 1960.

Thus a new branch of functional analysis—the theory of linear operators in spaces with an indefinite metric—takes its origin from 1944, although theoretical physicists had encountered such spaces somewhat earlier (see [XXI]). We emphasize that we are speaking of infinite-dimensional spaces, since linear transformations in finite-dimensional spaces with an indefinite metric were already being studied (Frobenius) at the end of the previous century, although there has been a revival of interest in them and their applications in our own time.

L. S. Pontryagin's work was continued, above all, by M. G. Krein and I. S. Iokhvidov. They axiomatized Pontryagin's approach to complex spaces with an indefinite metric, which they called $\Pi_\varkappa$-spaces, they considered various problems about the geometry of such spaces, and they obtained a number of new facts for operators in $\Pi_\varkappa$. M. G. Krein also studied real spaces Π_1 in connection with the so-called Lorentz transformation and also in connection with the theory of screw curves in infinite-dimensional Lobachevskiy spaces (M. G. Krein [3], see also [XV]). In M. G. Krein's paper [4], he developed an entirely new method, different from Pontryagin's, of proving theorems about invariant subspaces of plus-operators (to use the modern terminology), based on topological theorems about fixed points. I. Iokhvidov [1] suggested the application of the Cayley–Neyman transformation to the study of the connection between different classes of operators in $\Pi_\varkappa$. All this was subsequently summarized first in I. Iokhvidov's 'Kandidat's' dissertation [3], and later in a long article [XIV], written jointly by I. S. Iokhvidov and M. G. Krein, published in 1956. In 1959 its second part [XV] appeared, containing various applications including one to the indefinite problem of moments.

About this time, the theory began to grow not only in depth but also in width. V. P. Potapov became interested in its finite-dimensional analytic aspect, attracting Yu. P. Ginzburg later to working out infinite-dimensional analogues of his results. In these same years it appears that, independently of Soviet mathematicians, R. Nevanlinna in Finland began work on the general problems of an indefinite metric, and after him E. Pesonen and I. S. Louhivaara. Abroad in the fifties and sixties G. Langer (East Germany) joined in the investigations, basing himself on the work of both Soviet and Finnish mathematicians.

At about the same time the first survey was published on the geometry of infinite-dimensional spaces with an indefinite metric, carried out by Ginzburg and I. Iokhvidov [VIII]. This survey already included in part some results from the very beginning of the sixties, which later became years of rapid growth of the whole theory. One after the other appeared papers by Phillips, Langer, Ginzburg, M. Krein, I. Iokhvidov, Naymark, Shmul'yan, Bognar, Kuzhel, and many other mathematicians. The theory found more and more new applications—to dissipative hyperbolic and parabolic systems of differential equations (Phillips), to damped oscillations of infinite-dimensional elastic systems (M. Krein, Langer), to canonical systems of differential equations (M. Krein, Yakubovich, Derguzov), to the theory of group representations (Naymark), *etc.* M. Krein's remarkable lectures on indefinite metric [XVII] appeared (unfortunately in a very small edition), and in 1970 was published the book by Daletskiy and M. Krein [VI] in which the methods of indefinite metric found application and further development. At the end of this period the survey by Azizov and I. Iokhvidov [III] was published (1971), and finally Bognar's first book [V] entirely devoted to indefinite metric appeared in 1974.

In the sixties to the existing Odessa school (M. G. Krein) and the Moscow shcool (M. A. Naymark) occupied in this country with the problems of indefinite metric active new centres were added, among which should first be mentioned Voronezh were the investigations were grouped round I. S. Iokvidov's seminars at the Voronezh State University and the Scientific-Technical Mathematics Institute. Here a large collective of young mathematicians arose (T. Ya. Azizov, V. A. Khatskevich, V. A. Shtraus, E. I. Iokhvidov, E. B. Usvyatsova, Yu. S. Ektov, V. S. Ritsner, S. A. Khoroshavin, and many others), some of whose results are reflected in this monograph.

Now, however, when people literally throughout the world are occupied with the problems of indefinite metric, when courses on individual topics have begun to appear ([II], [XX]), when Azizov and I. Iokhvidov's survey [IV] carried out in 1979 by order of VINITI (the Institute for Scientific and Technical Information) already included about 400 names, there is a pressing need for an interpretation in the form of a monograph on at least the purely theoretical aspect of the accumulated material. Bognar's excellent book [V] illuminated only part of the theory (as it stood up to 1973). This also applies to the detailed monograph of I. Iokhvidov, Krein and Langer [XVI], containing important material, but only on $\Pi_\varkappa$-spaces and operators acting in them. All this provokes an urgent need for a book devoted to the fundamentals of the theory.

The monograph now offered to the reader differs considerably in its contents from the authors' original plan, which was to expound in detail with complete proofs the fundamentals of the theory of linear operators in spaces with an indefinite metric and its applications on approximately the same scale as was planned in their survey [IV]. However, our plan for such an extended treatment had to be abandoned because of the very limited size of the book, forcing us to almost unavoidable abridgement. And now the question arose: what should be sacrificed?

We could in no way sacrifice the rather extensive introductory first chapter setting out the geometry of spaces with an indefinite metric (we remark that in Bognar's work [V] geometry takes up the first half of the book). To do so would have deprived our book of a whole contingent of readers, in particular of students, post-graduate students, and specialists in natural science wishing to investigate the subject-matter but knowing little of its fundamentals. It is also clear from its very title that the book is supposed to shed light sufficiently fully on the central topics of operator theory. We mention at once that after the easy 'warming-up' pace of Chapter 1 on geometry, we allow ourselves a more and more compressed style of exposition in the later chapters, often leaving the reader to think out for himself many of the arguments and their details. With the same purpose many of the auxiliary propositions (sometimes very important ones) have been reduced to the category of exercises and problems, with which each section of the book ends. The range is such that the reader has to go from the quite simple initial problems to increasingly difficult ones. Some of the problems form important logical links in the text and without their solution it will be impossible to understand some of the proofs.

As a result, the aspect which suffered most turned out to be the third part of our intended plan (*cf.* [IV])—the application of the geometry and operator theory to actual problems; we touch on this only in Chapter 4,§3, and then only to a very limited extent, and the choice of applications is subordinated entirely to the authors' tastes.

As a justification for this may serve the fact that, on the one hand, it would in any case have been quite impossible to satisfy straight away all the wide and very varied interests of specialists in dissipative hyperbolic and parabolic systems of differential equations, of specialists in the problem of moments, in the problem of damped oscillations of mechanical systems, in the theory of group representations, of geometers, theoretical physicists and others. On the other hand, a number of monographs, extensive articles and surveys dealing with applications of indefinite metric to the domains mentioned here (and others) have already appeared. It suffices to mention M. Krein's lectures [XVII], Phillips's papers [1], [2], Sobolev's paper [1], Daletskiy and M. Krein's book [VI], the works of M. Krein and Langer [1], [2], I. Iokhvidov and M. Krein [XIV], [XV], of Naymark [2], of Kopachevskiy [1]–[3], the extensive cycle of articles by M. Krein and Langer [4]–[7] which represent an almost complete monograph, Nagy's book [XXI], *etc.*

But even with such a self-limitation we have been unable to include all the topics of the theory itself (not even in the form of problems). This applies first of all to topics in perturbation theory, various realizations of indefinite spaces, and many details relating to the spaces $\Pi_\varkappa$. In particular, we do not touch on the theory of characteristic functions for operators in $\Pi_\varkappa$, or questions connected with rigged spaces, variational theory of eigenvalues, etc.

The whole theory is set out within the framework of Krein and Pontryagin spaces, and therefore many generalizations to Banach spaces or simply normed spaces are also omitted. A little is said about them in [IV].

Each chapter is preceded by a short annotation, saving us the need to review here the structure of the book. At the end of each chapter there are remarks and bibliographical indications, but these in no way pretend to be complete.

Most of the difficult problems are accompanied by hints, sometimes rather detailed. At the same time, for many of the problems (including the difficult ones) no hints are given, but instead the source from which they were borrowed is indicated. Such problems have the purpose of extending the circle of readers and of introducing them to the contemporary state of the theory.

The five chapters of the book are divided into sections. All the special notation is introduced in the form of definitions.

In the citation of references Roman numerals indicate a reference to a monograph or survey listed in the first part of the bibliography. The other references, e.g., 'Jonas [2]' refer to the second part of the list—to Russian and foreign journals and other bibliography—arranged in the alphabetical order of the authors' names.

The book has been written rather quickly and therefore we have been unable to use the critical remarks and advice of people interested. An exception is V. A. Khatskevich, who read the manuscript of the book and made a number of important remarks for which we are extremely grateful to him.

We also thank M. G. Krein, G. Langer, A. V. Kuzhel, Ya. Bognar, N. D. Kopachevskiy, V. S. Shul'man, and many other mathematicians for their interest which manifested itself, in particular, in the systematic exchange of information.

Finally, without the patience, understanding and support shown to us by the members of our families, this book would not have see the light.

T. Ya. Azizov and I. S. Iokhvidov

Voronezh, March 1984.

The unhappy lot has fallen on me of informing the readers of this book that my dear teacher and co-author, Iosif Semeonovich lokhvidov, one of the founders of the theory of spaces with an indefinite metric, an eminent mathematician and a man of fine spirit, died on 1 July 1984. At this time the manuscript had already been put into production and it scarcely needed editorial correction later. This fact is due to the deep pedagogic talent and literary mastery of Iosif Semeonovich, who was the first scrupulous editor of the whole manuscript.

T. YA. AZIZOV

0727JW1...U1046

1 THE GEOMETRY OF SPACES WITH AN INDEFINITE METRIC

This chapter consists of ten sections. Its main substance (§§2–8, 10) is devoted to the geometry of Krein spaces—the principal arena of the action of the linear operators studied in this book. The central item is §8, in which the method of Ginzburg–Phillips angular operators is developed in detail, and some of its applications (for the time being, purely geometrical) are introduced. The whole presentation of the chapter is based on §1, in which we give a short sketch of the theory of linear spaces with an arbitrary indefinite metric (an Hermitian sesquilinear form). The most important particular case of Krein spaces, the Pontryagin spaces Π_k, are studied in particular in §9, though in fact they are encountered in examples much earlier (starting with §4).

In a more compact form than usual the theory is set out of orthogonal projection and the projection completeness of subspaces up to the maximum ones. The question of §10 of the decomposition of a subspace relative to a uniform dual pair also seems to be new.

As regards certain generalizations of Krein spaces and Pontryagin spaces, they are illustrated at the end of §§6 and 9 respectively; but preference is given here to those of them (W-spaces, G-spaces) which are used subsequently in the theory of operators in Krein and Pontryagin spaces.

§1 Linear spaces with an Hermitian form

1 Let $\mathscr{F}$ be a vector space over the field $\mathbb{C}$ of complex numbers, and let a sesquilinear Hermitian form $Q(x, y)$ be given on $\mathscr{F}$, i.e. the mapping $Q: \mathscr{F} \times \mathscr{F} \to \mathbb{C}$ is *linear* in the first argument:

$$Q(\lambda_1 x_1 + \lambda_2 x_2, y) = \lambda_1 Q(x_1, y) + \lambda_2 Q(x_2, y) \qquad (x_1, x_2, y \in \mathscr{F}_1 \lambda_1, \lambda_2 \in \mathbb{C}) \tag{1.1}$$

and *Hermitian symmetric:*

$$Q(y, x) = \overline{Q(x, y)} \qquad (x, y \in \mathscr{F}) \tag{1.2}$$

From (1.2) and (1.1) it follows that

$$q(x, \mu_1 y_1 + \mu_2 y_2) = \bar{\mu}_1 Q(x, y_1) + \bar{\mu}_2 Q(x, y_2) \qquad (x, y_1, y_2 \in \mathscr{F};\ \mu_1, \mu_2 \in \mathbb{C})$$

—the so-called *semi-linearity* (or *anti-linearity*) of the form $Q(x, y)$ in the second argument.

□ *Example 1.1:* Let $\mathscr{F}$ be the vector space over the field $\mathbb{C}$ consisting of all *finite* infinite sequences $x = (\xi_1, \xi_2, \ldots, \xi_n, \ldots)$ of complex numbers (with $\xi_n = 0$ for $n > N_x$) with the natural (co-ordinate-wise) definition of linear operations, and let $\{\alpha_1, \alpha_2, \ldots, \alpha_n, \ldots\}$ be an arbitrary infinite sequence of real numbers ($\alpha_n \in \mathbb{R}$, $n = 1, 2, \ldots$). We define a form $Q(x, y)$ on $\mathscr{F}$ in the following way: if $x = \{\xi_n\}_{n=1}^{\infty} \in \mathscr{F}$, $y = \{\eta_n\}_{n=1}^{\infty} \in \mathscr{F}$, then

$$Q(x, y) = \sum_{n=1}^{\infty} \alpha_n \xi_n \bar{\eta}_n \tag{1.3}$$

It is clear that $Q(x, y)$ satisfies the conditions (1.1) and (1.2), since for each concrete pair $x, y \in \mathscr{F}$ the formally infinite sum on the right hand side of (13) reduces to a finite sum having the usual properties of a sesquilinear form. ■

The continual analogue of Example 1.1 is

□ *Example 1.2:* Let $\mathscr{F}$ be the linear space of all finite, complex-valued, continuous functions defined on the whole real axis $\mathbb{R}$. We introduce the form Q by the formula

$$Q(x, y) = \int_{-\infty}^{\infty} x(t)\overline{y(t)}\, d\sigma(t) \qquad (x, y \in \mathscr{F}), \tag{1.4}$$

where $\sigma(t)$ is an arbitrary fixed real-valued function defined on $\mathbb{R}$ with bounded variation on each finite interval. ■

A Hermitian form $Q(x, y)$ with the properties (1.1) and (1.2) is called a *Q-metric*. We find it convenient to introduce a shorter notation for it:

$$[x, y] \equiv Q(x, y) \qquad (x, y \in \mathscr{F}). \tag{1.5}$$

2 In this section from now on $\mathscr{F}$ is to be understood to be a vector space with a Q-metric $[x, y]$. We remark that the form $Q(x, y)$ is, generally speaking, *indefinite*, i.e. (see (1.5)) the real number $[x, x] = Q(x, x)$ may have either sign. For this reason the Q-metric $[x, y]$ is also called an *indefinite metric*. We introduce the following classification of vectors and *lineals* (i.e. linear subsets) of the space $\mathscr{F}$; at the same time let us agree that throughout the rest of the book the cursive capital letter $\mathscr{L}$ (possibly with indices: $\mathscr{L}^+, \mathscr{L}^-, \mathscr{L}_1, \mathscr{L}^0, \mathscr{L}_\lambda(A)$) shall always denote a lineal.

Definition 1.3: A vector $x(\in\mathscr{F})$ is said to be *positive*, *negative*, or *neutral* depending on whether $[x, x] > 0$, $[x, x] < 0$, or $[x, x] = 0$ respectively.

It is clear, for example, that for the vector $x = \theta$ (the zero vector) we have $[\theta, \theta] = 0$, i.e., θ is a neutral vector; but the reverse implication is, in general, untrue: the neutrality of a vector x *does not imply* that $x = \theta$. The presence or absence in $\mathscr{F}$ of *non-zero* neutral vectors depends on the properties of the Q-metric, a point discussed below in paragraph 4.

Positive (respectively negative) vectors and neutral vectors are combined under the general term *non-negative* (respectively *non-positive*) vectors.

Remark 1.4: As follows from the properties of a Q-metric non-negative, non-positive, and neutral vectors preserve their non-negativeness, non-positiveness, and neutrality respectively on multiplication by an arbitrary scalar $\lambda \in \mathbb{C}$. Positive and negative vectors behave similarly when multiplied by a non-zero scalar λ.

We denote the sets of all positive, negative, and neutral vectors of a space $\mathscr{F}$ respectively by $\mathscr{P}^{++}(\mathscr{F}) \equiv \mathscr{P}^{++}$, $\mathscr{P}^{--}(\mathscr{F}) \equiv \mathscr{P}^{--}$, and $\mathscr{P}^{0}(\mathscr{F}) \equiv \mathscr{P}^{0}$, i.e.

$$\mathscr{P}^{++} = \{x \mid [x, x] > 0\} \qquad \mathscr{P}^{--} = \{x \mid [x, x] < 0\}, \ \mathscr{P}^{0} = \{x \mid [x, x] = 0\}. \tag{1.6}$$

We also denote by

$$\mathscr{P}^{+}(\mathscr{F}) \equiv \mathscr{P}^{+} \equiv \mathscr{P}^{++} \cup \mathscr{P}^{0}, \qquad \mathscr{P}^{-}(\mathscr{F}) \equiv \mathscr{P}^{-} \equiv \mathscr{P}^{--} \cup \mathscr{P}^{0} \tag{1.7}$$

the sets of all non-negative and non-positive vectors in $\mathscr{F}$ respectively. It is clear that $\mathscr{P}^{+} \cap \mathscr{P}^{-1} = \mathscr{P}^{0}$; however, it should not be forgotten that any one of these three sets may, depending on the properties of the Q-metric, reduce to $\{\theta\}$, and the set $\mathscr{P}^{++}$ or $\mathscr{P}^{--}$ may even be void.

Definition 1.5: A lineal λ $(\subset\mathscr{F})$ is said to be *non-negative*, *non-positive*, or *neutral* if $\mathscr{L} \subset \mathscr{P}^{+}$, $\mathscr{L} \subset \mathscr{P}^{-}$, or $\mathscr{L} \subset \mathscr{P}^{0}$ respectively. All these three types of lineal are combined under the general name of *same-definition lineals*. It is clear that neutral $\mathscr{L}$ are simultaneously non-negative and non-positive.

Definition 1.6: A lineal $\mathscr{L}$ is said to be *positive* (*or negative*) if $\mathscr{L} \subset \mathscr{P}^{++} \cup \{\theta\}$ (or $\mathscr{L} \subset \mathscr{P}^{--} \cup \{\theta\}$). Positive and negative lineals are included in the common term '*definite* lineals'. On the other hand, a lineal $\mathscr{L}$ which contains both positive and negative vectors ($\mathscr{L} \cap \mathscr{P}^{++} \neq \emptyset$, $\mathscr{L} \cap \mathscr{P}^{--} \neq \emptyset$) is said to be *indefinite*.

3 Suppose two vector spaces $\mathscr{F}_1$ and $\mathscr{F}_2$ are given with the Q_1-metric and the Q_2-metric respectively:

$$[x, y]_1 = Q_1(x, y) \qquad (x, y \in \mathscr{F}_1);$$
$$[u, v]_2 = Q_2(u, v) \qquad (u, v \in \mathscr{F}_2).$$

Definition 1.7: The spaces $\mathscr{F}_1$ and $\mathscr{F}_2$ are said to be (Q_1, Q_2)-*isometrically isomorphic* if there is a linear bijective mapping[1] $T: \mathscr{F}_1 \to \mathscr{F}_2$ such that

$$[x, y]_1 = [Tx, Ty]_2 \qquad (x, y \in \mathscr{F}_1).$$

From this definition it follows, in particular, that the sets $\mathscr{P}_k^+$, $\mathscr{P}_k^-$, $\mathscr{P}_k^0$, $\mathscr{P}_k^{++}$, $\mathscr{P}_k^{--}$ $(k = 1, 2)$ in the spaces $\mathscr{F}_1$ and $\mathscr{F}_2$ respectively are connected by the relations $\mathscr{P}_2^{\pm} = T\mathscr{P}_1^{\pm}$, $\mathscr{P}_2^0 = T\mathscr{P}_1^0$, $\mathscr{P}_2^{\pm\pm} = T\mathscr{P}_1^{\pm\pm}$. The same operator T, called a (Q_1, Q_2)-*isometric isomorphism*, sets up a one-to-one correspondence between all the non-negative (non-positive), and, in particular, between the positive (negative), lineals of the spaces $\mathscr{F}_1$ and $\mathscr{F}_2$ respectively.

The concept of a (Q_1, Q_2)-*skew-symmetric isomorphism* S between the same two spaces $\mathscr{F}_1$ and $\mathscr{F}_2$ is introduced in a similar way. This linear bijective operator $S: \mathscr{F}_1 \to \mathscr{F}_2$ acts so that

$$[x, y]_1 = -[Sx, Sy]_2 \qquad (x, y \in \mathscr{F}_1).$$

We shall frequently have to use these concepts in the following particular case:

Definition 1.8: If $\mathscr{F}$ is a space with the Q-metric $[x, y]$, then the spare $\mathscr{F}_1$ which is (Q, Q_1)-skew-symmetric to the space $\mathscr{F}$, and which coincides with $\mathscr{F}$ as a set, but which has the Q_1 $(\equiv -Q)$-metric

$$Q_1(x, y) = -Q(x, y) = -[x, y] \qquad (x, y \in \mathscr{F} = \mathscr{F}_1)$$

is called the *anti-space* of $\mathscr{F}$.

4 We return to the question of non-zero, neutral vectors.

1.9 If $\mathscr{L}$ $(\subset \mathscr{F})$ is indefinite, then it contains non-zero, neutral vectors.
□ Let $x, y \in \mathscr{L}$ and $[x, x] > 0$, but $[y, y] < 0$. We consider the function φ of the real variable τ $(-\infty < \tau < \infty)$

$$\varphi(\tau) = [(1-\tau)x + \tau y, (1-\tau)x + \tau y],$$

(i.e., a square trinomial).

Since φ is continuous everywhere on $\mathbb{R}$, and $\varphi(0) = [x, x] > 0$ but $\varphi(1) = [y, y] < 0$, there is a τ_0 $(0 < \tau_0 < 1)$ such that $\varphi(\tau_0) = 0$, i.e., the vector $z = (1-\tau_0)x + \tau_0 y$ $(\in \mathscr{L})$ is neutral. Moreover $z \neq \theta$, because the vectors x and y are linearly independent (see Remark 1.4). ■

Proposition 1.9 cannot be inverted, because non-zero neutral vectors are contained not only in an indefinite lineal but also in any semi-definite (but not definite!) lineal. In this case they play a particular rôle, which will be explained later in §1.6.

---

[1] Throughout the rest of Chapter 1 we adhere to the classical symbolism for mapping (*cf.* [XII], for example); i.e., $T: \mathscr{F}_1 \to \mathscr{F}_2$ means that the mapping T is defined on the whole of $\mathscr{F}_1$, and the corresponding images lie in $\mathscr{F}_2$. Later, starting in Chapter 2, it will be more convenient to us to treat the symbol $T: \mathscr{F}_1 \to \mathscr{F}_2$ in a rather wider sense (see Chapter 2, §1.1).

5 *Definition 1.10:* Vectors x, y ($\in \mathscr{F}$) are said to be *Q-orthogonal* if $[x, y] = 0$. This fact is denoted by the symbol $[\perp]$: $x[\perp]y$.

It is clear from (1.2) and (1.5) that the relation just now introduced is symmetric:

$$x[\perp]y \Leftrightarrow y[\perp]x.$$

Q-orthogonality of sets $\mathscr{M}, \mathscr{N}$ ($\subset \mathscr{F}$) is naturally defined by requiring that $x[\perp]y$ (for all $x \in \mathscr{M}$ and $y \in \mathscr{N}$), and is denoted by $\mathscr{M}[\perp]\mathscr{N}$ ($\Leftrightarrow \mathscr{N}[\perp]\mathscr{M}$). In particular, if $\mathscr{N}$ reduces to a single vector x ($\mathscr{N} = \{x\}$), then we write simply $x[\perp]\mathscr{M}$.

Definition 1.11: The *Q-orthogonal complement of a set* $\mathscr{M}$ ($\subset \mathscr{F}$) is the set

$$\mathscr{M}^{[\perp]} \equiv \{x \mid x[\perp]\mathscr{M}\}.$$

□ From the property (1.1) of a Q-metric $[x, y]$ it follows that $\mathscr{M}^{[\perp]}$ is always a lineal (even if $\mathscr{M}$ is not a lineal). For any sets $\mathscr{M}, \mathscr{N}$ ($\subset \mathscr{F}$) the implication

$$\mathscr{M} \subset \mathscr{N} \Rightarrow \mathscr{M}^{[\perp]} \supset \mathscr{N}^{[\perp]} \tag{1.8}$$

is obvious, and so are the relations

$$\mathscr{M}^{[\perp]} \cap \mathscr{N}^{[\perp]} \subset (\mathscr{M} + \mathscr{N})^{[\perp]}, \tag{1.9}$$

$$(\mathscr{M} \cap \mathscr{N})^{[\perp]} \supset \mathscr{M}^{[\perp]} + \mathscr{N}^{[\perp]}, \tag{1.10}$$

$$\mathscr{M}^{[\perp][\perp]} \supset \mathscr{M}, \tag{1.11}$$

where the sign + between sets means the algebraic sum of the sets $\mathscr{M}$ and $\mathscr{N}$, i.e., the result of addition elementwise of all possible pairs $\mathscr{M} + \mathscr{N} \equiv \{x + y \mid x \in \mathscr{M}, y \in \mathscr{N}\}$, and $\mathscr{M}^{[\perp][\perp]} \equiv (\mathscr{M}^{[\perp]})^{[\perp]}$. We note that if $\mathscr{L}_1, \mathscr{L}_2$ ($\subset \mathscr{F}$) are lineals, then (1.9) can be made more precise:

$$\mathscr{L}_1^{[\perp]} \cap \mathscr{L}_2^{[\perp]} = (\mathscr{L}_1 + \mathscr{L}_2)^{[\perp]}. \quad \blacksquare \tag{1.12}$$

6 *Definition 1.12:* A vector x_0 ($\in \mathscr{L}$) is said to be an *isotropic vector* $\mathscr{L}$ ($\subset \mathscr{F}$) if $x_0 \neq \theta$ and $x_0[\perp]\mathscr{L}$.

It follows from this definition that any non-zero linear combination of isotropic vectors of a lineal $\mathscr{L}$ is again an isotropic vector of $\mathscr{L}$.

Definition 1.13: The linear envelope $\text{Lin}\{x_0\}$ of all the isotropic vectors x_0 ($\in \mathscr{L}$) is called the *isotropic lineal for* $\mathscr{L}$ and is denoted by $\mathscr{L}^0$. In other words,

$$\mathscr{L}^0 \equiv \mathscr{L} \cap \mathscr{L}^{[\perp]}, \tag{1.13}$$

and the equality $\mathscr{L}^0 = \{\theta\}$ indicates the absence of isotropic vectors in $\mathscr{L}$. Speaking rather loosely, we shall sometimes call $\mathscr{L}^0$ the isotropic part of the lineal $\mathscr{L}$.

Definition 1.14: If $\mathscr{L}^0 = \{\theta\}$, the lineal $\mathscr{L}$ is said to be *non-degenerate*, and in the opposite case *degenerate*.

It is easy to see that *every definite lineal $\mathscr{L}$ is non-degenerate;* but an indefinite $\mathscr{L}$ can also be non-degenerate.

□ *Example 1.15:* We return to Example 1.1 and assume additionally that $\alpha_1 = -1$, and $\alpha_n = 1$ $(n \geqslant 2)$. Then the two-dimensional lineal

$$\mathscr{L} = \text{Lin}(e_1, e_2) \ (\subset \mathscr{F}), \text{ where } e_1 = \{1, 0, 0, \ldots\} \text{ and } e_2 = \{0, 1, 0, 0, \ldots\}$$

is indefinite: $[e_1, e_1] = -1$, $[e_2, e_2] = 1$ (see (1.3)). At the same time $\mathscr{L}$ is non-degenerate, for it would follow from the relation $x_0 = \lambda_1 e_1 + \lambda_2 e_2 \,[\perp]\, \mathscr{L}$ (because of (1.3)) that $0 = [x_0, e_1] = \lambda_1$ and $0 = [x_0, e_2] = \lambda_2$, i.e. $x_0 = \theta$. We leave the reader to verify that in this example the whole space $\mathscr{F}$ is also a non-degenerate (and, as we saw, an indefinite) lineal. ■

Examples (but by no means the only ones) of degenerate ideals are all $(\neq \{\theta\})$ semi-definite lineals which are different from definite lineals. To see this we note that the restriction of a Q-metric $[x, y]$ to any semi-definite lineal $\mathscr{L}$ is subject to the *Cauchy–Bunyakovski inequality*:

1.16 If $\mathscr{L}$ is semi-definite, then

$$|[x, y]|^2 \leqslant [x, x][y, y] \qquad (x, y \in \mathscr{L}). \tag{1.14}$$

□ The proof (in the case of $\mathscr{L}$ being non-negative) follows at once from consideration of the Hermitian form (non-negative for all $\xi, \eta \in \mathbb{C}$) $(0 \leqslant) \quad [\xi x + \eta y, \xi x + \eta y] = [x, x]\,|\xi|^2 + [x, y]\xi\bar{\eta} + [y, x]\bar{\xi}\eta + [y, y]\,|\eta|^2$, for which the inequality (1.14) is simply the expression of the non-negativity of its discriminant. In the case of a non-positive $\mathscr{L}$ we arrive at the same conclusion by considering the form $(-[x, y])$. ■

From 1.16 immediately follows the proposition:

1.17 In a semi-definite $\mathscr{L}$ every neutral vector x_0 $(\neq \theta)$ is isotropic. In particular, on a neutral $\mathscr{L}$ the form $[x, y] \equiv 0$.

□ If $(\theta \neq)$ $x_0 \in \mathscr{L}$ and if $[x_0, x_0] = 0$, then for any $y \in \mathscr{L}$ we have, by (1.14), $|[x_0, y]|^2 \leqslant [x_0, x_0][y, y] = 0$, i.e. $x_0 \,[\perp]\, \mathscr{L}$. ■

7 Each of the classes of lineals in $\mathscr{F}$ described in the preceding sections contains certain special lineals which will be of particular interest to us later.

Definition 1.18: A positive $\mathscr{L}$ $(\subset \mathscr{F})$ is said to be *maximal positive lineal* if for any positive $\mathscr{L}_1 \supset \mathscr{L}$ we have $\mathscr{L}_1 = \mathscr{L}$. *Maximal non-negative, maximal negative, maximal non-positive, maximal neutral, and maximal non-degenerative lineals* are defined similarly.

For all the classes of lineals mentioned one important principle holds; we formulate and prove this principle for positive lineals (the formulation and proofs for the other classes are entirely analogous).

Theorem 1.19: (The maximality principle). *Every positive lineal $\mathscr{L}$ is contained in a certain maximal positive lineal $\mathscr{L}_{\max}$.*

□ We consider the set $\mathcal{M} = \{\tilde{\mathscr{L}}\}$ of all positive lineals $\tilde{\mathscr{L}}$ of the space $\mathscr{F}$ which contain $\mathscr{L}$. This set is not empty: $\mathscr{L} \in \mathcal{M}$. We introduce in it a *partial ordering* ($<$) *by inclusion*, i.e., for $\mathscr{L}_1, \mathscr{L}_2 \in \mathcal{M}$, we put $\mathscr{L}_1 < \mathscr{L}_2$ if and only if $\mathscr{L}_1 \subset \mathscr{L}_2$. We shall show that in $\mathcal{M}$ any chain (i.e. *a linearly ordered subset*) is bounded above. Indeed, for such a chain $\mathfrak{Z}$ we obtain the upper bound by forming the union of all elements of the chain, i.e., of all the positive lineals containing $\mathfrak{Z}$ which enter into $\mathfrak{Z}$. Because $\mathfrak{Z}$ is linearly ordered, this union will again be a positive lineal containing $\mathscr{L}$, i.e., it will be an element of the set $\mathcal{M}$, and by construction it contains all the lineals of the chain $\mathfrak{Z}$, i.e., it is the upper bound for $\mathfrak{Z}$. By Zorn's lemma, the set $\mathcal{M}$ contains at least one maximal element $\mathscr{L}_{\max}$, which obviously will indeed be the required maximal positive lineal containing $\mathscr{L}$. ■

Remark 1.20: It will be shown later that in the particular case most important for us, when the space $\mathscr{F}$ with an indefinite metric $[x, y]$ is a so-called Krein space (see §2), realization of the maximality principle for semi-definite lineals does not require the application of Zorn's Lemma, but the construction of the corresponding 'maximal object' can be effectively carried out.

We also remark that any lineal $\mathscr{L}$ ($\subset \mathscr{F}$) can be regarded afresh us a space with Q-metric which contains, therefore, possibly positive, negative lineals and the like, for which in turn the maximality principle operates (locally, i.e. within $\mathscr{L}$).

8 We consider an arbitrary degenerate lineal $\mathscr{L}$ ($\subset \mathscr{F}$) and its isotropic lineal $\mathscr{L}^0$ ($\neq \{\theta^0\}$). As a well-known, there is an infinite set of decompositions of $\mathscr{L}$ into a direct sum

$$\mathscr{L} = \mathscr{L}^0 \dotplus \mathscr{L}_1, \tag{1.15}$$

where $\mathscr{L}_1$ is any (algebraic) complement to $\mathscr{L}^0$. From the fact that (1.15) is a direct sum, i.e. $\mathscr{L}^0 \cap \mathscr{L}_1 = \{\theta\}$, and that all the vectors isotropic for $\mathscr{L}$ are collected in $\mathscr{L}^0$ (*cf.* Definition 1.13), it follows that with any choice of the complement $\mathscr{L}_1$ this lineal is degenerate. Moreover, the following hold:

□ *1.21 In every decomposition* (1.15) *of an arbitrary lineal $\mathscr{L}$, where $\mathscr{L}^0$ is its isotropic lineal, $\mathscr{L}_1$ is the maximal* (*in $\mathscr{L}$*) *non-denegerate lineal.* ■

□ *1.22 Every maximal* (*in $\mathscr{L}$*) *semi-definite lineal $\mathscr{L}^1$ contains the isotropic lineal $\mathscr{L}^0$* (see (1.15)). ■

The decomposition (1.15) also leads to the construction of the factor lineal $\hat{\mathscr{L}} \equiv \mathscr{L}/\mathscr{L}^0$, whose elements (co-sets) will be denoted by $\hat{x}$, $\hat{y}$ and so on. Every class $\hat{x}$ ($\in \hat{\mathscr{L}}$) is defined, with an arbitrary vector $x \in \hat{x}$ ($x \in \mathscr{L}$), as usual by the formula $\hat{x} = x + \mathscr{L}^0$ (i.e. $\hat{x} = \{x + y\}_{y \in \mathscr{L}^0}$). The canonical homomorphism $\mathscr{L} \to \hat{\mathscr{L}}$ can be enriched with additional content if by means of it a certain $\hat{Q}$-metric is induced from $\mathscr{L}$ into $\hat{\mathscr{L}}$ according to the rule

$$\hat{Q}(\hat{x}, \hat{y}) = Q(x, y), \quad \text{where} \quad x \in \hat{x}, y \in \hat{y}. \tag{1.16}$$

It is not difficult to verify that the definition is a proper one, i.e., it does not depend on the chice of the vectors $x \in \hat{x}$, $y \in \hat{y}$.

Introducing again the short notation $[x, y] \equiv Q(x, y)$, and $[\hat{x}, \hat{y}]^\wedge \equiv \hat{Q}(\hat{x}, \hat{y})$, for any $x_1 \in \hat{x}$, $y_1 \in \hat{y}$ we have $x_1 - x = x_0 \in \mathscr{L}^0$, $y_1 - y = y_0 \in \mathscr{L}^0$, so that by virtue of the vectors x_0 and y_0 being isotopic we have $[\hat{x}, \hat{y}]^\wedge = [x, y] = [x + x_0, y + y_0] = [x_1, y_1]$. Hence, it follows immediately that

□ *1.23 The lineal* $\mathscr{L}_1$ *in the decomposition* (1.15) *is* $(Q, \hat{Q})$*-isometrically isomorphic to the factor lineal* $\hat{\mathscr{L}} = \mathscr{L}/\mathscr{L}_0$ *with the Q-metric* (1.16). ■

9 A decomposition (1.15) of an arbitrary lineal $\mathscr{L}$ $(\subset \mathscr{F})$ into an isotropic lineal $\mathscr{L}^0$ and a non-degenerate lineal $\mathscr{L}_1$ gives rise to the natural question concerning the possibility of the further decomposition of a *non-degenerate* lineal $\mathscr{L}_1$, if it is indefinite, into the direct sum of a positive and a negative lineal. Without going deeply into this difficult problem here (it has given rise to an extensive literature (see §6 below)), we shall for the moment establish only some comparatively simple facts.

First of all, although by no means all non-degenerate lineals $\mathscr{L}$ can be decomposed into the direct sum $\mathscr{L} = \mathscr{L}^+ \dotplus \mathscr{L}^-$ of a positive lineal $\mathscr{L}^+$ and a negative lineal $\mathscr{L}^-$ (see Example 1.33 below), the converse is true:

1.24 If $\mathscr{L}$ $(\subset \mathscr{F})$ *admits the decomposition*

$$\mathscr{L} = \mathscr{L}^+ \dotplus \mathscr{L}^{-1} \tag{1.17}$$

into the direct sum of the lineals $\mathscr{L}^+$ $(\subset \mathscr{P}^{++} \cup \{\theta\})$ *and* $\mathscr{L}^-$ $(\subset \mathscr{P}^{--} \cup \{\theta\})$, *then* $\mathscr{L}$ *is non-degenerate.*

□ If $x_0 \in \mathscr{L}$ and $x_0\, [\perp]\, \mathscr{L}$, then decomposing x_0 in accordance with (1.17) into the components $x_0 = x_0^+ + x_0^-$ $(x^\pm \in \mathscr{L}^\pm)$, we obtain

$$0 = [x_0, x_0^+] = [x_0^+, x_0^+] + [x_0^-, x^+] \Rightarrow [x_0^+, x_0^+] = -[x_0^-, x^+],$$
$$0 = [x_0, x_0^-] = [x_0^+, x_0^-] + [x_0^-, x_0^-] \Rightarrow [x_0^-, x_0^-] = -[x_0^+, x_0^-].$$

Hence it is clear that $[x_0^+, x_0^-]$ (and also $[x_0^-, x_0^+]$) is a real number, i.e. $[x_0^+, x_0^-] = [x_0^-, x_0^+]$, and therefore $[x_0^+, x_0^+] = [x_0^-, x_0^-]$, which is possible only if $x_0^+ = x_0^- = \theta$, i.e. $x_0 = \theta$. ■

The Proposition 1.24 can be reversed in a known sense; more precisely, we can prove the following proposition, rather stronger than the converse of 1.24.

1.25 Let the non-degenerate lineal $\mathscr{L}$ *be the algebraic sum* $\mathscr{L} = \mathscr{L}^+ + \mathscr{L}^- = \{x^+ + x^- \mid x^\pm \in \mathscr{L}^\pm\}$, *where* $\mathscr{L}^\pm \subset \mathscr{P}^\pm$. *Then* $\mathscr{L} = \mathscr{L}^+ \dotplus \mathscr{L}^-$ *and* $\mathscr{L}^\pm$ *are maximal in* $\mathscr{P}^\pm(\mathscr{L})$ *respectively. In particular, if either of* $\mathscr{L}^\pm$ *is definite, then it is a maximal definite lineal in* $\mathscr{L}$.

□ Supposing, for example, that there are $\tilde{\mathscr{L}}^+ \subset \mathscr{P}^+(\mathscr{L})$, $\tilde{\mathscr{L}}^+ \supset \mathscr{L}^+$

and $\tilde{x}^+ \in \tilde{\mathscr{L}}^+ \backslash \mathscr{L}^+$, we obtain $\tilde{x}^+ = x^+ + x^-$ $(x^{\pm} \in \mathscr{L}^{\pm})$, i.e. $x^- = \tilde{x}^+ - x^+ \in \tilde{\mathscr{L}}^+ \subset \mathscr{P}^+$, hence $[\bar{x}, \bar{x}] = 0$. But then $x_0 \equiv x^- = \tilde{x}^+ - x^+$ is an isotropic vector both in $\mathscr{L}^-$ and in $\tilde{\mathscr{L}}^+$ (see (1.17)); in particular $x_0 [\perp] \mathscr{L}^{\pm}$, i.e. $x_0 = \bar{x}^+ - x^+ [\perp] \mathscr{L}$, hence $\tilde{x}^+ = x^+ \in \mathscr{L}^+$, which contradicts the choice of $\tilde{x}^+$. ■

In view of Proposition 1.25 the result 1.24 can in turn be regarded as a simple consequence of a more general fact:

1.26 Every $\mathscr{L}$ $(\subset \mathscr{F})$ which contains a lineal $\mathscr{L}^+ \subset \mathscr{P}^{++} \cup \{\theta\}$ $(\mathscr{L}^- \subset \mathscr{P}^{--} \cup \{\theta\})$ and which is maximal in $\mathscr{P}^+(\mathscr{L})$ (in $\mathscr{P}^-(\mathscr{L})$) is non-degenerate.

□ If in $\mathscr{L}$ there is an isotropic vector x_0, then, as a consequence of its neutrality, we have $x_0 \notin \mathscr{L}^+$ (respectively $x_0 \in \mathscr{L}^-$), and $\mathrm{Lin}\{\mathscr{L}^+, x_0\}$ (resp. $\mathrm{Lin}\{\mathscr{L}^-, x_0\}$) is a non-negative (non-positive) lineal which is contained in $\mathscr{L}$ and which contains $\mathscr{L}^+$ $(\mathscr{L}^-)$ as a proper subset. We have obtained a contradiction. ■

A decomposition (1.17) in which $\mathscr{L}^{\pm} \subset \mathscr{P}^{\pm}$, as soon as it exists, generates two linear projectors $P^{\pm}$ which relate to any vector $x \in \mathscr{L}$ its components $x^{\pm} = P^{\pm}x$ $(\in \mathscr{L}^{\pm})$, where $x^+ + x^- = x$. In other words, $P^+ + P^- = I_{\mathscr{L}}$ is the identity operator in $\mathscr{L}$, and $P^+P^- = P^-P^+$ is an operator annihilating $\mathscr{L}$.

Lemma 1.27: *If $\mathscr{L}$ $(\subset \mathscr{L}^{\pm})$, admits a decomposition* (1.17) *in which $\mathscr{L}^-$ $(\mathscr{L}^+)$ is a definite lineal, and if $\mathscr{L}^- \subset \mathscr{P}^+(\mathscr{L})$ $(\mathscr{L}^- \subset \mathscr{P}^-(\mathscr{L}))$, then the mapping P^+: $\mathscr{L}' \to \mathscr{L}^+$ $(P^-: \mathscr{L}' \to \mathscr{L}^-)$ is injective. For arbitrary $\mathscr{L}^{\pm} \subset \mathscr{P}^{\pm}$ these mappings are also injective for $\mathscr{L}' \subset \mathscr{P}^{++}(\mathscr{L}) \cup \{\theta\}$ $(\mathscr{L}' \subset \mathscr{P}^{--}(\mathscr{L}) \cup \{\theta\})$.*

□ Suppose, for example, that $\mathscr{L}' \subset \mathscr{P}^+(\mathscr{L})$ and that for some $x \in \mathscr{L}'$ we have $P^+x = \theta$. Then $x = P^+x + P^-x = P^-x \in \mathscr{L}^-$; hence $x = \theta$. In the case $\mathscr{L}' \subset \mathscr{P}^{++}(\mathscr{L}) \cup \{\theta\}$, if for $x \in \mathscr{L}'$ we have $P^+x = \theta$, then $x = P^+x + P^-x = P^-x = \theta$, because P^-x is non-positive, and x is positive or equal to zero. ■

Corollary 1.28: *Under the conditions of Lemma* 1.27 *for maximal non-negativity (maximal non-positivity) in $\mathscr{L}$ of the lineal $\mathscr{L}' \subset \mathscr{P}^+(\mathscr{L})$ $(\mathscr{L}' \subset \mathscr{P}^-(\mathscr{L}))$ it is sufficient that* $P^+\mathscr{L}' = \mathscr{L}^+$ $(P^-\mathscr{L}' = \mathscr{L}^-)$.

□ Suppose, for example, that $\mathscr{L}' \subset \mathscr{P}^+(\mathscr{L})$ and that $P^+\mathscr{L}' = \mathscr{L}^+$. If also $\mathscr{L}_1 \subset \mathscr{P}^+(\mathscr{L})$ and $\mathscr{L}_1 \supset \mathscr{L}'$, then $\mathscr{L}_1 = \mathscr{L}'$; for otherwise for any $x_1 \in \mathscr{L}_1 \backslash \mathscr{L}'$ we would have $P^+x_1 = x_1^+ \in \mathscr{L}^+$ and at the same time $x_1 = P^+x$ for some $x \in \mathscr{L}'$ $(\subset \mathscr{L}_1)$, i.e., $P^+(x_1 - x) = \theta$. But $x_1 - x \in \mathscr{L}_1$ and therefore by Lemma 1.27 $x_1 = x \in \mathscr{L}'$, contrary to the choice of x_1. ■

10 We shall return later in paragraph 11 to decompositions of the form (17) considering them in another light, motivated by the concepts introduced in the present paragraph.

Definition 1.29: $\mathscr{L}_1, \mathscr{L}_2$ $(\subset \mathscr{F})$ are said to be *skewly linked* if $\mathscr{L}_1 \cap \mathscr{L}_2^{[\perp]} = \mathscr{L}_2 \cap \mathscr{L}_1^{[\perp]} = \{\theta\}$. To indicate that $\mathscr{L}_1$ and $\mathscr{L}_2$ are skewly linked we shall write $\mathscr{L}_1 \# \mathscr{L}_2$.

1.30 In order that $\mathscr{L}_1$ and $\mathscr{L}_2$, of which at least one is neutral, shall be skewly linked it is necessary that two conditions hold: a) $\mathscr{L}_1 \cup \mathscr{L}_2 = \{\theta\}$, b) $\mathscr{L}_1 \dot{+} \mathscr{L}_2$ be non-degenerate. If $\mathscr{L}_1$ and $\mathscr{L}_2$ are both neutral, these conditions are also sufficient.

□ Suppose, for example, that $\mathscr{L}_1 \subset \mathscr{P}^0$. If $\mathscr{L}_1 \# \mathscr{L}_2$, then since (see Proposition 1.17) $\mathscr{L}_1 \subset \mathscr{L}_1^{[\perp]}$, condition a) follows from $\mathscr{L}_2 \cap \mathscr{L}_1^{[\perp]} = \{\theta\}$. Further, if $x_0 \in \mathscr{L}_1 \dot{+} \mathscr{L}_2$, then $x_0 = x_1 + x_2$ $(x_k \in \mathscr{L}_k, k = 1, 2)$. If furthermore $x_0 [\perp] \mathscr{L}_1 \dot{+} \mathscr{L}_2$, then, in particular, $x_0 \in \mathscr{L}_1^{[\perp]}$ and $x_1 \in \mathscr{L}_1 \subset \mathscr{L}_1^{[\perp]}$, so that $x_2 \in \mathscr{L}_1^{[\perp]}$ also. But $x_2 \in \mathscr{L}_2$, and so $x_2 = \theta$. Hence $x_1 = x_0 [\perp] \mathscr{L}_2$, i.e., $x_1 \in \mathscr{L}_1 \cap \mathscr{L}_2^{[\perp]}$, and therefore $x_1 = \theta$. Thus $x_0 = \theta$, and condition b) has been proved.

Conversely, suppose that $\mathscr{L}_1, \mathscr{L}_2 \subset \mathscr{P}^0$ and that conditions a) and b) hold. We note that in this case condition a) automatically follows from condition b), which can be replaced by the even weaker condition b′) *the algebraic sum* $\mathscr{L}_1 + \mathscr{L}_2 = [x_1 + x_2 \mid x_1 \in \mathscr{L}_1, x_2 \in \mathscr{L}_2\}$ *be non-degenerate*. Since now $\mathscr{L}_1 \subset \mathscr{L}_1^{[\perp]}$ and $\mathscr{L}_2 \subset \mathscr{L}_2^{[\perp]}$, any non-null vector from $\mathscr{L}_1 \cap \mathscr{L}_2^{[\perp]}$ or $\mathscr{L} \cap \mathscr{L}_1^{[\perp]}$ would be isotropic for $\mathscr{L}_1 + \mathscr{L}_2$, contrary to condition b′). ■

The simplest example of skewly linked lineals are two arbitrary, *one-dimensional* lineals $\mathscr{L}_1 = \text{Lin}\{x_1\}$ and $\mathscr{L}_2 = \text{Lin}\{x_2\}$, if $[x_1, x_2] \neq 0$ $(x_1, x_2 \in \mathscr{F})$. The fact that in this example the dimensions of $\mathscr{L}_1$ and $\mathscr{L}_2$ are the same is not accidental, as the following lemma shows.

Lemma 1.31: *If* $\mathscr{L}_1 \# \mathscr{L}_2$ *and* $0 < \dim \mathscr{L}_1 = m < \infty$, *then* $\dim \mathscr{L}_1 = \dim \mathscr{L}_2$ *and for any basis* $\{e_j\}_1^m$ *in* $\mathscr{L}_1$ *a basis* $\{f_k\}_1^m$ *can be chosen in* $\mathscr{L}_2$ *such that* $[e_j, f_k] = \delta_{jk}$ $(k, k = 1, \ldots, m)$. (*'Q-biorthogonality'*).

□ Choose an arbitrary basis $\{e_j\}_1^m$ in $\mathscr{L}_1$ and suppose at first that $\dim \mathscr{L}_2 > m$. Let $x_1, \ldots, x_m, x_{m+1}$ be linearly independent vectors in $\mathscr{L}_2$. Consider the system of m linear homogeneous equations

$$\xi_1 [x_1, e_j] + \cdots + \xi_m [x_m, e_j] + \xi_{m+1} [x_{m+1}, e_j] = 0 \qquad (j = 1, 2, \ldots, m) \tag{1.18}$$

in the $m + 1$ unknowns $\xi_1, \ldots, \xi_m, \xi_{m+1}$; it has a non-trivial solution $\xi_1^0, \ldots, \xi_m^0, \xi_{m+1}^0$, i.e, the vector $(\theta \neq)\ x_0 = \sum_{k=1}^m \xi_k^0 x_k \in \mathscr{L}_2 \cap \mathscr{L}_1^{[\perp]}$, contrary to the condition. Therefore $\dim \mathscr{L}_2 \leqslant \dim \mathscr{L}_1 = m$, and by the same argument $\dim \mathscr{L}_1 \leqslant \dim \mathscr{L}_2$, i.e., $\dim \mathscr{L}_1 = \dim \mathscr{L}_2 = m$.

Further, since $\mathscr{L}_1 \in \mathscr{L}_2^{[\perp]} = \{\theta\}$, a vector f_1 in $\mathscr{L}_2$ can be found such that $[e_1, f_1] = 1$, and if $m = 1$, then the construction of the Q-biorthogonal bases $\{e_1\}$ and $\{f_1\}$ is complete. But if $m > 1$, then we first choose $(\theta \neq) f_1 \in \mathscr{L}_2$ such that $[e_2, f_1] = [e_3, f_1] = \cdots = [e_m, f_1] = 0$. This is possible, since $m = \dim \mathscr{L}_2 > \dim \text{Lin}\{e_2, e_3, \ldots, e_m\} = m - 1$ and the construction of f_1 is carried out by the method used at the beginning of the proof. Since $\mathscr{L}_2 \cap \mathscr{L}_1^{[\perp]} = \{\theta\}$, so $[e_1, f_1] \neq 0$ and we may assume $[e_1, f_1] = 1$. Then we choose a vector $(\theta \neq)\ f_2 \in \mathscr{L}_2$ such that $[e_1, f_2] = [e_3, f_2] = \cdots = [e_m, f_2] = 0$ and $[e_2, f_2] = 1$, and so on until the system $\{f_k\}_1^m$ is completely constructed. Linear independence (i.e., basicity in $\mathscr{L}_2$) of this system easily follows from the condition of Q-biorthogonality. ■

Corollary 1.32: *If $\mathscr{L}_1 \# \mathscr{L}_2$ and $\mathscr{L}_1, \mathscr{L}_2$ are finite-dimensional, then $\mathscr{L}_1 \dotplus \mathscr{L}_2^{[\perp]} = \mathscr{F}$.*

□ When $\mathscr{L}_1 = \mathscr{L}_2 = \{\theta\}$, the assertion follows from the fact that $\mathscr{L}_2^{[\perp]} = \mathscr{F}$. But if (see Lemma 1.31) $\dim \mathscr{L}_1 = \dim \mathscr{L}_2 = m \geqslant 1$, then, having chosen Q-biorthogonal bases $\{e_j\}_1^m$ and $\{f_j\}_1^m$ in $\mathscr{L}_1$ and $\mathscr{L}_2$ respectively, for any $x \in \mathscr{F}$ we put $x_1 = \sum_{j=1}^m [x, f_j] e_j$. Then $x_1 \in \mathscr{L}_1$ and $x - x_1 [\perp] f_k$ $(k = 1, 2, \ldots, m)$, i.e., $x_2 = x - x_1 \in \mathscr{L}_2^{[\perp]}$. ■

11 The proof in Lemma 1.31 of the coincidence of the (linear) dimensions of skewly linked $\mathscr{L}_1$ and $\mathscr{L}_2$ is based essentially on the finite dimensionality of $\mathscr{L}_1$ and $\mathscr{L}_2$. As regards infinite-dimensional $\mathscr{L}_1$ and $\mathscr{L}_2$, their linear isomorphism does not always follow from $\mathscr{L}_1 \# \mathscr{L}_2$.

□ *Example 1.33:* Let $\mathscr{F}$ be a linear space of infinite-dimensional (in both directions) *finite (on the left)* sequences $x = \{\xi_j\}_{j=-\infty}^{\infty}$, $\xi_j = 0$ for $j \leqslant j_0(x)$. We introduce an indefinite metric in $\mathscr{F}$ by the formula

$$[x, y] = \sum_{j=-\infty}^{\infty} \xi_j \bar{\eta}_{-j-1}, \tag{1.19}$$

where

$$x = \{\xi_j\}_{j=-\infty}^{\infty}, \qquad y = \{\eta_j\}_{j=-\infty}^{\infty} \in \mathscr{F}.$$

We now consider $\mathscr{L}_1$ $(\subset \mathscr{F})$ consisting of vectors x with $\xi_{-1} = \xi_{-2} = \cdots = 0$. It follows from (1.19) that for such vectors we have $[x, x] = 0$, i.e. $\mathscr{L}_1 \subset \mathscr{P}^0$. Similarly we define $\mathscr{L}_2$ as the set of all y $(\in \mathscr{F})$ with $\eta_0 = \eta_1 = \cdots = 0$. Again from (1.19) we have $[y, y] = 0$, so that $\mathscr{L}_2$ is also neutral. It is clear that: a) $\mathscr{L}_1 \cap \mathscr{L}_2 = \{\theta\}$, and b) $\mathscr{L}_1 \dotplus \mathscr{L}_2 = \mathscr{F}$ is non-degenerate, and so by 1.30 $\mathscr{L}_1 \# \mathscr{L}_2$. At the same time $\mathscr{L}_1$ and $\mathscr{L}_2$ are not isomorphic, because $\mathscr{L}_1$ is isomorphic to the space of all sequences $\{\xi_0, \xi_1, \ldots\}$, but $\mathscr{L}_2$ is isomorphic to the space of all finite sequences $\{\eta_{-1}, \eta_{-2}, \ldots\}$, which has, as is well known, a smaller (namely, a countable) linear dimensionality.

At the same time we show that $\mathscr{F}$ is an example of a *non-degenerate space which does not admit a decomposition of the form* (1.17). For, suppose that $\mathscr{F} = \mathscr{L}^+ \dotplus \mathscr{L}^-$, where $\mathscr{L}^+ \subset \mathscr{P}^{++} \cup \{\theta\}$, and $\mathscr{L}^- \subset \mathscr{P}^{--} \cup \{\theta\}$.

Consider the mapping T_2: $\mathscr{L}^+ \to \mathscr{L}_2$, which to every vector $x = \{\xi_j\}_{j=-\infty}^{\infty}$ $(\in \mathscr{L}^+)$ relates the vector $T_2 x = \{\ldots, \xi_{-2}, \xi_{-1}, 0, 0, \ldots\}$ $(\in \mathscr{L}_2)$. This linear mapping is injective, because by (1.19) the equality $T_2x = \theta$ implies the equality $[x, x] = 0$, and so $x = \theta$ (because $\mathscr{L}^+$ is positive). The injectivity of the mapping T_2: $\mathscr{L}^- \to \mathscr{L}_2$ is proved similarly. But this means that the linear dimensions of the lineals $\mathscr{L}^+$ and $\mathscr{L}^-$ do not exceed the linear dimension of the lineal $\mathscr{L}_2$, and therefore they are not more than countable, which is impossible because their direct sum $\mathscr{L}^+ \dotplus \mathscr{L} = \mathscr{F} = \mathscr{L}_1 \dotplus \mathscr{L}_2$ is isomorphic to the lineal $\mathscr{L}_1$ of higher dimension. ■

We point out that the foregoing argument contains essentially the proof of the following (abstract) theorem:

Theorem 1.34: *If a non-degenerate spare $\mathscr{F}$ with an indefinite metric is the direct sum $\mathscr{F} = \mathscr{L}_1 \dotplus \mathscr{L}_2$ of two non-isomorphic lineals $\mathscr{L}_1, \mathscr{L}_2 \subset \mathscr{P}^0$, then it does not admit decomposition into the direct sum $\mathscr{F} = \mathscr{L}^+ \dotplus \mathscr{L}^-$ of two definite lineals $\mathscr{L}^+$ and $\mathscr{L}^-$.*

□ The proof is the same as in Example 1.33 if we consider that the decomposition $\mathscr{F} = \mathscr{L}_1 \dotplus \mathscr{L}_2$ generates projectors P_k: $\mathscr{F} \to \mathscr{L}_k$ $(k = 1, 2)$, $P_1 + P_2 = I_{\mathscr{F}}$, and use again that one of these two projectors which correspond to the ideal (let us say, $\mathscr{L}_2$) of lesser dimension (than $\mathscr{L}_1$). ■

12 In conclusion we return to the situation in paragraph 3 when there are two complex linear spaces $\mathscr{F}_1$ and $\mathscr{F}_2$ with a Q_1-metric and a Q_2-metric $[\cdot, \cdot]_1$ and $[\cdot, \cdot]_2$ given on them respectively, and a linear mapping T: $\mathscr{F}_1 \to \mathscr{F}_2$. However, instead of the stringent requirements of isomorphism and (Q_1, Q_2)-isometricity here a more 'liberal' condition will be imposed on T which, in the notation of paragraph 3, reads $T\mathscr{P}_1^0 \subset \mathscr{P}_2^+$.

Lemma 1.35: *Suppose that the Q_1-metric in the space $\mathscr{F}_1$ is known to be indefinite, the Q_2-metric in $\mathscr{F}_2$ is arbitrary, and the linear mapping T: $\mathscr{F}_1 \to \mathscr{F}_2$ has the property $T\mathscr{P}_1^0 \subset \mathscr{P}_2^+$. Then, for all $y \in \mathscr{P}_1^{--}$ and $z \in \mathscr{P}_1^{++}$,*

$$\frac{[Ty, Ty]_2}{[y, y]_1} \leqslant \frac{[Tz, Tz]_2}{[z, z]_1}. \tag{1.20}$$

□ Restricting ourselves to 'normalized' vectors y: $[y, y]_1 = -1$ and z: $[z, z]_1 = 1$, we rewrite the relation (1.20) as $-[Ty, Ty]_2 \leqslant [Tz, Tz]_2$, and we assume the contrary. Namely, suppose that for some $y_0 \in \mathscr{F}_1$ with $[y_0, y_0]_1 = -1$ and $z_0 \in \mathscr{F}_1$ with $[z_0, z_0]_1 = 1$ we have $-[Ty_0, Ty]_2 > [Tz_0, Tz_0]_2$. We consider the vector $x_0 = \varepsilon y_0 + z_0$ where ε ($|\varepsilon| = 1$) is, for the time being, an arbitrary parameter. We have

$$[x_0, x_0]_1 = 2\,\mathrm{Re}(\varepsilon[y_0, z_0]_1), \qquad [Tx_0, Tx_0]_2 \leqslant 2\,\mathrm{Re}(\varepsilon[Ty_0, Tz_0]_2).$$

But since $\varepsilon = e^{i}\varphi$ $(0 \leqslant \varphi < 2\pi)$, so, by choosing the argument φ suitably, we can always arange that $\mathrm{Re}(\varepsilon[y_0, z_0]_1) = 0$ and $\mathrm{Re}(\varepsilon[Ty_0, Tz_0]_2) \leqslant 0$. For such an ε we obtain $[x_0, x_0]_1 = 0$ and $[Tx_0, Tx_0]_2 < 0$, which contradicts the condition $T\mathscr{P}_1^0 \subset \mathscr{P}_1^+$. ■

Corollary 1.36: *Under the conditions of Lemma* 1.35 *the finite limits*

$$\mu_+(T) \equiv \inf_{x \in \mathscr{P}_1^{++}} \frac{[Tx, Tx]_2}{[x, x]_1} \; (> -\infty), \; \mu_-(T) \equiv \sup_{x \in \mathscr{P}_1^{--}} \frac{[Tx, Tx]_2}{[x, x]_1} \; (< +\infty)$$

exist. Moreover, $\mu_-(t) \leqslant \mu_+(t)$ *and for any* μ *with* $\mu_-(T) < \mu < \mu_+(T)$ *the inequality*

$$[Tx, Tx]_2 > \mu[x, x]_1 \tag{1.21}$$

holds for all $x \in \mathscr{F}_1$.

□ For the finiteness of the limits $\mu_+(T)$ and $\mu_-(T)$ and the inequality $\mu_-(T) \leqslant \mu_+(T)$ follow immediately from (1.20). Further, for $x \in \mathscr{P}_1^0$ the inequality (1.21) follows from the condition $T\mathscr{P}_1^0 \subset \mathscr{P}_2^+$. In the remaining cases when $x \in \mathscr{P}_1^{++} \cup \mathscr{P}_1^{--}$, it is easily seen from the chain of inequalities (*cf.* (1.20))

$$(\mu_-(T) =) \; \sup_{x \in \mathscr{P}_1^{--}} \frac{[Tx, Tx]_2}{[x, x]_1} \leqslant \mu \leqslant \inf_{x \in \mathscr{P}_1^{++}} \frac{[Tx, Tx]_2}{[x, x]_1} \qquad (= \mu_+(T)).$$

■

Exercises and problems

1 Verify that for the form $[x, y]$ defined in para. 1 the normalization formula

$$[x, y] = \frac{1}{4}[x + y, x + y] - \frac{1}{4}[x - y, x - y] + \frac{i}{4}[x + iy, x + iy] - \frac{i}{4}[x - iy, x - iy] \qquad (x, y \in \mathscr{F})$$

holds.

2 Consider the linear space $\mathscr{F} = C[-1, 1]$ of all complex-valued continuous functions defined on the interval $[-1, 1]$. Verify that the form

$$[x, y] = \int_{-1}^{1} x(t)\overline{y(-t)}\, dt \qquad (x, y \in \mathscr{F}) \tag{1.22}$$

is Hermitian and defines an indefinite metric on $\mathscr{F}$. Give examples of positive and negative vectors in $\mathscr{F}$.

3 If a space $\mathscr{F}$ with Q-metric $[x, y]$ contains at least one positive (negative) vector, then $\mathscr{F} = \mathscr{P}^{++} + \mathscr{P}^{++}$ ($\mathscr{F} = \mathscr{P}^{--} + \mathscr{P}^{--}$), where the symbol denotes the algebraic sum of sets (see M. Krein and Shmul'yan [2]).
Hint: For a positive vector $x_+ \in \mathscr{F}$ (for instance) and an arbitrary $x \in \mathscr{F}$ consider the quadratic form $[x + \alpha x_+, x + \alpha x_+]$ for large $\alpha \in \mathbb{R}$.

4 In order that $\mathscr{F}$ (see Exercise 3) should be indefinite it is necessary that not one of the sets $\mathscr{P}^{++}, \mathscr{P}^{--}, \mathscr{P}^{+}, \mathscr{P}^{-}$ should be a lineal, and it is sufficient that at least neither of the sets $\mathscr{P}^{+}, \mathscr{P}^{-}$ should not be a lineal. (*Cf.* M. Krein & Shmul'yan [2].)
Hint: Use the result of Exercise 3.

5 Prove that for any set $\mathcal{M}\ (\subset \mathcal{F})$ it is always true that $\mathcal{M}^{[\perp][\perp][\perp]} = \mathcal{M}^{[\perp]}$ (E. Scheibe [1]).

6 Prove that the relation (1.12) holds not only for lineals but for any sets $\mathcal{M}_1, \mathcal{M}_2\ (\subset \mathcal{F})$ provided only that they both contain the vector θ.

7 If $\mathcal{L} \subset \mathcal{P}^0$, then $\mathcal{L}^{[\perp][\perp]} \subset \mathcal{P}^0$ ([V]).

8 If $\mathcal{L}$ is degenerate, then $\mathcal{L}^{[\perp][\perp]}$ is also degenerate ([V]). A similar proposition for non-degenerate lineals is false.

9 If $\mathcal{L} \subset \mathcal{P}^+$, then $\mathcal{L}^{[\perp][\perp]}$ may turn out to be indefinite (and may even coincide with the whole space) ([V]).

10 Prove the converse of Proposition 1.17.
Hint: Use the decomposition (1.15) and Proposition 1.9.

11 Obtain the converse of Proposition 1.21.

12 Verify that the inequality (1.21) may fail when $\mu \notin [\mu_-(T), \mu_+(T)]$; construct an example.

§2 Krein spaces (axiomatics)

1 We consider a linear space $\mathcal{F}$ with a Q-metric $Q(x, y) = [x, y]$. Suppose that $\mathcal{F}$ admits decomposition into the direct sum of a positive ($\mathcal{F}^+$) and a negative ($\mathcal{F}^-$) lineal: $\mathcal{F} = \mathcal{F}^+ \dotplus \mathcal{F}^-$, from which it follows by 1.24 that the whole space $\mathcal{F}$ is non-degenerate.

We take the following step and suppose that $\mathcal{F}^+ [\perp] \mathcal{F}^-$, and so we can write

$$\mathcal{F} = \mathcal{F}^+ [\dotplus] \mathcal{F}^{-1}. \tag{2.1}$$

where the symbol $[\dotplus]$ denotes the *Q-orthogonal direct sum*.

Definition 2.1: The decomposition (1.21) is caled a *canonical decomposition of the space $\mathcal{F}$.*

The lineals $\mathcal{F}^+$ and $\mathcal{F}^-$ in (2.1) are pre-Hilbert spaces: $\mathcal{F}^+$ with the scalar product (a positive-definite form) $[x, y]$ $(x, y \in \mathcal{F}^+)$, and $\mathcal{F}^-$ with the scalar product $(-[x, y])$ $(x, y \in \mathcal{F}^{-1})$.

Definition 2.2: A space $\mathcal{F}$ with a Q-metric $[x, y]$, which admits a canonical decomposition (2.1) in which $\mathcal{F}^+$ and $\mathcal{F}^-$ are *complete, i.e. Hilbert, spaces* relative to the norms $\|x\| = [x, x]^{1/2} (x \in \mathcal{F}^+)$ and $\|x\| = (-[x, x]^{1/2})\ (x \in \mathcal{F})$ respectively, is called a *Krein space*.

Remark 2.3: Here that case in which one or both of the spaces $\mathcal{F}^{\pm}$ is *finite-dimensional* is not excluded. §9 is devoted entirely to this important particular case. But if, in particular, $\mathcal{F}^- = \{\theta\}$ (or $\mathcal{F}^+ = \{\theta\}$), i.e., the whole of $\mathcal{F}$ is definite, then this Krein space is simply a Hilbert space with the scalar product $[x, y]$ (or $-[x,]$).

2 In considering a Krein space, we shall from the very first time start from the fixed canonical decomposition (2.1) which features in its definition (2.2), although in the case where $\mathscr{F}$ is indefinite, as may easily be understood (see theorem 8.17 below), this decomposition is *not unique*.

A canonical decomposition (2.1) enables a (positive definite) scalar product to be introduced into the whole Krein space (i.e. for all $x, y \in \mathscr{F}$) according to the formula

$$(x, y) = [x^+, y^+] - [x^-, y^-],$$
$$x = x^+ + x^-, \qquad y = y^+ + y^-; \qquad x^+, y^+ \in \mathscr{D}^+; \qquad x^-, y^- \in \mathscr{F}^-. \tag{2.2}$$

In particular, for vectors $u, v \in \mathscr{F}^+$ we have, clearly $(u, v) = [u, v]$, and for $u, v \in \mathscr{F}^-$ we have $(u, v) = -[u, v]$. But if $u \in \mathscr{F}^+$ and $v \in \mathscr{F}^-$, then it follows from (2.2) that $(u, v) = [u, \theta] - [\theta, v] = 0$, i.e. $\mathscr{F}^+ \perp \mathscr{F}^-$, where the symbol $\perp$ denotes, as usual, orthogonality relative to the scalar product (2.2).

□ Thus, the space with the scalar product (2.2) can be regarded as the *orthogonal sum of the two Hilbert spaces* $\mathscr{F}^+$ *and* $\mathscr{F}^-$, *from which follows*

2.4 *The space* $\mathscr{F}$ *is complete, i.e. it is a Hilbert space relative to the norm*

$$\|x\| = (x, x)^{1/2} \tag{2.3}$$

generated by the scalar product (2.2). ■

Remark 2.5: Since our main purpose later will be to study linear operators acting in Krein spaces, the topology of this space (and it is defined by the norm (2.3), i.e., by the scalar product (2.2)) is important for questions connected with the boundedness and closure of operators, with their spectral theory, and so on. At the same time the Definitions (2.2) and (2.3) may create the impression that the topology about which we are speaking depends essentially on the actual choice of the canonical decomposition (2.1). Later (see Theorem 7.19) it will be proved that this impression is erroneous: *all norms defined by different canonical decompositions of the form* (2.1) *turn out to be equivalent*.

3 In connection with the above definitions it is expedient to distinguish Krein spaces from all spaces $\mathscr{F}$ with an indefinite metric by means of a special notation. For this we shall adopt the traditional notation for Hilbert spaces (as we saw in paragraph 2, a Krein space is a Hilbert space), namely $\mathscr{H}$. The canonical decomposition (2.1) will now be written in the form

$$\mathscr{H} = \mathscr{H}^+ [\dot{+}] \mathscr{H}^-; \tag{2.4}$$

we recall, moreover, that not only is $\mathscr{H}^+ [\perp] \mathscr{H}^-$, but also $\mathscr{H}^+ \perp \mathscr{H}^-$, i.e. the decomposition (2.4) could also be written in the form

$$\mathscr{H} = \mathscr{H}^+ \oplus \mathscr{H}_1^- \tag{2.5}$$

where $\oplus$ is the symbol for a (Hilbert) *orthogonal sum*.

Returning to formula (2.2) we can, in a known sense, invert it, i.e., express the metric (the Q-form) $[x, y]$ in terms of the scalar product (2.2). Keeping for arbitrary vectors $x, y \in \mathscr{H}$ the notation $x^{\pm}, y^{\pm}$ for their components lying in $\mathscr{H}^{\pm}$ respectively (in accordance with (2.4) or (2.5)), we have

$$[x, y] = [x^+ + x^-, y^+ + y^-] = [x^+, y^+] + [x^-, y^-] = (x^+, y^+) - (x^-, y^-),$$

or finally

$$[x, y] = (x^+, y^+) - (x^-, y^-). \tag{2.6}$$

In particular,

$$[x, x] = \| x^+ \|^2 - \| x^- \|^2 \tag{2.7}$$

□ From the formula (2.7) we obtain

2.6 Positivity, negativity, or neutrality of a vector $x \in \mathscr{H}$ are equivalent to the relations $\| x^+ \| > \| x^- \|$, $\| x^+ \| < \| x^- \|$, or $\| x^+ \| = \| x^- \|$ *respectively.* ■

4 The results of paragraph 3 show that a Krein space can be looked on as an *arbitrary* Hilbert space $\mathscr{H}$ decomposed into the usual orthogonal sum (2.5) of the two subspaces $\mathscr{H}^+$ and $\mathscr{H}^-$ ($\mathscr{H}^+ \perp \mathscr{H}^-$) and equipped in addition to the original Hilbert metric (i.e., the scalar product (x, y)) with the additional indefinite metric $[x, y]$ given by the relation (2.6). We suggest to the reader that, as a simple exercise, he should convince himself, that with such a definition of the form $Q(x, y) = [x, y]$ the subspaces $\mathscr{H}^+$ and $\mathscr{H}^-$ turn out to be positive and negative respectively relative to it, that they are Q-orthogonal ($\mathscr{H}^+ [\perp] \mathscr{H}^-$), and therefore the decomposition (2.5) is simultaneously also a canonical decomposition of the form (2.4). Moreover, the original scalar product, as can easily be calculated, is again expressed in terms of the form $[x, y]$ by means of the relation (2.2).

In the next section it will be shown (see §3, para. 5) that the point of view presented here on Krein spaces can be further modified and that to this concept can be given an extremely flexible form most convenient for the theory of operators.

Exercises and problems

1 The anti-space (see Definition 1.8) of a Krein space is again a Krein space.

2 Suppose that in a canonical decomposition (2.1) of a space $\mathscr{F}$ at least one of the terms $\mathscr{F}^+$ or $\mathscr{F}^-$ is not complete relative to the norm $\| x \| = | [x, x] |^{1/2}$; we complete $\mathscr{F}^{\pm}$ relative to this norm to Hilbert spaces $\tilde{\mathscr{F}}^{\pm}$. Show that the orthogonal sum $\tilde{\mathscr{F}}^+ \oplus \tilde{\mathscr{F}}^-$ is a Krein space which contains $\mathscr{F}$ as a dense part. We obtain the same result if we at once complete the whole of $\mathscr{F}$ relative to the norm (2.3).

3 In Exercise 2 of §1 the space $\mathscr{F}$ admits a canonical decomposition $\mathscr{F} = \mathscr{F}^+ [\dot{+}] \mathscr{F}^-$, where $\mathscr{F}^+$ and $\mathscr{F}^-$ are the spaces of all continuous even and odd functions on $[-1, 1]$ respectively. Completing it in accordance with the procedure considered in the previous Exercise leads to the space $\tilde{\mathscr{F}} = L^2[-1, 1]$ with the indefinite metric (1.22), in which the Riemann integral should be replaced by a Lebesgue integral. The interval $(-1,1)$ can, of course, be replaced by any symmetric interval $(-a, a)$.

4 We modify Example 1.1 by considering the space $\mathscr{F}$ of infinite (in both directions) and finite (also in both directions) sequences $x = \{\ldots, \xi_{-2}, \xi_{-1}, \xi_1, \xi_2, \ldots\}$, with co-ordinatewise definition of the linear operations, and introducing for $x, y \in \mathscr{F}$ $(y = \{\eta_k\}_{k=-\infty}^{\infty})_{k \neq 0}$ the indefinite metric

$$Q(x, y) = [x, y] = \sum_{k=1}^{\infty} \xi_k \bar{\eta}_k - \sum_{k=1}^{\infty} \xi_{-k} \bar{\eta}_{-k}. \tag{2.8}$$

Find a canonical decomposition of the space $\mathscr{F}$ and show that completion of $\mathscr{F}$ in the manner of Exercise 2 leads to a Krein space which is Q-isometrically isomorphous to $l^2 \oplus l^2$, with the indefinite form (2.8) ([XV]).

5 In Example 1.2 we restrict ourselves to a finite interval $[a, b] \subset \mathbb{R}$ and we denote by $\omega(t)$ the total variation of a function $\sigma(t)$ on the interval $[a, t]$ $(a \leqslant t \leqslant b)$. Then the indefinite form

$$[x, y] = \int_a^b x(t)\overline{y(t)}\, d\sigma(t) \qquad (x, y \in L^2_\omega(a, b))$$

gives in the $L^2_\omega(a, b)$ space of all ω-measurable and ω-summable squared functions the structure of a Krein space. Find its canonical decomposition (I. Iokhvidov and Ektov [2]).

Hint: Use Hahn's decomposition theorem (see, e.g. [VII], [VIII], §5, Theorem 1).

6 Generalize the example in the preceding exercise to arbitrary $L^2(S, \Sigma, \mu)$-spaces (see [XII]), where $\mu = \mu_+ - \mu_-$ is a measure with bounded variation $|\mu| = \mu_+ + \mu_-$ ([XII]).

7 A model of a *real* Krein space (see, at the end of Chapter 1, the remarks on §1, paragraphs 1, 2) is the orthogonal sum $\mathscr{H} = \mathscr{H}^+ \oplus \mathscr{H}^-$ of two *real* Hilbert $\mathscr{H}^+$ and $\mathscr{H}^-$ with a form

$$Q(x, y) = [x, y] = (x^+, y^+)_+ - (x^-, y^-)_-,$$

where $x = x^+ + x^-$, $y = y^+ + y^-$, and $(\cdot, \cdot)_+$, $(\cdot, \cdot)_-$ are the scalar products in $\mathscr{H}^+$ and $\mathscr{H}^-$ respectively.

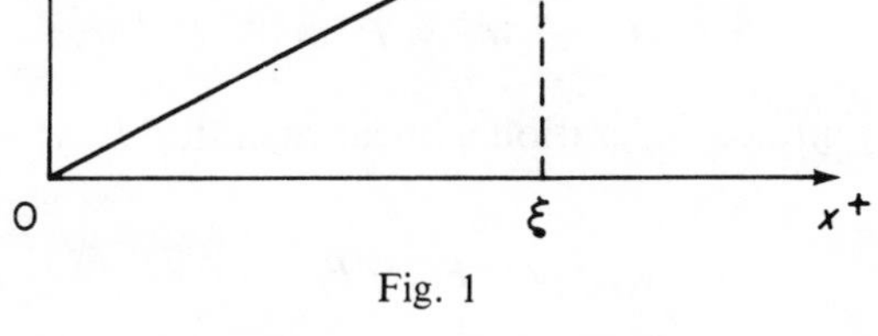

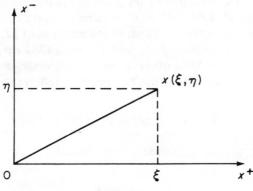

Fig. 1

Show that in the particular case when $\mathscr{H}^+ = \mathscr{H}^- = l^2$ (a real l^2), such a Krein space is Q-isometrically isomorphic to the completion relative to the l^2 norm of the analogue of the space $\mathscr{F}$ in Exercise 4 for real sequences with a real form (2.8). Verify that this analogue for the case $\dim \mathscr{H}^+ = \dim \mathscr{H}^- = 1$ is Q-isometrically isomorphic to the Cartesian plane $\mathbb{R}^2$ (see Fig. 1) with rectangular Cartesian axes $(\mathscr{H}^+, \mathscr{H}^-)$, with vectors $x_1 = (\xi_1, \eta_1)$, $x_2 = (\eta_2, \eta_2)$, and with the forms: indefinite form $Q(x_1, x_2) = [x_1, x_2] = \xi_1\xi_2 - \eta_1\eta_2$, and the Hilbert form $(x_1, x_2) = \xi_1\xi_2 + \eta_1\eta_2$ (the scalar product).

In the case of real $\mathscr{H}^+, \mathscr{H}^-$ with $\dim \mathscr{H}^+ = 1$, $\dim \mathscr{H}^- = 3$, the Krein space $\mathscr{H} = \mathscr{H}^+ \oplus \mathscr{H}^-$ is Q-isometrically isomorphic to the Minkowski space of the special theory of relativity with the Q-metric (see (2.8))

$$Q(x, y) = [x, y] = \xi_0\eta_0 - \sum_{k=1}^{3} \xi_k\eta_k.$$

8 Let $\mathscr{H} = \mathscr{H}^+ \oplus \mathscr{H}^-$ be a Krein space and let $x^\pm \in \mathscr{H}^\pm$. Then the conditions a) $x^\pm \perp y$ and b) $x^\pm\, [\perp]\, y$ are equivalent for any $y \in \mathscr{H}$.
Hint: Use formula (2.2).

9 Under the conditions of the previous exercise

$$\{x_0\, [\perp]\, x,\, x_0 \perp x\} \Leftrightarrow \{x_0^+ \perp x^+,\, x_0^- \perp x^-)$$

for any

$$x_0 = x_0^+ + x_0^- \quad \text{and} \quad x = x^+ + x^- \text{ in } \mathscr{H} \qquad (x_0^\pm, x^\pm \in \mathscr{H}^\pm).$$

Hint: Use formula (2.2).

10 Let $\mathscr{L}$ be a subspace of a Krein space, and let $\mathscr{L}^0$ be its isotropic part. The space $\hat{\mathscr{L}} = \mathscr{L} | \mathscr{L}^0$ will be Krein space if and only if $\mathscr{L}$ admits the decomposition $\mathscr{L} = \mathscr{L}^0\, [\dot{+}]\, \mathscr{L}^1$ into the direct sum of the isotopic subspace $\mathscr{L}^0$ and a Krein space $\mathscr{L}^1$. Moreover, if $\mathscr{L} = \mathscr{L}_0\, [\dot{+}]\, \mathscr{L}''$ is another direct decomposition, then $\mathscr{L}''$ is also a Krein space (Langer [9]).
Hint: Use the Definition 2.2 of a Krein space and Proposition 1.23.

§3 Canonical projectors $P^\pm$ and canonical symmetry J

1 Suppose a Krein space $\mathscr{H}$ is given, with a canonical decomposition $\mathscr{H} = \mathscr{H}^+\, [\dot{+}]\, \mathscr{H}^-$. This decomposition generates two mutually complementary projectors P^+ and P^- ($P^+ + P^- = I$, the identity operator in $\mathscr{H}$) mapping $\mathscr{H}$ on to $\mathscr{H}^+$ and $\mathscr{H}^-$ respectively. Thus, in the notation of §2, for any $x \in \mathscr{H}$ we have $P^\pm x = x^\pm$. The projectors P^+ and P^- are called *canonical projectors*. Recalling the relation (2.5) we observe that $P^\pm$ are *ortho-projectors*, i.e., they are orthogonal (self-adjoint) projection operators relative to the scalar product (2.2):

$$\mathscr{H} = \mathscr{H}^+ \oplus \mathscr{H}^- = P^+\mathscr{H} \oplus P^{-1}\mathscr{H}.$$

We now bring into consideration a linear operator $J\colon \mathscr{H} \to \mathscr{H}$ defined by the formula

$$J = P^+ - P^-, \tag{3.1}$$

and call it the *canonical symmetry* of the Krein space $\mathscr{H}$. The justification for this name is that, firstly, the operator J is defined by the canonical decomposition (2.4), and secondly, that it is a bounded symmetric (self-adjoint) operator in $\mathscr{H}$, since it is the difference between two self-adjoint operators P^+ and P^-.

3.1 The canonical symmetry J has the following properties:

$$J^* = J \qquad (\textit{self-adjointness}), \tag{3.2}$$

$$J^2 = I \ (J^{-1} = J) \qquad (\textit{the property of being involutory}) \tag{3.3}$$

$$J^{-1} = J^* \qquad (\textit{the property of being unitary}). \tag{3.4}$$

□ The relation (3.2) was established above. Further, for $x = x^+ + x^- \in \mathscr{H}$ we have, by (3.1), $Jx = x^+ - x^-$, and so $J^2x = J(x^+ - x^-) = x^+ + x^- = x$, i.e., $J^2 = I$. Finally, (3.4) follows from (3.2) and (3.3). ■

2 We trace the action of the operator J on the subspaces $\mathscr{H}^+$ and $\mathscr{H}^-$. It follows from (3.4) that

□ *3.2 $\mathscr{H}^+$ is an eigen-subspace of the operator J, corresponding to the eigenvalue $\lambda = 1$.* ■

□ *3.3 $\mathscr{H}^-$ is an eigen-subspace of the operator J, corresponding to the eigenvalue $\lambda = -1$.* ■

Since the operator J is simultaneously both self-adjoint and unitary (see 3.1), its whole spectrum lies on the intersection of the real axis and the unit circle, i.e., on the join of the points $\lambda_1 = 1$ and $\lambda_2 = -1$. As we see from 3.2 and 3.3, it actually (provided only that $\mathscr{H}^+ \neq \{\theta\}$ and $\mathscr{H}^- \neq \{\theta\}$) contains both these two numbers, which are eigenvalues of the operator J: $\sigma(J) = \sigma_P(J) = (-1, 1)$. We recall that $\sigma(\cdot)$ denotes the whole spectrum, and $\sigma_P(\cdot)$ the point spectrum, of an operator.

We notice further that the definition $J(x^+ + x^-) = x^+ - x^-$ of the canonical symmetry operator J enables us to treat the result of its action as the 'mirror reflection' of the space $\mathscr{H}$ in the subspace $\mathscr{H}^+$.

3 The introduction of the operator J makes it possible to reunite in a new and more compact way the fundamental relations derived in §2 between the indefinite metric ($[x, y]$) and the Hilbert metric ((x, y)) in a Krein space, namely:

3.4

$$(x, y) = [Jx, y], \qquad [x, y] = (Jx, y) \qquad (x, y \in \mathscr{H}) \tag{3.5}$$

□ The first of these relations is the equivalent of formula (2.2), and the second is the equivalent of formula (2.6). ■

It follows from 3.4 that, firstly,

$$\| x \|^2 = [Jx, x], \qquad [x, x] = (Jx, x) \qquad (x \in \mathscr{H}), \tag{3.6}$$

and secondly that:

3.5 The form $[x, y]$ is a bounded (and therefore continuous over all the variables) Hermitian sequilinear functional on $\mathscr{H} \times \mathscr{H}$ with the bound (since $\| J \| = 1$)

$$|[x, y]| \leqslant \| J \| \, \| x \| \, \| y \| = \| x \| \, \| y \| \qquad (x, y \in \mathscr{H}). \quad ■ \tag{3.7}$$

□ The Proposition 3.5 enables an important deduction to be made about the closures (with respect to the norm $\| x \|$) of semi-definite lineals of a Krein space:

3.6 If $\mathscr{L} \subset \mathscr{P}^+$ $(\mathscr{P}^-)$, then also $\bar{\mathscr{L}} \subset \mathscr{P}^+ (\mathscr{P}^-)$. If $\mathscr{L} \subset \mathscr{P}^0$, then also $\bar{\mathscr{L}} \subset \mathscr{P}^0$. ■

□ From this follows, in turn,

3.7 In a Krein space every maximal lineal $\mathscr{L}$ $(\subset \mathscr{P}^+)$ (or $\mathscr{L}$ $(\subset \mathscr{P}^-)$, $\mathscr{L}$ $(\subset \mathscr{P}^0)$) is closed, i.e., it is a subspace: $\mathscr{L} = \bar{\mathscr{L}}$ ■

We note that for maximal definite lineals a similar assertion is certainly not true (see Example 4.12 below).

4 □ To the properties of the operator J of being symmetric

$$((Jx, y) = (x, Jy), \qquad x, y \in \mathscr{H}$$

and of being unitary, and therefore isometric

$$(Jx, Jy) = (x, y), \qquad x, y \in \mathscr{H},$$

which reflect its properties relative to the scalar product (x, y), can be added its entirely analogous properties relative to the indefinite metric $[x, y]$:

$$[Jx, y] = [x, Jy], \; [Jx, Jy] = [x, y] \qquad (x, y \in \mathscr{H}), \tag{3.8}$$

which follow from (3.5):

$$[Jx, y] = (x, y) = \overline{(y, x)} = \overline{[Jy, x]} \qquad (x, y \in \mathscr{H});$$

$$[Jz, Jy] = (x, Jy = (Jx, y) = [x, y] \qquad (x, y \in \mathscr{H}). \quad ■$$

These properties have been named the *J-symmetry* and the *J-isometricity* respectively of the operator J; they are discussed in detail later in Chapter 2, Sections 3.5. As regards the indefinite metric $[x, y]$ itself, in view of the second formula in (3.5) it is also known as the *J-metric*, and a Krein space is called a (Hilbert) *space with a J-metric* or, shortly, a *J-space*.

For the same reason the Q-form $Q(x, y) = [x, y] = (Jx, y)$ is called the *J-form*, and correspondingly we speak of *J-orthogonality* ($x\,[\perp]\,y$) of vectors $x, y \in \mathscr{H}$, of the *J-orthogonal complement* $\mathscr{M}^{[\perp]}$ to a set $\mathscr{M}$ ($\subset \mathscr{H}$), and so on. We shall adhere to this latest terminology throughout all the rest of the exposition unless otherwise stipulated (see, e.g. §9, para. 4).

5 In the theory of linear operators in Krein spaces (see Chapter 2 and later) it is often convenient to regard $\mathscr{H}$ as a Hilbert space in which an indefinite metric is defined not by a canonical decomposition $\mathscr{H} = \mathscr{H}^+ \oplus \mathscr{H}^-$ given in advance, i.e., by canonical orthoprojectors $P^{\pm}$ ($P^+ + P^- = I$) (see §2.4), but by some *preassigned symmetry operator J*. For this purpose it is expedient to modify the definition (as will be seen in a moment, to an equivalent form) of the symmetry operator J and, at the same time, also of a canonical decomposition and of canonical projectors by putting as a basis not the form (3.1) but the relations (3.2)–(3.4), any two of which imply the third (see the Proof of Proposition 3.1).

Definition 3.8: Any linear operator J: $\mathscr{H} \to \mathscr{H}$ which is simultaneously unitary and self-adjoint:

$$J^{-1} = J^* = J$$

is called a *canonical symmetry* in the Hilbert space $\mathscr{H}$.

□ A canonical symmetry J immediately generates *orthogonal canonical orthoprojectors* $P^{\pm}$ according to the formulae

$$P^+ = \tfrac{1}{2}(I + J), \qquad P^- = \tfrac{1}{2}(I - J),$$

and a *canonical decomposition*

$$\mathscr{H} = \mathscr{H}^+ \oplus \mathscr{H}^-, \qquad \mathscr{H}^{\pm} = P^{\pm}\mathscr{H},$$

and also the *J-metric*

$$[x, y] = (Jx, y) \qquad (x, y \in (x, y \in \mathscr{H}),$$

where (x, y) is the original scalar product in $\mathscr{H}$, and the subspaces $\mathscr{H}^{\pm}$, relative to the J-metric, are definite (of different 'signs') and J-orthogonal. We mention that here the 'limiting' particular cases where $J = \pm I$, i.e., either $\mathscr{H}^- = \{\theta\}$ or $\mathscr{H}^+ = \{\theta\}$, and the J-metric $[x, y]$ is identical with either (x, y) or $-(x, y)$, are not excluded. ■

Both in the general theory and (in particular) in applications the canonical symmetry operator J may be defined in the Hilbert space in very different ways. We illustrate this fact here by two examples (the reader will encounter others in the later sections and chapters).

□ *Example 3.9:* We consider a Hilbert space $\tilde{\mathscr{H}} = \mathscr{H}_1 \oplus \mathscr{H}_2$ which is the orthogonal sum of two copies of a Hilbert space $\mathscr{H}$ ($= \mathscr{H}_1 = \mathscr{H}_2$), and we define in $\tilde{\mathscr{H}}$ an operator J in the 'operator-matrix' form

$$\tilde{J} = \begin{Vmatrix} G & (I-G^2)^{1/2}V \\ V^*(I-G^2)^{1/2} & -V^*GV \end{Vmatrix}, \tag{3.9}$$

where G is a bounded self-adjoint operator, and I is the identity operator in $\mathscr{H}$, with $0 \leqslant G \leqslant I$, and V is a semi-unitary operator in $\mathscr{H}$ mapping $\mathscr{H}$ on to $V\mathscr{H} \supset [\operatorname{Ker}(I-G^2)]^\perp$ (actual examples of such a pair G, V will be met later in Example 1.4 in Chapter 3 and elsewhere), or V is an arbitrary unitary operator ($V^*V = VV^* = I$).

In the matrix form (3.9) the operators $(I-G^2)^{1/2}V$ and $V^*(I-G^2)^{1/2}$ appearing on the secondary diagonal are to be understood as operating, the first from $\mathscr{H}_2$ into $\mathscr{H}_1$, and the second from $\mathscr{H}_1$ into $\mathscr{H}_2$. Thus, strictly speaking, they should be written in the form $(I-G^2)^{1/2}Vj_{12}$ and $j_{21}V^*(I-G^2)^{1/2}$, where j_{12} and j_{21} are the operators of 'canonical imbedding' (of identification of elements) of $\mathscr{H}_2$ into $\mathscr{H}_1$ and of $\mathscr{H}_1$ into $\mathscr{H}_2$ respectively. In order not to complicate the notation we shall not do this either now or later, hoping that no misunderstanding will arise in the reader's mind.

Clearly $\tilde{J}^* = \tilde{J}$, and so it remains to verify that $\tilde{J}^2 = \tilde{I}$ (the identity operator in $\tilde{\mathscr{H}}$). But it is easy to see this by squaring the matrix (3.9) and noting that $(I-G^2)\mathscr{H} \subset [\operatorname{Ker}(I-G^2)]^\perp$ and $GV\mathscr{H} \subset [\operatorname{Ker}(I-G^2)]^\perp$. ■

The most important cases of the canonical symmetry operator (3.9), those most often used in applications are the 'extreme cases':

$$G = I, \qquad V = I\colon \tilde{J} = \begin{Vmatrix} I & 0 \\ 0 & -I \end{Vmatrix}; \tag{3.10a}$$

$$G = 0, \qquad V = I\colon \tilde{J} = \begin{Vmatrix} 0 & I \\ I & 0 \end{Vmatrix}. \tag{3.10b}$$

We note that only for the first of these is the original representation $\tilde{\mathscr{H}} = \mathscr{H}_1 \oplus \mathscr{H}_2$ also a canonical decomposition $\tilde{\mathscr{H}} = \tilde{\mathscr{H}}^+ [+] \tilde{\mathscr{H}}^-$ ($\tilde{\mathscr{H}}^+ = \mathscr{H}_1, \tilde{\mathscr{H}}^- = \mathscr{H}_2$). In other cases the canonical decomposition has to be constructed afresh (see Theorem 8.17 below).

□ *Example 3.10:* Again let $\tilde{\mathscr{H}} = \mathscr{H}_1 \oplus \mathscr{H}_2$ ($\mathscr{H}_1 = \mathscr{H}_2 = \mathscr{H}$), but this time $\mathscr{H}$ itself is a Hibert space with a J-metric, i.e., it is a Krein space. We introduce in $\tilde{\mathscr{H}}$ the canonical symmetry operator

$$\tilde{J} = \begin{Vmatrix} 0 & \varepsilon J \\ \bar{\varepsilon} J & 0 \end{Vmatrix}, \qquad |\varepsilon| = 1 \tag{3.11}$$

Here again the self-adjointness of $\tilde{J}$ in $\tilde{\mathscr{H}}$ is obvious, and the formula $\tilde{J}^2 = \tilde{I}$ can be verified immediately. In the particular case when $J = I$, $\varepsilon = 1$ the formula (3.11) turns into (3.10b). ■

□ *Example 3.11:* Consider the Hilbert space $\mathscr{H} = l^2$ of all infinite (in both directions) sequences $x = \{\xi_j\}_{j=-\infty}^{\infty}$, $y = \{\eta_j\}_{j=-\infty}^{\infty}$ ($x, y \in l^2$) with square-summable moduli, and with a scalar product $(\cdot, \cdot)$ and an infinite form $[\cdot, \cdot]$ given respectively by the formulae (cf. Example 1.33)

$$(x, y) = \sum_{j=-\infty}^{\infty} \xi_j \bar{\eta}_j, \qquad [x, y] = \sum_{j=-\infty}^{\infty} \xi_j \bar{\eta}_{-j-1}.$$

Here the operator $J\{\xi_j\}_{j=-\infty}^{\infty} = \{\xi_{-j-1}\}_{j=-\infty}^{\infty}$ is an involution and $J^* = J$. ■

Exercises and problems

1 In Exercise 3 of §2 construct the canonical symmetry operator J from the given canonical decomposition.

2 Construct the canonical symmetry operator J for the space in §2, Exercise 4.

3 Do the same for the spares in §2, Exercises 5 and 6.

4 In a Krein space the lineals $\mathscr{L}$ and $J\mathscr{L}$ are always skewly linked: $\mathscr{L} \# J\mathscr{L}$ (see §1, para 10) ([V]) (*cf.* Exercise 6b below).

5 Find the canonical decompositions of the Krein spaces in Examples 3.10 and 3.11 corresponding to the canonic symmetry operator $\tilde{J}$ indicated therein.

6 The concept of a canonical symmetry operator makes sense also in any space $\mathscr{F}$ with a Hermitian form $Q(x, y) = [x, y]$ $(x, y \in \mathscr{F})$ if $\mathscr{F}$ admits a decomposition $\mathscr{F} = \mathscr{F}^+ [\dot{+}] \mathscr{F}^-$, which, clearly, defines projectors $P^{\pm}$ from $\mathscr{F}$ on to $\mathscr{F}^{\pm}$ respectively $(P^+ + P^- = I)$. Verify that (*cf.* Scheibe [1]):

a) $J = P^+ - P^-$ is an involution: $J^2 = I$, and a Q-symmetry: $[Jx, y] = [x, Jy]$ $(x, y \in \mathscr{F})$, and is a Q-isometry: $[Jx, Jy] = (x, y)$ $(x, y \in \mathscr{F})$;

b) For any lineal $\mathscr{L}$ $(\subset \mathscr{F})$ the lineals $\mathscr{L}$ and $J\mathscr{L}$ are skewly linked ($\mathscr{L} \# J\mathscr{L}$) and are Q-isometrically isomorphic (see §1, para. 3); here a Q-metric is induced in $\mathscr{L}$ and $J\mathscr{L}$ from $\mathscr{F}$;

c) $J\mathscr{P}^{\pm} = \mathscr{P}^{\pm}$, $J\mathscr{P}^0 = \mathscr{P}^0$, $J\mathscr{P}^{\pm\pm} = \mathscr{P}^{\pm\pm}$, $(J\mathscr{L})^{[\perp]} = J\mathscr{L}^{[\perp]}$ $(\mathscr{L} \subset \mathscr{F})$; in particular, if $\mathscr{L}$ is maximal in the class of non-negative (positive, neutral, non-positive, or negative) lineals, then $J\mathscr{L}$ is also maximal in the same class; a similar assertion holds for non-degenerate lineals, and if $\mathscr{L}^{[\perp][\perp]} = \mathscr{L}$, then $(J\mathscr{L})^{[\perp][\perp]} = J\mathscr{L}$;

d) for a neutral $\mathscr{L}$ the lineal $\mathscr{L} + J\mathscr{L}$ admits the canonical decomposition $\mathscr{L} + J\mathscr{L} = P^+\mathscr{L} [\dot{+}] P^-\mathscr{L}$, and moreover $P^+\mathscr{L}$ is Q-isometrically isomorphic to the anti-space for $P^-\mathscr{L}$ (see §1, para. 3).

7 If $\mathscr{H} = \mathscr{L} \oplus \mathscr{L}^{\perp}$ is the orthogonal sum of two subspaces of a Hilbert space $\mathscr{H}$, then in order that the matrix

$$J = \begin{Vmatrix} J_{11} & J_{12} \\ J_{12}^* & J_{22} \end{Vmatrix}$$

with bounded elements should give (relative to the given decomposition of $\mathscr{H}$) a canonical symmetry operator in $\mathscr{H}$, it is necessary and sufficient that

a) $J_{11} = J_{11}^*$, $J_{22} = J_{22}^*$; b) $J_{11}^2 + J_{12}J_{12}^* = I_{\mathscr{L}}$;

c) $J_{22}^2 + J_{12}^*J_{12} = I_{\mathscr{L}^{\perp}}$; d) $J_{11}J_{12} + J_{12}J_{22} = 0$.

Hint: Verify the condition $J^2 = I$.

§4 Semi-definite and definite lineals and subspaces in a Krein space

1 Let $\mathscr{H} = \mathscr{H}^+ [\oplus] \mathscr{H}^-$ be a canonical decomposition of a Krein space, and let $P^\pm$ be the corresponding canonical projectors. The symbol $[\oplus]$ is used here (for the first time) as a brief remainder of the 'two-fold orthogonality' (the usual one and also in the J-metric) of the canonical decomposition we have fixed on. We consider an arbitrary semi-definite (for definiteness, a non-negative) lineal $\mathscr{L}$ $(\subset \mathscr{H})$ and we study the mapping $P^+ : \mathscr{L} \to \mathscr{H}^+$. Speaking more accurately, we actually study here the restriction $P^+ \mid \mathscr{L}$ of the operator P^+ on to $\mathscr{L}$, and the image of $\mathscr{L}$ in this mapping is the lineal $P^+ \mathscr{L}$ $(\subset \mathscr{H}^+)$.

Theorem 4.1: *The mapping $P^+ \mid \mathscr{L}$: $\mathscr{L} \to P^+ \mathscr{L}$ of a non-negative lineal $\mathscr{L}$ $(\subset \mathscr{H})$ is a linear homeomorphism, i.e., it is bijective, continuous, and the inverse mapping $(P^+ \mid \mathscr{L})^{-1}$: $P^+ \mathscr{L} \to \mathscr{L}$ is also continuous.*

□ The linearity and continuity of the ortho-projector P^+, and also of its restriction $P^+ \mid \mathscr{L}$, are obvious. The bijectivity follows from Lemma 1.27. Further, for $x \in \mathscr{L}$ we have (see 2.6) $\| x^+ \| > \| x^- \|$, and so

$$\| (P^+ \mid \mathscr{L}) x \|^2 = \| x^+ \| \geqslant \tfrac{1}{2}(\| x^+ \|^2 + \| x^- \|^2) = \tfrac{1}{2} \| x \|^2 \qquad (x \in \mathscr{L}). \qquad (4.1)$$

Putting $(P^+ \mid \mathscr{L}) x = y$, we have $x = (P^+ \mid \mathscr{L})^{-1} y$, and the relation (4.1) gives

$$\| (P^+ \mid \mathscr{L})^{-1} y \|^2 \leqslant 2 \| y \|^2 \qquad (y \in P^+ \mathscr{L}), \text{ i.e. } \| (P^+ \mid \mathscr{L})^{-1} \| \leqslant \sqrt{2}. \quad ■$$

□ **Corollary 4.2:** *Under the conditions of Theorem 4.1 the lineals $\mathscr{L}$ and $P^+ \mathscr{L}$ are either both closed or both are not closed.* ■

□ **Corollary 4.3:** *The dimension of any non-negative subspace $\mathscr{L}$ $(\subset \mathscr{H} = \mathscr{H}^+ [\dot{+}] \mathscr{H}^-)$ does not exceed the dimension of $\mathscr{H}^+$:* $\dim \mathscr{L} \leqslant \dim \mathscr{H}^+$. ■

Remark 4.4: By virtue of Lemma 1.27 the bijectivity of the mapping $(P^+ \mid \mathscr{L})$: $\mathscr{L} \to P^+ \mathscr{L}$ in Theorem 4.1 holds for any linear space $\mathscr{F}$ with a Q-form $[x, y]$ (see §1, para. 1) which admits decomposition into a direct sum (not necessarily even a Q-orthogonal sum!) $\mathscr{F} = \mathscr{F}^+ \dot{+} \mathscr{F}^-$ of a positive lineal $(\mathscr{F}^+)$ and a negative lineal $(\mathscr{F}^-)$, if $P^\pm$ are the projectors corresponding to this decomposition ($P^+ + P^- = I_{\mathscr{F}}$, the identity operator in $\mathscr{F}$). So in this case Corollary 4.3 still holds if in the inequality $\dim \mathscr{L} \leqslant \dim \mathscr{L}^+$ we take the symbol dim to be not the Hilbert dimension but the *linear dimension*, i.e. the cardinality of the algebraic basis. Finally, even if $\mathscr{F}$ is degenerate, if it still admits decomposition into the direct sum $\mathscr{F} = \mathscr{F}^0 + \mathscr{F}^+ + \mathscr{F}^-$ of an isotropic lineal $\mathscr{F}^0$, a positive lineal $\mathscr{F}^+$, and a negative lineal $\mathscr{F}^-$ (see 1.24), then similar arguments show that the linear dimension of any *positive* $\mathscr{L}$ $(\subset \mathscr{F})$ does not exceed the linear dimension of $\mathscr{F}^+$. ■

In conclusion we point out that we have, only for the sake of definiteness, been dicussing non-negative and positive lineals. The reader will easily be able to formulate and prove analogues of all these assertions for non-positive and negative lineals. But, actually, repetition of the proofs is unnecessary here because a simple transition from the space $\mathscr{H}$ (respectively $\mathscr{F}$) to the anti-space (see Definition 1.8) changes the rôles of non-negative (positive) and non-positive (negative) lineals, in particular of $\mathscr{H}^+(\mathscr{F}^+)$ and $\mathscr{H}^-(\mathscr{F}^-)$ and correspondingly of the projectors P^+ and P^-.

2 The results of paragraph 1 open the way to the establishment of criteria for maximality of semi-definite lineals in a Krein space.

Theorem 4.5: *In order that $\mathscr{L}$ $(\subset \mathscr{P}^+)$ $(\mathscr{L}$ $(\subset \mathscr{P}^-))$ in the Krein space $\mathscr{H} = \mathscr{H}^+ [\oplus] \mathscr{H}^-$ should be a maximal non-negative lineal (maximal non-positive lineal) it is necessary and sufficient that $P^+\mathscr{L} = \mathscr{H}^+$ $(P^-\mathscr{L} = \mathscr{H}^-)$.*

□ We carry out the proof for a non-negative $\mathscr{L}$.

Necessity: Let $\mathscr{L}$ be the maximal lineal from $\mathscr{P}^+$. Then (see 3.5) $\mathscr{L}$ is closed, i.e., it is a subspace, and therefore (see Corollary 4.2) $P^+\mathscr{L}$ $(\subset \mathscr{H}^+)$ is also a subspace. Let us assume that $P^+\mathscr{L} \neq \mathscr{H}^+$. Since on $\mathscr{H}^+$ the metrics (x, y) and $[x, y]$ are identical (see (2.2)), we can find a vector $(\theta \neq) z_0 \in \mathscr{D}_{\mathscr{L}}^+ \equiv \mathscr{H}^+ \ominus P^+\mathscr{L}$ which is simultaneously orthogonal and J-orthogonal to $P^+\mathscr{L}$. $\mathscr{D}_{\mathscr{L}}^+$ is called the *deficiency subspace* for $\mathscr{L}$.

Thus $z_0 \perp P^+\mathscr{L}$ and $z_0 [\perp] P^+\mathscr{L}$. At the same time $z_0 [\perp] \mathscr{L}$, because for any $x \in \mathscr{L}$ we have

$$[z_0, x] = [z_0, x^+ + x^-] = [z_0, x^+] = (z_0, x^+) = (z_0, Px^+) = 0.$$

We note further that $z_0 \notin \mathscr{L}$, for otherwise it would be isotropic for $\mathscr{L}$ and therefore a neutral vector, but z_0 is positive $(\theta \neq z_0 \in \mathscr{H}^+)$. But then the linear envelope $\text{Lin}\{\mathscr{L}, z_0\} \equiv \tilde{\mathscr{L}} \subset \mathscr{P}^+$, and also $\tilde{\mathscr{L}} \supset \mathscr{L}$ and $\tilde{\mathscr{L}} \neq \mathscr{L}$, which contradicts the maximal non-negativity of $\mathscr{L}$.

Sufficiency was established earlier in Corollary 1.28. ■

Corollary 4.6: *If $\mathscr{L}$ is a maximal non-negative (maximal non-positive) subspace of the Krein space $\mathscr{H} = \mathscr{H}^+ [\oplus] \mathscr{H}^-$, then* $\dim \mathscr{L} = \dim \mathscr{H}^+$ $(\dim \mathscr{L} = \dim \mathscr{H}^-)$.

□ This follows immediately from Theorems 4.1 and 4.5. ■

3 The situation is more complicated with maximal positive (maximal negative) lineals, which, as we shall see below (Example 4.12), can also be non-closed. We start, however, from a simple fact:

Theorem 4.7: *In order that $\mathscr{L} \subset \mathscr{P}^{++} \cup \{\theta\}$ $(\mathscr{L} \subset \mathscr{P}^{--} \cup \{\theta\})$ should be a* *maximal positive (maximal negative) lineal, it is necessary, and in the case when $\mathscr{L}$ is closed, it is also sufficient, that $\overline{P^+\mathscr{L}} = \mathscr{H}^+$ ($\overline{P^-\mathscr{L}} = \mathscr{H}^-$).*

□ *Necessity*. Suppose for definiteness that $\mathscr{L}$ is a maximal positive lineal. If $\overline{P^+\mathscr{L}} \neq \mathscr{H}^+$, then, again choosing an arbitrary vector $z_0 \neq \theta$ in the deficiency space $\mathscr{D}^+_{\mathscr{L}} = \mathscr{H} \ominus \overline{P^+\mathscr{L}}$, we discover, exactly as in the proof of Theorem 4.5, that $\tilde{\mathscr{L}} = \text{Lin}\{\mathscr{L}, z_0\}$ is positive, and also $\tilde{\mathscr{L}} \supset \mathscr{L}$ and $\tilde{\mathscr{L}} \neq \mathscr{L}$, which is impossible.

Sufficiency: Let $\overline{P^+\mathscr{L}} = \mathscr{H}^+$. If the positive $\mathscr{L}$ is closed ($\mathscr{L} = \bar{\mathscr{L}}$), then by Corollary 4.2 $P^+\mathscr{L} = \overline{P^+\mathscr{L}} = \mathscr{H}^+$ is also closed. By Theorem 4.5 $\mathscr{L}$ is a maximal non-negative subspace and by the same token a maximal positive subspace. ■

□ In the course of proving the 'sufficiency part' of Theorem 4.7 we have incidentally established.

Corollary 4.8: *A maximal positive lineal $\mathscr{L}$ in the case when it is closed is also a maximal non-negative lineal.* ■

Remark 4.9: In the case when $\mathscr{L} \subset \mathscr{P}^{++} \cup \{\theta\}$ ($\mathscr{L} \subset \mathscr{P}^{--} \cup \{\theta\}$) is not closed, the condition $\overline{P^+\mathscr{L}} = \mathscr{H}^+$ ($\overline{P^-\mathscr{L}} = \mathscr{H}^-$)) is certainly *not sufficient* for the maximality of $\mathscr{L}$, because it is also possible that $\bar{\mathscr{L}} \subset \mathscr{P}^{++} \cup \{\theta\}$ ($\bar{\mathscr{L}} \subset \mathscr{P}^{--} \cup \{\theta\}$). As the simplest example in the case, say, of an infinite dimensional $\mathscr{H}^+$ we may take any lineal $\mathscr{L}$ ($\neq \mathscr{H}^+$) which is dense in $\mathscr{H}^+$ ($\bar{\mathscr{L}} = \mathscr{H}^+$). ■

In view of Remark 4.9 the criteria (i.e., the necessary and sufficient conditions) for maximality of definite lineals have to be sought in other terms.

Theorem 4.10: *In order that in the Krein space $\mathscr{H} = \mathscr{H}^+ [\oplus] \mathscr{H}^-$ the lineal $\mathscr{L} \subset \mathscr{P}^{++} \cup \{\theta\}$ ($\mathscr{L} \subset \mathscr{P}^{--} \cup \{\theta\}$) should be a maximal positive (maximal negative) lineal, it is necessary and sufficient that the following two conditions hold:*

(I) *The closure $\bar{\mathscr{L}}$ of the lineal $\mathscr{L}$ is a maximal non-negative (non-positive) subspace.*
(II) *$\bar{\mathscr{L}} = \mathscr{L} + (\bar{\mathscr{L}})^0$, where $(\bar{\mathscr{L}})^0$ is the isotropic lineal for $\bar{\mathscr{L}}$ (not excluding the possibility $(\bar{\mathscr{L}})^0 = \{\theta\}$).*

□ *Necessity*. Suppose, for definiteness, that $\mathscr{L}$ is a maximal positive lineal. By virtue of Theorem 4.7 we have $\overline{P^+\mathscr{L}} = \mathscr{H}^+$ and *a fortiori* $P^+\bar{\mathscr{L}} = \mathscr{H}^+$, and therefore (Theorem 4.5) $\bar{\mathscr{L}}$ is a maximal non-negative subspace; so (I) has been proved. Further, $\mathscr{L} \subset \bar{\mathscr{L}}$, $(\bar{\mathscr{L}})^0 \subset \bar{\mathscr{L}}$, and in addition $\mathscr{L} \cap (\bar{\mathscr{L}})^0 = \{\theta\}$, because $\mathscr{L}$ is positive and $(\bar{\mathscr{L}})^0$ is neutral. But then $\bar{\mathscr{L}} = \mathscr{L} \dotplus (\bar{\mathscr{L}})^0 \dotplus \mathscr{L}_1$, where $\mathscr{L}_1$ is positive or equal to $\{\theta\}$. But since the positive lineal $\mathscr{L} \dotplus \mathscr{L}_1 \supset \mathscr{L}$, so, by virtue of the maximality of $\mathscr{L}$, we have $\mathscr{L}_1 = \{\theta\}$, i.e. (II) has been proved.

Sufficiency: Let the conditions (I)–(II) hold for a positive $\mathscr{L}$. We consider the maximal positive lineal $\mathscr{L}_{\max}$ ($\supset \mathscr{L}$); it exists by virtue of the maximality principle (Theorem 1.19). Then, on the one hand, $\bar{\mathscr{L}}_{\max} \supset \bar{\mathscr{L}}$ and because of (I) $\bar{\mathscr{L}}_{\max} = \bar{\mathscr{L}}$, but because of (II)

$$\bar{\mathscr{L}}_{\max} = \bar{\mathscr{L}} = \mathscr{L} \dot{+} (\bar{\mathscr{L}})^0. \tag{4.2}$$

Now suppose there is a $z \in \mathscr{L}_{\max} \backslash \mathscr{L} (\subset \mathscr{L}_{\max} \subset \bar{\mathscr{L}}_{\max})$. Then, in accordance with (4.2), we have $z = x + x_0$ $(x \in \mathscr{L}, x_0 \in (\bar{\mathscr{L}})^0)$ and $x_0 = z - x \in \mathscr{L}_{\max} \cap (\bar{\mathscr{L}})^0$, i.e. $x_0 = \theta$ and $z = x \in \mathscr{L}$, contrary to the choice of z. Thus, $\mathscr{L}_{\max} = \mathscr{L}$. ■

Before giving an example of a non-closed maximal positive lineal, we recall one lemma of a general character, which we shall have occasion to use more than once later.

Lemma 4.11: *Let $\mathscr{N}$ be an infinite-dimensional normed linear space and let $e \in \mathscr{N}$, $e \neq \theta$. Then there is a linear $\mathscr{L}$ $(\subset \mathscr{N})$ such that*

$$\bar{\mathscr{L}} = \mathscr{N}, \qquad \mathscr{N} = \mathscr{L} \dot{+} \operatorname{Lin}\{e\}. \tag{4.3}$$

□ Without loss of generality we shall suppose that $\| e \| = 1$. As is well known, there is an unbounded (discontinuous) linear functional φ: $\mathscr{N} \to \mathbb{C}$ such that $\varphi(e) = 1$. Let $\mathscr{L} = \operatorname{Ker} \varphi$, i.e., the set of zeros (the kernel) of the functional φ. Then (see, for example, [XIII]) $\bar{\mathscr{L}} = \mathscr{N}$. At the same time for any $x \in \mathscr{N}$, having chosen $\lambda = \varphi(x)$, for $x_0 = x - \lambda e$ we have $\varphi(x_0) = \varphi(x) - \lambda = 0$, i.e. $x_0 \in \mathscr{L}$ and $x = x_0 + \lambda e$. ■

□ *Example 4.12:* We consider a Krein space $\mathscr{H} = \mathscr{H}^+ [\oplus] \mathscr{H}^-$, where $\dim \mathscr{H}^+ = \infty$ and $\{e_\alpha^+\}_{\alpha \in A}$ is a complete orthonormal system in $\mathscr{H}^+$, but $\dim \mathscr{H}^- = 1$, i.e. $\mathscr{H}^- = \operatorname{Lin}\{e^-\}$, $\| e^- \| = 1$. Thus

$$\begin{aligned} &[e_\alpha^+, e_\beta^+] = (e_\alpha^+, e_\beta^+) = \delta_{\alpha\beta}, \qquad (\alpha, \beta \in A), \\ &[e_\alpha^+, e^-] = (e_\alpha^+, e^-) = 0, \\ &[e^-, e^-] = -\| e^- \|^2 = -1, \end{aligned} \tag{4.4}$$

where A is some (infinite) set of indices. We fix in A a certain index α_0 and consider the subspace $\tilde{\mathscr{L}} = C \operatorname{Lin}\{e_{\alpha_0}^+ + e^-;\ e_\alpha^+\}_{\alpha \in A, \alpha \neq \alpha_0}$. By (4.4) this subspace is non-negative, and moreover it is maximal non-negative, because $P^+ \tilde{\mathscr{L}} = C \operatorname{Lin}\{e_\alpha^+\}_{\alpha \in A} = \mathscr{H}^+$. The vector $e = e_{\alpha_0}^+ + e^-$ $(\neq \theta)$ is neutral, and so $\operatorname{Lin}\{e\} = \tilde{\mathscr{L}}^0$ is an isotropic subspace for $\tilde{\mathscr{L}}$ (see 1.17 and the analogue of Corollary 4.3 for non-positive subspaces). We now apply Lemma 4.11 and express $\tilde{\mathscr{L}}$ in the form $\tilde{\mathscr{L}} = \mathscr{L} \dot{+} \operatorname{Lin}\{e\} = \mathscr{L} \dot{+} \tilde{\mathscr{L}}^0$, where $\bar{L} = \tilde{\mathscr{L}}$. Further, $\mathscr{L}$ is positive, $\bar{\mathscr{L}} (= \tilde{\mathscr{L}})$ is a maximal non-negative subspace, $(\bar{\mathscr{L}})^0 = \tilde{\mathscr{L}}^0$ and $\mathscr{L} = \mathscr{L} + (\bar{\mathscr{L}})^0$. By Theorem 4.10 $\mathscr{L}$ is a maximal positive lineal (and is, moreover, non-closed) (see Exercise 2 below). ■

4 We now consider another particular case of semi-definite lineals in a Krein space, that of neutral lineals $\mathscr{L}$ $(\subset \mathscr{P}^0)$. Since such an $\mathscr{L}$ is simultaneously non-negative and non-positive, *both the mappings*

$$P^+ \mid \mathscr{L}: \mathscr{L} \to P^+ \mathscr{L} \ (\subset \mathscr{H}^+) \quad \text{and} \quad P^- \mid \mathscr{L}: \mathscr{L} \to P^- \mathscr{L} \ (\subset \mathscr{H}^-)$$

are linear homeomorphisms (Theorem 4.1).

Theorem 4.13: *In order that a lineal $\mathscr{L}$ ($\subset \mathscr{P}^0$) should be a maximal neutral subspace it is necessary and sufficient that one of the conditions*

$$\text{(I)} \ P^+ \mathscr{L} = \mathscr{H}^+; \qquad \text{(II)} \ P^- \mathscr{L} = \mathscr{H}^-.$$

should hold.

□ *Necessity:* Let $\mathscr{L}$ be a maximal lineal from $\mathscr{P}^0$ and therefore closed (see 3.5). If both the conditions (I) and (II) are infringed, then in the (non-zero) deficiency spaces $\mathscr{D}^+_{\mathscr{L}} = \mathscr{H}^+ \ominus P^- \mathscr{L}$ and $\mathscr{D}^-_{\mathscr{L}} = \mathscr{H}^- \ominus P^- \mathscr{L}$ we choose vectors x^+ and x^- respectively (with $\| x^+ \| = \| x^- \| > 0$). Then (see the proof of Necessity in Theorem 4.5) $x^{+ [\perp]} \mathscr{L}$ and $x^{- [\perp]} \mathscr{L}$, and so $x = x^+ + x^{- [\perp]} \mathscr{L}$. Moreover, the vector x is neutral, because (see (2.7)) $[x, x] = \| x^+ \|^2 - \| x^- \|^2 = 0$. At the same time $x \notin \mathscr{L}$, because otherwise it would follow from the relation $x^+ \perp P^+ \mathscr{L}$ for instance (with x^+ chosen in the same way) that $\| x^+ \|^2 = (x^+, x^+) = (x^+, P^+ x) = 0$, contrary to the condition $\| x^+ \| > 0$. It remains to consider the neutral lineal $\widetilde{\mathscr{L}} = \text{Lin}(\mathscr{L}, x) \supset \mathscr{L}$ ($\widetilde{\mathscr{L}} \neq \mathscr{L}$) to arrive at a contradiction with the maximal neutrality of $\mathscr{L}$.

Sufficiency: If, for example, condition (I) holds, then the neutral subspace $\mathscr{L}$ is, by Theorem 4.5, maximal non-negative, and therefore is also a maximal neutral lineal. ■

Corollary 4.14: *Every maximal neutral subspace is either a maximal non-negative subspace, or a maximal non-positive subspace, or is simultaneously both the one and the other.*

□ This follows from the conditions (I), (II) of Theorem 4.13, and Theorem 4.5. ■

Definition 4.15: If a maximal neutral lineal $\mathscr{L}$ is simultaneously both maximal non-negative and maximal non-positive, then it is called a *hyper-maximal neutral linear.*

□ *Remark 4.16:* From the Definition 4.15 and Corollary 4.6 it follows that $\dim \mathscr{L} = \dim \mathscr{H}^{\pm}$, and so maximal neutral subspaces can exist only in those Krein spaces $\mathscr{H} = \mathscr{H}^+ [\oplus] \mathscr{H}^-$ in which $\dim \mathscr{H}^+ = \dim \mathscr{H}^-$. ■

5 To conclude this section we indicate other criteria for neutrality, maximal and hyper-maximal neutrality of a lineal $\mathscr{L}$ ($\subset \mathscr{H}$) which can be formulated and proved without reference to a canonical decomposition $\mathscr{H} = \mathscr{H}^+ [\dot{+}] \mathscr{H}^-$, i.e., in a form valid for arbitrary spaces $\mathscr{F}$ with an indefinite metric (see §1).

4.17 For neutrality of $\mathscr{L}$ it is necessary and sufficient that $\mathscr{L} \subset \mathscr{L}^{[\perp]}$.

□ If $\mathscr{L}$ is neutral, then all its non-zero vectors are isotropic (see 1.17), and therefore $\mathscr{L} \subset \mathscr{L}^{[\perp]}$. Conversely, it follows from this inclusion that the lineal $\mathscr{L}^0 = \mathscr{L} \cap \mathscr{L}^{[\perp]}$ isotropic for $\mathscr{L}$ coincides with $\mathscr{L}$ itself, i.e., $\mathscr{L}$ is neutral. ■

4.18 In order that $\mathscr{L}$ $(\subset \mathscr{P}^0)$ should be maximal in $\mathscr{P}^0$, it is necessary and sufficient that $\mathscr{L}^{[\perp]}$ be semi-definite and $\mathscr{L}^{[\perp][\perp]} = \mathscr{L}$.

□ If $\mathscr{L}$ is a maximal linear in $\mathscr{P}^0$, then by proposition 4.17 $\mathscr{L} \subset \mathscr{L}^{[\perp]}$, and therefore $\mathscr{L}^{[\perp]}$ cannot be indefinite, since otherwise in any decomposition $\mathscr{L}^{[\perp]} = \mathscr{L} \dotplus \mathscr{L}_1$ the lineal $\mathscr{L}_1$ would also be indefinite (since $\mathscr{L} [\perp] \mathscr{L}_1$) and so, by virtue of 1.9, it would contain neutral vectors not falling within $\mathscr{L}$, and this would contradict the maximality of $\mathscr{L}$ $(\subset \mathscr{P}^0)$. Thus $\mathscr{L}^{[\perp]}$ is semi-definite. Further, since from Proposition 1.17 we have that any neutral vector from $\mathscr{L}^{[\perp]}$ is isotropic for $\mathscr{L}^{[\perp]}$, by virtue of its maximality $\mathscr{L}$ $(\subset \mathscr{P}^0)$ must coincide with the isotropic lineal $(\mathscr{L}^{[\perp]})^0$, i.e., $\mathscr{L} = \mathscr{L}^{[\perp]} \cap \mathscr{L}^{[\perp][\perp]}$. But $\mathscr{L}^{[\perp][\perp]} \supset \mathscr{L}$ (see (1.11)), i.e., $\mathscr{L}^{[\perp][\perp]} = \mathscr{L}$.

Conversely, let $\mathscr{L} \subset \mathscr{L}^{[\perp]}$, let $\mathscr{L}^{[\perp]}$ be semi-definite, and $\mathscr{L}^{[\perp][\perp]} = \mathscr{L}$. Let $\tilde{\mathscr{L}} \supset \mathscr{L}$, where $\tilde{\mathscr{L}} \subset \mathscr{P}^0$, and there is an $x_0 \in \tilde{\mathscr{L}} \setminus \mathscr{L}$. By Proposition 1.17 the vector x_0 is isotropic for $\tilde{\mathscr{L}}$ and *a fortiori* $x_0 [\perp] \mathscr{L}$, i.e., $x_0 \in \mathscr{L}^{[\perp]}$. But because $\mathscr{L}^{[\perp]}$ is semi-definite the natural vector x_0 is also isotropic for $\mathscr{L}^{[\perp]}$, i.e., $x_0 \in \mathscr{L}^{[\perp]} \cap \mathscr{L}^{[\perp][\perp]} = \mathscr{L}^{[\perp]} \cap \mathscr{L}$, and therefore $x_0 \in \mathscr{L}$, contrary to the choice of x_0. ■

4.19 From hyper-maximal neutrality of $\mathscr{L}$ it is necessary and sufficient that $\mathscr{L} = \mathscr{L}^{[\perp]}$.

□ From maximality neutrality of $\mathscr{L}$ alone it follows, by 4.18, that $\mathscr{L}^{[\perp]}$ $(\supset \mathscr{L})$ is semi-definite, and from the hyper-maximal neutrality of $\mathscr{L}$ it follows, by virtue of the Definition 4.15, that $\mathscr{L} = \mathscr{L}^{[\perp]}$.

Conversely, if $\mathscr{L} = \mathscr{L}^{[\perp]}$, then, by Proposition 4.17, $\mathscr{L} \subset \mathscr{P}^0$. If now the lineal $\tilde{\mathscr{L}} \supset \mathscr{L}$ is semi-definite, then $\mathscr{L} [\perp] \tilde{\mathscr{L}}$, and therefore $\tilde{\mathscr{L}} \subset \mathscr{L}^{[\perp]} = \mathscr{L}$, i.e., $\tilde{\mathscr{L}} = \mathscr{L}$. ■

For neutral subspaces $\mathscr{L}$ in a Krein space $\mathscr{H} = \mathscr{H}^+ [\oplus] \mathscr{H}^-$ Proposition 4.19 enables us to obtain the curious formula:

4.20 *If the subspace $\mathscr{L} \subset \mathscr{P}^0$, then*

$$\mathscr{L}^{[\perp]} = \mathscr{L} [\dotplus] \mathscr{D}_{\mathscr{L}}^+ [\dotplus] \mathscr{D}_{\mathscr{L}}^-, \tag{4.5}$$

where $\mathscr{D}_{\mathscr{L}}^\pm = \mathscr{H}^\pm \ominus P^\pm \mathscr{L}$ are the deficiency subspaces for $\mathscr{L}$.

□ The inclusion $\mathscr{L} [\dotplus] \mathscr{D}_{\mathscr{L}}^+ [\dotplus] \mathscr{D}_{\mathscr{L}}^- \subset \mathscr{L}^{[\perp]}$ is obvious (*cf.* Proposition 4.17 and the proof of Theorem 4.13). Conversely, let $z \in \mathscr{L}^{[\perp]}$. Having expressed $\mathscr{H}^\pm$ in the form $\mathscr{H}^\pm = P^\pm \mathscr{L} [\oplus] \mathscr{D}_{\mathscr{L}}^\pm$, we have $\mathscr{H} = \mathscr{H}' [\oplus](\mathscr{D}_{\mathscr{L}}^+ [\oplus] \mathscr{D}_{\mathscr{L}}^-)$, where $\mathscr{H}' = P^+ \mathscr{L} [\oplus] P^- \mathscr{L}$ is obviously again a Krein space, and $\mathscr{L}$ $(\subset h')$ is a hyper-maximal neutral subspace in $\mathscr{H}'$. Now for $x = x^+ + x^-$ $(x^\pm \in \mathscr{H}^\pm)$ we have $x^\pm = x_1^\pm + x_2^\pm$, where $x_1^\pm \in P^\pm \mathscr{L}$ $(\subset \mathscr{H}')$ and $x_2^\pm \in \mathscr{D}_{\mathscr{L}}^\pm$. But $x = (x_1^+ + x_1^-) + (x_2^+ + x_2^-) [\perp] \mathscr{L}$ and $x_2^+ + x_2^- [\perp] \mathscr{L}$. Therefore $(\mathscr{H}' \ni)$ $x_1^+ + x_1^- [\perp] \mathscr{L}$, and since $\mathscr{L}$ is hyper-maximal in $\mathscr{H}'$, so (*cf.* 4.19) $x_1^+ + x_1^- \in \mathscr{L}$. ■

With formula (4.5) and the argument we have just given there is closely connected the following Proposition, which will be extremely useful later:

4.21 If under the conditions of Proposition 4.10 $\mathscr{L}'$ is a maximal non-negative (non-positive) subspace in the Krein space $\mathscr{D}^+_{\mathscr{L}} [\dot{+}] \mathscr{D}^-_{\mathscr{L}}$, then $\mathscr{L}' + \mathscr{L}$ is a maximal non-negative (maximal non-positive) subspace in $\mathscr{H}$.

□ We restrict ourselves to the case $\mathscr{L}' \subset \mathscr{P}^+$ and (see Theorem 4.5) $P^+\mathscr{L}' = \mathscr{D}^+_{\mathscr{L}}$. Then $P^+(\mathscr{L}' + \mathscr{L}) = \text{Lin}\{P^+\mathscr{L}', P^+\mathscr{L}\} = \text{Lin}\{\mathscr{D}^+_{\mathscr{L}}, {}^+\mathscr{L}\} = \mathscr{H}^+$. It remains to use Theorem 4.5. ■

Exercises and problems

1 If $\mathscr{L}_1$ and $\mathscr{L}_2$ are maximal non-negative (maximal non-positive) lineals, and $\mathscr{L} = \text{Lin}\{\mathscr{L}_1, \mathscr{L}_2\}$, then for an isotropic lineal $\mathscr{L}^0$ the inclusion $\mathscr{L}^0 \subset \mathscr{L}_1 \cap \mathscr{L}_2$ holds ([V]).

2 Prove that a non-closed positive lineal $\mathscr{L}_1$ $(\subset \mathscr{H})$ exists with $\overline{P^+\mathscr{L}_1} = \mathscr{H}^+$ which admits a proper extension into a positive, but again non-closed, lineal $\mathscr{L}$ $(\supset \mathscr{L}_1)$. *Hint:* As $\mathscr{L}$ use, for example, the maximal positive lineal in Example 4.12, and then, having again applied Lemma 4.11, construct on $\mathscr{L}_1 \subset \mathscr{L}$.

3 We consider for $\mathscr{L}_1$ $(\subset \mathscr{H})$ its deficiency subspaces $\mathscr{D}^{\pm}_{\mathscr{L}_1} = \mathscr{H}^{\pm} \ominus \overline{P^{\pm}\mathscr{L}_1}$. Let $\mathscr{L}$ be the maximal lineal from $\mathscr{P}^+$ and $\mathscr{L}_1 \subset \mathscr{L}$. Then $\dim(\mathscr{L} \ominus \overline{\mathscr{L}}_1) = \dim \mathscr{D}^+_{\mathscr{L}_1}$ ([V]).

4 Give examples of hyper-maximal neutral lineals $\mathscr{L}$ in the space $\mathscr{H}$ from Example 3.11.

5 In the real, two-dimensional, Krein space (see Exercise 7 to §2) distinguish by shading in Fig. 1 the sets $\mathscr{P}^+$ and $\mathscr{P}^-$ (see (1.7)), and find positive, negative, and neutral lineals. Verify that they are all maximal (in their respective classes) and that the neutral lineals are even hyper-maximal.

6 Show that in a Krein space Proposition 4.19 and 4.20 (i.e., Formula (4.5)) are equivalent.

§5 Uniformly definite (regular) lineals and subspaces. Subspaces of the classes $h^{\pm}$

1 In this section we shall mainly discuss certain subclasses of definite lineals and subspaces in a Krein space $\mathscr{H} = \mathscr{H}^+ [\oplus] \mathscr{H}^-$. In this connection the basic concept by means of which these subclasses will be distinguished is the so-called *intrinsic metric*.

Suppose, for definiteness, that the lineal $\mathscr{L}$ is positive. Since the form $[x, y]$ is positive definite ($[x,] > 0$ for all $(\theta \neq)\ x \in \mathscr{L}$), it can be adopted as a *new scalar product* on $\mathscr{L}$; by so doing, $\mathscr{L}$ is converted into, generally speaking, a pre-Hilbert space.

Definition 5.1: Without introducing new designations we shall call the restriction of the metric $[x, y]$ on to $\mathscr{L}$ the *intrinsic metric* on $\mathscr{L}$, and the corresponding norm

$$|x|_{\mathscr{L}} = [x, x]^{1/2} \qquad (x \in \mathscr{L}) \tag{5.1}$$

will be called the *intrinsic norm* on $\mathscr{L}$.

□ From (3.7) it follows that the estimate

$$|x|_{\mathscr{L}} \leqslant \|x\| \qquad (x \in \mathscr{L}) \tag{5.2}$$

holds for the intrinsic norm $|x|_{\mathscr{L}}$ on $\mathscr{L}$ introduced by formula (5.1), i.e., $|x|_{\mathscr{L}}$ is less than the basic ('exterior') norm $\|x\|$ on $\mathscr{L}$. ■

Our first aim is to distinguish the case when these two norms are equivalent.

Definition 5.2: A positive lineal $\mathscr{L}$ is said to be *uniformly positive* if there is a constant $\alpha > 0$ such that

$$|x|_{\mathscr{L}} \geqslant \alpha \|x\| \qquad (x \in \mathscr{L}) \tag{5.3}$$

or, what is the same thing,

$$[x, x] \geqslant \alpha^2 \|x\|^2 \qquad (x \in \mathscr{L}). \tag{5.1}$$

It is clear that (5.3) combined with (5.2) means that the norms $|x|_{\mathscr{L}}$ and $\|x\|$ are equivalent on $\mathscr{L}$.

Entirely similarly—that is, by the same requirement (5.3)—we introduce the definition of a *uniformly negative lineal* $\mathscr{L}$; this time the intrinsic norm $|x|_{\mathscr{L}}$ on a negative lineal $\mathscr{L}$ is defined by the equality

$$|x|_{\mathscr{L}} = |[x, x]|^{1/2} = (-[x, x])^{1/2} \qquad (x \in \mathscr{L}). \tag{5.5}$$

In this case the relation (5.4) becomes

$$-[x, x] \geqslant \alpha^2 \|x\|^2 \qquad (x \in \mathscr{L}). \tag{5.6}$$

Uniformly positive and uniformly negative lineals $\mathscr{L}$ are combined under the general name *uniformly definite lineals.*

□ Uniformly definite lineals have an obvious 'inheritance' property:

5.3 *Every lineal which is contained in a uniformly positive (uniformly negative) lineal is itself uniformly positive (uniformly negative).* ■

The simplest examples of uniformly definite lineals are the components $\mathscr{H}^{\pm}$ of the canonical decomposition $\mathscr{H} = \mathscr{H}^{+} [\oplus] \mathscr{H}^{-}$, and also, by virtue of Proportion 5.3, all lineals which are contained in $\mathscr{H}^{+}$ ($\mathscr{H}^{-}$).

5.4 *All finite-dimensional definite lineals are uniformly definite.*

□ This follows from the fact that in a finite-dimensional space $\mathscr{L}$ all norms (including $\|\cdot\|$ and $|\cdot|_{\mathscr{L}}$) are equivalent. ■

5.5 *$\mathscr{L}$ is uniformly definite if and only if its closure, the subspace $\bar{\mathscr{L}}$, is uniformly definite.*

□ This follows immediately from 5.3 and the fact that the inequalities (5.4) and (5.6) continue to hold on passage from $\mathscr{L}$ to its closure $\bar{\mathscr{L}}$ (see proposition 3.5). ■

2 The question of the connection between the *closedness* of a definite lineal $\mathscr{L}$ ($\subset \mathscr{H}$) relative to the exterior norm and its *completeness* relative to the intrinsic norm $|\cdot|_{\mathscr{L}}$ (*intrinsic completeness*) is of particular interest. When $\mathscr{L}$ is uniformly definite, i.e., when the norms $\|\cdot\|$ and $|\cdot|_{\mathscr{L}}$ are equivalent, it is clear that the two properties just mentioned either hold or do not hold simultaneously. This assertion admits a partial inversion:

5.6 A definite subspace $\mathscr{L}$ $(=\bar{\mathscr{L}})$ is uniformly definite if and only if it is intrinsically complete.

□ Because $\mathscr{H}$ is complete, closure of $\mathscr{L}$ relative to $\|\cdot\|$ implies the completeness of $\mathscr{L}$ relative to the same norm. Now let $\mathscr{L}$ be also intrinsically complete. Since the norms $\|\cdot\|$ and $|\cdot|_{\mathscr{L}}$ are connected by the inequality (5.2), these norms are equivalent, by a well-known Banach theorem.

Conversely, uniform definiteness of $\mathscr{L}$ implies the equivalence of the norms $\|\cdot\|$ and $|\cdot|_{\mathscr{L}}$ on $\mathscr{L}$. Therefore from the completeness of $\mathscr{L}$ in the norm $\|\cdot\|$ follows its intrinsic completeness. ■

As regard *non-closed* definite lineals $\mathscr{L}$, however, the situation is more complicated; such a lineal $\mathscr{L}$ may be either complete or not complete relative to the intrinsic norm $|\cdot|_{\mathscr{L}}$. We have actually already met the first of these two cases (the lineal $\mathscr{L}$ in Example 4.12) (see §9, Exercise 11 below). However, it is clear that in the case of intrinsic completeness such a (non-closed) lineal cannot be uniformly definite.

3 Later, in §6, *inter alia* we establish criteria (necessary and sufficient conditions) for the uniform definiteness of closed lineals (of subspaces—see Proposition 6.11), and in §8 (Lemma 8.4) a method of describing all such subspaces is indicated. But here we return for a moment to the first examples of uniformly definite lineals—to the subspaces $\mathscr{H}^+$ and $\mathscr{H}^-$ in the canonical decomposition $\mathscr{H} = \mathscr{H}^+ [\oplus] \mathscr{H}^-$. In a certain sense these examples can be called models, as the next theorem indicates.

Theorem 5.7: *In order that a linear $\mathscr{L}$ $(\neq \{\theta\})$ of a Krein space $\mathscr{H}$ should be a Krein space (relative to the original metric $[x, y]$) it is necessary and sufficient to admit a J-orthogonal decomposition $\mathscr{L} = \mathscr{L}^+ [\dotplus] \mathscr{L}^-$ into two uniformly definite subspaces: a positive one $\mathscr{L}^+$ and a negative one $\mathscr{L}^-$. Here it is not excluded that $\mathscr{L}^+ \{\theta\}$ or $\mathscr{L}^- = \{\theta\}$.*

□ *Necessity:* The necessary condition follows from Definition 2.2 of a Krein space, from the Definitions (2.2) and (2.3) of the scalar product and norm $\|\cdot\|$ in a Krein space, and from the Definitions (5.4) and (5.6).

Sufficiency follows from Proposition 5.6 and the Definition (2.2). ■

Corollary 5.8: *If $\mathscr{L}$ is a neutral subspace of a Krein space, then the factor-space $\hat{\mathscr{H}} \equiv \mathscr{L}^{[\perp]}/\mathscr{L}$ is again a Krein space* (cf. Exercise 10 in §2).

□ Recalling formula (3.5) we have $\mathscr{L}^{[\perp]} = \mathscr{L}\,[\dot{+}]\,\mathscr{D}^+_{\mathscr{L}}\,[\dot{+}]\,\mathscr{D}^-_{\mathscr{L}}$, where $\mathscr{D}^+_{\mathscr{L}}\,[\dot{+}]\,\mathscr{D}^-_{\mathscr{L}}$ is a Krein space by 5.3 and Theorem 5.7. It remains to apply 1.23. ■

4 The concept of a uniformly definite subspace admits a certain generalization which will later play an important part in the theory of certain classes of linear operators in Krein spaces (see Chapter 3, §5).

Definition 5.9: A non-negative (non-positive) subspace $\mathscr{L}$ of a Krein space $\mathscr{H}$ is called a *subspace of class h^+* (*class h^-*) if it admits a decomposition $\mathscr{L} = \mathscr{L}^0\,[\dot{+}]\,\mathscr{L}^+$ ($\mathscr{L} = \mathscr{L}^0\,[\dot{+}]\,\mathscr{L}^-$) into a direct J-orthogonal sum of a *finite-dimensional isotropic* subspace $\mathscr{L}^0$ ($\dim \mathscr{L}^0 < \infty$) and a *uniformly positive* (*uniformly negative*) subspare $\mathscr{L}^+$ ($\mathscr{L}^-$).

A caution here is necessary: the fact that a subspace $\mathscr{L}$ belongs, say, to the class h^+ does not in any way imply that in *every* decomposition $\mathscr{L} = \mathscr{L}^0\,[\dot{+}]\,\mathscr{L}^+_1$ the positive component (the lineal $\mathscr{L}^+_1$) will be uniformly positive.

Example 5.10: In Example 4.12, the subspace $\tilde{\mathscr{L}} \in h_1$, because it admits the decomposition $\tilde{\mathscr{L}} = \mathrm{Lin}(e^+_{\alpha_0} + e^-)\,[\dot{+}]\,C \quad \mathrm{Lin}\{e^+_\alpha\}_{\alpha \in A, \alpha \neq \alpha_0}$, in which $(\tilde{\mathscr{L}})^0 = \mathrm{Lin}\{e^+_{\alpha_0} + e^-\}$ is a one-dimensional isotropic subspace, and $(\mathrm{Lin}\{e^+_\alpha\}_{\alpha \in A, \alpha \neq \alpha_0}$ is, by 5.3, uniformly positive.

On the other hand, the same $\tilde{\mathscr{L}}$ can be represented (see Example 4.12) in the form $\tilde{\mathscr{L}} = (\tilde{\mathscr{L}})^0[\dot{+}]\tilde{\mathscr{L}}$, where $\mathscr{L}$ is by no means uniformly positive—for if it were, then by 5.5 its closure $\bar{\mathscr{L}}$ would also be uniformly positive, but in Example 4.12 it was degenerate!

Exercises and problems

1 Any Krein $\mathscr{H} = \mathscr{H}^+\,[\oplus]\,\mathscr{H}^-$ with infinite-dimensional $\mathscr{H}^\pm$ contains definite subspaces which are not uniformly definite (Ginzburg [4]).

In the case of separable $\mathscr{H}^\pm$ with orthonormalized bases $\{e^+_k\}^\infty_{k=1}$ and $\{e^-_k\}^\infty_{k=1}$, consider, for example, the subspace $\mathscr{L} = C\,\mathrm{Lin}\{e^+_k + (k/k+1)e^-_k\}^\infty_{k=2}$; prove that it is positive but not uniformly positive (I. Iokhvidov [13]). Modify this example for negative subspaces and for non-separable $\mathscr{H}^\pm$.

2 Let $\mathscr{L}$ be an arbitrary lineal in a Krein space, and let $\mathscr{D}_{\mathscr{L}} = \mathscr{L}^{[\perp]} \cap \mathscr{L}^\perp$. Then $\mathscr{D}_{\mathscr{L}}$ is a Krein space (see Ritsner [4]).

Hint. Prove that $\mathscr{D}_{\mathscr{L}} = \mathscr{D}^+_{\mathscr{L}}\,[\dot{+}]\,\mathscr{D}^-_{\mathscr{L}}$, where $\mathscr{D}^\pm_{\mathscr{L}}$ are the deficiency subspaces for $\mathscr{L}$ (see §4), and then apply 5.3 and Theorem 5.7.

3 Prove that all maximal, uniformly definite lineals are closed, and obtain the maximality principle for them (*cf.* Theorem 1.19).

4 Every maximal uniformly definite lineal $\mathscr{L}$ is a maximal semi-definite lineal ([V]).
Hint: If $\mathscr{L}$ is positive, for example, then, assuming the contrary, consider its deficiency subspace $\mathscr{D}^+_{\mathscr{L}}$ and prove that $\mathscr{L}_1 = \mathscr{L} [\dot{+}] \mathscr{D}^+_{\mathscr{L}}$ is uniformly positive.

5 Let $\mathscr{L}$ ($\subset \mathscr{H}$) be a definite lineal, and let $y_0 \in \mathscr{H}$. In order that the functional $\varphi_{y_0} = [x, y_0]$ should be continuous relative to the intrinsic norm $|\cdot|_{\mathscr{L}}$ it is necessary and sufficient that

$$m(y_0) = \inf_{x \in \mathscr{L}} [x - y_0, x - y_0] > -\infty \quad \text{(if } \mathscr{L} \text{ is positive)},$$

$$M(y_0) = \sup_{x \in \mathscr{L}} [x, y_0, x - y_0] < \infty \quad \text{(is } \mathscr{L} \text{ is negative)}.$$

If these conditions hold for the (intrinsic) norm $|\varphi_0|_{\mathscr{L}}$ of the functional φ_{y_0} we have $|\varphi_{y_0}|^2_{\mathscr{L}} = [y_0, y_0] - m(y_0)$ and $|\varphi_{y_0}|_{\mathscr{L}} = M(y_0) - [y_0, y_0]$ respectively (I. Iakhvidov, see [VIII].

6 A definite lineal $\mathscr{L}$ ($\subset \mathscr{H}$) is said to be *regular* if for every $y \in \mathscr{H}$ the linear functional $\varphi_y(x) = [x, y]$ ($x \in \mathscr{L}$) is continuous relative to the intrinsic norm $|\cdot|_{\mathscr{L}}$, and to be *singular* otherwise (I. Iokhvidov [7]; see also [VIII]. Prove that regularity of a definite lineal is equivalent to its being uniformly definite ([VIII]).
Hint: Extend by the Hahn–Banach theorem every linear functional which is continuous (relative to the external norm $\|\cdot\|$) on $\mathscr{L}$ into a functional continuous on the whole of $\mathscr{H}$, and then apply F. Riesz's theorem, and thus show that the stock of linear functionals on $\mathscr{L}$ which are continuous relative to the norms $\|\cdot\|$ and $|\cdot|_{\mathscr{L}}$ is the same, from which the equivalence of these norms will follow.

7 Let the subspace $\mathscr{L} \subset \mathscr{P}^0$, and let $\hat{\mathscr{H}} = \mathscr{L}^{[\perp]}/\mathscr{L}$ be the corresponding Krein space (see Corollary 5.8). Then maximal semi-definiteness in $\hat{\mathscr{H}}$ of the subspace $\hat{\mathscr{L}}$ is equivalent to the corresponding semi-definiteness in $\mathscr{H}$ of the lineal $\text{Lin}\{x \mid x \in \hat{x}, \hat{x} \in \hat{\mathscr{L}}\}$.
Hint. Use (4.5), 1.23 and 4.21.

§6 Decomposability of lineals and subspaces of a Krein space. The Gram operator of a subspace. *W*-spaces and *G*-spaces

1 In §1, paragraph 9 we touched on, and in Theorem 1.34 considered in rather more detail, the question whether an arbitrary non-degenerate lineal $\mathscr{L}$ can be decomposed into the direct sum $\mathscr{L} = \mathscr{L}^+ \dot{+} \mathscr{L}^-$ of positive and negative lineals $\mathscr{L}^\pm$. Here we shall narrow down the problem somewhat, since we shall discuss only lineals $\mathscr{L}$ in a Krein space $\mathscr{H}$ and decompositions of the form

$$\mathscr{L} = \mathscr{L}^+ [\dot{+}] \mathscr{L}^-, \tag{6.1}$$

in which $\mathscr{L}^+ [\perp] \mathscr{L}^-$.

Definition 6.1: If a *J*-orthogonal decomposition (6.1) exists for a lineal $\mathscr{L}$ ($\subset \mathscr{H}$), then $\mathscr{L}$ is called a *decomposable lineal*; otherwise $\mathscr{L}$ is called an *indecomposable lineal*.

However, even the transition to Krein spaces does not reduce the problems of decomposability of non-degenerate lineals $\mathscr{L}$ (non-degeneracy is a necessary condition, see 1.24).

□ As an example we consider in the Krein space $\tilde{\mathscr{F}}$ of Exercise 3 to §2 the lineal $\mathscr{L}$ consisting of all those functions $\varphi \in \tilde{\mathscr{F}}$ which, on the interval $[-1,0]$ coincide with polynomials. It is clear that $\mathscr{L}$ is dense in $\tilde{\mathscr{F}}$, and therefore it is a non-degenerate lineal. On the other hand, $\mathscr{L} = \mathscr{L}_1 \dotplus \mathscr{L}_2$, where $\mathscr{L}_1$ consists of all those functions $\varphi \in \mathscr{L}$ which are equal to 0 on $[-1,0]$, and $\mathscr{L}_2$ consists of all those $\varphi \in \mathscr{L}$ which are equal to 0 on $[0,1]$. $\mathscr{L}_1$ and $\mathscr{L}_2$ are neutral (see the metric (1.20)) and by construction they are not isomorphic (the linear dimension of $\mathscr{L}_1$ is greater than that of $\mathscr{L}_2$). Hence (see Theorem 1.34) follows the impossibility in general of representing $\mathscr{L}$ in the form $\mathscr{L} = \mathscr{L}^+ \dotplus \mathscr{L}^-$ where $\mathscr{L}^\pm$ are definite, and *a fortiori* the indecomposability of $\mathscr{L}$. ■

2 At the same time, the transition to Krein spaces enables us to solve completely (and moreover in the positive sense) the problem of the decomposability of its closed lineals, i.e., of subspaces. We start from an elementary fact:

6.2 An isotropic lineal $\mathscr{L}^0$ of any subspace $\mathscr{L}$ in a Krein space is closed, i.e., it is an isotropic subspace.

□ This follows immediately from the fact that, for any $x \in \mathscr{L}$, the passage to the limit as $n \to \infty$ in the equality $[x_n^0, x] = 0$, which holds for any sequence of vectors isotropic in $\mathscr{L}$, $x_n^0 \to x^0$ $(\in \mathscr{L})$ gives, by 3.5, $[x^0, x] = 0$, i.e., the vector x^0 is isotropic in $\mathscr{L}$ (if $x^0 \neq \theta$).

Later is this section (and in a number of other questions) the concept of the *Gram operator* of a given subspace will play a leading part. Let $\mathscr{L}$ be a subspace of the Krein space $\mathscr{H} = \mathscr{H}^+ [\oplus] \mathscr{H}^-$ with the canonical symmetry operator $J = P^+ - P^-$ and the J-metric $[x, y] = (Jx, y)$ (see §3, paragraph 1).

We consider the restriction of this J-metric on to $\mathscr{L}$: $[x, y]_{\mathscr{L}}$, i.e., simply the form $[x, y]$ with the arguments x, y traversing only $\mathscr{L}$. Since $\mathscr{L}$, like the whole of $\mathscr{H}$, is a Hilbert space with the scalar product (x, y) $(x, y \in \mathscr{L})$ and the norm $\|x\| = (x, x)^{1/2}$ $(x \in \mathscr{L})$, and $[x, y]_{\mathscr{L}}$ is a sesquilinear (and moreover a bounded (see 3.5)) functional in $\mathscr{L}$ (or, more precisely, in $\mathscr{L} \times \mathscr{L}$), there is a unique bounded self-adjoint operator $G_{\mathscr{L}}$ operating in $\mathscr{L}$ such that

$$([x, y]_{\mathscr{L}} =)\ [x, y] = (G_{\mathscr{L}} x, y) \qquad (x, y \in \mathscr{L}).$$

Let $P_{\mathscr{L}}$ be the (Hilbert) orthoprojector on to $\mathscr{L}$. Then for any $x, y \in \mathscr{L}$

$$(G_{\mathscr{L}} x, y) = [x, y] = (Jx, y) = (Jx, P_{\mathscr{L}} y) = (P_{\mathscr{L}} Jx, y),$$

and so $G_{\mathscr{L}} = P_{\mathscr{L}}(J \mid \mathscr{L})$, where $J \mid \mathscr{L}$ is, as usual, the restriction of J on to $\mathscr{L}$.

Definition 6.3: The linear operator $G_{\mathscr{L}} = P_{\mathscr{L}} J \mid \mathscr{L}$ is called the *Gram operator* of the subspace $\mathscr{L}$.

3 Noting that $\| G_{\mathscr{L}} \| = \| P_{\mathscr{L}}(J \mid \mathscr{L}) \| \leqslant 1$, we now use the spectral decomposition of the bounded self-adjoint operator $G_{\mathscr{L}}$ in $\mathscr{L}$:

$$G_{\mathscr{L}} = \int_{-1}^{1} \lambda \, dE_{\mathscr{L}}(\lambda), \tag{6.2}$$

where $E_{\mathscr{L}}(\lambda)$ is the *spectral function* (*resolution of the identity*) of the operator $G_{\mathscr{L}}$, and the integral in (6.2) is understood as the limit of the corresponding integral sums in the uniform operator topology (i.e., with respect to the operator norm; see e.g. [XXIII]).

Suppose, for definiteness, that the spectral function $E_{\mathscr{L}}(\lambda)$ is *strongly continuous on the right*, i.e. $E_{\mathscr{L}}(\lambda + 0) = E_{\mathscr{L}}(\lambda)$, where $E_{\mathscr{L}}(\lambda + 0) = s - \lim_{\mu \downarrow \lambda} E_{\mathscr{L}}(\mu)$ (strong limit). We bring into consideration three orthoprojectors in the Hilbert space $\mathscr{L}$:

$$P_{\mathscr{L}}^{-} = \int_{-1}^{-0} dE_{\mathscr{L}}(\lambda) = E_{\mathscr{L}}(-0), \qquad P_{\mathscr{L}}^{0} = E_{\mathscr{L}}(0) - E_{\mathscr{L}}(-0),$$

$$P_{\mathscr{L}}^{+} = \int_{0}^{1} dE_{\mathscr{L}}(\lambda) = I_{\mathscr{L}} - E_{\mathscr{L}}(0), \tag{6.3}$$

where the symbol $\int_{-1}^{-0}$ means that in setting up the corresponding integral sums the value of $E_{\mathscr{L}}(\lambda)$ at zero is here taken to be equal to $E_{\mathscr{L}}(-0)$ (*cf.* [XXIII]).

It follows from the properties of the spectral function that the three orthoprojectors defined in (6.3) are pairwise orthogonal, and, by their definition,

$$P_{\mathscr{L}}^{-} + P_{\mathscr{L}}^{0} + P_{\mathscr{L}}^{+} = \int_{-1}^{1} dE_{\mathscr{L}}(\lambda) = I_{\mathscr{L}},$$

i.e., they generate an orthogonal (in the Hilbert metric) decomposition of $\mathscr{L}$:

$$\mathscr{L} = \mathscr{L}^{-} \oplus \mathscr{L}^{0} \oplus \mathscr{L}^{+} \qquad (\mathscr{L}^{\pm} = P_{\mathscr{L}}^{\pm}\mathscr{L},\ \mathscr{L}^{0} = P_{\mathscr{L}}^{0}\mathscr{L}) \tag{6.4}$$

Theorem 6.4: *In the decomposition* (6.4) $\mathscr{L}^{-}$ *and* $\mathscr{L}^{+}$ *are pairwise J-orthogonal negative and positive subspaces respectively, and* $\mathscr{L}^{0}$ *is the isotropic subspace for* $\mathscr{L}$.

□ It should be made clear at once that, even when $\mathscr{L} \neq \{\theta\}$, any one (or even any two) of the subspaces on the right-hand side of the decomposition (6.4) may reduce to $\{\theta\}$; we shall not mention this fact again in future.

We start by proving that $\mathscr{L}^{0}$ is an isotropic subspace for $\mathscr{L}$. By definition $\mathscr{L}^{0} = P_{\mathscr{L}}^{0}\mathscr{L} = (E_{\mathscr{L}}(0) - E_{\mathscr{L}}(-0))\mathscr{L}$, i.e., $\mathscr{L}_0$ is the eigen-subspace of the operator $G_{\mathscr{L}}$ corresponding to the eigenvalue $\lambda = 0$ (more shortly, $\mathscr{L}^{0} = \operatorname{Ker} G_{\mathscr{L}}$), and this is equivalent to $\mathscr{L}^{0}$ being isotropic in $\mathscr{L}$. For, if $x_0 \in \mathscr{L}$ the relation $x_0 [\perp] \mathscr{L}$ is equivalent to the fact that, for all $x \in \mathscr{L}$

$$([x_0, x] =)\ (G_{\mathscr{L}} x_0, x) = 0, \qquad \text{i.e., } G_{\mathscr{L}} x_0 = \theta.$$

We now prove that $\mathscr{L}^-$ is non-positive. If $(\theta\neq)x\in\mathscr{L}^-=P_{\mathscr{L}}^-\mathscr{L}$, then, in accordance with (6.1)–(6.3)

$$[x,x]=(G_{\mathscr{L}}x,x)=\int_{-1}^{1}\mathscr{L}\,\mathrm{d}(E_{\mathscr{L}}(\lambda)x,x)=\int_{-1}^{1}\lambda\,\mathrm{d}(E_{\mathscr{L}}(\lambda)x,P_{\mathscr{L}}^-x)$$

$$=\int_{-1}^{1}\lambda\,\mathrm{d}(E_{\mathscr{L}}(\lambda)x,E_{\mathscr{L}}(-0)x)=\int_{-1}^{1}\lambda\,\mathrm{d}(E_{\mathscr{L}}(-0)E_{\mathscr{L}}(\lambda)x,x). \tag{6.5}$$

But

$$(E_{\mathscr{L}}(-0)E_{\mathscr{L}}(\lambda)x,x)=\begin{cases}(e_{\mathscr{L}}(\lambda)x,x) & \text{when } \lambda<0,\\ (E_{\mathscr{L}}(-0)x,x)\equiv\text{const} & \text{when } \lambda\geqslant 0.\end{cases}$$

Returning to (6.5) we obtain

$$[x,x]=\int_{-1}^{1}\lambda\,\mathrm{d}(E_{\mathscr{L}}(-0)E_{\mathscr{L}}(\lambda)x,x)=\int_{-1}^{-0}\lambda\,\mathrm{d}(E_{\mathscr{L}}(\lambda)x,x)\leqslant 0, \tag{6.6}$$

since $-1\leqslant\lambda<0$, and $(E_{\mathscr{L}}(\mathscr{L})x,x)$ is a non-decreasing function.

By means of similar calculations the reader will be able without difficulty to show that $\mathscr{L}^+$ is non-negative and $\mathscr{L}^-\,[\perp]\,\mathscr{L}^+$.

From what has been proved it follows in turn that $\mathscr{L}^-$ and $\mathscr{L}^+$ are definite, because a neutral vector x_0 contained in either of them would be an isotropic vector for this subspace (when $x\neq\theta$) and therefore, in view of the J-orthogonality $\mathscr{L}^0\,[\perp]\,\mathscr{L}^\pm$ and $\mathscr{L}^+\,[\perp]\,\mathscr{L}^-$ already proved it would be isotropic for the whole of $\mathscr{L}=\mathscr{L}^-\,[\oplus]\,\mathscr{L}^0\,[\oplus]\,\mathscr{L}^+$. But then $x_0\in\mathscr{L}^0$ and by (6.4) $x_0=\theta$. ■

We point out that in the course of proving Theorem 6.4 we have on the way established the proposition.

6.5 *$\mathscr{L}$ is non-degenerate if and only if* Ker $G_{\mathscr{L}}=\{\theta\}$.

□ This follows from the fact that $\mathscr{L}^0=\operatorname{Ker}G_{\mathscr{L}}$. ■

4 By analogy with Definition 2.1 and generalizing it slightly we adopt the

Definition 6.6. If $\mathscr{L}$ is a subspace of a Krein space, then any representation of $\mathscr{L}$ in the form of a J-orthogonal direct sum

$$\mathscr{L}=\mathscr{L}^0\,[\dotplus]\,\mathscr{L}^+\,[\dotplus]\,\mathscr{L}^- \tag{6.7}$$

of three subspaces, of which $\mathscr{L}^0$ is isotropic for $\mathscr{L}$ and $\mathscr{L}^+$ ($\mathscr{L}^-$) are positive (negative), is called a *canonical decomposition* of $\mathscr{L}$.

The existence of such a decomposition (see (6.4)) for any subspace $\mathscr{L}$ is established in Theorem 6.4. Moreover, it is clear that the (isotropic) component $\mathscr{L}^0$ of a canonical decomposition is defined uniquely (see 6.5). As regards the components $\mathscr{L}^\pm$, their construction in Theorem 6.4 was carried out in a special way by means of the spectral decomposition of the Gram operator $G_{\mathscr{L}}$, i.e., it was based on the scalar product (x, y). But the latter, as is well known, was defined (see (2.2)) by a fixed choice of the canonical decomposition $\mathscr{H} = \mathscr{H}^+ [\dot{+}] \mathscr{H}^-$ (or, equivalently, by the choice of the canonical symmetry operator J; see §3, paragraph 5). Hence it is already clear *a priori* that other canonical decompositions for $\mathscr{L}$ are possible, for example,

$$\mathscr{L} = \mathscr{L}^0 [\dot{+}] \mathscr{L}_1^+ [\dot{+}] \mathscr{L}_1^-, \tag{6.8}$$

in which, generally speaking, $\mathscr{L}_1^+ \neq \mathscr{L}^+$ and $\mathscr{L}_1^- \neq \mathscr{L}^-$ (see, e.g., Theorem 8.17 below). So the next theorem has all the more interest:

Theorem 6.7: (*Law of inertia*). *Whatever the canonical decomposition* (6.8) *of the subspace* $\mathscr{L}$ *may be, it is always the case that*

$$\dim \mathscr{L}_1^+ = \dim \mathscr{L}^+, \qquad \dim \mathscr{L}_1^- = \dim \mathscr{L}^-, \tag{6.9}$$

where $\mathscr{L}^+$ *and* $\mathscr{L}^-$ *are understood to be the corresponding components of any other canonical decomposition* (6.7) *of the same subspace* $\mathscr{L}$.

□ Starting from the canonical decomposition (6.7), say, we use the deduction made earlier at the end of Remark 4.4, by virtue of which we have that, for the linear dimensions, and also in the present case for the Hilbert dimensions,

$$\dim \mathscr{L}_1^+ \leqslant \dim \mathscr{L}^+, \qquad \dim \mathscr{L}_1^- \leqslant \dim \mathscr{L}^-.$$

But in this argument the rôles of the decompositions (6.7) and (6.8) can be interchanged; hence (6.9) follows. ■

So, let a certain canonical decomposition (6.7) of the subspace $\mathscr{L}$ be fixed.

Definition 6.8: The trio $(\pi_{\mathscr{L}}, \nu_{\mathscr{L}}, \omega_{\mathscr{L}})$ of cardinal numbers

$$\pi_{\mathscr{L}} = \dim \mathscr{L}^+, \qquad \nu_{\mathscr{L}} = \dim \mathscr{L}^-, \qquad \omega_{\mathscr{L}} = \dim \mathscr{L}^0$$

is called the *inertia index of the subspace* $\mathscr{L}$ and it is denoted by In $\mathscr{L}$.

It follows from Theorem 6.7 that In $\mathscr{L}$ is an invariant of the subspace $\mathscr{L}$ relative to all its possible canonical decompositions (6.7).

The concept of the inertia index makes it possible to introduce for an arbitrary subspace $\mathscr{L}$ of a Krein space $\mathscr{H}$ one further important characteristic, which will in particular come into the foreground in §9.

Definition 6.9: Let $\mathscr{L}$ be a subspace of a Krein space $\mathscr{H}$ and let In $\mathscr{L} = (\pi_{\mathscr{L}}, \nu_{\mathscr{L}}, \omega_{\mathscr{L}})$ be its inertia index. Then the cardinal number $\varkappa_{\mathscr{L}} = \min\{\pi_{\mathscr{L}}, \nu_{\mathscr{L}}\}$ is called the *rank of indefiniteness* of the subspace $\mathscr{L}$. In particular, the rank of indefiniteness of the whole space $\mathscr{H} = \mathscr{H}^+ [\oplus] \mathscr{H}^-$ is denoted simply by $\varkappa$, i.e., $\varkappa = \min\{\dim \varkappa^+, \dim \varkappa^-\}$.

5 ☐ From the results of §6, para. 3 (*cf.* (6.1) and (6.2)) this obvious proposition follows:

6.10 A subspace $\mathscr{L}$ of a Krein space is non-negative (non-positive) if and only if its Gram operator $G_{\mathscr{L}}$ is non-negative (non-positive), i.e., $(G_{\mathscr{L}}x, x) \geqslant 0$ $(\leqslant 0)$ for all $x \in \mathscr{L}$. A subspace $\mathscr{L}$ is positive (negative) if and only if $(G_{\mathscr{L}}x, x) > 0$ (<0) for all $(\theta \neq)$ $x \in \mathscr{L}$. ■

We also obtain without difficulty a criterion for uniform definiteness of a subspace $\mathscr{L}$ in terms of its Gram operator $G_{\mathscr{L}}$:

6.11 In order that a subspace $\mathscr{L}$ should be uniformly positive (uniformly negative) it is necessary and sufficient that its Gram operator $G_{\mathscr{L}}$ should, for some $\alpha > 0$, satisfy the inequality $G_{\mathscr{L}} \geqslant \alpha^2 I_{\mathscr{L}}$ $(G_{\mathscr{L}} \leqslant -\alpha^2 I_{\mathscr{L}})$.

☐ Since $[x, y] = (G_{\mathscr{L}}x, y)$ $(x, y \in \mathscr{L})$, the inequality $G_{\mathscr{L}} \geqslant \alpha^2 I_{\mathscr{L}}$ $(\leqslant \alpha^2 I_{\mathscr{L}})$ is equivalent to the condition $[x, x] > \alpha^2(x, x)$ $(\leqslant -\alpha^2(x, x))$ $(x \in \mathscr{L})$ for the subspace $\mathscr{L}$ to be uniformly positive (uniformly negative) (see (5.4)). ■

Corollary 6.12: *For a subspace $\mathscr{L}$ of a Krein space $\mathscr{H}$ to be itself a Krein space relative to the J-metric $[x, y]$ it is sufficient that the point $\lambda = 0$ should be a regular point for its Gram operator $G_{\mathscr{L}}$: $0 \in \rho(G_{\mathscr{L}})$.*

☐ The condition $0 \in \rho(G_{\mathscr{L}})$ is equivalent to the existence of an $\alpha > 0$ such that

$$G_{\mathscr{L}} = \int_{-1}^{-\alpha^2} \lambda \, dE_{\mathscr{L}}(\lambda) + \int_{\alpha^2}^{1} \lambda \, dE_{\mathscr{L}}(\lambda)$$

(the so-called 'spectral gap' $(-\alpha^2, \alpha^2)$ at zero). But then $\mathscr{L}^0 = \{\theta\}$, and in the canonical decomposition given by Theorem 6.4, $\mathscr{L} = \mathscr{L}^+ [\dotplus] \mathscr{L}^-$, the conditions of Proposition 6.11 hold for the Gram operator $G_{\mathscr{L}^\pm} = G_{\mathscr{L}} \mid \mathscr{L}^\pm$, and so the $\mathscr{L}^\pm$ are uniformly definite. It only remains to apply Theorem 5.7. ■

Later (see Theorem 7.16) it becomes clear that the condition $0 \in \rho(G_{\mathscr{L}})$ is not only sufficient, but also *necessary*, for $\mathscr{L}$ to be a Krein space.

6 We became acquainted above with the concept of the Gram operator $G_{\mathscr{L}}$ of an arbitrary subspace $\mathscr{L}$ of a Krein space $\mathscr{H}$. Starting from this concept we can consider a space more general than a Krein space, namely, a Hilbert space with an indefinite metric.

Let $\mathscr{H}$ be a Hilbert space with a scalar product (x, y) $(x, y \in \mathscr{H})$ and norm $\|x\| = (x, x)^{1/2}$ $(x \in \mathscr{H})$, and let W be an arbitrary bounded selfadjoint operator $(W^* = W)$ given on $\mathscr{H}$. Then the Hermitian sesquilinear form $[x, y] = (Wx, y)$ defines in $\mathscr{H}$ an *indefinite metric* which we shall call the *W-metric*, and we shall call the space $\mathscr{H}$ itself with the W-metric a *W-space*. W is called the *Gram operator of the space* $\mathscr{H}$.

It is clear that all the results of §1 hold for a W-space (for $Q(x, y) = (Wx, y) = [x, y]$). However, it is also possible to assert that many (though by no means all) of the facts proved for a Krein space remain true for W-spaces.

□ The most important of them is the inequality

$$|[x, y]| \leqslant \|W\| \, \|x\| \, \|y\| \tag{6.10}$$

which establishes the *continuity relative to the norm* $\|\cdot\|$ *of the Hermitian sesquilinear functional* $[x, y]$ (*the W-metric*) *over the set of variables* (*cf.* Proposition 3.5). ■

Further, it is easy to see that the *non-degeneracy of a W-space* $\mathscr{H}$ (*or the non-degeneracy of the W-metric*) *is equivalent to the condition* Ker $W = \{\theta\}$, *or, what is the same thing*, $0 \notin \sigma_p(W)$ (*cf.* 6.5). In this particular case we shall denote the Gram operator W by the letter G and speak of the *G-metric* and the *G-space* $\mathscr{H}$.

It is not difficult to understand that *if* $\mathscr{H}$ *is a W-space then the factor-space* $\hat{\mathscr{H}} = \mathscr{H}|(\text{Ker } W)$ *with the induced* $\hat{W}$*-metric* (see 1.23) *is a G-space.*

A G-metric may be either *regular* ($0 \notin \sigma(G) \Leftrightarrow 0 \in \rho(G)$) or *singular* ($0 \in \sigma(G)$).

6.13 *A Hilbert space* $\mathscr{H}$ *with a regular G-metric is a Krein space. If* $\mathscr{H}$ *is a W-space, then* $\hat{\mathscr{H}} = \mathscr{H}|(\text{Ker } W)$ *is a Krein space if and only if* $\mathscr{R}_W = \bar{\mathscr{R}}_W$.

□ The first assertion was established, essentially, in Corollary 6.12. The second assertion follows from the decomposition $\mathscr{H} = \bar{\mathscr{R}}_W [\oplus] \text{Ker } W$ and Proportion 1.23, by virtue of which $\hat{\mathscr{H}}$ is $\hat{W}$-isometrically isomorphic to $\hat{R}_W$, and therefore the condition $\bar{\mathscr{R}}_W = \mathscr{R}_W$ is equivalent to the regularity of the G-metric (i.e., the $\hat{W}$-metric) in $\hat{\mathscr{H}}$. ■

We note that a singular G-space too can be converted into a certain Krein space by a completion method. The precise result is formulated thus:

6.14 *Let* $\mathscr{H}$ *be a singular G-space, let* $mI \leqslant G \leqslant MI$, *let* $G = \int_{m-0}^{M} \lambda \, dE_\lambda$ ($E_{\lambda+0} = E_\lambda$) *be the spectral decomposition of its Gram operator G, and let* $P^- = \int_{m-0}^{-0} dE_\lambda$ *and* $P^+ = \int_0^M dE_\lambda$ *be the orthoprojectors defined by it* ($P^+ + P^- = I$: *cf.* (6.3)). *For any* $x \in \mathscr{H}$ *we write* $x^\pm = P^\pm x$ *and we introduce in* $\mathscr{H}$ *a new scalar product.*

$$(\widetilde{x, y}) = [x^+, y^+] - [x^-, y^-] \qquad (x, \quad y \in \mathscr{H})$$

and a corresponding norm $\|x\|^\sim = (\widetilde{x, x})^{1/2}$ ($x \in \mathscr{H}$).

Then the form $[x, y]$ *can be continuously extended* (*i.e. continuously relative to the norm* $\|\cdot\|^\sim$) *on to the completion* $\tilde{\mathscr{H}}$ *of the space* $\mathscr{H}$ *relative to this norm;* $\tilde{\mathscr{H}}$ *is a Krein space relative to the extended form* $[x, y]$.

□ We note that the projectors $P^\pm$, as also the Gram operator G, are bounded relative to the original norm; hence for the new norm we obtain the estimate

$\|x\|^{\sim} \leqslant c\|x\|$ $(c>0)$. The form $[x, y]$ remains continuous in the new norm $\|\cdot\|^{\sim}$ too:

$$|[x,y]| = |[x^+,y^+]+[x^-,y^-]| = |\widetilde{(x^+,y^+)} - \widetilde{(x^-,y^-)}| = |\widetilde{(x^+-x^-,y)}| \leqslant \|x^+-x^-\|^{\sim}\|y\|^{\sim} \leqslant \|x\|^{\sim}\|y\|^{\sim} \qquad (x,y\in\mathscr{H}) \tag{6.11}$$

(here we have used the G-orthogonality of the subspaces $\mathscr{H}^{\pm} = P^{\pm}\mathscr{H}$; *cf.* Theorem 6.4). Since $\mathscr{H}$ is dense in its completion $\tilde{\mathscr{H}}$ relative to the norm $\|\cdot\|^{\sim}$, so the form $[x, y]$ can be continuously extended on the whole of $\tilde{\mathscr{H}}$ (we keep the same notation for it). We denote by $\tilde{\mathscr{H}}^{\pm}$ the closures of the lineals $\mathscr{H}^{\pm}$ $(\subset\tilde{\mathscr{H}})$ relative to the norm $\|\cdot\|^{\sim}$ respectively. Moreover, by (6.11), $\tilde{\mathscr{H}}^{+}[\perp]\tilde{\mathscr{H}}^{-}$ relative to the extended form $[x, y]$ and, obviously, $\tilde{\mathscr{H}}^{+}\perp\tilde{\mathscr{H}}^{-}$ relative to the scalar product $\widetilde{(x, y)}$, so that $\tilde{\mathscr{H}} = \tilde{\mathscr{H}}^{+}[\oplus]\tilde{\mathscr{H}}^{-}$ is a Krein space relative to the form $[x, y]$, or, what is the same thing, it is a $\tilde{J}$-space, where $\tilde{J} = \tilde{P}^{+} - \tilde{P}^{-}$ and $\tilde{P}^{\pm}$ are the orthoprojectors from $\tilde{\mathscr{H}}$ on to $\tilde{\mathscr{H}}^{\pm}$ respectively. It is clear also that $\tilde{J}$ is the extension, by continuity relative to the norm $\|\cdot\|^{\sim}$, of the operator $P^{+} - P^{-}$ from $\mathscr{H}$ on to the whole of $\tilde{\mathscr{H}}$ (continuity follows from (6.11)). ■

□ The question examined in Proposition 6.14 about the imbedding of G-spaces in a Krein space has also another aspect, which reveals, so to speak, the *universality* of Krein spaces among all W-spaces. In fact, if $\mathscr{H}$ is a W-space, then without loss of generality we may suppose that $\|W\|\leqslant 1$ (for otherwise we can renorm $\mathscr{H}$ equivalently, by replacing the original scalar product $(\cdot,\cdot)$ by a new one: $(\cdot,\cdot)_1 = \alpha(\cdot,\cdot)$, where $\alpha \geqslant \|W\|$; and then for the new Gram operator $W_1 = (1/\alpha)W$ we obtain $\|W_1\|_1 = (1/\alpha)\|W\| \leqslant 1$). Now the space $\tilde{\mathscr{H}} = \mathscr{H}\oplus\mathscr{H}$ is turned by means of the canonical symmetry operator (see Example 3.9)

$$\tilde{J} = \begin{Vmatrix} W & (I-W^2)^{1/2} \\ (I-W^2)^{1/2} & -W \end{Vmatrix}$$

into a Krein space, and $\mathscr{H}$ is a subspace of it (more precisely, it is $(W,\tilde{\mathscr{H}})$-isometrically isomorphic to a certain subspace of it). Since $\tilde{\mathscr{H}}$, in its turn, is $(\tilde{J}, J_1)$-isometrically isomorphic to a certain subspace of an arbitrary J_1-space $\mathscr{H}_1 = \mathscr{H}_1^{+}\oplus\mathscr{H}_1^{-}$ with subspaces $\mathscr{H}_1^{\pm}$ of sufficiently large dimension, so $\mathscr{H}$ too is (W, J_1)-isometrically isomorphic to a certain subspace of the space $\mathscr{H}_1$. Thus, *an arbitrary Krein space* $\mathscr{H}_1$ (*provided the subspaces* $\mathscr{H}_1^{\pm}$ *in its canonical decomposition are of sufficiently large dimension*) *contains a subspace which is* (W, J_1)*-isometrically isomorphic to a given* W*-space* $\mathscr{H}$. ■

The theory of semi-definite lineals and subspaces (in particular, the concepts of the intrinsic metric and uniform definiteness, of regular and singular lineals) which has been developed in §§4–6 for Krein spaces also remains valid to a considerable extent for W-spaces and, in particular, for G-spaces.

Exercises and problems

1 Consider the Krein space $\mathscr{H}$ from Example 3.11 and find a lineal which is dense in this space and indecomposable.
Hint: Use the idea of Example 1.33.

2 Prove that the indecomposable lineal (space) $\mathscr{F}$ constructed in Example 1.33 not only is not a Krein space, but (in contrast to the situation in Exercise 1) it cannot in general be imbedded in any Hilbert space $\mathscr{H}$ with a scalar product $(\cdot, \cdot)$ 'majorizing' the form $[\cdot, \cdot]$: $|[x, y]|^2 \leqslant (x, x)(y, y)$, i.e., $|[x, y]| \leqslant \|x\| \|y\|$ $(x, y \in \mathscr{F})$ [VIII].
Hint. Arguing from the contrary, consider vectors $x = \{\xi_j\}_{j=-\infty}^{\infty}$ with $\xi_{-k} = 1$ and $\xi_j = 0$ when $j \neq -k$ $(k = 1, 2, \ldots)$, and also a vector $y = (\ldots, 0, \ldots, 0, \eta_0, \eta_1, \ldots)$, where $\eta_{k-1} = (x_k, x_k) + k$ $(k = 1, 2, \ldots)$, and obtain the contradictory inequality $\eta_{k-1} \leqslant \|y\|^2$ for all $k = 1, 2, \ldots$ (Ginzburg).

3 Let $G_{\mathscr{L}}$ be the Gram operator of a subspace $\mathscr{L}$ $(\subset \mathscr{H})$. Then $\mathscr{L}$ is non-degenerate if and only if $\|x\|_{\mathscr{L}} = \|G_{\mathscr{L}} x\|$ is a norm in $\mathscr{L}$ ([VI]; see Definition 7.13 below).

4 A sequence $\{x_n\}_1^{\infty}$ is said to be *asymptotically isotropic in a lineal* $\mathscr{L}$ $(\subset \mathscr{H})$ if $[x_n, y] \to 0$ $(n \to \infty)$ uniformly relative to all $y \in \mathscr{L}$ with $\|y\| = 1$. Let $G_{\mathscr{L}}$ be the Gram operator of the subspace $\mathscr{L}$. The sequence $\{x_n\}_1^{\infty}$ is asymptotically isotropic in $\mathscr{L}$ if and only if $\lim_{n \to \infty} \|G_{\mathscr{L}} x_n\| = 0$ (Ginzburg; see [VIII]).

5 A definite subspace $\mathscr{L}$ $(\subset \mathscr{H})$ is uniformly definite if and only if an arbitrary asymptotically isotropic sequence in $\mathscr{L}$ converges to zero (Ginzburg; see [VIII], *cf.* Theorem 7.16 below).

6 The result of Exercise 4 can be substantially generalized: the closure $\bar{\mathscr{L}}$ of a lineal $\mathscr{L}$ $(\subset \mathscr{H})$ is a Krein space if and only if every asymptotically isotropic sequence in $\mathscr{L}$ converges to zero (I. Iokhidov, see [VIII], Proposition 5.5, and Theorem 7.16 below.)
Hint: Use the results of Exercises 3 and 4.

7 Let $\mathscr{L} = \mathscr{L}^+ [\dotplus] \mathscr{L}^- [\dotplus] \mathscr{L}^0$ be the canonical decomposition of a subspace $\mathscr{L}$ of a Krein space, and let $G_{\mathscr{L}^{\pm}}$ be the Gram operators for $\mathscr{L}^{\pm}$ respectively. In order that every uniformly positive (uniformly negative) subspace in $\mathscr{L}$ should be finite-dimensional it is necessary and sufficient that $G_{\mathscr{L}^+} \in \mathscr{S}_{\infty}$ $(G_{\mathscr{L}^-} \in \mathscr{S}_{\infty})$ (Azizov).
Hint: In the 'necessary' part use the fact that $(I_{\mathscr{L}^+} - E_{\mathscr{L}^+}(\lambda))G_{\mathscr{L}^+} \to G_{\mathscr{L}^+}$ when $(0 <)$ $\lambda \to 0$ in the uniform operator topology, and $\dim(I_{\mathscr{L}^+} - E_{\mathscr{L}^+}(\lambda)) < \infty$ when $\lambda > 0$ (here $E_{\mathscr{L}^+}(\lambda)$ is the spectral function of the operator $G_{\mathscr{L}^+}$). In the 'sufficient' part use the injectiveness of the orthoprojector P_+ from $\mathscr{L}$ on to $\mathscr{L}^+$ on positive subspaces (see Lemma 1.27) and the fact that, when $x \in \mathscr{M}^+$ (any uniformly positive subspace from $\mathscr{L}$) $(G_{\mathscr{L}^+} P_+ x, P_+ x) \geqslant c \|x\|^2$, i.e., $G_{\mathscr{L}^+}$ is invertible on $P_+ \mathscr{M}^+$.

§7 J-orthogonal complements and projections. Projectional completeness

1 We recall that we have already encountered from time to time J-orthogonal complements (and even earlier Q-orthogonal complements), starting in §1.5 (see also §1.10, §3.4, §4.5, etc.) But here we begin their systematic study as applied to lineals and, in particular, to subspaces of a Krein space $\mathscr{H}$ with a fixed canonical symmetry operator J in $\mathscr{H}$ (see §3.5).

As before, we indicate J-orthogonality by the symbol $[\perp]$ and a J-

orthogonal complement by the same symbol raised as an index, while the usual (Hilbert) orthogonality with respect to the scalar product $(x, y) = [Jx, y]$ is denoted by the symbol $\perp$ and the usual orthogonal complement is indicated by the same symbol used as an index. It is not difficult to prove the Propositions 7.1–7.5, listed below, by direct verification using the relation (3.5) or Propositions 3.5 and 3.7; we leave the reader to do this.

□ 7.1 *For any subset $\mathcal{M}$ ($\subset \mathcal{H}$) its J-orthogonal complement ($\mathcal{M}^{[\perp]}$) and its orthogonal complement ($\mathcal{M}^{\perp}$) are subspaces, connected by the formulae*

$$\mathcal{M}^{[\perp]} = J\mathcal{M}^{\perp}, \tag{7.1a}$$

$$\mathcal{M}^{\perp} = J\mathcal{M}^{[\perp]}. \quad \blacksquare \tag{7.1b}$$

□ 7.2 *For any set $\mathcal{M}$ ($\subset \mathcal{H}$)*

$$\mathcal{M}^{[\perp]} = \bar{\mathcal{M}}^{[\perp]} = (\text{Lin } \mathcal{M})^{[\perp]}. \quad \blacksquare$$

□ 7.3 *For any $\mathcal{M}$ ($\subset \mathcal{H}$)*

$$(J\mathcal{M})^{[\perp]} = J\mathcal{M}^{[\perp]}. \quad \blacksquare \tag{7.2}$$

□ 7.4 *For any set $\mathcal{M}$ ($\subset \mathcal{H}$)*

$$\mathcal{M}^{[\perp][\perp]} = C \text{ Lin } \mathcal{M}$$

(cf. (1.11)), and, in particular, for any lineal $\mathcal{L}$ ($\subset \mathcal{H}$)

$$\mathcal{L}^{[\perp][\perp]} = \bar{\mathcal{L}}. \quad \blacksquare \tag{7.3}$$

□ 7.5 *For a subspace $\mathcal{L}$ ($= \bar{\mathcal{L}}$) it is always true that $\mathcal{L}^{[\perp][\perp]} = \mathcal{L}$, and therefore the isotropic subspaces $\mathcal{L}^0$ of $\mathcal{L}$ and $\mathcal{L}^{[\perp]}$ coincide:*

$$\mathcal{L}^0 = \mathcal{L} \cap \mathcal{L}^{[\perp]} = \mathcal{L}^{[\perp]} \cap \mathcal{L}^{[\perp][\perp]}. \quad \blacksquare$$

2 In this paragraph we restrict our attention to the case mentioned in 7.5 above, where $\mathcal{L} = \bar{\mathcal{L}}$ is a subspace of a Krein space. Although it is always true that $\mathcal{L} \cap \mathcal{L}^{\perp} = \{\theta\}$ and $\mathcal{L}^{\perp} \oplus \bar{\mathcal{L}} = \mathcal{H}$ relative to the Hilbert metric, yet the analogous equalities with respect to the indefinite metric are, generally, speaking, not true if only because $\mathcal{L}$ may be a degenerate subspace, and then $\mathcal{L} \cap \mathcal{L}^{[\perp]} = \mathcal{L}^0 \neq \{\theta\}$. In this case the algebraic sum $\mathcal{L} + \mathcal{L}^{[\perp]}$, i.e., the lineal $\mathcal{L} + \mathcal{L}^{[\perp]} = \{x, y \mid x \in \mathcal{L}, y \in \mathcal{L}^{[\perp]}\}$ is no longer a direct sum.

7.6 *For any subspace $\mathcal{L}$ the relation $(J\mathcal{L}^0)^{\perp} = \overline{(\mathcal{L} + \mathcal{L}^{[\perp]})}$ holds.*

□ Using (1.12) and Proposition 7.5 we have $(\mathcal{L} + \mathcal{L}^{[\perp]})^{[\perp]} = \mathcal{L}^{[\perp]} \cap \mathcal{L}^{[\perp][\perp]} = \mathcal{L}^0$, whence, by (7.16), $(\mathcal{L} + \mathcal{L}^{[\perp]})^{\perp} = J\mathcal{L}^0$ and $(J\mathcal{L}^0)^{[\perp]} = (\mathcal{L} + \mathcal{L}^{[\perp]})^{\perp\perp} = \overline{(\mathcal{L} + \mathcal{L}^{[\perp]})}$. ■

The fact proved in Proposition 7.6 implies that the whole space $\mathscr{H}$ can be expressed in the form of an orthogonal sum:

$$\mathscr{H} = (\overline{\mathscr{L} + \mathscr{L}^{[\perp]}}) \oplus J\mathscr{L}^0. \tag{7.4}$$

This enables us to replace the relation $\mathscr{L} \oplus \mathscr{L}^\perp = \mathscr{H}$, which is true only in the Hilbert metric by the following more precise form of it:

Lemma 7.7: *In order that*

$$\overline{\mathscr{L} + \mathscr{L}^{[\perp]}} = \mathscr{H} \tag{7.5}$$

it is necessary and sufficient that the subspace $\mathscr{L}$ be non-degenerate.

$\square$ The proof follows directly from formula (7.4) if we take into account that, since the operator J is unitary (see (3.4)), the relations $J\mathscr{L}^0 = \{\theta\}$ and $\mathscr{L}^0 = \{\theta\}$ are equivalent. $\blacksquare$

3   Returning to the question discussed in Paragraph 2, but from a more general viewpoint, we introduce the following

Definition 7.8:   A lineal $\mathscr{L}$ in a Krein space is said to be *projectively complete* if

$$\mathscr{L} + \mathscr{L}^{[\perp]} = \mathscr{H}. \tag{7.6}$$

In seeking conditions for projective completeness we discover that, as in the analogous question in the case of a definite (Hilbert) metric, the following proposition holds:

7.9   For a linear $\mathscr{L} (\subset \mathscr{H})$ to be projectively complete it must be closed: $\mathscr{L} = \bar{\mathscr{L}}$. In particular, the subspaces $\mathscr{L}$ and $\mathscr{L}^{[\perp]}$ can be projectively complete only simultaneously.

$\square$ Suppose (7.6) holds, but there is a vector $y_0 \in \bar{\mathscr{L}} \| \mathscr{L}$. We express the vector y_0 in accordance with (7.6) in the form $y_0 = x_0 + z_0$ ($x_0 \in \mathscr{L}, z_0 \in \mathscr{L}^{[\perp]}$). Thus, $x_0, y_0 \in \bar{\mathscr{L}}$, $(y_0 - x_0 =) z_0 \in \bar{\mathscr{L}} \cap \mathscr{L}^{[\perp]} = \mathscr{L}^{[\perp][\perp]} \cap \mathscr{L}^{[\perp]}$, i.e., $z_0 [\perp] \mathscr{L} + \mathscr{L}^{[\perp]} = \mathscr{H}$ and $z_0 = \theta$; hence $y_0 = x_0 \in \mathscr{H}$ contrary to the choice of y_0. The second assertion in 7.9 follows from the equality $\mathscr{L}^{[\perp][\perp]} = \mathscr{L}_0$ (see Proposition 7.5). $\blacksquare$

However, although in the 'definite' situation the condition $\mathscr{L} = \bar{\mathscr{L}}$ is not only necessary but also sufficient for the existence of the decomposition $\mathscr{H} = \mathscr{L} \oplus \mathscr{L}_0$, the indefinite metric, as already mentioned at the beginning of paragraph 7.2, here shows its specific character. In particular, we have

7.10   In order that a subspace $\mathscr{L} (\subset \mathscr{H})$ should be projectively complete it must be non-degenerate: $\mathscr{L} \cap \mathscr{L}^{[\perp]} = \{\theta\}$.

$\square$ This is a simple consequence of Lemma 7.7, just as (7.5) follows *a fortiori* from (7.6). $\blacksquare$

But even both the conditions $\mathscr{L} = \bar{\mathscr{L}}$ and $\mathscr{L} \cap \mathscr{L}^{[\perp]} = \{\theta\}$ together are *not sufficient* for $\mathscr{L}$ to be projectively complete. We can quickly see this by approaching the problem from a rather different standpoint.

4 A vector x is called the *J-orthogonal projection* of a vector $y \in \mathscr{H}$ on to a subspace $\mathscr{L}$ $(\subset \mathscr{H})$ if

$$1)\ x \in \mathscr{L} \qquad 2)\ y - x\,[\perp]\,\mathscr{L}. \tag{7.7}$$

For every (so to say, 'individual') vector y_0 a criterion for the existence of its J-orthogonal projection on to a given subspace $\mathscr{L}$ $(\subset \mathscr{H})$ is obtained simply in terms of the Gram operator $G_{\mathscr{L}}$ of the subspace $\mathscr{L}$ (see Definition 6.3).

Lemma 7.11: *Let $P_{\mathscr{L}}$ be the (Hilbert) orthogonal projector on to a subspace $\mathscr{L}$ $(\neq \{\theta\})$ of a J-space $\mathscr{H}$ and let $G_{\mathscr{L}} = P_{\mathscr{L}}(J \mid \mathscr{L})$ be the Gram operator of the subspace $\mathscr{L}$. Then in order that a vector y_0 $(\in \mathscr{H})$ should have a J-orthogonal projection x_0 on to $\mathscr{L}$ it is necessary and sufficient that $P_{\mathscr{L}}Jy_0 \in \mathscr{R}_{G_{\mathscr{L}}}$.*

□ The existence of the required J-orthogonal projection $x_0(\in \mathscr{L})$ of the vector y_0 is equivalent, by (7.7) to

$$[y_0 - x_0, x] = 0 \Leftrightarrow [y_0, x] = [x_0, x] \Leftrightarrow (Jy_0, P_{\mathscr{L}}x) = (G_{\mathscr{L}}x_0, x)$$

holding for all $x \in \mathscr{L}$, i.e., $(P_{\mathscr{L}}Jy_0, x) = (G_{\mathscr{L}}x_0, x)$, whence $P_{\mathscr{L}}Jy_0 = G_{\mathscr{L}}x_0 \in \mathscr{R}_{G_{\mathscr{L}}}$. Since the whole argument is reversible, Lemma 7.11 is proved. ■

Lemma 7.11 does not answer the question whether the J-orthogonal projection is unique when its projection exists.

Lemma 7.12: *In order that, even if only for a single vector y_0 $(\in \mathscr{H})$ with a J-orthogonal projection x_0 on to a subspace $\mathscr{L}$ $(\neq \{\theta\})$, this J-orthogonal projection should be unique, it is necessary that $\mathscr{L}$ be non-degenerate. This same condition is sufficient to ensure that any vector y $(\in \mathscr{H})$ shall have not more than one J-orthogonal projection on to $\mathscr{L}$.*

□ The existence of an isotropic vector z_0 $(\neq \theta)$ in $\mathscr{L}$ would bring it about that, for the vector y_0 $(\in \mathscr{H})$ which has the J-orthogonal projection x_0 on $\mathscr{L}_1$ the vector $x_0 + z_0$ $(\neq x_0)$ would also be a J-orthogonal projection on to $\mathscr{L}$, because $x_0 + z_0 \in \mathscr{L}$ and $y_0 - (x_0 + z_0)\,[\perp]\,\mathscr{L}$.

Conversely, if for some vector y $(\in \mathscr{H})$ there are two orthogonal projections x_1 and x_2 on to $\mathscr{L}$ with $x_1 \neq x_2$, then the vector $(\theta \neq)z_0 = x_1 - x_2(\in \mathscr{L})$ has the property: $z_0 = (y - x_1) - (y - x_2)\,[\perp]\,\mathscr{L}$, i.e., it is an isotropic vector for $\mathscr{L}$. ■

5 We now return to the search started in para. 7.3 for criteria for the projective completeness of a subspace $\mathscr{L}$ $(\subset \mathscr{H})$, and for this purpose we introduce some further definitions.

Definition 7.13: In a non-degenerate subspace $\mathscr{L}$ having the Gram operator $G_{\mathscr{L}}$ we introduce the *norm* $\| x \|_{\mathscr{L}} = \| G_{\mathscr{L}} x \|$ $(x \in \mathscr{L})$.

Definition 7.14: A non-generate subspace $\mathscr{L}$ is said to be *regular* if the norms $\| \cdot \|_{\mathscr{L}}$ and $\| \cdot \|$ are equivalent on it.

7.15 A subspace $\mathscr{L}$ with a Gram operator $G_{\mathscr{L}}$ is regular if and only if $0 \in \rho(G_{\mathscr{L}})$.

□ The estimate $\| G_{\mathscr{L}} x \| = \| x \|_{\mathscr{L}} \geqslant c \| x \|$ for all $x \in \mathscr{L}$ $(c > 0)$ is equivalent to the continuous invertiblity of the bounded selfadjoint operator $G_{\mathscr{L}}$. ■

Now we can establish a fundamental theorem, containing a set of criteria for projective completeness.

Theorem 7.16: *For a subspace $(\{\theta\} \neq)$ $\mathscr{L}$ $(\subset \mathscr{H})$ with a Gram operator $G_{\mathscr{L}}$ the following four assertions are equivalent:*

a) *$\mathscr{L}$ is projectively complete;*
b) *$\mathscr{L}$ is regular, i.e., $0 \in \rho(G_{\mathscr{L}})$* (see 7.15);
c) *$\mathscr{L}$ is a Krein space;*
d) *any vector $y \in \mathscr{H}$ has at least one J-orthogonal projection on to $\mathscr{L}$.*

□ a) ⇒ b). Since $\mathscr{L}$ is projectively complete, it follows from 7.10 that (7.6) represents a decomposition into a direct sum: $\mathscr{H} = \mathscr{L} [\dot{+}] \mathscr{L}^{[\perp]}$. To this decomposition correspond the bounded projectors Q and $(I - Q)$ respectively. Therefore for any $(\theta \neq)$ $x \in \mathscr{L}$ and $y = (1/\| x \|)Jx$ we have

$$\begin{aligned} \| x \| = [x, y] = [x, Qy] = (G_{\mathscr{L}} x, y) \leqslant \| G_l x \| \, \| Q \| \\ = \| Q \| \, \| x \|_{\mathscr{L}} \leqslant \| G_{\mathscr{L}} \| \, \| Q \| \, \| x \|, \end{aligned}$$

i.e., $\mathscr{L}$ is regular.

b) ⇒ c). This implication was established earlier in Corollary 6.12.

c) ⇒ d). By Theorem 5.7 $\mathscr{L} = \mathscr{L}^+ [\dot{+}] \mathscr{L}^-$, where $\mathscr{L}^{\pm}$ are uniformly definite. We show that any vector $y \in \mathscr{H}$ has a J-orthogonal projection on to $\mathscr{L}^+$ and $\mathscr{L}^-$. Consider, for example, a linear functional $\varphi_y(x) = [x, y]$ $(x \in \mathscr{L}^+)$. Since $| \varphi_y(x) | = | (x, Jy) | \leqslant \| x \| \, \| y \|$, so φ_y is continuous relative to the norm $\| \cdot \|$, and therefore also relative to the intrinsic norm $| \cdot |_{\mathscr{L}^+}$ which is equivalent to it (see (5.2) and (5.3)). Since $\mathscr{L}^+$ is complete relative to $| \cdot |_{\mathscr{L}^+}$ (see 5.6), there is, by Riesz's theorem, an $x_1 \in \mathscr{L}^+$ such that $\varphi_y(x) = [x, x_1]$ $(x \in \mathscr{L}^+)$; hence $[x, y] = [x, x_1]$ for all $x \in \mathscr{L}^+$ and $y - x [\perp] \mathscr{L}^+$. Thus, x_1 is the *J*-orthogonal projection of y on to $\mathscr{L}^+$, and similarly we find an x_2, the *J*-orthogonal projection of y on to $\mathscr{L}^-$. But then $x_0 = x_1 + x_2$ is the *J*-orthogonal projection of y on to $\mathscr{L}$.

d) ⇒ a). Since any y $(\in \mathscr{H})$ has a *J*-orthogonal projection x on to $\mathscr{L}$, so $y = x + z$ $(x \in \mathscr{L}, z \in \mathscr{L}^{[\perp]})$ and $\mathscr{H} = \mathscr{L} [\dot{+}] \mathscr{L}^-$, i.e., $\mathscr{L}$ is projectively complete. ■

Corollary 7.17: *If a subspace $\mathscr{L}$ is definite, then its projective completeness is equivalent to uniform definiteness.*

□ This follows, for example, from 7.10, 6.11, and the equivalence a) ⇐ b) in Theorem 7.16. ■

Corollary 7.18: *Every finite-dimensional non-degenerate subspace $\mathscr{L}$ is projectively complete.*

□ This follows from assertion b) in Theorem 7.16 and the equivalence of all norms in a finite-dimensional space.

6 To conclude this section we again return to the question about different canonical decompositions of a Krein space $\mathscr{H}$ and the Hilbert topologies determined by them (*cf.* Remark 2.4 above).

We consider, as well as a particular fixed canonical decomposition

$$\mathscr{H} = \mathscr{H}^+ [\dot{+}] \mathscr{H}^- \tag{7.8}$$

with the canonical projectors $P^{\pm}$, the canonical symmetry operator $J = P^+ - P^-$, and the scalar product $(x, y) = [Jx, y]$ and norm $\| x \| = (x, x)^{1/2}$ $(x, y \in \mathscr{H})$ generated by them (see §3), some other canonical decomposition

$$\mathscr{H} = \mathscr{H}_1^+ [\dot{+}] \mathscr{H}_1^-, \tag{7.9}$$

i.e., we carry out, as it were, 'a rotation of the co-ordinate axes'. It is clear (7.9) also generates corresponding canonical projectors $P_1^{\pm}$: $P_1^{\pm}\mathscr{H} = \mathscr{H}_1^{\pm}$, $P_1^+ + P_1^- = I$, a new canonical symmetry operator $J_1 = P_1^+ - P_1^-$, and also a new scalar product $(x, y)_1 = [J_1 x, y]$ and norm $\| x \|_1 = (x, x)_1^{1/2}$ $(x, y \in \mathscr{H})$.

Theorem 7.19: *The norms* $\| \cdot \|$ and $\| \cdot \|_1$ *generated by different canonical decompositions* (7.8) *and* (7.9) *of a Krien space* $\mathscr{H}$ *are equivalent.*

□ We start from the fact that one of the canonical decompositions, let us say (7.8), generates in $\mathscr{H}$ (in accordance with §2.2) the structure of a Hilbert space with the norm $\| x \| = (x, x)^{1/2}$ $(x \in \mathscr{H})$. The presence of the other canonical decomposition (7.9), where $\mathscr{H}_1^{\pm}$ are definite lineals and $\mathscr{H}_1^+ [\perp] \mathscr{H}_1^-$, shows that they are projectively complete, and closed in the norm $\| \cdot \|$ (see 7.9), that the projectors $P_1^{\pm}$ are closed in this norm, and finally that $\mathscr{H}_1^{\pm}$ are uniformly definite in this norm (see Corollary 7.17). If we now take into account that the norm $\| \cdot \|_1$ is the *intrinsic norm* on $\mathscr{H}_1^{\pm}$ (*cf.* (5.1)), then it is clear that on $\mathscr{H}_1^{\pm}$ the norms $\| \cdot \|$ and $\| \cdot \|_1$ are equivalent. Therefore $\mathscr{H}_1^{\pm}$, and with them also the whole of $\mathscr{H}$ (see 2.3), are complete relative to the norm $\| \cdot \|_1$.

We remark further that for $x_1 \in \mathscr{H}_1^{\pm}$ the norms $\| x_1^{\pm} \|_1$ (as intrinsic norms) are simply subordinate to the original (external) norms: $\| x_1^{\pm} \|_1 \leqslant \| x_1^{\pm} \|$. Therefore, for any $x \in \mathscr{H}$, $x = x_1^+ + x_1^-$ $(x_1^{\pm} \in \mathscr{H}_1^{\pm})$ we have

$$\begin{aligned} \| x \|_1 &= \| x_1^+ + x_1^- \| \leqslant \| x_1^+ \|_1 + \| x_1^- \|_1 \leqslant \| x_1^+ \| + \| x_1^- \| = \| P_1^+ x \| + \| P_1^- x \| \\ &\leqslant \| P_1^+ \| \, \| x \| + \| P_1^- \| \, \| x \| = \beta \| x \|, \end{aligned}$$

where

$$\beta = \| P_1^+ \| + \| P_1^- \| > 0.$$

Hence by virtue of the completeness relative to both the norms $\| \cdot \|$ and $\| \cdot \|_1$ it follows that these norms are equivalent. ■

Exercises and problems

1 A semi-definite lineal $\mathscr{L}$ is uniformly definite if and only if $\bar{\mathscr{L}}$ is projectively complete ([V]).

2 Let $\mathscr{L}$ $(\subset \mathscr{H})$ contain a maximal uniformly positive (uniformly negative) lineal (see Exercises 3 and 4 to §5). Then $\bar{\mathscr{L}}$ is projectively complete and $\mathscr{L}^{[\perp]}$ is uniformly negative (uniformly positive) ([V]).
Hint: Use Corollary 7.17, Exercise 4 to §5, Proposition 5.3, and also 7.9 and 7.4.

3 A lineal $\mathscr{L}$ is a maximal uniformly definite lineal if and only if there is a cononical decomposition $\mathscr{H} = \mathscr{H}_1^+ [\dotplus] \mathscr{H}_1^-$ such that $\mathscr{L} = \mathscr{H}_1^+$ (or $\mathscr{L} = \mathscr{H}_1^-$) ([VIII]).
Hint: Use the preceding Exercise and the result of Exercise 3 to §5.

4 A lineal $\mathscr{L}$ is projectively complete if and only if a) $\mathscr{L}$ is closed b) $\mathscr{L}$ is non-degenerate c) for any decomposition of it $\mathscr{L} = \mathscr{L}^+ [\dotplus] \mathscr{L}^-$ $(\mathscr{L}^\pm \subset \mathscr{P}^{\pm\pm} \cup \{\theta\})$ there is a canonical decomposition $\mathscr{H} = \mathscr{H}_1^+ [\dotplus] \mathscr{H}_1^-$ such that $\mathscr{L}^\pm \subset \mathscr{H}_1^\pm$ ([V]).
Hint: In the 'necessary' part use 7.9, 7.10, Theorem 7.16, the maximality principle for uniformly definite lineals (see §5, Exercise 3), and the result of Exercise 3 above.

5 In Theorem 7.16 deduce the equivalence d) $\Leftrightarrow$ b) from Lemma 7.1.

6 Without using Theorem 7.16 prove that for subspaces $\mathscr{L}$ which are definite, regularity is equivalent to uniform definiteness.
Hint: Compare the norms $\| x \|_{\mathscr{L}} = \| G_{\mathscr{L}} x \|$ and $| x |_{\mathscr{L}} = \| G^{1/2} x \|$ $(x \in \mathscr{L}$, $\mathscr{L}$ is positive).

7 Let $\mathscr{H}$ be a Krein space, and $\mathscr{L}$ be a subspace of it which admits the decomposition $\mathscr{L} = \mathscr{L}_0 [\dotplus] \mathscr{L}'$, where $\mathscr{L}'$ is positively complete, and $\mathscr{L}_0$ is the isotropic part of $\mathscr{L}$. Then in any decomposition $\mathscr{L}^{[\perp]} = \mathscr{L}_0 [\dotplus] (\mathscr{L}^{[\perp]})'$ into a direct sum of subspaces the second term will be a projectively complete subspace.
Hint: Use the result of Exercise 10 to §2, taking into account that $(\mathscr{L}^{[\perp]})'$ is J-isometrically isomorphic to the subspace $\mathscr{L}^{[\perp]}/\mathscr{L}_0$.

8 Prove that the sum $\mathscr{L}_1 [\dotplus] \mathscr{L}_2$ of projectively complete subspaces $\mathscr{L}_1$ and $\mathscr{L}_2$ is projectively complete.

§8 The method of angular operators

1 Let $\mathscr{L} \subset \mathscr{P}^+$ in a Krein space $\mathscr{H} = \mathscr{H}^+ [\oplus] \mathscr{H}^-$. If $P^\pm$ are the canonical projectors, then by Theorem 4.1 the mapping $P^+ | \mathscr{L}$: $\mathscr{L} \to P^+ \mathscr{L}$ $(\mathscr{H}^+)$ is a linear homeomorphism, and $\| (P^+ | \mathscr{L})^{-1} \| \leqslant \sqrt{2}$. We now consider the bounded linear operator

$$K = P^- (P^+ | \mathscr{L})^{-1}, \qquad K: P^+ \mathscr{L} \to \mathscr{H}^-. \tag{8.1}$$

Definition 8.1: The operator $\mathscr{K}$ defined by the relation (8.1) is called the *angular operator for $\mathscr{L}$ with respect to $\mathscr{H}^+$.*

The meaning of this nomenclature is explained a little later (see Exercises 1 and 2 to §8).

□ For an arbitrary vector $x (\in \mathscr{L})$ we have $x = x^+ + x^-$ $(x^\pm \in \mathscr{H}^\pm)$, $x^+ = P^+ x \in P^+ \mathscr{L}$, $Kx^+ = P^-(P^+ \mid \mathscr{L})^{-1} x^+ = P^- x = x^-$. Thus every vector $x\ (\in \mathscr{L})$ has the form $x = x^+ + Kx^+$ $(x^+ \in P^+ \mathscr{L})$, and since $\mathscr{L}$ is non-negative, for all $x^+ \in P^+ \mathscr{L}$ we have (see Proposition 2.5) $\| x^+ \| \geqslant \| x^- \| = \| Kx^+ \|$, i.e., $\| K \| \leqslant 1$ (K is a compression).

Conversely, given an arbitrary $\mathscr{L}^+$ $(\subset \mathscr{H}^+)$ and an arbitrary compression $K: \mathscr{L}^+ \to \mathscr{H}^-$ $(\| K \| \leqslant 1)$, we consider the set of vectors

$$\mathscr{L} = \{x^+ + Kx^+\}_{x^+ \in \mathscr{L}^+} \tag{8.2}$$

Since K is a linear operator, $\mathscr{L}$ is a lineal and $\mathscr{L} \subset \mathscr{P}^+$ because, for any $x = x^+ + Kx^+ (\in \mathscr{L})$ we have $x^- = Kx^+$, hence $\|x^-\| = \|Kx^+\| \leqslant \|x^+\|$. Moreover it is clear that $P^+ \mathscr{L} = \mathscr{L}^+$ and K is an angular operator for $\mathscr{L}$ with respect to $\mathscr{H}^+$.

Summing up, we have obtained a complete description of all non-negative lineals of a Krein space, and, in fact, we have proved

Theorem 8.2: *The formula* (8.2) *in which $\mathscr{L}^+$ is an arbitrary lineal from $\mathscr{H}^+$, and K: $\mathscr{L}^+ \to \mathscr{H}^-$ is an arbitrary compression* $(\| K \| \leqslant 1)$, *gives the general form of all $\mathscr{L} \subset \mathscr{P}^+$ of the Krein space $\mathscr{H} = \mathscr{H}^+ [\oplus] \mathscr{H}^-$, and $\mathscr{L}^+ = P^+ \mathscr{L}$ and K is the angular operator for $\mathscr{L}$ with respect to $\mathscr{H}^+$.* ■

All non-positive lineals $\mathscr{L}$ $(\subset \mathscr{H})$ are described similarly.

Definition 8.1': If $\mathscr{L}$ is a non-positive lineal in the Krein space $\mathscr{H} = \mathscr{H}^+ [\oplus] \mathscr{H}^-$, then the operator

$$Q = P^+(P^- \mid \mathscr{L})^{-1}, \qquad Q\colon P^- \mathscr{L} \to \mathscr{H}^+ \tag{8.3}$$

is called the *angular operator for $\mathscr{L}$ with respect to $\mathscr{H}^-$.*

□ The reader will be able himself to establish the following analogue of Theorem 8.2:

Theorem 8.2': *The formula*

$$\mathscr{L} = \{Qx^- + x^-\}_{x^- \in \mathscr{L}^-},$$

in which $\mathscr{L}^-$ is an arbitrary lineal form $\mathscr{H}^-$, and Q: $\mathscr{L}^- \to \mathscr{H}^+$ $(\| Q \| < 1)$ *is an arbitrary compression, gives the general form for all $\mathscr{L} \subset \mathscr{P}^-$ in the Krein space $\mathscr{H} = \mathscr{H}^+ [\oplus] \mathscr{H}^-$, and $\mathscr{L}^- = P^{-1} \mathscr{L}$, and Q is the angular operator for $\mathscr{L}$ with respect to $\mathscr{H}^-$.* ■

2 From now on (Sections .2 and .3) we restrict ourselves to the case of lineals $\mathscr{L} \subset \mathscr{P}^+$ and their angular operators K, since for $\mathscr{L} \subset \mathscr{P}^-$ the whole theory is entirely analogous (the reader can himself formulate the corresponding Proposition 8.3′, Lemma 8.4′, etc.).

In spite of its simple construction the angular operator K for $\mathscr{L}$ (the brevity we shall often omit the words 'with respect to $\mathscr{H}$') contains quite a lot of information about the properties of the lineal itself.

8.3 If K is the angular operator for $\mathscr{L} \subset \mathscr{P}^+$, then $\mathscr{L} \subset \mathscr{P}^{++} \cup \{\theta\}$ if and only if $\| Kx^+ \| \leqslant \| x^+ \|$ for all $(\theta \neq)\ x^+ \in P^+\mathscr{L}$, and $\mathscr{L} \subset \mathscr{P}^0$ if and only if K is an isometric operation: $\| Kx^+ \| = \| x^+ \|$ $(x^+ \in P^+\mathscr{L})$.

□ Since for $x \in \mathscr{L}$ we have $x = x^+ + x^- = x^+ + Kx^+$, all the assertions follow immediately from the formula (see (2.7)) $[x, x] = \| x^+ \|^2 - \| x^- \|^2 = \| x^+ \|^2 - \| Kx^+ \|^2$. ■

We note that for $\mathscr{L} \subset \mathscr{P}^0$ the description can be obtained by means of the analogue 8.3′ of Proposition 8.3 in terms of the isometric operator Q: $P^-\mathscr{L} \to \mathscr{H}^+$ which is the angular operator of $\mathscr{L}$ with respect to $\mathscr{H}^-$ (see (8.3)). Here, obviously, $Q = K^{-1}$.

A rather more subtle fact than Proposition 8.3 is revealed by

Lemma 8.4: *$\mathscr{L}$ $(\subset \mathscr{P}^+)$ with the angular operator K is uniformly positive if and only if $\| K \| < 1$.*

□ Let $\mathscr{L}$ be uniformly positive, i.e., for all $x \in \mathscr{L}$

$$\| x^+ \|^2 - \| x^- \|^2 = [x, x] \geqslant \alpha^2 \| x \|^2 = \alpha^2(\| x^+ \|^2 + \| x^- \|^2) \qquad (\alpha > 0), \tag{8.4}$$

and by the meaning of Definition 5.2 α can always be chosen so that $0 < \alpha < 1$, and this we do. The relation (8.4) is equivalent to the inequality

$$(1 - \alpha^2) \| x^+ \|^2 \geqslant (1 + \alpha^2) \| x^- \|^2, \tag{8.5}$$

whence

$$\| Kx^+ \|^2 = \| x^- \|^2 \leqslant \frac{1 - \alpha^2}{1 + \alpha^2} \| x^+ \|^2, \qquad \text{i.e., } \| K \| \leqslant \sqrt{\frac{1 - \alpha^2}{1 + \alpha^2}} < 1.$$

Conversely, let $\| K \| = \rho < 1$. Put

$$\alpha = \sqrt{\frac{1 - \rho^2}{1 + \rho^2}} \qquad (0 < \alpha < 1).$$

Then $\rho^2 = (1 - \alpha^2)/(1 + \alpha^2)$, and for all $x^+ \in P^+\mathscr{L}$ and $x = x^+ + x^- = x^+ + Kx^+$ we have

$$\| x^- \|^2 = \| Kx^+ \|^2 \leqslant \rho^2 \| x^+ \|^2 = \frac{1 - \alpha^2}{1 + \alpha^2} \| x^+ \|^2,$$

or $(1 - \alpha^2) \| x^+ \|^2 \geqslant (1 + \alpha^2) \| x^- \|^2$, which, as we have seen (*cf.* (8.5)), is equivalent to (8.4). ■

Lemma 8.4 together with Proposition 8.3 shows an effective method of constructing both definite and uniformly definite lineals.

3 The concept of an angular operator proves to be particularly useful for work with *maximal semi-definite* subspaces. For brevity of exposition we introduce the following two sets of subspaces:

$(\mathcal{M}^+(\mathcal{H}) =)\mathcal{M}^+$—the set of all maximal non-negative subspaces of a Krein space $\mathcal{H}$, (8.6)

$(\mathcal{M}^-(\mathcal{H}) =)\mathcal{M}^-$—the set of all maximal non-positive subspaces of a Krein space $\mathcal{H}$. (8.7)

8.5 In order that $\mathcal{L}$ $(\subset \mathcal{P}^+)$ be contained in $\mathcal{M}^+$ it is necessary and sufficient that its angular operator K be defined everywhere in $\mathcal{H}^+$ $(K: \mathcal{H}^+ \to \mathcal{H}^-)$.

□ This follows from Theorem 4.5 and 8.2. ■

Matters are naturally more complicated with *maximal positive* lineals $\mathcal{L}$ because as we have already seen (see Example 4.12) such an $\mathcal{L}$ may be non-closed and in this case $\mathcal{L} \notin \mathcal{M}^+$.

8.6 In order that $\mathcal{L}$ $(\subset \mathcal{P}^{++} \cup \{\theta\})$ with the angular operator K should be a maximal positive lineal it is necessary, and if $\mathcal{L}$ is closed $(\mathcal{L} = \bar{\mathcal{L}})$ it is also sufficient, that the domain of definition of the operator be dense in $\mathcal{H}^+$: $\bar{\mathcal{D}}_K = \mathcal{H}^+$.

□ This is simply a re-formulation of Theorem 4.7. ■

Propositions 8.5 and 8.6 outline a way of constructing extensions of semi-definite and definite lineals to maximal lineals of the corresponding classes, and, moreover, a way of describing all such extensions.

8.7 Let $\mathcal{L} \subset \mathcal{P}^+$, and let its angular operator K be defined on the lineal $\mathcal{D}_K$ $(\subset \mathcal{H}^+)$. Then:

a) *non-negative extensions $\tilde{\mathcal{L}}$ $(\supset \mathcal{L})$ are described in accordance with Theorem 8.2 by all possible contractions $\tilde{K}$ $(\supset K)$, i.e., by extensions of the operator K with $\|\tilde{K}\| < 1$ for which $\mathcal{D}_{\tilde{K}} \subset \mathcal{H}^+$ and $\tilde{K}$: $\mathcal{D}_{\tilde{K}} \to \mathcal{H}^-$;*
b) *in particular, all $\tilde{\mathcal{L}}$ $(\supset \mathcal{L})$ from $\mathcal{M}^+$ are described by such contractions $\tilde{K}$ $(\subseteq K)$ for which $\mathcal{D}_{\tilde{K}} = \mathcal{H}^+$.*

□ This follows from Theorem 8.2 and Proposition 8.5. ■

□ *Remark 8.8:* The procedure for constructing $\tilde{\mathcal{L}}$ $(\in \mathcal{M}^+)$ from assertion b) of Proposition 8.7 begins with the closure of $\mathcal{L}$, i.e., consideration of the non-negative subspace $\bar{\mathcal{L}}$, with the angular operator $\bar{K}$ (the closure of the operator K). If $\mathcal{D}_{\bar{K}} = \mathcal{H}^+$, then it is clear that $\bar{\mathcal{L}} \in \mathcal{M}^+$ and there are no other $\mathcal{L} \in \mathcal{M}^+$ such that $\tilde{\mathcal{L}} \supset \mathcal{L}$. But if $\mathcal{D}_{\bar{K}} \neq \mathcal{H}^+$, then one of the extensions $\tilde{\mathcal{L}}$ $(\supset \mathcal{L})$ of the class $\mathcal{M}^+$ can be constructed (without resorting to Zorn's lemma—see Remark 1.20) for example by means of the *trivial* extension $\tilde{K}$ of the contraction K on to the whole of $\mathscr{H}^+$, namely by defining the *linear* operator $\tilde{K}$ by the formulae

$$\tilde{K}x^+ = \begin{cases} \bar{K}x^+, & x^+ \in \mathscr{D}_{\bar{K}}, \\ \theta, & x^+ \in \mathscr{D}_{\bar{\mathscr{L}}}^{+} \ (= \mathscr{H}^+ \ominus \mathscr{D}_{\bar{K}}). \end{cases} \tag{8.8}$$

But as regards assertion a) of Proposition 8.7, here intermediate extensions $\tilde{\mathscr{L}}$ of the lineal $\mathscr{L}$ between $\mathscr{L}$ and $\tilde{\mathscr{L}}$ are possible ($\mathscr{L} \subset \tilde{\mathscr{L}} \subset \bar{\mathscr{L}}$). ■

Remark 8.9: The procedure described in formula (8.8) of Remark 8.9 is applicable also to the construction of maximal *positive* extensions $\tilde{\mathscr{L}}$ of a *positive* $\mathscr{L}$ with the angular operator K, but only in the case when its closure $\bar{\mathscr{L}}$ is also positive.

4 Turning to the consideration of neutral extensions of neutral lineals we need (in accordance with what was said after the proof of Proposition 8.2) the Theorem 8.2′ and also the analogue 8.7′ (which we have not formulated) of Proposition 8.7 for non-positive lineals.

Theorem 8.10: *Let $\mathscr{H} = \mathscr{H}^+ [\oplus] \mathscr{H}^-$ be a Krein space, let $\mathscr{L} \subset \mathscr{P}^0$, and let K and Q be its angular operators with respect to $\mathscr{H}^+$ and $\mathscr{H}^-$ respectively. Then:*

a) *all neutral extensions $\tilde{\mathscr{L}}$ of the lineal $\mathscr{L}$ are described in accordance with Theorems* 8.2 *and* 8.2′ *by all the angular operators from the union $\{\tilde{K}\} \cup \{\tilde{Q}\}$ of the set $\{\tilde{K}\}$ of all isometric extensions $\tilde{K}\,(\supset K)$ of the isometric operator K ($\mathscr{D}_{\tilde{K}} \subset \mathscr{H}^+$, $\tilde{K}$: $\mathscr{D}_{\tilde{K}} \to \mathscr{H}^-$) and the set $\{\tilde{Q}\}$ of all isometric extensions $\tilde{Q}\,(\supset Q)$ of the isometric operator Q ($\mathscr{D}_{\tilde{Q}} \subset \mathscr{H}^-$, $\tilde{Q}$: $\mathscr{D}_{\tilde{Q}} \to \mathscr{H}^+$);*

b) *in particular, all maximal neutral extensions $\tilde{\mathscr{L}}\ (\supset \mathscr{L})$ in the case* dim $\mathscr{H}^+ <$ dim $\mathscr{H}^-$ *are described by all the maximal isometric operators $\tilde{K}$ ($\mathscr{D}_{\tilde{K}} = \mathscr{H}^+$) described in assertion* a) *of the class $\{\tilde{K}\}$, and in the case* dim $\mathscr{H}^+ >$ dim $\mathscr{H}^-$ *by all the maximal isometric operators $\tilde{Q}$ ($\mathscr{D}_{\tilde{Q}} = \mathscr{H}^-$) described in assertion* a) *of the class $\{\tilde{Q}\}$;*

c) *in the case* dim $\mathscr{H}^+ =$ dim $\mathscr{H}^- = \infty$ *the description of all maximal neutral extensions $\tilde{\mathscr{L}}(\supset \mathscr{L})$ is obtained by considering all the maximal isometric extensions of the operators again from the union $\{\tilde{K}\} \cup \{\tilde{Q}\}$ in assertion* a); *but in the case* dim $\mathscr{H}^+ =$ dim $\mathscr{H}^- < \infty$ *it suffices to use the maximal isometric operators only from $\{\tilde{K}\}$ or only from $\{\tilde{Q}\}$;*

d) *hyper-maximal neutral extensions $\tilde{\mathscr{L}}\ (\supset \mathscr{L})$, which exist only when* dim $\mathscr{H}^+ =$ dim $\mathscr{H}^-$ *are described by the set of all unitary extensions $\tilde{K}$ ($\supset K$) (or, what is equivalent, $\tilde{Q}\ (\supset Q)$), i.e., by isometric extensions such that $\tilde{K}\mathscr{H}^+ = \mathscr{H}^-$ ($\tilde{Q}\mathscr{H}^- = \mathscr{H}^+$).*

□ Assertion a) follows from 8.3 and 8.3′.

b) If $\dim \mathscr{H}^+ < \dim \mathscr{H}^-$, then every maximal neutral subspace $\tilde{\mathscr{L}}$ ($\supset \mathscr{L}$) belongs to the class $\mathscr{M}^+$. In the opposite case in accordance with Corollary 4.14 we would have $\tilde{\mathscr{L}} \in \mathscr{M}^-$, $P^-\tilde{\mathscr{L}} = \mathscr{H}^-$, and on the basis of the analogue 8.7′ of Proposition 8.7 b) the angular operator $\tilde{Q}$ of the subspace $\tilde{\mathscr{L}}$ with respect to $\mathscr{H}^-$ would be defined everywhere in $\mathscr{H}^-$ and would be isometric, and $\tilde{Q}$: $\mathscr{H}^- \to \mathscr{H}^+$; but this is impossible when $\dim \mathscr{H}^+ < \dim \mathscr{H}^-$. It remains only to refer to 8.7 b). In the case $\dim \mathscr{H}^+ > \dim \mathscr{H}^-$ the argument is similar.

c) In view of the isometricity, when $\dim \mathscr{H}^+ = \dim \mathscr{H}^-$, of the Hilbert spaces $\mathscr{H}^+$ and $\mathscr{H}^-$, in the case *when they are finite-dimensional* all the maximal isometric (unitary) extensions $\tilde{K}$: $\mathscr{H}^+ \to$ on to $\mathscr{H}^-$ of the operator K and the maximal isometric (unitary) extensions $\tilde{Q}$: $\mathscr{H}^- \to$ on to $\mathscr{H}^+$ of the inverse operator Q can be obtained from one another by the formulae $\tilde{Q} = \tilde{K}^{-1}$ (or $\tilde{K} = \tilde{Q}^{-1}$). But if the dimension $\dim \mathscr{H}^+ = \dim \mathscr{H}^-$ is infinite, then maximal isometric extensions $\tilde{K}$: $\mathscr{H}^+ \to \mathscr{H}^-$ for which $\tilde{K}\mathscr{H}^+ \neq \mathscr{H}^-$ are possible, and similarly $\tilde{Q}$: $\mathscr{H}^- \to \mathscr{H}^+$ with $\tilde{Q}\mathscr{H}^- \neq \mathscr{H}^+$ are possible. Therefore a complete description of all maximal neutral extensions $\tilde{\mathscr{L}}$ ($\supset \mathscr{L}$) is achieved only by a sorting-out of all the maximal isometric operators from the union $\{\tilde{K}\} \cup \{\tilde{Q}\}$. Of course, with regard to those of them which are unitary, we can limit ourselves in this sorting-out, as in the 'finite-dimensional' case, to the unitary operators from $\{\tilde{K}\}$ alone (or those from $\{\tilde{Q}\}$ which are inverse to them).

d) The truth of the last assertion of the theorem follows from 8.7 (or from 8.7′) and Corollary 4.14. ■

5 In this section we consider only subspaces $\mathscr{L}$ from the $\mathscr{M}^+$ class and the $\mathscr{M}^-$ class. Therefore their angular operators K (resp. Q) are defined everywhere in $\mathscr{H}^+$ (resp. $\mathscr{H}^-$).

We now need the well-known concept of the operator T^* adjoint to an operator T: $\mathscr{H}_1 \to \mathscr{H}_2$ acting from one Hilbert space ($\mathscr{H}_1$) to another ($\mathscr{H}_2$) (see, for example, [VII]). Later (see Chapter 2, §1.1) this concept will be significantly generalized, but in the simplest particular case, which is what we now need, when *the operator T is bounded and defined everywhere in $\mathscr{H}_1$*, the *adjoint adjuster T^**: $\mathscr{H}_2 \to \mathscr{H}_1$ is naturally defined on *all* $y_2 \in \mathscr{H}_2$ by the relation

$$(x_1, T^*y_2)_1 = (Tx_1, y_2)_2 \qquad (x_1 \in \mathscr{H}_1), \tag{8.9}$$

where $(\cdot, \cdot)_1$ and $(\cdot, \cdot)_2$ are the scalar products in $\mathscr{H}_1$ and $\mathscr{H}_2$ respectively. It is clear from (8.9) that T^* is also bounded and that it has the usual properties:

$$(S+T)^* = S^* + T^*, \qquad (ST)^* = T^*S^*, \qquad T^{**} = T.$$

Theorem 8.11: *Let $\mathscr{L} \in \mathscr{M}^+$ and let K be the angular operator of $\mathscr{L}$ with respect to $\mathscr{H}^+$. Then $\mathscr{L}^{[\perp]} \in \mathscr{M}^-$ and the angular operator Q of the subspace $\mathscr{L}^{[\perp]}$ with respect to $\mathscr{H}^-$ is K^*:*

$$Q = K^*. \tag{8.10}$$

□ Let $y \in \mathscr{L}^{[\perp]}$. Then, for all $x \in \mathscr{L}$, we have

$$0 = [x, y] = [x^+ + Kx^+, y] = (x^+, y^+) - (Kx^+, y^-) = (x^+, y^+) - (x^+, K^*y^-), \tag{8.11}$$

i.e., for all $x^+ \in \mathscr{H}^+$

$$(x^+, y^+) = (x^+, K^*y^-) \tag{8.12}$$

(here the rôles of the spaces $\mathscr{H}_1$ and $\mathscr{H}_2$ in the definition (8.9) are played by $\mathscr{H}^+$ and $\mathscr{H}^-$ respectively). But since (*cf.* (8.9)) $K^*y^- \in \mathscr{H}^+$, it follows from (8.12) that $y^+ = K^*y^-$ and so $y = K^*y^- + y^-$.

Conversely, for all $y^- \in \mathscr{H}^-$ the vector $y = K^*y^- + y^-$ $(y^+ = K^*y^-)$ satisfies the relation (8.12), from which, taking the steps in the reverse order, the chain of equalities (8.11) is established. Thus $y \in \mathscr{L}^{[\perp]}$.

So

$$\mathscr{L}^{[\perp]} = \{K^*y^- + y^-\}_{y^- \in \mathscr{H}^-} \quad \text{and} \quad \| K^* \| = \| K \| < 1,$$

so that $\mathscr{L}^{[\perp]} \in \mathscr{M}^-$ and its angular operator $Q = K^*$. ■

Theorem 8.11′ (the dual of Theorem 8.11) the reader will be able to formulate for himself without difficulty.

Remark 8.12: For any subspace $\mathscr{L}$ $(=\bar{\mathscr{L}})$, as is well-known (see 7.5), we always have $\mathscr{L}^{[\perp][\perp]} = \mathscr{L}$. In the particular case when $\mathscr{L} \in \mathscr{M}^+$ $(\mathscr{M}^-)$, to this fact the following relations, deriving from Theorems 8.11 and 8.11′, for the corresponding angular operators are equivalent: $K^{**} = K$ $(Q^{**} = Q)$.

Corollary 8.13 (8.13′): *If $\mathscr{L} \in \mathscr{M}^+$ $(\mathscr{M}^-)$ and is positive (negative), then the subspace $\mathscr{L}^{[\perp]} \in \mathscr{M}^-$ $(\mathscr{M}^+)$ is negative (positive).*

□ Suppose, for example, that $\mathscr{L} \in \mathscr{M}^+$ and is positive. Then, by Theorem 8.11, $\mathscr{L}^{[\perp]} \in \mathscr{M}^-$. But also $\mathscr{L}^{[\perp]}$ is non-degenerate, for otherwise any isotropic vector z_0 $(\neq \theta)$ in it would be J-orthogonal to $\mathscr{L}$ and then $\tilde{\mathscr{L}} = \text{Lin}\{\mathscr{L}, z_0\}$ $(\neq \mathscr{L})$ would be non-negative and wider than $\mathscr{L}$ $(z_0 \notin \mathscr{L}$, because $\mathscr{L}$ is positive), which contradicts Corollary 4.8. ■

Corollary 8.14 (8.14′): *If $\mathscr{L} \in \mathscr{M}^+$ $(\mathscr{M}^-)$ and is uniformly definite, then $\mathscr{L}^{[\perp]} (\in \mathscr{M}^-)$ $(\mathscr{L}^{[\perp]} \in \mathscr{M}^+)$ is also uniformly definite. Moreover $\mathscr{L} [\dot{+}] \mathscr{L}^{[\perp]} = \mathscr{H}$.*

□ This follows from the fact that $\| K^* \| = \| K \| < 1$ $(\| Q^* \| = \| Q \| < 1)$ for the angular operators K and K^* (resp. Q and Q^*) of the subspaces $\mathscr{L}$ and $\mathscr{L}^{[\perp]}$ respectively. The last assertion follows from Corollary 7.17 ■

6 For the pair of subspaces $\mathscr{L}_\pm \in \mathscr{M}^\pm$ it is possible to obtain a curious characterization (widely applicable later) of the mutual disposition of these subspaces in terms of the spectra produced by their angular operators K and Q respectively. As a preliminary we recall that the operators QK and KQ act in the Hilbert spaces $\mathscr{H}^+$ and $\mathscr{H}^-$ respectively.

Theorem 8.15: *The following implications hold:*

a) $\mathscr{L}_+ \cap \mathscr{L}_- \neq \{\theta\} \Leftrightarrow 1 \in \sigma_P(KQ) \cup \sigma_P(QK)$;
b) $\overline{\mathscr{L}_+ [\dot{+}] \mathscr{L}_-} = \mathscr{H} \Leftrightarrow \sigma_P(KQ) \cup \sigma_P(QK)$;
c) $\mathscr{L}_+ [\dot{+}] \mathscr{L}_- = \mathscr{H} \Leftrightarrow \rho(KQ) \cap \rho(QK)$.

□ a) For any vector $x_0 \in \mathscr{L}_+ \cap \mathscr{L}_-$ we have two representations:

$$x_0 = x_0^+ + Kx_0^+ = Qx_0^- + x_0^- \qquad (x_0^\pm \in \mathscr{H}^\pm, x_0^+ + x_0^- = x_0),$$

from which it follows that

$$Qx_0^- = x_0^+, \qquad Kx_0^+ = x_0^-.$$

Hence (8.13)

$$KQx_0^- = x_0^- \quad \text{and} \quad Qkx_0^+ = x_0^+.$$

If $x_0 \neq \theta$ it follows from (8.13) that $x_0^\pm \neq \theta$ also, and so

$$1 \in \sigma_p(KQ) \cap \sigma_{p_-}(QK)(\subset \sigma_p(KQ) \cup \sigma_p(QK)). \tag{8.14}$$

Conversely, if, for example, $QKx_0^+ = x_0\ (\neq \theta)$, then

$$(\theta \neq)\ x_0^+ + Kx_0^+ = QKx_0^+ + Kx_0^+ \in \mathscr{L}_+ \cap \mathscr{L}_-.$$

b) The inequality $\overline{\mathscr{L}_+ + \mathscr{L}_-} \neq \mathscr{H}$ is equivalent to the existence of a vector $(\theta \neq)y_0 \perp \mathscr{L}_+ + \mathscr{L}_-$, or, what is the same thing, $x_0 = Jy_0 [\perp] \mathscr{L}_+ + \mathscr{L}_-$. Since $x_0 [\perp] \mathscr{L}_+$, so $[x_0, x_0] \leqslant 0$; and it follows from $x_0 [\perp] \mathscr{L}_-$ that $[x_0, x_0] \geqslant 0$, i.e., x_0 is a neutral vector. Since $\mathscr{L}_\pm$ are maximally semi-definite, it follows that $x_0 \in \mathscr{L}_+ \cap \mathscr{L}_-$, and the whole of this chain of reasoning is reversible. Thus $\overline{\mathscr{L}_+ + \mathscr{L}_-} \neq \mathscr{H} \Leftrightarrow \mathscr{L}_+ \cap \mathscr{L}_- \neq \{\theta\}$, and it remains to use assertion a) (in particular, the 'principal part' of the relation (8.14)).

c) Let $\mathscr{L}_+ [\dot{+}] \mathscr{L}_- = \mathscr{H}$. Then for every $x^+ \in \mathscr{H}^+$ there are $x_0^\pm \in \mathscr{H}^\pm$ such that $x^+ = x_0^+ + Kx_0^+ + Qx_0^- + x_0$. It is clear that $x^+ = x_0^+ + Qx_0^-$ but $Kx_0^+ + x_0^- = \theta_1$ and therefore for any $x^+ \in \mathscr{H}^+$ the equation $x_0^+ - QKx_0^+ = x^+$ is soluble (with respect to $x_0^+ \in \mathscr{H}^+$). Therefore the number $\lambda = 1$ cannot enter into either the continuous spectrum nor the residual spectrum of the operator QK: $1 \in \sigma_p(QK) \cup \rho(QK)$. But $\mathscr{L}_+ \cap \mathscr{L}_- = \{\theta\}$, so it follows from a) that $1 \in \rho(QK)$. Similarly we can prove that $1 \in \rho(KQ)$, i.e., $1 \in \rho(QK) \cap \rho(KQ) \subset \rho(QK) \cup \rho(KQ)$.

Conversely, suppose, for example, that $1 \in \rho(QK)$. Since an arbitrary vector from $\mathscr{L}^+ + \mathscr{L}^-$ can be written in the form $x^+ + Kx^+ + x^- + Qx^-$, where $x^\pm \in \mathscr{H}^\pm$, it is sufficient to verify that, for any $y^\pm \in \mathscr{H}^\pm$, the system $x^+ + Qx^- = y^+$, $x^- + Kx^+ = y^-$ is soluble. From the second equation $x^- = y^- - Kx^+$, and so we can rewrite the first equation in the form

$x^+ - QKx^+ = y^+ - Qy^-$, and therefore $x^+ = (I^+ - QK)^-(y^+ - Qy^-)$. Hence $x^- = y^- - K(I^+ - QK)^{-1}(y^+ - Qy^-)$. ■

Corollary 8.16: *If in the pair $\mathscr{L}_\pm$ $(\subset \mathscr{H}^\pm)$ at least one of the subspaces $\mathscr{L}_\pm$ is uniformly definite, then $\mathscr{L}_+ [\dotplus] \mathscr{L}_- = \mathscr{H}$.*

□ This assertion follows from Lemma 8.4 and assertion b) in Theorem 8.15. ■

7 We now return to the investigation started in §7.6 of the question of a 'rotation of the co-ordinate axes' in a Krein space $\mathscr{H}$, i.e., of a transition from one of its canonical decompositions

$$\mathscr{H} = \mathscr{H}^+ [\dotplus] \mathscr{H}^- \tag{8.15}$$

to another decomposition $\mathscr{H} = \mathscr{H}_1^+ [\dotplus] \mathscr{H}_1^-$. We shall regard the decomposition (8.15) with the canonical projectors $P^\pm$ $(\neq 0)$ and $J = P^+ - P^-$ as fixed and as defining the scalar product $(x, y) = [Jx, y]$ $((x, y) \in \mathscr{H})$, and we pose the problem of calculating the new canonical projectors $P_1^\pm: \mathscr{H} \to \mathscr{H}_1^\pm$. We recall that $\mathscr{H}_1^\pm$ are uniformly definite and for their angular operators K and K^* respectively (see Corollary 8.14) we have $\|K\| = \|K^*\| < 1$.

Theorem 8.17: *The new canonical projectors $P^\pm$ are represented in matrix form relative to the decomposition* (8.15) *as*

$$\begin{aligned} P_1^+ &= \begin{Vmatrix} (I^+ - K^*K)^{-1} & -(I^+ - K^*K)^- K^* \\ K(I^+ - K^*K)^{-1} & -K(I^* - K^*K)^{-1}K^* \end{Vmatrix}, \\ P_1^- &= \begin{Vmatrix} -K^*(I^- - KK^*)^{-1}K & K^*(I^- - KK^*)^{-1} \\ -(I^- - KK^*)^{-1}K & (I^- - KK^*)^{-1} \end{Vmatrix}, \end{aligned} \tag{8.16}$$

where $I^\pm$ are the identity operators in $\mathscr{H}^\pm$ respectively.

□ The existence of the operators represented by the right-hand sides of formula (8.16) is obvious. The fact that they are idempotent, i.e., $(P_1^\pm)^2 = P_1^\pm$, is verified immediately by squaring the corresponding matrices. Further, the construction of these matrices is such that for any vector-column

$$x = \begin{Vmatrix} x^+ \\ x^- \end{Vmatrix} \in \mathscr{H} \qquad (x^\pm \in \mathscr{H}^\pm)$$

we have

$$P_1^+ x = P_1^+ \begin{Vmatrix} x^+ \\ x^- \end{Vmatrix} = \begin{Vmatrix} (I^+ - K^*K)^{-1}x^+ - (I^+ - K^*K)^{-1}K^*x^- \\ K(I^* - K^*K)^{-1}x^+ - K(I^+ - K^*K)^{-1}K^+x^- \end{Vmatrix}$$

$$= \begin{Vmatrix} y^+ \\ Ky^+ \end{Vmatrix} \equiv y \in \mathscr{H}_1^+,$$

where $y^+ = (I^+ - K^*K)^{-1}x^+ - (I^+ - K^*K)^{-1}K^*x^- \in \mathscr{H}^+$.

In exactly the same way

$$P_1^- x = P_1^- \begin{pmatrix} x^+ \\ x^- \end{pmatrix} = \begin{pmatrix} K^* z^- \\ z^- \end{pmatrix} \equiv z \in \mathscr{H}_1^-,$$

where $z^- = -(I^- - KK^*)Kx^+(I^- - KK^*)^{-1}x^- \in \mathscr{H}^-$.

It remains to verify that y and z are the J-orthogonal projection of the vector x on to $\mathscr{H}_1^+$ and $\mathscr{H}_1^-$ respectively. We have

$$x - y = \left\| \begin{matrix} x^+ - y^+ \\ x^- - Ky^+ \end{matrix} \right\|$$

$$= \left\| \begin{matrix} x^+ - (I^+ - K^*K)^{-1}x^+ - (I^+ - K^*K)^{-1}K^*z^- \\ x^- - K(I^+ - K^*K)^{-1}x^+ + K(I^+ - K^*K)^{-1}K^*x^- \end{matrix} \right\|,$$

so that for any vector

$$u = \left\| \begin{matrix} u^+ \\ ku^+ \end{matrix} \right\| \in \mathscr{H}_1^+,$$

carrying out straightforward calculations, we obtain

$$\begin{aligned}
[x - y, y] &= (x^+ - y^+, u^+) - (x^- - Ky^+, Ku^+) \\
&= (x^+ - (I^+ - K^*K)^{-1}x^+ + (I^+ - K^*K)^{-1}K^*x^-, u^+) \\
&\quad - (x^- - K(I^+ - K^*K)^{-1}x^+ + K(I^+ - K^*K)^{-1}K^*x^-, Ku^+) \\
&= (\{I^+ - (I^+ - K^*K)^{-1}\}x^+ + (I^+ - K^*K)^{-1}K^*x^-, u^+) \\
&\quad - (-K^*K(I^+ - K^*K)^{-1}x^+ + \{I^+ + K^*K(I^+ - K^*K)^{-1}\}K^*x^-, u^+) \\
&= (\{I^+ - (I^+ - K^*K)^{-1} + K^*K(I^+ - K^*K)^{-1}\}x^+ \\
&\quad + \{(I^+ - K^*K)^{-1} - I^+ - K^*K(I^+ - K^*K)^{-1}\}K^*x^-, u^+) \\
&= (\theta, u^+) = 0,
\end{aligned}$$

i.e., $x - y\,[\perp]\,\mathscr{H}_1^+$, and in exactly the same way one proves that $x - z\,[\perp]\,\mathscr{H}_1^-$. ■

8 We shall now indicate some useful generalizations of theorems using angular operators, and at the same time some generalizations of the concept 'angular operator' itself.

Suppose, for example, that in a non-degenerate subspace $\mathscr{L}$ of a Krein space $\mathscr{H}$ the Gram operator G generates the canonical decomposition (see Theorem 6.4)

$$\mathscr{L} = \mathscr{L}^+ [\oplus] \mathscr{L}^-, \tag{8.17}$$

and that the Gram operator $G_{\mathscr{L}}$ itself relating to (8.17) is represented by the matrix

$$G_{\mathscr{L}} = \left\| \begin{matrix} I & 0 \\ 0 & -A \end{matrix} \right\|, \tag{8.18}$$

where I ($\equiv I_{\mathscr{L}^+}$) is the identity operator in $\mathscr{L}^+$, and A (>0) is a bounded positive operator in the Hilbert space $\mathscr{L}^-$, and it is possible that $0 \in \sigma_c(A)$, so that $\mathscr{L}^-$ is a negative, but not, generally speaking, a uniformly negative subspace (*cf.* Theorem 6.12).

8.18 If the subspace $\mathscr{L}$ satisfies the conditions (8.17) *and* (8.18), *then any non-positive lineal $\mathscr{L}_1$ ($\subset \mathscr{L}$) can be expressed in the form $\mathscr{L}_1 = \{F_1 x^- + x^-\}_{x^- \in P_{\mathscr{L}^-}\mathscr{L}_1}$, where $P_{\mathscr{L}^-}$ is the orthoprojector on to $\mathscr{L}^-$ and F_1 is a linear operator, F_1: $P_{\mathscr{L}^-}\mathscr{L}_1 \to \mathscr{L}^+$.*

The subspace $\mathscr{L}_1$ is a maximal non-positive subspace in $\mathscr{L}$ if and only if $P_{\mathscr{L}^-}\mathscr{L}_1 = \mathscr{L}^-$.

□ We note that, in accordance with (8.17), (8.18) for any $x = x^+ + x^- \in \mathscr{L}(x^\pm \in \mathscr{L}^\pm)$ we have

$$[x, x] = (G_{\mathscr{L}}x, x) = (x^+, x^+) - (Ax^-, x^-) = (x^+, x^+) - (A^{1/2}x^-, A^{1/2}x^-),$$

and therefore, in particular, for $x \in \mathscr{L}_1$, where $[x, x] \leqslant 0$,

$$\| x^+ \| = (x^+, x^+) \leqslant (A^{1/2}x^-, A^{1/2}x^-) = \| A^{1/2}x^- \|^2 \qquad (x^- \in P_{\mathscr{L}^-}\mathscr{L}_1). \tag{8.19}$$

The operator Q_1 which relates to any vector $A^{1/2}x$ ($\in A^{1/2}P_{\mathscr{L}^-}\mathscr{L}_1$) the vector $x^+ \in \mathscr{L}^+$ is correctly defined as a linear operator, for in accordance with Lemma 1.27 the vector x^+ is uniquely regenerated by x^- ($\in P_{\mathscr{L}^-}\mathscr{L}_1$), and it follows from (8.19) that when $A^{1/2}x^- = \theta$ we have $x^+ = \theta$. Thus, $x^+ = Q_1 A^{1/2}x^-$ ($x^- \in P_{\mathscr{L}^-}\mathscr{L}_1$), and further, again by virtue of 18.19), $\| Q_1 \| \leqslant 1$, so that $F_1 = Q_1 A^{1/2}$ is indeed the required operator.

Further, when $P_{\mathscr{L}^-}\mathscr{L}_1 = \mathscr{L}^-$ the maximal non-positivity of $\mathscr{L}_1$ in $\mathscr{L}^-$ follows from Corollary 1.28. Conversely, let $\mathscr{L}_1$ be the maximal non-positive space (in $\mathscr{L}$). Then, by what has been proved, $\mathscr{L}_1 = \{x^- + Q_1 A^{1/2}x^-\}_{x^- \in P_{\mathscr{L}^-}\mathscr{L}_1}$. But the contraction Q_1 defined on $A^{1/2}P_{\mathscr{L}^-}\mathscr{L}_1$ ($\subset \mathscr{L}^-$) admits extension into a contraction defined on $A^{1/2}\mathscr{L}^-$ which maps $A^{1/2}\mathscr{L}^-$ into $\mathscr{L}^+$; this would imply a non-positive extension of $\mathscr{L}_1$ in $\mathscr{L}$. But $\mathscr{L}_1$ is maximal, and therefore Q_1 is defined on $A^{1/2}\mathscr{L}^-$. Hence it follows that indeed $P_{\mathscr{L}^-} = \mathscr{L}^-$. For, let us now consider

$$\tilde{\mathscr{L}}_1 = \{x^- + Q_1 A^{1/2}x^-\}_{x^- \in \mathscr{L}^-} \qquad (\subset \mathscr{L});$$

by virtue of (8.19) this is again non-positive and it contains $\mathscr{L}_1$, i.e., $\tilde{\mathscr{L}}_1 = \mathscr{L}_1$ and $P_{\mathscr{L}^-}\mathscr{L}_1 = P_{\mathscr{L}^-}\tilde{\mathscr{L}}_1 = \mathscr{L}^-$. ■

It is clear that a proposition analogous to 8.18 can be established for the case when

$$G_{\mathscr{L}} = \begin{Vmatrix} A & 0 \\ 0 & -I \end{Vmatrix}, \qquad A > 0, \quad \text{and } \mathscr{L}_1 \ (\subset \mathscr{L}) \text{ is non-negative.}$$

□ We note further that it is possible by means of angular operators to describe not only semi-definite lineals but a far more general type of lineals of a Krein space $\mathscr{H} = \mathscr{H}^+ [\dot{+}] \mathscr{H}^-$. With this purpose let us consider, for example, *the class* $\mathscr{A}^+$ *of all lineals* $\mathscr{L}$ *from* $\mathscr{H}$ *which are mapped injectively into* $\mathscr{H}^+$ *by a projector* P^+. It is not difficult to figure out that $\mathscr{A}^+$ consists precisely of those $\mathscr{L}$ for which $\mathscr{L} \cap \mathscr{H}^- = \{\theta\}$ (compare with the proof of Lemma 1.26). It is clear that all such $\mathscr{L}$ admit the following description:

$$\mathscr{L} = \{x^+ + Kx^+\}_{x^+ \in P^+\mathscr{L}},$$

where $K = P^-(P^+ \mid \mathscr{L})^{-1}$, but on this occasion the 'angular operator' K is no longer obliged, generally speaking, to be a contraction. In fact, it is a contraction if and only if $\mathscr{L}$ is non-negative (it is obvious that all non-negative $\mathscr{L}$ are contained in $\mathscr{A}^+$).

The class $\mathscr{A}^-$ *of all* $\mathscr{L}$ *for which* $\mathscr{L} \cap \mathscr{H}^+ = \{\theta\}$ is defined analogously. They are described by the formula

$$\mathscr{L} = \{Qx^- + x^-\}_{x^- \in P^-\mathscr{L}},$$

where $Q = P^+(P^- \mid \mathscr{L})^{-1}$. The class $\mathscr{A}^-$ contains, in particular, all non-positive $\mathscr{L}$. ■

9 In conclusion we consider 'in the large' the sets $\mathscr{K}^{\pm}$ of all the angular operators $K(Q)$ of subspaces of the class $\mathscr{M}^{\pm}$ respectively.

□ From Theorems 8.2, 8.2′ and Proposition 8.5 it follows that:

8.19

$$(\mathscr{K}^+ \mid \mathscr{H}^+, \mathscr{H}^-) \equiv)\mathscr{K}^+ = \{K \mid K: \mathscr{H}^+ \to \mathscr{H}^-, \ \|K\| \leqslant 1\},$$
$$(\mathscr{K}^-(\mathscr{H}^-, \mathscr{H}^+) \equiv)\mathscr{K}^- = \{Q \mid Q: \mathscr{H}^- \to \mathscr{H}^+, \ \|Q\| < 1\}. \quad ■$$

□ Thus, $\mathscr{K}^+$ ($\mathscr{K}^-$) represents an *operator ball* in the Banach space $\mathscr{B}(\mathscr{H}^+, \mathscr{H}^-)$ ($\mathscr{B}(\mathscr{H}^-, \mathscr{H}^+)$) of all bounded operators $T: \mathscr{H}^+ \to \mathscr{H}^-$ ($T: \mathscr{H}^- \to \mathscr{H}^+$). ■

□ In what follows we shall often use the following well-known fact (see, e.g., [VII]):

8.20 The balls $\mathscr{K}^{\pm}$ *are bicompact in the weak operator topology.* ■

We recall that a system $\{\mathscr{M}_\alpha\}_{\alpha \in A}$ of subsets of a certain set is said to be *centralized* if every finite sub-system of it $\{\mathscr{M}_{\alpha_i}\}_{i=1}^n$ has a non-empty intersection $\bigcap_{i=1}^n \mathscr{M}_{\alpha_i} \neq \emptyset$.

□ As is well-known (see, e.g., [VII]) Proposition 8.20 is equivalent to the following:

8.21 Every centralized system $\{\mathscr{K}_\alpha^{\pm}\}_{\alpha \in A}$ *of sets* $\mathscr{K}_\alpha^{\pm} \in \mathscr{K}^{\pm}(\alpha \in A)$ *which are closed in the weak topology has a non-empty intersection.* ■

Let $\mathscr{L}_+$ ($\mathscr{L}_-$) be a non-negative (non-positive) lineal with the angular operator K_0 (Q_0), and let $\mathscr{K}_+(K_0)$ ($\mathscr{H}_-(Q_0)$ be the set of angular operators of all the subspaces from $\mathscr{M}^+$ ($\mathscr{M}^-$) which are extensions of $\mathscr{L}_+$ ($\mathscr{L}_-$), i.e., (*cf.* 8.7)

$$\mathscr{K}_+(K_0) = \{K \mid K \in \mathscr{K}^+, K \supset K_0\}, \tag{8.20}$$

$$\mathscr{K}_-(Q_0) = \{Q \mid Q \in \mathscr{K}^-, Q \supset Q_0\}. \tag{8.21}$$

In particular, if $\mathscr{L}_+ = \{\theta\}$ ($\mathscr{L}_- = \{\theta\}$), then $\mathscr{K}_+(K_0) = \mathscr{K}^+$ ($\mathscr{K}_-(Q_0) = \mathscr{K}^-$). We bring into consideration (see (8.9)) the sets

$$\mathscr{K}^*_-(Q_0) = \{K \mid K = Q^*, Q \in \mathscr{K}_-(Q_0)\}, \tag{8.22}$$

$$\mathscr{K}^*_+(K_0) = \{Q \mid Q = K^*, K \in \mathscr{K}_+(K_0)\}. \tag{8.23}$$

8.22 *The set $\mathscr{K}^*_-(Q_0)$ (respectively $\mathscr{K}^*_+(K_0)$) coincides with the set of all angular operators of subspaces from $\mathscr{M}^+$ (resp. from $\mathscr{M}^-$) which are J-orthogonal to the lineal $_-$ ($\subset \mathscr{P}^-$) (resp. $\mathscr{L}_+$ ($\supset \mathscr{P}^+$)) with the angular operator Q_0 (resp. K_0).*

□ The assertion follows directly from Theorems 8.11 and 8.11′. ■

Theorem 8.23: *The sets $\mathscr{K}_+(K_0)$, $\mathscr{K}_-(Q_0)$, $\mathscr{K}^*_-(Q_0)$ and $\mathscr{K}^*_+(K_0)$ are convex and bicompact in the weak operator topology.*

□ We begin with $\mathscr{K}_+(K_0)$. Let $K_1, K_2 \in \mathscr{K}^+(K_0)$. Since $\mathscr{K}_+(K_0) \subset \mathscr{K}^+$, so by virtue of the convexity of $\mathscr{K}^+$ when $\alpha \in [0,1]$ we have $\alpha K_1 + (1-\alpha)K_2 \in \mathscr{K}^+$ for any $K_1, K_2 \in \mathscr{K}^+$. Further, if $x^+ \in \mathscr{D}_{K_0}$ and $K_1, K_2 \in \mathscr{K}_+(K_0)$, then

$$(\alpha K_1 + (1-\alpha)K_2)x^+ = \alpha K_1 x^+ + (1-\alpha)K_2 x^+ = \alpha K_0 x^+ + (1-\alpha)K_0 x^+ = K_0 x^+,$$

i.e., $\alpha K_1 + (1-\alpha)K_2 \supset K_0$, and the convexity of $\mathscr{K}_+(K_0)$ is proved.

Now let a generalized sequence $\{K_\gamma\}_{\gamma \in \Gamma}$ ($\subset \mathscr{K}_+(K_0)$), where Γ is a certain directed set (see [VII]) converge in the weak operator topology to a certain operator $\tilde{K} \in \mathscr{B}(\mathscr{H}^+, \mathscr{H}^-)$. By virtue of 8.20 the ball $\mathscr{K}^+$ is closed in this topology, and so $\tilde{K} \in \mathscr{K}^+$. Since for any $\varepsilon > 0$ and for any $x^\pm \in \mathscr{H}^\pm$ a $\gamma_0 \in \Gamma$ can be found such that $|((K - K_\gamma)x^+, x^-)| < \varepsilon$ for all $\gamma > \gamma_0$, so we have

$$|((\tilde{K} - K_0)x^+, x^-)| < \varepsilon \text{ for all } \varepsilon > 0, \text{ for all } x^+ \in \mathscr{D}_{K_0} \text{ and for all } x^- \in \mathscr{H}^-;$$

hence

$$(\tilde{K} - K_0)x^+ = \theta \ (x^+ \in \mathscr{D}_{K_0}), \text{ i.e., } \tilde{K} \supset K_0 \text{ and } \tilde{K} \in \mathscr{K}_+(K_0).$$

Thus (see [VII]) $\mathscr{K}_+(K_0)$ is a closed (in the weak operator topology) subset of the bicompact ball $\mathscr{K}^+$ and therefore it is itself bicompact.

The arguments are analogous for $\mathscr{K}_-(Q_0)$, $\mathscr{K}^*_-(Q_0)$, and $\mathscr{K}^*_+(K_0)$. ■

Corollary 8.24: *The sets $\mathscr{K}_+(K_0) \cap \mathscr{K}^*_-(Q_0)$ and $\mathscr{K}_-(Q_0) \cap \mathscr{K}^*_+(K_0)$ are convex and bicompact in the weak operator topology.* ■

Remark 8.25: In Corollary 8.24 it is not excluded, generally speaking, that the intersections mentioned in it may be empty. The non-emptiness of either of them implies, by 8.22, the relation $\mathscr{L}_+ [\perp] \mathscr{L}_-$. Later (see Theorem 10.2) it will be proved that this is not only a necessary, but also a *sufficient*, condition for both the intersections mentioned in Corollary 8.24 to be non-empty.

Exercises and problems

1 Consider the two-dimensional real Krein space $\mathscr{H}$ in Exercise 7 to §2 (see Fig. 1 there). In Fig. 2 is shown a non-negative (even positive) subspace $\mathscr{L}(\subset \mathscr{H})$. For any $x \in \mathscr{L}$ we have $x = x^+ + x^-$ $(x^{\pm} \in \mathscr{H}^{\pm})$, and $x^- = Kx^+$, where K is the operator of rotating the vector x^+ through an angle $\pi/2$ (in the positive direction), and then multiplying by the scalar $k = \tan \varphi$—the angular coefficient of the 'line' $\mathscr{L}$. Compare this with the definition 8.1. Verify that $\| K \| = | k |$.

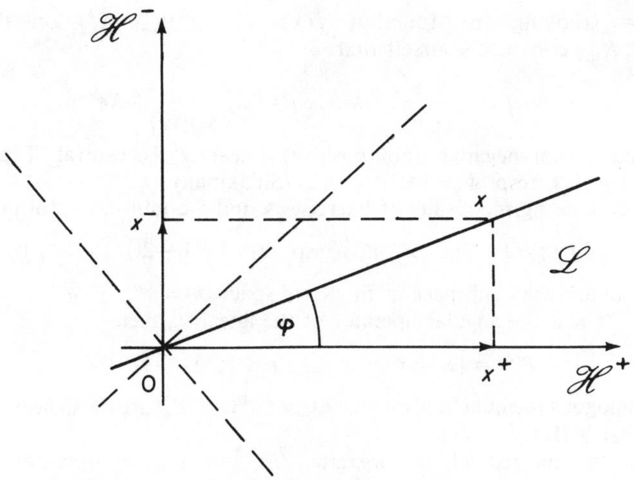

Fig. 2.

2 If in Exercise 1 φ is always understood to be the *minimal* angle between $\mathscr{L}$ and 'the axis' $\mathscr{H}^+$, then $\tan \varphi = \| K \|$. Prove that in the general case too ($\dim \mathscr{H} \leqslant \infty$) for the angular operator K of a non-negative subspace $\mathscr{L}$ we have $\tan \varphi(\mathscr{L}, \mathscr{H}^+) = \| K \|$ if the (minimal) angle φ is defined by the equality $\sin \varphi(\mathscr{L}, \mathscr{H}^+) = \sup_{e \in S(\mathscr{L})} \| e - ZP^+ e \|$ where $S(\mathscr{L})$ is the unit sphere of the lineal $\mathscr{L}$ ($\| e \| = 1$) (M. Krein [XVII]).

3 It follows from the result of Exercise 2 and Lemma 8.4 that uniform positivity of the subspace $\mathscr{L}$ is equivalent to the inequality $\varphi < \pi/4$. Hence it is possible to derive in terms of the *aperture* $\Theta(\mathscr{L}_1, \mathscr{L}_2) = \max\{\sin \varphi(\mathscr{L}_1, \mathscr{L}_2), \sin \varphi(\mathscr{L}_2, \mathscr{L}_1)\}$ *of two subspaces* $\mathscr{L}_1$ *and* $\mathscr{L}_2$ (see, e.g., [I]) the following criterion: *in order that all subspaces* $\mathscr{L}'$ *with a sufficiently small aperture* $\Theta(\mathscr{L}, \mathscr{L}')$ *be non-negative it is necessary and sufficient that the subspace* $\mathscr{L}$ *be uniformly positive* (M. Krein [XVII]).

4 Let $\mathscr{L}$ be a non-negative subspace with angular operator K and let $\Theta(\mathscr{L}, \mathscr{H}^+) < 1$. Then

$$\Theta(\mathscr{L}, \mathscr{H}^+) = \frac{\| K \|}{\sqrt{1 + \| K \|^2}} \quad \text{(Shlakman)}.$$

Hint: Use the properties of the aperture (see [I]) and the result of Exercise 2.

5 On the set of all semi-definite lineals $\mathscr{L}$ of a Krein space $\mathscr{H}$ we introduce the functional

$$\Phi(l) = \begin{cases} \sup\limits_{e \in S(\mathscr{L})} [e, e], & \text{if } \mathscr{L} \text{ is non-negative,} \\ \inf\limits_{e \in S(\mathscr{L})} [e, e], & \text{if } \mathscr{L} \text{ is non-positive.} \end{cases}$$

Suppose, for definiteness, that $\mathscr{L}$ is non-negative, and let K be its angular operator. It is clear that $0 \leqslant \Phi(\mathscr{L}) \leqslant 1$ and that $\{\Phi(\mathscr{L}) = 0\} \Leftrightarrow \{\mathscr{L} \text{ is neutral}\}$. Prove that

$$\Phi(\mathscr{L}) = \begin{cases} \dfrac{\| K^{-1} \|^2 - 1}{\| K^{-1} \|^2 + 1} & \text{if } K \text{ is bounded invertible,} \\ 1 & \text{otherwise} \end{cases} \quad \text{(Shlakman)}$$

Hint: By studying the function $f(x) = (1 - x^2)/(1 + x^2)$ on the interval $0 \leqslant x \leqslant \| K \|$, convince yourself that

$$\Phi(\mathscr{L}) = \frac{1 - \mu^2}{1 + \mu^2}, \quad \text{where } \mu = \inf_{e^+ \in S(P^+ \mathscr{L})} \| K e^+ \|.$$

6 Prove that a non-negative (non-positive) lineal $\mathscr{L}$ is neutral if and only if $\Phi(\mathscr{L}, \mathscr{H}^-) = 1/\sqrt{2}$ (resp. $\Phi(\mathscr{L}, \mathscr{H}^+) = 1/\sqrt{2}$) (Shlakman).
Hint: By comparing the results of Exercises 4 and 5 establish the formula

$$\Phi(\mathscr{L}) = 1 - 2\Theta^2(\mathscr{L}, \mathscr{H}^-) (\text{resp. } \Phi(\mathscr{L}) = 1 - 2\Theta^2(\mathscr{L}, \mathscr{H}^+)).$$

7 Let $\mathscr{L}$ be an arbitrary subspace of the Krein space $\mathscr{H} = \mathscr{H}^+ \oplus \mathscr{H}^-$, and let $\mathscr{L} \in \mathscr{A}^+$ (see §8.8). If K is the angular operator of the lineal $\mathscr{L}$, then

$$\mathscr{L}^{[\perp]} = \{K^{(*)} x^- + x^-\}_{x^- \in P^- \mathscr{L}} [\oplus] \mathscr{D}^+_{\mathscr{L}} [\oplus] \mathscr{D}^-_{\mathscr{L}}, \tag{8.24}$$

and an analogous formula holds when $\mathscr{L} \in \mathscr{A}^-$. Here $\mathscr{D}^{\pm}_{\mathscr{L}}$ are the deficiency subspaces of the lineal $\mathscr{L}$ (Ritsner [4]).
Hint. The 'generalized adjoint operator' $T^{(*)}$ of a linear operator $T\colon \mathscr{D}_T \to \mathscr{H}$ ($\mathscr{D}_T \subset \mathscr{H}$) should here be understood to be the operator whose graph $\Gamma_{T^{(*)}}$ is given by the formula $\Gamma_{T^{(*)}} = \{\langle u, v \rangle \in \mathscr{H} \times \mathscr{D}_T \mid (Tx, u) = (x, v) \text{ for all } x \in \mathscr{D}_T\}$ (*cf.* (8.10)). For any bounded T such a $T^{(*)}$ exists and $\| T^{(*)} \| = \| T \|$.

8 Derive the formula (4.5) and Theorem 8.11 from the formula (8.24).

9 Suppose a linear operator E in the Krein space $\mathscr{H} = \mathscr{H}^+ [\oplus] \mathscr{H}^-$ satisfies the conditions $EP^+ = E$, $P^+ E = P^+$. Then E is a projector ($E^2 = E$), and further $E\mathscr{H} \subset \mathscr{P}^+$ if and only if $\| E \| \leqslant \sqrt{2}$. An analogous result is true if P^+ and $\mathscr{P}^+$ are replaced by P^- and $\mathscr{P}^-$ respectively (Langer [2]).

10 Prove that the projector E in Exercise 9 and the angular operator K of the lineal $E\mathscr{H}$ ($\subset \mathscr{P}^+$) are connected by the formula $E' = P^+ + KP^+$ (in the case $E\mathscr{H} \subset \mathscr{P}^-$ the corresponding formula is $E = P^- + QP^-$, where Q is the angular operator of the lineal $E\mathscr{H}$) ([XVII]).

11 Prove that for the projector E in Exercise 9 the condition $\| E \| < \sqrt{2}$ is necessary and sufficient for the lineal $E\mathscr{H}$ to be uniformly definite.
Hint. Use the result of Exercise 10 and Lemma 8.4.

12 Prove that for a subspace $\mathscr{L}_+$ $(\in\mathscr{M}^+)$ of the class h^+ (see Definition 5.9) we have $\mathscr{L}_- \equiv \mathscr{L}_+^{[\perp]} \in \mathscr{M}^- \cap h^-$.

13 Let $\mathscr{L} \in \mathscr{M}^-$, let K be the corresponding angular operator, and let $P_{\mathscr{L}}$ be the orthoprojector on to $\mathscr{L}$. Prove that

$$P_{\mathscr{L}} = \left\| \begin{matrix} (I^+ + K^*K)^{-1} & (I^+ + K^*K)^{-1}K^* \\ K(I^+ + K^*K)^{-1} & K(I^+ + K^*K)K^* \end{matrix} \right\| \quad \text{(Azizov).}$$

Hint: See the proof of Theorem 8.17.

14 Under the conditions of Exercise 13 prove that for the operator $T = P^+ \mid \mathscr{L}: \mathscr{L} \to \mathscr{H}^+$ the adjoint operator $T^*: \mathscr{H}^+ \to \mathscr{L}$ has the form $T^* = T^{-1}(I^+ + K^*K)^{-1}$ (Azizov).

15 Under the conditions of Exercise 13 prove that the Gram operator $G_{\mathscr{L}} (= P_{\mathscr{L}} J \mid \mathscr{L})$ has the form $G_{\mathscr{L}} = V^*(I^+ - K^*K)V$, where $V: \mathscr{L} \to \mathscr{H}^+$ is a linear homeomorphism, i.e., $G_{\mathscr{L}}$ and $(I^+ - K^*K)$ are 'congruent' (Azizov).
Hint: Use the results of Exercises 13 and 14.

16 Let $\mathscr{L}^- \in \mathscr{M}^-$, let Q be its angular operator, and let $G_{\mathscr{L}^-}$ be the Gram operator for $\mathscr{L}^-$. Then the operators $G_{\mathscr{L}^-}$ and $I - Q^*Q$ are congruent, i.e., there is a linear homeomorphism $V: \mathscr{L}^- \to \mathscr{H}^-$ such that $G_{\mathscr{L}} = V^*(I^+ - Q^*Q)V$ (Azizov).
Hint: $P^- \mid \mathscr{L}^-$ can be taken as V.

17 Under the conditions of the preceding problem suppose $\mathscr{L}^+$ $(\mathscr{L}^- = \mathscr{L}^{+[\perp]})$ has no infinite-dimensional uniformly definite subspaces. Then

$$|K_{\mathscr{L}^+}| = I^+ + S \qquad (|K_{\mathscr{L}^-}| = I^- + S_1), \quad \text{where } S(S_1) \in \mathscr{S}_\infty. \quad \text{(Azizov).}$$

Hint: Use the results of the preceding exercise and of Exercise 7 in §6.

18 Let $\mathscr{L}^+ \subset \mathscr{P}^+$ and $K_{\mathscr{L}^+} \in \mathscr{S}_\infty$. Then $\mathscr{L}^+ \in h^+$ (J.W. Helton [2]).

19 Let $\mathscr{L}^+ \in \mathscr{M}^+ \cap h^+$ and $\mathscr{L}^- \in \mathscr{M}^-$. Then $\mathscr{L}^+ \cap \mathscr{L}^- = \{\theta\} \Leftrightarrow \mathscr{L}^+ \dotplus \mathscr{L}^- = \mathscr{H}$ (Azizov).
Hint: Use Theorem 8.15.

20 Let $\mathscr{H}$ be a Krein space and let $\mathscr{L}^+ \subset \mathscr{P}^+$. Prove that $\mathscr{L}^+ \in \mathscr{M}^+$ if and only if $\mathscr{L}^{+[\perp]} \subset \mathscr{P}^-$.
Hint: Use Theorem 8.11.

21 Suppose $\mathscr{H}$ is a Krein space, $\mathscr{L}^+ \subset \mathscr{P}^+$, $\mathscr{L}_0 \subset \mathscr{L}^+ \cap \mathscr{L}^{+[\perp]}$, and $\hat{\mathscr{H}} = \mathscr{L}_0^{[\perp]}/\mathscr{L}_0$. Prove that $\mathscr{L}^+ \in \mathscr{M}^+(\mathscr{H})$ if and only if $\mathscr{L}^+/\mathscr{L}_0 \in \mathscr{M}^+(\hat{\mathscr{H}})$.
Hint: Use the definition of the factor space $\hat{\mathscr{H}}$ and Exercise 20.

22 Suppose $\mathscr{H}$ is a Krein space, $\mathscr{L}_\pm \subset \mathscr{P}_\pm$, and $\mathscr{L}_0 \subset \mathscr{L}_\pm \cap \mathscr{P}^0$. Prove that, if $\dim \mathscr{L}_0 < \infty$, then $\mathscr{L}_\pm \in h^\pm$ if and only if $\mathscr{L}_\pm/\mathscr{L}_0 \in h^\pm$ (Azizov).
Hint: Verify that the subspaces $\mathscr{L}_\pm \in h^\pm$ if and only if their angular operators can be expressed in the form of a direct sum of a finite-dimensional isometric operator and a uniform contraction, i.e., an operator T with $\|T\| < 1$.

23 We introduce on $\mathscr{M}^+$ a functional $\theta(\mathscr{L}, \mathscr{N}) = \sup\{\theta(x, y) \mid x \in \mathscr{L}, y \in \mathscr{N};$ $\mathscr{L}, \mathscr{N} \in \mathscr{M}^+\}$, where

$\theta(x, y) = 1$ if $\mathrm{Re}^2[x, y] \leqslant [x, x][y, y]$;

$\theta(x, y) = 2\delta^2 + 2\delta\sqrt{\delta^2 - 1} - 1$ if $\mathrm{Re}^2[x, y] \geqslant [x, x][y, y] > 0$

and $\delta^2 = \mathrm{Re}^2[x, y]/[x, x][y, y]$;

$\theta(x, y) = 0$ in the remaining cases.

Prove that the relation $\mathscr{L} \sim \mathscr{N}$ which is expressed by the inequality $\theta(\mathscr{L}, \mathscr{N}) < \infty$ is an equivalence relation on $\mathcal{M}^+$, that on each equivalence class the function $\rho(\mathcal{L}, \mathcal{N}) = \ln \theta(\mathcal{L}, \mathcal{N})$ is a metric, and that each equivalence class is a complete metric space (A. V. Sobolev and V. A. Khatskevich, [1], [2]).

24 Prove that the set $\mathcal{M}_0^+$ of maximal uniformly positive subspaces in an equivalence class (Khatskevich, [13], [14]).

25 Using the connection between operators from $\mathcal{K}^+(\mathcal{H}^+, \mathcal{H}^-)$ and subspaces from $\mathcal{M}^+$ introduce an equivalence relation on $\mathcal{K}^+(\mathcal{H}^+, \mathcal{H}^-)$:

$$K_1 \sim K_2 \quad \text{if } \theta(\mathcal{L}_1, \mathcal{L}_2) < \infty, \quad \text{where } \mathcal{L}_i = \{x^+ + K_i x^+ \mid x^+ \in \mathcal{H}^+\}, \qquad i = 1, 2.$$

Verify that the interior of the ball $\mathcal{K}^+(\mathcal{H}^+, \mathcal{H}^-)$ is an equivalence class (Khatskevich, [13], [14]).

§9 Pontryagin spaces $\Pi_\varkappa$. $W^{(\varkappa)}$-spaces and $G^{(\varkappa)}$-spaces

1 One subclass of Krein spaces is particularly important in operator theory.

Definition 9.1: A Krein space $\mathcal{H} = \mathcal{H}^+ [\dot{+}] \mathcal{H}^-$ with a finite rank of indefiniteness

$$K = \min\{\dim \mathcal{H}^+, \dim \mathcal{H}^-\} \text{ (see Definition 6.5)}$$

is called a *Pontryagin space* and is denoted by $\Pi_\varkappa$. Admitting a certain freedom of speech, we shall sometimes call $\varkappa$ *the number of positive or negative squares.*

In accordance with this definition we shall denote a fixed canonical decomposition of a Pontryagin space $\Pi_\varkappa$ by

$$\Pi_\varkappa = \Pi^+ [\dot{+}] \Pi^-, \tag{9.1}$$

and without loss of generality *we shall in future assume, if nothing to the contrary is stated, that* $(\infty >)\varkappa = \dim \Pi^+$. Otherwise we would have changed over to the anti-space (see Definition 1.8). Thus the equality $\varkappa = 0$ corresponds to the (trivial) case when $\Pi_\varkappa = \Pi^-$ is a Hilbert space with the scalar product $(x, y) = -[x, y]$ $(x, y \in \Pi_\varkappa)$ (see Remark 2.3). Examples of spaces $\Pi_\varkappa$ are given in the Exercises for this section.

It follows from Definition 9.1 that all the general geometric facts about Krein spaces are true for a space $\Pi_\varkappa$. However, and additional condition, namely, the inequality $\varkappa < \infty$ naturally gives rise to a number of additional properties of Pontryagin spaces. It is with these properties that we shall concern ourselves. The simplest of them is contained in the following proposition.

9.2 *All non-negative lineals of the* $\Pi_\varkappa$ *are finite-dimensional.*

□ By Corollary 4.3 we have for any non-negative $\mathcal{L}$ $(\subset \Pi_\varkappa)$ that $\dim \mathcal{L} \leqslant \varkappa (< \infty)$. ■

Corollary 9.3: *In the space* $\Pi_\varkappa$ *every positive lineal is uniformly positive.*

□ This follows from 9.2 and 5.4. ■

9.4 *A non-negative (positive, neutral) lineal* $\mathscr{L}$ *in the space* $\Pi_\varkappa$ *is a maximal non-negative* (*a maximal positive, a maximal neutral*) *lineal if and only if* dim $\mathscr{L} = \varkappa$.

□ This follows from Corollaries 4.6 and 4.8 and Theorem 4.13 (taking 9.2 into account). In applying Corollary 4.8 it should be borne in mind, that in contrast to the general case, *in the space* $\Pi_\varkappa$ *all maximal positive lineals are closed* (because they are finite-dimensional). ■

To conclude this section we establish a lemma which will often be applied in the theory of operators acting in the space $\Pi_\varkappa$.

Lemma 9.5: *Any lineal* $\mathscr{L}$ *dense in* $\Pi_\varkappa$ ($\overline{\mathscr{L}} = \Pi_\varkappa$) *contains a maximal* (*i.e. a* $\varkappa$*-dimensional*) *positive subspace.*

□ We consider the canonical decomposition (9.1), where dim $\Pi^+ = \varkappa$, and a certain orthonormalized basis $\{e_1, e_2, \ldots, e_\varkappa\}$ in Π^+. It is clear that $[e_j, e_k] = (e_j, e_k) = \delta_{jk}$, $(j, k = 1, 2, \ldots, \varkappa)$. For any $\varepsilon > 0$, because of the condition $\overline{\mathscr{L}} = \Pi_\varkappa$, vectors $f_1, f_2, \ldots, f_\varkappa \in \mathscr{L}$ can be found with the property $\| f_k - e_k \| <$ $(k = 1, 2, \ldots, \varkappa)$. Since all the Gram determinants of the system $\{e_1, e_2, \ldots, e_\varkappa\}$ are positive:

$$\det \| [e_j, e_k] \|_{j,k=1}^{m} = \det \| \delta_{jk} \|_{j,k=1}^{m} = 1 > 0 \qquad (m = 1, 2, \ldots, \varkappa),$$

it follows that, for sufficiently small $\varepsilon > 0$, all the determinants

$$\det \| [f_j, f_k] \|_{j,k=1}^{m} \ (m = 1, 2, \ldots, \varkappa) \qquad \text{are also positive,}$$

i.e., the Gram matrix $\| [f_j, f_k] \|_{j,k=1}^{\varkappa}$ of the subspace $\mathscr{F} = \text{Lin}\{f_1, f_2, \ldots, f_\varkappa\}$ $(\subset \mathscr{L})$ is positive definite: thus the vectors $f_1, f_2, \ldots, f_\varkappa$ are linearly independent, and $\mathscr{F}$ is a $\varkappa$-dimensional positive subspace. ■

2 The condition $\varkappa < \infty$, apart from its natural influence (as explained in section 9.1) on the character of non-negative lineals (subspace) of the space $\Pi_\varkappa$, also has an influence on the negative subspaces of $\pi_\varkappa$.

Theorem 9.6: *All negative subspaces* $\mathscr{L}$ *of the space* $\Pi_\varkappa$ *are uniformly negative.*

□ By virtue of the obvious analogue 8.9′ of Remark 8.9 there is a maximal negative subspace $\mathscr{L}_{\max} \supset \mathscr{L}$. By Corollary 8.13′, $\mathscr{L}_{\max}^{[\perp]}$ is a maximal positive subspace and, more than that (see Proposition 9.2 and Corollary 9.3), it is a uniformly positive $\varkappa$-dimensional subspace. By virtue of Corollary 7.15 it is projectively complete, and since (see Proposition 7.9) $\mathscr{L}_{\max} = \mathscr{L}_{\max}^{[\perp][\perp]}$ is also projectively complete, it is also uniformly negative. Therefore (see 5.3) $\mathscr{L}$ too is uniformly negative. ■

Corollary 9.7: *All definite subspaces $\mathscr{L}$ of $\Pi_\varkappa$ are projectively complete.*

□ This follows from Corollary 7.15 and Corollary 9.3 (for positive $\mathscr{L}$) and from Theorem 9.6 (for negative $\mathscr{L}$).

Remark 9.8: The assertion of Theorem 9.6 cannot by any means be extended to *non-closed* negative lineals. The example 4.12, which we have already used more than once, bears witness to this (with accuracy up to a transition to the anti-space). The space $\mathscr{H}$ considered in this example is (up to a sign, i.e., up to a transtion to the anti-space) the space Π_1, and the non-closed positive lineal $\mathscr{L}$ constructed in it is not uniformly positive (as was explained later in Example 5.10), i.e., $\mathscr{L}$ is singular (*cf.* Exercise 6 to §5). We shall return later, in Exercises 9–12, to the analysis of similar situations.

Theorem 9.9: *In a space $\Lambda_\varkappa$ a subspace $\mathscr{L}$ $(=\bar{\mathscr{L}})$ is projectively complete if and only if it is non-degenerate.*

□ We recall (Proposition 7.10) that non-degeneracy of a subspace $\mathscr{L}$ in the general case of a Krein space is a necessary condition for its projective completeness. We show that in our case it is also sufficient. If $\mathscr{L}$ is non-degenerate, then in any canonical decomposition of it $\mathscr{L}=\mathscr{L}^+[+]\mathscr{L}^-$ (which exists by virtue of Theorem 6.4) the definite subspaces $\mathscr{L}^+$ and $\mathscr{L}^-$ are uniformly definite (Corollary 9.3 and Theorem 9.6), and therefore (Theorems 7.16 and 5.7) $\mathscr{L}$ is projectively complete. ■

Corollary 9.10: *Every non-degenerate subspace $\mathscr{L}$ in $\Pi_\varkappa$ is itself a Pontryagin space $\Pi_{\varkappa'}$ with a certain rank of indefiniteness $\varkappa'$: $0 \leqslant \varkappa' \leqslant \varkappa$.*

□ $\mathscr{L}$ is projectively complete, and in its canonical decomposition $\mathscr{L}=\mathscr{L}^+[\dotplus]\mathscr{L}^-$, by virtue of Proposition 9.2 $0 \leqslant \dim \mathscr{L}^+ \leqslant \varkappa$, so that when $\dim \mathscr{L}^+ \leqslant \dim \mathscr{L}^-$ we have $\varkappa' = \dim \mathscr{L}^+ \leqslant \varkappa$, and when $\dim \mathscr{L}^- \leqslant \dim \mathscr{L}^+$ *a fortiori* $\varkappa' = \dim \mathscr{L}^- < \varkappa$. ■

3 Comparing some of the facts discovered in §9.2 we remark that they are characteristic for picking our Pontryagin spaces in the class of all Krein spaces.

Theorem 9.11: *For a Krein space $\mathscr{H}=\mathscr{H}^+[\dotplus]\mathscr{H}^-$ the following three assertions are equivalent:*

a) *the space $\mathscr{H}$ is a Pontryagin space ($\mathscr{H}=\Pi_\varkappa$);*

b) *all the definite subspaces* $\mathscr{L}$ $(\subset \mathscr{H})$ *are uniformly definite (projectively complete)*;

c) *all the non-degenerate subspaces* $\mathscr{L}$ $(\subset \mathscr{H})$ *are projectively complete.*

□ That a) ⇒ b) ⇒ c) was established in §9.2. It remains to prove the implication c) ⇒ a). To do this it is sufficient to adduce an example of a non-degenerate but not projectively complete subspace $\mathscr{L}$ in any Krein space $\mathscr{H} = \mathscr{H}^+ [\dotplus] \mathscr{H}^-$ of infinite rank of indefiniteness: $\varkappa = \min\{\dim \mathscr{H}^+, \dim \mathscr{H}^-\} = \infty$. But as such an example any subspace $\mathscr{L}$ which is, say, positive but not uniformly positive will serve (see Exercise 1 to §5). ■

4 In studying the spaces $\Pi_\varkappa$ (and later the operators acting in them), instead of the prefix '*J*-' (*J*-orthogonality, *J*-isometry, etc.) which is traditional for Krein spaces, the prefix 'π-' or '$\pi_\varkappa$-' is more expressive since it indicates at once that *everything takes place in a Pontryagin space*, and the second form indicates the rank of indefiniteness. In particular, we shall in future use these symbols, naming, for example, vectors x, y $(\in \Pi_\varkappa)$ for which $x[\perp]y$, and sets $\mathscr{L}, \mathscr{M}$ $(\subset \Pi_\varkappa)$ for which $\mathscr{L}[\perp]\mathscr{M}$ as *π-orthogonal*, and we shall call the lineal $\mathscr{M}^{[\perp]}$ the *π-orthogonal complement of the set* $\mathscr{M}$.

We recall (see Proposition 9.2) that the isometric lineal $\mathscr{L}^0 = \mathscr{L} \cap \mathscr{L}^{[\perp]}$ of a lineal $\mathscr{L}$ in a space $\Pi_\varkappa$ is always finite-dimensional ($\dim \mathscr{L}^0 < \varkappa$). For degenerate subspaces $\mathscr{L}$ ($\mathscr{L} = \bar{\mathscr{L}}, \mathscr{L}^0 = \{\theta\}$) in $\Pi_\varkappa$ instead of Theorem 9.9 we have to be content with a more complicated decomposition (generated by $\mathscr{L}$) of $\Pi_\varkappa$ into a direct sum of subspaces more complicated than the simple π-orthogonal sum $\mathscr{L} [\dotplus] \mathscr{L}^{[\perp]}$.

Theorem 9.12: *Let* $\mathscr{L}$ *be a subspace in* $\Pi_\varkappa$, let $\mathscr{M} = \mathscr{L}^{[\perp]}$ *be its* π*-orthogonal complement, and* $\mathscr{L}^0 = \mathscr{L} \cap \mathscr{M}$ *be the isotropic part of* $\mathscr{L}$. *Then the following decomposition holds:*

$$\Pi_\varkappa = \mathscr{L}_1 [\dotplus] \mathscr{M}_1 [\dotplus] (\mathscr{L}^0 + \mathscr{N}), \tag{9.2}$$

where $\mathscr{L}_1$ *and* $\mathscr{M}_1$ *are non-degenerate subspaces connected with* $\mathscr{L}$ *and* $\mathscr{M}$ *respectively by the relations*

$$\mathscr{L} = \mathscr{L}_1 [\dotplus] \mathscr{L}^0, \qquad \mathscr{M} = \mathscr{M}_1 [\dotplus] \mathscr{L}^0, \tag{9.3}$$

and $\mathscr{N}$ *is a certain subspace skewly linked with* $\mathscr{L}^0$ *(see Definition 1.29)*: $\mathscr{N} \# \mathscr{L}^0$. *For any choice of* $\mathscr{L}_1$ *and* $\mathscr{M}_1$ *in* (9.3) *the subspace* $\mathscr{N}$ *skewly linked with* $\mathscr{L}^0$ *may be chosen arbitrarily in the subspace* $(\mathscr{L}_1 [\dotplus] \mathscr{M}_1)^{[\perp]}$.

On the other hand, for any choice in $\Pi_\varkappa$ *of a subspace* $\mathscr{N}$ $(\# \mathscr{L}^0)$ *satisfying the conditions* (9.3) *the subspaces* $\mathscr{L}_1$ *and* $\mathscr{M}_1$ *in* (9.2) *are defined uniquely:*

$$\mathscr{L}_1 = \mathscr{L} \cap \mathscr{N}^{[\perp]}, \qquad \mathscr{M}_1 = \mathscr{M} \cap \mathscr{N}^{[\perp]}.$$

□ The subspaces $\mathscr{L}_1$ and $\mathscr{M}_1$ satisfying the conditions (9.3) (by Proposition 7.3 $\mathscr{L}^0$ is the common isotropic subspace for $\mathscr{L}$ and $\mathscr{M}$) are not-degenerate, and therefore (Theorem 9.11c) they are projectively complete, and the same, as is easy to understand (see Exercise 8 to §7), applies also to their π-orthogonal sum $\mathscr{L}_1 [\dot{+}] \mathscr{M}_1$ (we recall that $\mathscr{L}_1 \subset \mathscr{L}$ and $\mathscr{M}_1 \subset \mathscr{M} = \mathscr{L}^{[\perp]}$). Thus, both $\mathscr{L}_1 [\dot{+}] \mathscr{M}_1$ and $(\mathscr{L}_1 [\dot{+}] \mathscr{M}_1)^{[\perp]}$ are certain Pontryagin spaces (see Corollary 9.10). The relations (9.3) show that the second of them contains $\mathscr{L}^0$ ($\dim \mathscr{L}^0 \leqslant \varkappa$).

As in every Krein space, in the space $(\mathscr{L}_1 [\dot{+}] \mathscr{M}_1)^{[\perp]}$ there are (see Exercise 4 to §3) subspaces skewly linked with $\mathscr{L}^0$. We choose any one of them and call it $\mathscr{N}$: $\mathscr{N} \# \mathscr{L}^0$. Then by virtue of Lemma 1.31 and Proposition 1.30 $\dim \mathscr{N} = \dim \mathscr{L}_0$ and the direct sum $\mathscr{L}^0 \dot{+} \mathscr{N}$ is non-degenerate, i.e., it is projectively complete. Thus (see Exercise 8 to §7) the subspace

$$\mathscr{R} = \mathscr{L}_1 [\dot{+}] \mathscr{M}_1 [\dot{+}] (\mathscr{L}^0 + \mathscr{N})$$

is also projectively complete, i.e., it is non-degenerate.

Now let x be such that $x [\perp] \mathscr{R}$. In particular, $x \in (\mathscr{L}^0 + \mathscr{L}_1)^{[\perp]} = \mathscr{L}^{[\perp]}$ and $x \in (\mathscr{L}^0 + \mathscr{M}_1)^{[\perp]} = \mathscr{M}^{[\perp]} = \mathscr{L}^{[\perp][\perp]} = \mathscr{L}$, i.e., $x \in \mathscr{L} \cap \mathscr{L}^{[\perp]} = \mathscr{L}^0$. On the other hand, $x [\perp] \mathscr{N}$ and therefore, since $\mathscr{N} \# \mathscr{L}^0$, we have by virtue of Definition 1.2 that $x = \theta$. Thus, $\mathscr{R} = \Pi_\varkappa$ and the equalities (9.3) are proved.

We pass on to the last assertion of the theorem. We fix in $\Pi_\varkappa$ an arbitrary $\mathscr{N}$ ($\# \mathscr{L}^0$) and we put $\mathscr{L}_1 = \mathscr{L} \cap \mathscr{N}^{[\perp]}$. The lineal $\mathscr{L}_1$ is obviously closed; $\mathscr{L}^0 [\perp] \mathscr{L}_1$ (since $\mathscr{L}^0 [\perp] \mathscr{L}$) and $\mathscr{L}_0 \cap \mathscr{L}_1 = \{\theta\}$ since $\mathscr{L}^0 \# \mathscr{N}$. Further, $\mathscr{L}^0 \dot{+} \mathscr{N}^{[\perp]} = \Pi_\varkappa$ (see Corollary 1.32), $\mathscr{L}^0 \subset \mathscr{L}$, and therefore

$$\mathscr{L} = \mathscr{L} \cap \Pi_\varkappa = \mathscr{L} \cap (\mathscr{L}^0 \dot{+} \mathscr{N}^{[\perp]}) = \mathscr{L}^0 \dot{+} (\mathscr{L} \cap \mathscr{N}^{[\perp]}) = \mathscr{L}^0 [\dot{+}] \mathscr{L}_1.$$

Similarly one verifies that for $\mathscr{M}_1 = \mathscr{M} \cap \mathscr{N}^{[\perp]}$ the relation $\mathscr{M} = \mathscr{L}^0 [\dot{+}] \mathscr{M}_1$ holds. Thus, for the constructed subspaces $\mathscr{L}_1$ and $\mathscr{M}_1$ the relations (9.3) hold and $\mathscr{L}_1 \cap \mathscr{M}_1 = \{\theta\}$. Since by construction $\mathscr{L}_1^{[\perp]} = (\mathscr{L} \cap \mathscr{N}^{[\perp]})^{[\perp]} \supset \mathscr{N}^{[\perp][\perp]} = \mathscr{N}$ and similarly $\mathscr{M}_1^{[\perp]} = (\mathscr{M} \cap \mathscr{N}^{[\perp]})^{[\perp]} \supset \mathscr{N}$, so $\mathscr{N} \subset (\mathscr{L}_1 \dot{+} \mathscr{M}_1)^{[\perp]}$, and by virtue of the first part (already proved) of the theorem the decomposition (9.2) holds.

We now prove that, in the decomposition (9.2) just constructed, the subspaces $\mathscr{L}_1$ and $\mathscr{M}_1$ satisfying the conditions (9.3) cannot, for a fixed $\mathscr{N}$ ($\# \mathscr{L}_0$), be chosen differently. For, if $\mathscr{L}_1$ and $\mathscr{M}_1$ satisfy the conditions (9.2) and (9.3), then

$$\mathscr{L}_1 \subset \mathscr{L} \cap \mathscr{N}^{[\perp]}, \qquad \mathscr{M}_1 \subset \mathscr{M} \cap \mathscr{N}^{[\perp]}, \tag{9.4}$$

However, we saw earlier that the subspaces $\mathscr{L} \cap \mathscr{N}^{[\perp]}$ and $\mathscr{M} \cap \mathscr{N}^{[\perp]}$ also satisfy the conditions of the form (9.3): $\mathscr{L} = (\mathscr{L} \cap \mathscr{N}^{[\perp]}) [\dot{+}] \mathscr{L}^0$, $\mathscr{M} = (\mathscr{M} \cap \mathscr{N}^{[\perp]}) [\dot{+}] \mathscr{L}^0$; therefore it follows from (9.4) and 1.21 that $\mathscr{L}_1 = \mathscr{L} \cap \mathscr{N}^{[\perp]}$ and $\mathscr{M}_1 = \mathscr{M} \cap \mathscr{N}^{[\perp]}$. ■

5 We present one more simple proposition, which is often applicable in investigations.

9.13 If $\mathscr{L}$ ($\subset \Pi_\varkappa$) is a space with inertia index (π, ν, ω), *and if* $\mathscr{L}^0$ *is its isotropic subspace* ($\dim \mathscr{L}^0 = \omega$), *then the factor space* $\mathscr{L}/\mathscr{L}_0$ *with the indefinite metric induced from* $\mathscr{L}$ *according to the rule* (1.16) *is a Pontryagin space* $\Pi'_{\varkappa'}$, *where* $\varkappa' = \min\{\pi, \nu\}$.

We express $\mathscr{L}$ in the form $\mathscr{L} = \mathscr{L}_1 [\dot{+}] \mathscr{L}^0$, where $\mathscr{L}_1$ is a non-degenerate subspace. By virtue of Corollary 9.10 $\mathscr{L}_1$ is a Pontryagin space $\Pi'_{\varkappa'}$: $0 \leqslant \varkappa' \leqslant \varkappa$. Let a canonical decomposition of $\mathscr{L}_1$ have the form $\mathscr{L}_1 = \mathscr{L}_1^+ [\dot{+}] \mathscr{L}_1^-$, where, clearly, $\min\{\dim \mathscr{L}_1^+, \dim \mathscr{L}_1^-\} = \varkappa'$. But $\mathscr{L} = \mathscr{L}_1^+ [\dot{+}] \mathscr{L}_1^- [\dot{+}] \mathscr{L}^0$ is the canonical decomposition of the whole of $\mathscr{L}$, and so $\varkappa' = \min\{\pi, \nu\}$ (Theorem 6.7). It remains to apply 1.23. ■

Corollary 9.14: *Let* $\mathscr{L}^0$ *be a neutral linear in* $\Pi_\varkappa$, $\dim \mathscr{L}^0 = \omega$ ($\leqslant \varkappa$), *and* $\mathscr{L} = (\mathscr{L}^0)^{[\perp]}$. *Then* $\mathscr{L}/\mathscr{L}^0$ *is a space* $\Pi'_{\varkappa'}$ *with* $\varkappa' = \varkappa - \omega$.

□ Since $\mathscr{L}^0$ is an isotropic subspace for $\mathscr{L}$ (see Proposition 7.5), so In $\mathscr{L} = \{\pi, \nu, \omega\}$ with certain π and ν, and therefore by virtue of Proposition 9.13 $\mathscr{L}/\mathscr{L}_0 = \Pi'_{\varkappa'}$, where $\varkappa' = \min\{\pi, \nu\}$. But it is easy to see that here $\pi = \dim \Pi^+ - \omega$, because $\mathscr{L} \supset \mathscr{D}^+_{\mathscr{L}^0}$ (the deficiency subspace for $\mathscr{L}^0$) and $\dim \mathscr{D}^+_{\mathscr{L}^0} = \varkappa - \omega$, and $\nu = \dim \Pi^-$, so that $\nu \geqslant \varkappa > \varkappa - \omega = \pi$ and therefore $\varkappa' = \varkappa - \omega$. ■

6 Among W-spaces (see §6.6) and in particular G-spaces the analogues and generalizations of Pontryagin spaces $\Pi_\varkappa$ are of particular interest in the theory of operators. We shall *denote by* $\varkappa$ *the maximal dimension of the non-negative subspaces contained in a W-space* $\mathscr{H}$. If $\varkappa < \infty$ we shall call $\mathscr{H}$ a $W^{(\varkappa)}$*-space*, and if it is non-degenerate a $G^{(\varkappa)}$*-space*.

9.15 A W-space $\mathscr{H}$ *is a* $W^{(\varkappa)}$*-space if and only if the whole positive part of the spectrum* $\sigma(W)$ *of the Gram operator W consists of eigenvalues and the sum of the multiplicities of all non-negative eigenvalues is equal to* $\varkappa$. *Moreover,* $\hat{\mathscr{H}} = \mathscr{H}/\mathrm{Ker}\ \mathscr{W}$ *is a* $G^{(\varkappa - \varkappa_0)}$*-space, where* $\varkappa_0 = \dim \mathrm{Ker}\ W$.

□ If the spectrum $\sigma(W)$ satisfies the requirement indicated, then (*cf.* Theorem 6.4), in the canonical decomposition

$$\mathscr{H} = \mathscr{H}^- [\oplus] \mathscr{H}^0 [\oplus] \mathscr{H}^+ \qquad (\mathscr{H}^\pm = P^\pm \mathscr{H}, \mathscr{H}^0 = P^0 \mathscr{H}) \tag{9.5}$$

defined by the spectral decomposition $W = \int_{m-0}^{M} \lambda \, dE_\lambda$ ($E_{\lambda+0} = E_\lambda$) with the aid of the projectors

$$P^- = \int_{-m}^{-0} dE_\lambda, \qquad P_0 = E_0 - E_{-0}, \qquad P^+ = \int_0^M dE_\lambda,$$

the (isotropic) subspaces $\mathscr{H}^0$ (= Ker W) is neutral, and also (see Proposition 6.5) $\dim \mathscr{H}^0 = \varkappa_0$ is the multiplicity of the number $0 \in \sigma_p(W)$, and the maximal dimension of the non-negative subspaces contained in $\mathscr{H}^- [\oplus] \mathscr{H}^+$ is by Lemma 1.27 equal to $\dim \mathscr{H}^+ = \varkappa_1$ the sum of the multiplicities of the positive eigenvalues of the operator W. Thus the maximal dimensionality of the non-negative subspaces in $\mathscr{H}$ is equal to $\varkappa_0 + \varkappa_1 = \varkappa$.

Conversely, if $\mathscr{H}$ is a $W^{(\varkappa)}$-space, then in its canonical decomposition (9.5) $\dim(\mathscr{H}^+ [\oplus] \mathscr{H}^+) = \varkappa$ (again by virtue of Proposition 6.5 and Lemma 1.27). The last assertion regarding $\hat{\mathscr{H}}$ follows from Proposition 1.23. ■

Corollary 9.16: *A G-space $\mathscr{H}$ is a $\Pi_\varkappa$ space if and only if it is regular and* $\dim \mathscr{L}^+(G) = \varkappa < \infty$, *where $\mathscr{L}^+(G)$ is the direct (and orthogonal) sum of all the eigen-subspaces of operator G corresponding to its positive eigenvalues. If $\mathscr{H}$ is a $W^{(\varkappa)}$-space, then $\hat{\mathscr{H}} = \mathscr{H}/\mathrm{Ker}\ W^{(\varkappa)}$ is a $\Pi_{\varkappa-\varkappa_0}$ space* ($\varkappa_0 = \dim \mathrm{Ker}\ W^{(\varkappa)}$) *if and only if $\mathscr{R}_{W^{(\varkappa)}} = \bar{\mathscr{R}}_{W^{(\varkappa)}}$.*

□The condition for regularity ($0 \in \rho(G)$) is connected with the fact that a $\Pi_\varkappa$ space is a Krein space (see Proposition 6.13), and all the rest-follows from Proposition 6.13. ■

But if a G-space $\mathscr{H}$ is singular and is a $G^{(\varkappa)}$-space, then it can be 'densely imbedded' in a $\Pi_\varkappa$ space by means of the procedure described in Proposition 6.14.

To conclude this paragraph 6.9 we indicate another useful criterion for a G-space to be a $G^{(\varkappa)}$-space.

9.17 Let $\mathscr{H} = \mathscr{L} \dot{+} \mathscr{N}$ be a G-space, $\mathscr{L}$ and $\mathscr{N}$ subspaces of it with $\mathscr{L} \subset \mathscr{P}^+$, $\dim \mathscr{L} = \varkappa < \infty$, *and $\mathscr{N} \subset \mathscr{P}^{--} \cup \{\theta\}$. Then $\mathscr{H}$ is a $G^{(\varkappa)}$-space.*

□ It follows from Proposition 1.25 that $\mathscr{L}$ is a maximal non-negative subspace. It remains to verify that $\mathscr{H}$ does not contain non-negative subspaces $\mathscr{L}'$ of higher dimension than $\varkappa = \dim \mathscr{L}$. Let P^+ and P^- ($P^+ + P^- = I$) be the projectors generated by the direct decomposition $\mathscr{H} = \mathscr{L} \dot{+} \mathscr{N}$, and let $\mathscr{L}' \subset \mathscr{P}^+$. Then by Lemma 1.27 the mapping $P^+: \mathscr{L}' \to \mathscr{L}$ is injective, hence $\dim \mathscr{L}' \leqslant \varkappa$. ■

Exercises and problems

1 Let $\mathscr{F}$ be a linear space of finite sequences $x = \{\xi_j\}_{j=1}^{\infty}$ ($\xi_j = 0, j > N_x$), and let $A = \| a_{jk} \|_{j,k=1}^{\infty}$ be an infinite Hermitian matrix defining on $\mathscr{F}$ an Hermitian metric

$$[x, y] = \sum_{j,k=1}^{\infty} a_{jk}\xi_j\bar{\eta}_k \qquad (x = \{\xi_j\}_{j=1}^{\infty}, y = \{\eta\}_{k=1}^{\infty} \in \mathscr{F}).$$

Let the matrix A be such that $\mathscr{F}$ does not contain isotropic vectors ($\mathscr{F}^0 = \{\theta\}$), and each of the forms $[x, x] = \sum_{j,k=1}^{N_x} a_{jk}\xi_j\bar{\xi}_k$ ($x \in \mathscr{F}$) contains not more than $\varkappa$, and at least one, positive squares. Prove that $\mathscr{F}$ after completion (*cf.* Exercise 4 to §2) becomes a $\Pi_\varkappa$ space ([XIV]).

2 Prove that if the condition on the number of positive squares for the forms in Exercise 1 is observed, then the condition for non-degeneracy of the metric $[x, y]$ on $\mathscr{F}$ (i.e. $\mathscr{H}^0 = \{\theta\}$) is equivalent to the requirement $\det \| a_{jk} \|_{j,k=1}^{n} \neq 0$ for all sufficiently large n. Without the condition on the number of positive squares, the given requirement is

sufficient, but not necessary, for the non-degeneracy of the form $[x, y]$ (I, Iokhvidov [19]).

3 Modifying the example in Exercise 1, consider the linear space $\mathscr{F}$ of two-sided and finite (on both sides) sequences $x = \{\ldots, \xi_{-1}, \xi_0, \xi_1, \xi_2, \ldots\}$, and an infinite (in all four directions) Hermitian matrix $A = \| a_{jk} \|_{j,k=-\infty}^{\infty}$; formulate the corresponding conditions that will ensure that $\mathscr{F}$ (after completion) becomes a $\Pi_\varkappa$ space ([XV], *cf.* [V]).

4 Verify that in the examples in Exercises 2 and 3, if non-degeneracy of the metric $[x, y]$ is not demanded (i.e., if $\mathscr{F}^0 \neq \{\theta\}$ is allowed), then the factor space $\mathscr{F}/\mathscr{F}^0$ after completion becomes a $\Pi_\varkappa$ space ([*XV*], [V]).
Hint: Compare with Proposition 9.13.

5 Let $\mathscr{H} = l^2$ be the space of all sequences $x = \{\xi_\varkappa\}_{k=1}^{\infty}, \Sigma_{k=1}^{\infty} |\xi_\varkappa|^2 < \infty$; the form $[x, y]$ for $x, y \in l^2$, $y = \{\eta_k\}_{k=1}^{\infty}$ is given by the formula

$$[x, y] = \Sigma_{k=1}^{\varkappa} \xi_k \bar{\eta}_k - \Sigma_{k=\varkappa+1}^{\infty} \xi_k \bar{\eta}_k.$$

Compare this with the example in Exercise 4 to §2, and show that $\mathscr{H}$ is a $\Pi_\varkappa$ space (Pontryagin [1]).

6 Let $\mathscr{F}$ be an abstract set on which a complex Hermitian kernel $K(p, q) = \overline{K(q, p)}$ $(p, q \in \mathscr{F})$ is defined such that the form $\Sigma_{j,k=1}^{n} K(p_j, p_k)\xi_j\xi_k$ for sufficiently large n contains exactly $\varkappa$ positive squares. The lineal consisting of all functions of the form

$$\varphi(p) = \sum_{j=1}^{n} \xi_j K(p, q_j), \qquad \psi(p) = \sum_{j=1}^{n} \eta_j K(p, q_j)$$

with the form $[\varphi, \psi] = \Sigma_{j,k=1}^{n} K(q_j, q_k)\xi_j\bar{\eta}_k$ becomes, after taking the factor-space by the isotropic lineal and completion, a $\Pi_\varkappa$ space; this is the continual analogue of the space in Exercise 1 (M. Krein; example published in I. Iokhvidov's report [4]).

7 In the example in Exercise 6 to §2 the space $L^2(S, \Sigma, \mu)$ will be a $\Pi_\varkappa$ space if in the canonic Jordan decomposition $\mu = \mu_+ - \mu_-$ of the measure into the difference of two positive measures the component μ_+ is concentrated in $\varkappa$ points of the set S not belonging to the carrier of the measure μ_- (see I. Iokhvidov [2] and Azizov [8], where particular cases of this situation, analogous to the example in Exercise 5 to §2, were considered, and also [IV]).

8 *In the real space* Π_1 for any two *non-negative vectors* x, y $(\in\Pi_1)$ establish the inequality $[x, y]^2 \geqslant [x, x][y, y]$, in which the inequality sign holds only when the vectors x and y are collinear (M. Krein, see [XV], see also M. Krein and Rutman [1]).

9 Consider $\Pi_\varkappa = \Pi^+ \oplus \Pi^-$, where $0 < \dim \Pi^- = \varkappa < \infty$, and Π^+ is infinite dimensional. Let $y_0 \in \Pi_\varkappa$ and let $\mathscr{L}$ be a positive lineal in $\Pi_\varkappa$. Then for continuity of the functional $\varphi_{y_0}(x) = [x, y_0]$ $(x \in \mathscr{L})$ relative to the internal norm $|\cdot|_{\mathscr{L}}$ it is necessary and sufficient that $y_0 [\perp] (\bar{\mathscr{L}})^0$ (I. Iokhvidov [11]).
Hint. In the proof use the canonical decomposition of the closure $\bar{\mathscr{L}}$: $\bar{\mathscr{L}} = \mathscr{L}_1 + (\bar{\mathscr{L}})^0$ and the fact that a positive subspace $\mathscr{L}_1$ is projectively complete (Theorem 9.11,b).

10 A definite lineal $\mathscr{L}$ $(\subset\Pi_\varkappa)$ is singular if and only if its closure $\bar{\mathscr{L}}$ is degenerate (I. Iokhvidov [11]; [VIII]).
Hint: Use the result of Exercise 9.

11 Let $\mathscr{L}$ $(\subset\Pi_\varkappa)$ be a singular lineal. Then for the equality $\bar{\mathscr{L}} = \mathscr{L} [\dot{+}] (\bar{\mathscr{L}})^0$ it is necessary that $\mathscr{L}$ should be complete (i.e., a Hilbert) space relative to the internal metric (I. Iokhurdov [11]).
Hint. Use Proposition 9.13.

12 The condition of completeness of $\mathscr{L}$ with respect to the internal norm $|\cdot|_{\mathscr{L}}$ is sufficient for the representation $\overline{\mathscr{L}} = \mathscr{L}[\dot{+}](\overline{\mathscr{L}})^0$ of the closure of a singular lineal $\mathscr{L}$ in any Krein space $\mathscr{H}$ (I. Iokhvidov [11]).
Hint: Prove that the lineal $\mathscr{L}[\dot{+}](\overline{\mathscr{L}})^0$ is closed relative to $\|\cdot\|$.

13 Consider the Pontryagin space $\Pi_\varkappa = \Pi^+[\oplus]\Pi^-$ ($\dim \Pi^+ < \infty$). Prove that a sequence $\{x_n\}$ strongly converges to a vector x_0 when $n \to \infty$ if $[x_n, x_n] \to [x_0, x_0]$ and $[x_n, y] \to [x_0, y]$ for y running through a set $\mathscr{D}$ dense in $\Pi_\varkappa$ (M. Krein and Langer [3]).
Hint: Without loss of generality it may be supposed that $\mathscr{D}$ is a lineal and $\Pi^+ \subset \mathscr{D}$. Verify first that $x_n^+ \to x_0^+$, and then use the ordinary criterion for strong convergence for $\{x_n^-\}_1^\infty$.

14 Let $\mathscr{H}_1 = \mathscr{H}_1^+[\oplus]\mathscr{H}_1^-$ be a J-space, and let $\mathscr{H}_2 = \mathscr{H}_2^+[\oplus]\mathscr{H}_2^0[\oplus]\mathscr{H}_2^-$ be the canonical decomposition of a W-space $\mathscr{H}_2$, where $W = T^*JT$, and $T: \mathscr{H}_2 \to \mathscr{H}_1$ is a bounded operator. Then $\dim \mathscr{H}_2^\pm \leqslant \dim \mathscr{H}_1^\pm$, and therefore, in particular, if $\mathscr{H}_1$ is a Pontryagin space with $\varkappa$ positive (negative) squares, then the positive (negative) part of the spectrum of the operator W consists of not more than $\varkappa$ (taking multiplicity into account) eigenvalues (Azizov).

§10 Dual pairs. J-orthonormalized systems and bases

1 *Definition 10.1:* A pair of subspaces $(\mathscr{L}_+, \mathscr{L}_-)$ of a Krein space $\mathscr{H} = \mathscr{H}^+[\oplus]\mathscr{H}^-$ is called a *dual pair* if $\mathscr{L}_\pm \subset \mathscr{P}^\pm$ and $\mathscr{L}_+[\perp]\mathscr{L}_-$. If, in addition, $\mathscr{L}_\pm \in \mathscr{M}^\pm$, then $(\mathscr{L}_+, y\mathscr{L}_-)$ is called a *maximal dual pair*. A dual pair $(\mathscr{L}'_+, \mathscr{L}'_-)$ is called *an extension of the dual pair* $(\mathscr{L}_+, \mathscr{L}_-)$ if $\mathscr{L}'_\pm \supset \mathscr{L}_\pm$, and it is called a *maximal extension of the pair* $(\mathscr{L}_+, \mathscr{L}_-)$ if this extension $(\mathscr{L}'_+, \mathscr{L}'_-)$ is a maximal dual pair.

Theorem 10.2: *Every dual pair $(\mathscr{L}_+, \mathscr{L}_-)$ has at least one maximal extension $(\mathscr{L}_+^{\max}, \mathscr{L}_-^{\max})$.*

Moreover, every maximal in $\mathscr{L}_+^{[\perp]}$ non-positive extension $\widetilde{\mathscr{L}}_-$ of the subspace $\mathscr{L}_-$ belongs to $\mathscr{M}^-$ and the pairs $(\widetilde{\mathscr{L}}_-^{[\perp]}, \widetilde{\mathscr{L}}_-)$ exhaust all the maximal extensions of the dual pair $(\mathscr{L}_+, \mathscr{L}_-)$.

□ Clearly it suffices to prove the second assertion of the theorem. Consider the sub-space $\lambda_+^{[\perp]}$ $(\supset \mathscr{L}_-)$. Let $(\mathscr{D}^+_{\mathscr{L}_+} \equiv)\mathscr{D}^+ = (P^+\mathscr{L}_+)^\perp \cap \mathscr{L}^+$ be the deficiency subspace for $\mathscr{L}_+$ in $\mathscr{K}^+$. We recall (see the proof of Theorem 4.5) that $\mathscr{D}^+[\perp]\mathscr{L}_+$ and $\mathscr{L}_+[\dot{+}]\mathscr{D}^+ \in \mathscr{M}^+$. Therefore $\mathscr{N} \equiv \mathscr{L}_+[\dot{+}]\mathscr{D}^+)^{[\perp]} \in \mathscr{M}^{-1}$ (see Theorem 8.11) and $\mathscr{N} = \mathscr{N}^0[\oplus]\mathscr{N}_1$, where $\mathscr{N}^0$ is the isotropic subspace for $\mathscr{N}$, and $\mathscr{N}_1$ is the negative subspace. Since $\mathscr{N} \subset \mathscr{L}_+^{[\perp]}$ and $\mathscr{D}^+ \subset \mathscr{L}_+^{[\perp]}$, it is clear that $\mathscr{L}_+^{[\perp]} \supset \mathscr{N}_0[\oplus]\mathscr{N}_1[\oplus]\mathscr{D}^+$.

Conversely, if $z[\perp]\mathscr{L}_+$, then $z = z^+ + z^-$ $(z^\pm \in \mathscr{H}^\pm)$, and $z^+ = z_1^+ + z_2^+$, where $z_1^+ \in \mathscr{D}^+$, and $z_2^+ \in (\mathscr{D}^+)^{[\perp]} \cap \mathscr{H}^+ = P^+\mathscr{L}_+$. Thus $z = z_1^+ + (z_2^+ + z^-)$, i.e., $z_2^+ + z^- = z - z_1^+[\perp]\mathscr{L}_+$, and, at the same time, $z_2^+ + z^- \in \mathscr{L}_+^{[\perp]} \cap (\mathscr{D}^+)^{[\perp]} = (\mathscr{L}_+[\dot{+}]\mathscr{D}^+)^{[\perp]} = \mathscr{N}$ (see (1.12)), and so

$z=(z_2^+ + z^-)+z_1^+ \in \mathcal{N}[\dotplus]\mathcal{D}^+$. Finally we obtain for $\mathcal{L}_+^{[\perp]}$ the canonical decomposition

$$\mathcal{L}_+^{[\perp]} = \mathcal{D}^+[\oplus]\mathcal{N}_1[\oplus]\mathcal{N}^0. \tag{10.1}$$

Now let $\tilde{\mathcal{L}}_-$ be a maximal in $\mathcal{L}_+^{[\perp]}$ non-positive extension of the subspace $\mathcal{L}_-$ ($\subset \mathcal{L}_+^{[\perp]}$). Then (see Proposition 1.22) $\tilde{\mathcal{L}}_- \supset \mathcal{N}^0$ and $\tilde{\mathcal{L}}_- = \mathcal{N}^0[\oplus]\mathcal{N}'_-$, where $\mathcal{L}'_-$ is the maximal, in the subspace $\mathcal{D}^+[\oplus]\mathcal{N}_1$, non-positive subspace (see (10.1)). We have turned out to be in the situation considered in Proposition 8.18, by virtue of which $\tilde{\mathcal{L}}'_- = \{Fx+x\}_{x\in\mathcal{N}_1}$, where the linear operator F: $\mathcal{N}_1 \to \mathcal{D}^+$ is defined on the whole of $\mathcal{N}_1$ in view of the maximal non-positivity of $\tilde{\mathcal{L}}'_-$ in $\mathcal{D}^+[\oplus]\mathcal{N}_1$. Therefore

$$\begin{aligned} P^-\tilde{\mathcal{L}}_- &= \operatorname{Lin}\{P^-\mathcal{N}^0, P^-\tilde{\mathcal{L}}'_-\} = \operatorname{Lin}\{P^-\mathcal{N}^0, P^-\mathcal{N}_1\} = P^-\{\mathcal{N}^0[\oplus]\mathcal{N}_1\} \\ &= P^-\mathcal{N} = \mathcal{H}^-, \end{aligned}$$

i.e., $\tilde{\mathcal{L}}_- \in \mathcal{M}^-$, and $^{[\perp]}_-$, $_-$) is one of the extensions of the dual pair $(\mathcal{L}_+, \mathcal{L}_-)$ of the form $(\mathcal{L}_+^{\max}, \mathcal{L}_-^{\max})$.

The last assertion of the theorem is obvious. ■

2 *Definition 10.3:* Let $\mathcal{H}$ be a Krein space, let $(\mathcal{L}_+, \mathcal{L}_-)$ be a maximal uniformly definite dual pair (i.e., $\mathcal{L}_\pm$ are uniformly definite), and let $\mathcal{D}$ be a subspace in $\mathcal{H}$. We call the decomposition $\mathcal{D}=\mathcal{D}_0[\dotplus]\mathcal{D}_1[\dotplus]\mathcal{D}_2$ of the subspace $\mathcal{D}$ its $(\mathcal{L}_+, \mathcal{L}_-)$—*decomposition* if $\mathcal{D}_0 = \mathcal{D}\cap\mathcal{D}^{[\perp]}$, $\mathcal{D}_1 = (\mathcal{D}\cap\mathcal{L}_+)[\dotplus](\mathcal{D}\cap\mathcal{L}_-)$, and $\mathcal{D}_2$ is the orthogonal complement to $\mathcal{D}_0[\dotplus]\mathcal{D}_1$ relative to the scalar product $(x,y)_1 = [x_+, y_+] - [x_-, y_-]$, where $x = x_+ + x_-$, $y = y_+ + y_-$, $x_\pm$, $y_\pm \in \mathcal{L}_\pm$, and therefore also relative to the form $[x, y]$ (see Exercise 8 to §2).

We note that *any subspace $\mathcal{D}$ admits an $(\mathcal{L}_+, \mathcal{L}_-)$-decomposition relative to any uniformly definite dual pair and this decomposition is unique.*

Theorem 10.4: *Let $\mathcal{D}=\mathcal{D}_0[\dotplus]\mathcal{D}_1[\dotplus]\mathcal{D}_2$ be the $(\mathcal{L}_+, \mathcal{L}_-)$-decomposition of a subspace $\mathcal{D}$, and let $\mathcal{D}_2 = \mathcal{D}'_2[\dotplus]\mathcal{D}''_2$, where $\mathcal{D}'_2$ is a projectively complete subspace. Then there is a maximal uniformly definite dual pair $(\mathcal{L}_1^+, \mathcal{L}_1^-)$ such that the decomposition $\mathcal{D}=\mathcal{D}_0[\dotplus](\mathcal{D}'_1[\dotplus]\mathcal{D}'_2)[\dotplus]\mathcal{D}''_2$ of the subspace $\mathcal{D}$ is its $(\mathcal{L}_1^+, \mathcal{L}_1^-)$-decomposition.*

□ Let $\mathcal{D}'_2 = \mathcal{D}'_+[\dotplus]\mathcal{D}'_-$ be a canonical decomposition. We prove our assertion for the case $\mathcal{D}'_2 = \mathcal{D}'_+$ and by so doing reduce it to the case $\mathcal{D}'_2 = \mathcal{D}'_-$, which is treated according to the same scheme.

Thus, $\mathcal{D}'_2 = \mathcal{D}'_+$. Without loss of generality we shall suppose that $\mathcal{H}^\pm = \mathcal{L}_\pm$, and we put $\mathcal{L}_1^+ = \mathcal{D}'_2[\dotplus](\mathcal{H}^+ \cap \mathcal{D}'_2{}^{[\perp]})$. By construction $\mathcal{L}_1^+$ is a uniformly definite subspace and, $\mathcal{D}''_2 \cap \mathcal{L}_1^+ = \{\theta\}$. Moreover, $\mathcal{L}_1^+ \in \mathcal{M}^+$; for if this were not so, then in $\mathcal{H}^+$ there would be a vector $x_0^+ \neq \theta$ J-orthogonal to $\mathcal{L}_1^+$ and therefore to $\mathcal{D}'_2$, i.e., x_0^+ would lie in $\mathcal{H}^+ \cap \mathcal{D}'_2{}^{[\perp]}$ and therefore $(x_0^+, x_0^+) = 0$; we would have a contradiction.

We put $\mathscr{L}_1^- = (\mathscr{L}_1^+)^{[\perp]}$. Then by (1.12) $\mathscr{L}_1^- = (\mathscr{H}^- + \mathscr{D}_2') \cap \mathscr{D}_2'^{[\perp]}$. Since $\mathscr{D}_2' \subset \mathscr{D}$ and $\mathscr{D} \cap \mathscr{H}^{\pm} \subset \mathscr{D}_1 \subset \mathscr{D}_2'^{[\perp]}$, so $\mathscr{D} \cap \mathscr{L}_1^+ = \mathscr{L}_2' [\dot{+}] (\mathscr{D} \cap \mathscr{H}^+)$ and $\mathscr{D} \cap \mathscr{L}_1^- = \mathscr{D} \cap [(\mathscr{H}^- + \mathscr{D}_2') \cap \mathscr{D}_2'^{[\perp]}] = \mathscr{D} \cap \mathscr{H}^-$, i.e., $(\mathscr{D} \cap \mathscr{L}_1^+) [\dot{+}] (\mathscr{D} \cap \mathscr{L}_1^-) = \mathscr{D}_1 [\dot{+}] \mathscr{D}_2$. Hence $\mathscr{D}_2'' \cap \mathscr{L}_1^- = \{\theta\}$. It therefore remains to verify that $\mathscr{D}_2''$ is orthogonal to $\mathscr{D}_0 [\dot{+}] \mathscr{D}_1 [\dot{+}] \mathscr{D}_2$ in the scalar product $(x, y)_1$ indicated in Definition 10.3.

Let $x \in \mathscr{D}_2''$. In accordance with Exercise 8 to §2 $(x, y)_1 = 0$ when $y \in \mathscr{D}_1 [\dot{+}] \mathscr{D}_2'$, and therefore the canonical projections $x_1^{\mp}$ of the vector x on to $\mathscr{L}_1^{\pm}$ lie in $\mathscr{H}^{\pm} \cap \mathscr{D}_2'^{[\perp]}$ respectively. Consequently $\mathscr{D}_2'' \subset (\mathscr{H}^+ \cap \mathscr{D}_2'^{[\perp]}) [\dot{+}] (\mathscr{H}^- \cap \mathscr{D}_2'^{[\perp]}) \equiv \mathscr{H}_1$. From Exercise 9 to §2 we conclude that $\mathscr{D}_0 \subset \mathscr{H}_1$. It remains to notice that the scalar products $(x, y)_1$ and (x, y) coincide on $\mathscr{H}_1$, and therefore also that $(x, z)_1 = (x, z) = 0$ when $x \in \mathscr{D}_2''$, $z \in \mathscr{D}_0$, i.e., the decomposition $\mathscr{D} = \mathscr{D}_0 [\dot{+}] (\mathscr{D}_1 [\dot{+}] \mathscr{D}_2') [\dot{+}] \mathscr{D}_2''$ is an $(\mathscr{L}_1^+, \mathscr{L}_1^-)$—decomposition of the subspace $\mathscr{D}$. ■

3 Later in Chapter 3 we shall encounter more than once dual pairs and maximal dual pairs of subspaces invariant relative to one or other operators or even whole families of operators. But in this section dual pairs arise in a natural way in the consideration of the so-called *J-orthonormalized systems* in a Krein space.

Definition 10.5: A system of vectors $\mathscr{F} = \{e_\alpha\}_{\alpha \in A}$ $(\subset \mathscr{H})$, where A is an arbitrary set of indices, is called a *J-orthonormalized system* if $[e_{\alpha_1}, e_{\alpha_2}] = \pm \delta_{\alpha_1 \alpha_2}$ for all $\alpha_1, a_2 \in A$, where $\delta_{\alpha_1 \alpha_2}$ is the Kronecker delta.

The simplest example of a J-orthonormalized system of $\mathscr{H} = \mathscr{H}^+ [\oplus] \mathscr{H}^-$ is the union of two arbitrary orthonormalized (in the usual sense) systems from the subspaces $\mathscr{H}^+$ and $\mathscr{H}^-$ respectively.

10.6 Any J-orthonormalized system $\mathscr{F} = \{e_\alpha\}_{\alpha \in A}$ is algebraically free, i.e., it does not contain linearly dependent finite-dimensional sub-systems; moreover, it is minimal in the sense that not one vector $e_\beta \in \mathscr{F}$ belongs to $C \operatorname{Lin}\{e_\alpha\}_{\alpha \in A, \alpha \neq \beta}$.

□ It is clear that the first assertion follows from the second. But the second is obvious, since (see Proposition 3.5) for any $x \in C \operatorname{Lin}\{e_\alpha\}_{\alpha \in A, \alpha \neq \beta}$ we have $[x, e_\beta] = 0$ and $[e_\alpha, e_\beta] = \pm 1$. ■

Definition 10.7: A J-orthonormalized system is said to be *maximal* if it is not contained in any wider J-orthonormalized system, and to be *J-complete* if there is no non-zero vector J-orthonormalized to this system (see Remark 10.11 below).

10.8 Any J-orthonormalized system is contained in a certain maximal J-orthonormalized system.

□ The proof is obtained in the usual way using Zorn's lemma (see Theorem 1.19). ■

4 Every J-orthonormalized system $\mathscr{F} = \{e_\alpha\}_{\alpha \in A}$ in a Krein space $\mathscr{H}$ generates a dual pair of subspaces $(\mathscr{L}_+, \mathscr{L}_-)$, where

$$\mathscr{L}_\pm = C \operatorname{Lin} \mathscr{F}_\pm, \qquad \mathscr{F}_\pm = \{e_\alpha \mid [e_\alpha, e_\alpha] = \pm 1\}_{\alpha \in A}. \tag{10.2}$$

10.9 If a dual pair $(\mathscr{L}_+, \mathscr{L}_-)$ is generated by a J-complete J-orthonormalized system $\mathscr{F}$ (see (10.2)), *then the subspaces $\mathscr{L}_\pm$ are definite.*

□ It must be shown that $\mathscr{L}_\pm$ are non-degenerate. But if an isotropic vector were found in $\mathscr{L}_+$ (or $\mathscr{L}_-$) is would be J-orthogonal to the whole system $\mathscr{F}$, which contradicts the J-completeness of $\mathscr{F}$. ■

At the same time maximality, or even J-completeness, of the system $\mathscr{F}$ does not, generally speaking, imply the maximality of the corresponding dual pair. What is more, the following proposition holds:

10.10 In any Krein space $\mathscr{H}$ distinct from $\Pi_\varkappa$ there are J-complete J-orthonormalized which generate dual pairs $(\mathscr{L}_+, \mathscr{L}_-)$ that are not maximal dual pairs.

□ We choose in $\mathscr{H}$ an arbitrary positive but singular, i.e., not a uniformly positive, subspace $\mathscr{L}_+ \in \mathscr{M}^+$ with the angular operator K, $\|K\| = 1$ (see Lemma 8.4). The corresponding Gram operator $G_{\mathscr{L}_+}$ is invertible $(0 \notin \sigma_p(G_{\mathscr{L}_+}))$, but $G_{\mathscr{L}_+}^{-1}$ is not bounded and defined on a domain $\mathscr{R}_{G_{\mathscr{L}_+}}$ $(\neq \mathscr{L}_+)$ dense in $\mathscr{L}_+$ (see Propositions 6.5 and 6.11). We consider an arbitrary vector $x_0 \in \mathscr{L}_+ \backslash \mathscr{R}_{G_{\mathscr{L}_+}}$ and the subspace $\mathscr{L}'_+ = \mathscr{L}_+ \ominus \operatorname{Lin}\{x_0\}$. We notice at once that $\overline{G_{\mathscr{L}_+}\mathscr{L}'_+} = \mathscr{L}_+$, for it would follow from $z_0 \in \mathscr{L}_+$ and $z_0 \perp G_{\mathscr{L}_+}\mathscr{L}'_+$ that $(G_{\mathscr{L}_+}z_0, v) = (z_0, G_{\mathscr{L}_+}, v) = 0$ for all $v \in \mathscr{L}'_+$, i.e., $G_{\mathscr{L}_+}z_0 = \lambda x_0$, and since $x_0 \notin \mathscr{R}_{G_{\mathscr{L}_+}}$, so $\lambda = 0$, and hence $G_{\mathscr{L}_+} = \theta$ and $z_0 = \theta$.

Now in the space $\mathscr{L}'_+$ we choose any complete orthonormalized (relative to the scalar product $(\cdot, \cdot)$) system (a Hilbert basis), and we orthonormalize it by the Schmidt process relative to the intrinsic metric $[\cdot, \cdot]$ which is positive on the whole of $\mathscr{L}_+$ and, in particular, on $\mathscr{L}'_+$. It is asserted that the J-orthonormalized system $\mathscr{F}_+$ so obtained, for which obviously $C \operatorname{Lin} \mathscr{F}_+ = \mathscr{L}'_+$, is J-complete and therefore maximal in $\mathscr{L}_+$.

For, if a vector $y_0 \in \mathscr{L}_+$ could be found such that $y_0 \, [\perp] \, \mathscr{F}_+$, then
$(y_0, G_{\mathscr{L}_+}u) = (G_{\mathscr{L}_+}y_0, y) = [y_0, u] = 0$ for all $u \in \mathscr{F}_+$ and
$y_0 \, [\perp] \, G_{\mathscr{L}_+}$ $(C \operatorname{Lin} \mathscr{F}_+ = G_{\mathscr{L}_+}\mathscr{L}'_+$, i.e., $y_0 \perp \overline{G_{\mathscr{L}_+}\mathscr{L}'_+}$ and $y_0 = 0$.

Thus, $C \operatorname{Lin} \mathscr{F}_+ = \mathscr{L}'_+$ and $\mathscr{L}'_+ \notin \mathscr{M}^+$. Now it remains to choose in the subspace $\mathscr{L}_- = \mathscr{L}_+^{[\perp]}$ $(\in \mathscr{M}^-)$ an arbitrary complete orthonormalized system and reconstruct it by means of the Schmidt process into a J-orthogonal system $\mathscr{F}_-$. The union $\mathscr{F} = \mathscr{F}_+ \cup \mathscr{F}_-$ is a J-orthonormalized system, and is also

J-complete, for

$$x[\perp]\mathscr{F} \Rightarrow x[\perp]\mathscr{F}_- \Rightarrow x[\perp]\, C\,\mathrm{Lin}\,\mathscr{F}_-\ (=\mathscr{L}_-),\ \text{i.e., } x \in \mathscr{L}_+.$$

But $x[\perp]\mathscr{F}_-$, from which it follows that $x=\theta$. At the same time the dual pair $(\mathscr{L}'_+, \mathscr{L}_-)$ generated by the system is not maximal. ■

Remark 10.11: In the course of the proof we have incidentally discovered the difference (for subspaces) between the concept of the J-completeness of a system and its completeness in the usual (Hilbert) sense: the system $\mathscr{F}_+$ we constructed at the beginning of the proof is J-complete in the subspace $\mathscr{L}_+$ but is by no means complete in it: $\theta \neq x_0 \in \mathscr{L}_+ \ominus C\,\mathrm{Lin}\,\mathscr{F}_+$. This is explained in the present case by the singularity of the subspace $\mathscr{L}_+$, regarding which we spoke (allowing a certain freedom of speech and interpreting Definition 10.7 broadly) of the J-completeness and completeness of the system $\mathscr{F}_+$. But as regards a J-orthonormalized system in the whole Krein space it is obvious that its J-completeness and ordinary completeness are equivalent.

5 We now consider *J-orthonormalized bases* in a separable Krein space $\mathscr{H} = \mathscr{H}^+ [\oplus] \mathscr{H}^-$, i.e., J-orthonormalized systems which are (Schauder) bases in $\mathscr{H}$.

10.12 If a J-orthonormalized system $\mathscr{F} = \{e_k\}_{k=1}^{\infty}$ *is a basis in* $\mathscr{H}$, *then the pair* $(\mathscr{L}_+, \mathscr{L}_-)$ *generated by it is a maximal dual pair.*

□ We prove, for example, the maximality of $\mathscr{L}_+$. Let us assume the contrary: $\mathscr{L}_+ \notin \mathscr{M}^+$. Then $P^+\mathscr{L}_4 \neq \mathscr{H}^+$, and in the deficiency space $\mathscr{L}_+^{[\perp]} \cap \mathscr{H}^+$ there is a non-zero vector $x_0 = \sum_{k=1}^{\infty} \xi_k e_k$. Let $\{e_{ki}\}_{i=1}^{\infty}$ be the set of all vectors of the basis $\{e_k\}_{k=1}^{\infty}$ for which $[e_k, e_k] = 1$. Then $0 = [x_0, e_{ki}] = \xi_{ki}$ $(i = 1, 2, \ldots)$, from which $[x_0, x_0] \leqslant 0$, i.e., $x_0 = \theta$, contrary to hypothesis. ■

We recall that a system of vectors $\{f_k\}_{k=1}^{\infty}$ in a Hilbert space is said to be *normalized* if $\| f_k \| = 1$ $(k = 1, 2, \ldots)$ and *almost normalized* if there are positive constants μ and M such that $(0<)\mu \leqslant \| f_k \| \leqslant M$ $(k = 1, 2, \ldots)$. The systems $\{f_k\}_{k=1}^{\infty}$ and $\{g_k\}_{k=1}^{\infty}$ are said to be *biorthogonal* if $(f_j, g_k) = \delta_{jk}$ $(j, k = 1, 2, \ldots)$. It is well-known (see, e.g., [XI]) that every system biorthogonal to a normalized system is almost normalized.

10.13 Every J-orthonormalized basis is almost normalized.

□ Let $\{e_k\}_{k=1}^{\infty}$ be a J-orthonormalized basis. Then the system $\{e_k^0 = \| e_k \|^{-1} e_k\}_{k=1}^{\infty}$ is normalized, and, as may easily be verified, the system $\{h_k = \operatorname{sign}[e_k, e_k]\, \| e_k \|\, J e_k\}_{k=1}^{\infty}$ is biorthogonal to it, and is therefore almost normalized: $0 < \mu \leqslant \| h_k \| \leqslant M$ $(k = 1, 2, \ldots)$. But since $\| e_k \| e_k = \operatorname{sign}[e_k, e_k]\, J h_k$, so $\| e_k \|^2 = \| J h_k \| = \| h_k \|$, and therefore $0 < \sqrt{\mu} \leqslant \| e_k \| \leqslant \sqrt{M}$. ■

Exercises and problems

1 Let $(\mathscr{L}_+, \mathscr{L}_-)$ be a dual pair of subspaces, let K and Q be the corresponding angular operators, and let $\mathscr{D}_{\overline{\mathscr{L}}_\pm}^{\mp}$ be the deficiency subspaces for $\mathscr{L}_\pm$ respectively (see §6.1, §6.2). Then

$$Kx = Q^{(*)}x + W^+(I^+ - QQ^{(*)})^{1/2}x \quad \text{for all } x \in P^+\mathscr{L}_+,$$
$$Qy = K^{(*)}y + W^-(I^- - KK^{(*)})^{1/2}y \quad \text{for all } y \in P^-\mathscr{L}_-,$$

where

$$W^+ : (I^+ - QQ^{(*)})^{1/2}P^+\mathscr{L}_+ \to \mathscr{D}_{\overline{\mathscr{L}}_-}^-$$

and

$$W^- : (I^- - KK^{(*)})^{1/2}P^-\mathscr{L}_- \to \mathscr{D}_{\overline{\mathscr{L}}_+}^+$$

are certain contractions. Moreover,
1) $\| Kx \| = \| x \| \Leftrightarrow \| W^+(I^+ - QQ^{(*)})x \| = \| (I^+ - QQ^{(*)})x \|$,
$\| Qy \| = \| y \| \Leftrightarrow \| W^-(I^- - KK^{(*)}y \| = \| (I^- - KK^{(*)}y \|$;
2) $\| K \| < 1 \Leftrightarrow \| W^+ \| < 1$, $\| Q \| < 1 \Leftrightarrow \| W^- \| < 1$;
3) $(\mathscr{L}_+, \mathscr{L}_-)$ is a maximal dual pair if and only if $W^+ : \mathscr{R}_{I^+ - QQ^{(*)}} \subset \operatorname{Ker} W^+$ and $W^- : \mathscr{R}_{I^- - KK^{(*)}} \subset \operatorname{Ker} W^-$ (Ritsner [4]).
Hint: See the definition of the symbol $T^{(*)}$ in Exercise 7 to §8.

2 In a Krein space $\mathscr{H} = \mathscr{H}^+ [\oplus] \mathscr{H}^-$ with infinite-dimensional $\mathscr{H}^\pm$ construct a dual pair $(\mathscr{L}_+, \mathscr{L}_-)$ which is not a maximal dual pair and is such that $\overline{\mathscr{L}_+ [\dot{+}] \mathscr{L}_-} = \mathscr{H}$ (Langer [7]).
Hint: See 10.10.

3 Choose the dual pair $(\mathscr{L}_+, \mathscr{L}_-)$ in Exercise 2 so that it has at least two different maximal extensions (Langer [7]).

4 Let $\mathscr{D} = \mathscr{D}_0 [\dot{+}] \mathscr{D}_1 [\dot{+}] \mathscr{D}_2$ and $\mathscr{D}^\perp = (\mathscr{D}^\perp)_0 [\dot{+}] (\mathscr{D}^\perp)_1 [\dot{+}] (\mathscr{D}^\perp)_2$ be the $(\mathscr{H}^+, \mathscr{H}^-)$-decomposition of the subspaces $\mathscr{D}$ and $\mathscr{D}^\perp$ of a J-space $\mathscr{H}$. Then $\mathscr{D}_0 \oplus (\mathscr{D}^\perp)_0$, $\mathscr{D}_1$, $(\mathscr{D}^\perp)_1$ and $\mathscr{D}_2 \oplus (\mathscr{D}^\perp)_2$ are pairwise orthogonal and J-orthogonal subspaces completely invariant relative to the operator J:

$$J[\mathscr{D}_0 \oplus (\mathscr{D}^\perp)_0] = \mathscr{D}_0 \oplus (\mathscr{D}^\perp)_0, \qquad J\mathscr{D}_1 = \mathscr{D}_1, \text{ etc.}$$

Hint: Use the definition 10.3 of an $(\mathscr{H}^+, \mathscr{H}^-)$-decomposition and the properties (7.1).

5 Under the conditions of Exercise 4

$$\mathscr{H} = [\mathscr{D}_0 [\dot{+}] \mathscr{D}_1 [\dot{+}] \mathscr{D}_2] \oplus [(\mathscr{D}^\perp)_0] + [(\mathscr{D}^\perp)_1] + [(\mathscr{D}^\perp)_2],$$

and the operator J is represented by the matrix $J = \| J_{ij} \|_{i,j=1}^6$. Prove that: $J_{41}^* = J_{14}$ is an isometry, mapping $(D^\perp)_0$ on to D_0; $J_{22} = J_{22}^* = J_{22}^{-1}$; $J_{33} = J_{33}^*$, $0 < J_{33}^2 < I$; $J_{63}^* = J_{36} = (I - J_{33}^2)V_{36}$, where V_{36} is an isometry, mapping $(\mathscr{D}^\perp)_2$ on to $\mathscr{D}_2$; $J_{55} = J_{55}^* = J_{55}^{-1}$; $J_{66} = -V_{36}^* J_{33} V_{36}$; and $J_{ij} = 0$ for the remaining i, j (Azizov).
Hint: Use the result of Exercise 4 and the properties of the operator J: $J = J^* = J^{-1}$.

6 Describe the maximal dual pairs $(\mathscr{L}_+, \mathscr{L}_-)$ of subspaces from $\mathscr{H}^\pm$ respectively.
Hint: Use the result of Exercise 12 to §8.

7 It is well-known that a basis $\{e_k\}_{k=1}^\infty$ of a separable Hilbert space $\mathscr{H}$ is said to be *unconditional* if it remains a basis under any rearrangement (i.e., renumbering) of its elements, and *conditional* in the opposite case. An almost normalized unconditional basis is called a *Riesz basis* (for an equivalent definition, see Chapter IV, §2.1).

Prove that a J-orthonormalized system $\mathscr{F} = \{e_k\}_{k=1}^{\infty}$ in a separable Krein space $\mathscr{H}$ is a Riesz basis if and only if the dual pair $(\mathscr{L}_+, \mathscr{L}_-)$ generated by it is a maximal dual pair of uniformly definite subspaces (i.e., $\mathscr{L}_+ [\dotplus] \mathscr{L}_- = \mathscr{H}$) ([VIII]).

8 On the basis of the result of Exercise 7 give a complete description of all J-orthonormalized Riess bases in a separable Krein space.
Hint. By means of Lemma 8.4 and Theorem 8.17 describe all the canonical decompositions $\mathscr{H} = \mathscr{H}^+ [\dotplus] \mathscr{H}^-$ of the Krein space and then use J-orthonormalized bases in $\mathscr{H}^+$ and $\mathscr{H}^-$ ([VIII]).

Remarks and bibliographical indications on Chapter I

§1.1, §1.2. The systematic study of infinite-dimensional linear spaces with an Hermitian Q-metric was undertaken for the first time in a general formulation, it seems, by Nevanlinna [1]–[5]. Particular cases of such spaces ($\Pi_\varkappa$ spaces) were considered earlier in the basic work of Pontryagin [1], M. Krein's articles [3], [4], I. Iokvidov's articles [1], [2], and their joint papers [XIV], [XV]. General complex spaces $\mathscr{F}$ with a Q-metric were examined after Nevanlinna by Pesonen [1], Louhivaara [1]–[3], Scheibe [1], Aronszajn [1], Ginzburg and Iokhvidov [VIII], M. Krein and Shmul'yan [1]–[4]. See the book [V] for a detailed exposition of the theory of these spaces, together with an extensive bibliography. We restrict ourselves here to only the necessary minimum of information. Hilbert spaces with a real Q-form together with concrete applications are encountered in M. Krein's [3] (see [XV]), and also M. Krein and Rutman [1], and in Hestenes [1].

§1.3. The concepts of a (Q_1, Q_2)-isometric isomorphism and of anti-space are well-known though formally introduced in [V].

§1.4. For another proof of Proposition 1.9 see M. Krein and Shmul'yan [2]. The proof in the text is due to I. Iokhvidov.

§1.5, §1.6. Here we follow [V] in our presentation.

§1.7. The maximality principle (Theorem 1.19) we encounter first in Scheibe [1], it seems.

§1.8. Propositions 1.21 and 1.22 we find in Scheibe [1], and Proposition 1.23 in [V] but essentially also in [VIII].

§1.9. Propositions 1.24 and 1.25 belong to 'mathematical folklore'; the proofs of them in the text belong to I. Iokhvidov and Azizov respectively. Proposition 1.26 is encountered in Pontryagin [1] and Lemma 1.27 in Ginzburg [3].

§1.10. The simplest case of finite-dimensional neutral skew-connected lineals and the theorem on the existence of Q-biorthogonal bases in them (*cf.* Lemma 1.31) we find in Pontryagin (1), and then in I. Iokhvidov [1] and in [XIV]. There Proposition 1.30 (*cf.* Theorem 9.12) is proved and used systematically. The term 'skew-connected' lineals itself, which has become widely used, is due to M. Krein in [XIV]. Our exposition follows Bognar's [V], which, unfortunately, uses a different name 'dual pairs', which now has in the literature (and in this book—see §10) quite a different meaning. Corollary 1.32 and the notation $\mathscr{L}_1 \# \mathscr{L}_2$ have been borrowed from [V].

§1.11. The Example 1.33, due to Mackie, was published by Savage [1]; Theorem 1.34 is due to Ovchinnikov [1]. In this connection see also Markin [1] and Bognar [4]. The example of a non-degenerate lineal (1.17) which does not admit decomposition, due to M. L. Brodskiy, and also constructed on the difference of dimensions, but using different ideas as well, was published in [VIII].

§1.12. Lemma 1.35, which in such a general form is presented here for the first time, has a long prehistory, briefly traced in [XVI]. Our proof follows the pattern in [V], where $\mathscr{F}_1 = \mathscr{F}_2$ and $T = I$.

§2.1–2.3. Essentially (although in an implicit form) Krein spaces were first considered in 1954 by Nevanlinna [4] and then in 1956 by Pesonen [1], Louhivaara [1], and (independently) by I. Iokhvidov and M. Krein (see [XIV], end of §13). Ginzburg [1], [2] began their systematic investigation (see below the remark on §2.4). In our exposition of the axiomatics and simplest properties of Krein spaces we follow Scheibe [1] and Bognar ([V]); the term 'Krein space' itself is due to the latter. It should, however, be pointed out that in the book [V] (as also in the basic works of Pontryagin [1] and also [XIV], [XV] and some others–*cf.* [XVI]) the rôle of square brackets in the notation for the indefinite metric $[x, y]$ is played by round brackets: (x, y). But we use round brackets to denote the (Hilbert) scalar product. This notation has been generally adopted in the last decades, especially in the Soviet literature.

§2.4. The approach to Krein space considered here was first presented by Ginzburg [1], [2].

§3. All the results of this section, which here already become 'folklore', can be found, essentially, in Ginzburg [1], [2]. To him also is due the extension of the term 'J-metric', introduced earlier by Pontryagin for finite-dimensional spaces, to the general case. Example 3.11 is borrowed from [V].

§4.1, §4.2. The results of these paragraphs are due to Ginzburg [3] (see also [VIII]). The deficiency subspaces $\mathscr{D}^{\pm}_{\mathscr{L}} = \mathscr{H}^{\pm} \ominus \overline{P^{\pm}\mathscr{L}} = \mathscr{H}^{\pm} \cap (P^{\pm}\mathscr{L})^{\perp}$ were considered in such a general formulation (for arbitrary lineals $\mathscr{L}$) by Ritsner [4]. In this connection see Exercises 7 and 8 to §8.

§4.3. The results of this paragraph are due to I. Iokhvidov, in particular, in connection with Theorem 4.10 see I. Iokhvidov [13]. Lemma 4.11 is well-known and is applied widely (see, e.g., [XV], from which the construction of Example 4.12 has been borrowed; there it served quite a different purpose; in our context Example 4.12 is in a certain sense 'universal'—it is mentioned in §§5, 8, 9.

§4.4, §4.5. The results of these two paragraphs, moreover in a more general form (in particular—for arbitrary spaces $\mathscr{G}$ with a Hermitian Q-metric, and the 'sufficient' part of Theorem 4.13—for spaces $\mathscr{F}$ with a canonical decomposition $\mathscr{F} = \mathscr{F}^+ [\dot{+}] \mathscr{F}^-$; see Exercise 6 to §3) were obtained by Scheibe (see also [VIII]). The proof given in the text of Proposition 4.19, which is suitable for arbitrary spaces $\mathscr{F}$ with a Q-metric, is due to I. Iokhvidov [VIII]. For a Krein space this Proposition follows from Ginzburg's results [3]. The formula (4.5) was published by Shmul'yan [4] with reference to an article of Bennevitz [1].

§5.1–5.3. The concept of a uniformly definite lineal (in the original terminology a 'regular definite lineal') was introduced in Ginzburg's investigation [4] (see also [VIII]). The term 'uniformly definite' was introduced by M. Krein ([XVII]). Our presentation here follows the book [V]. Corollary 5.8 is due to Langer [9]. We remark that the definitions (5.1) and (5.2) of the intrinsic norm $|\cdot|_{\mathscr{L}}$ of a definite lineal $\mathscr{L}$ make sense in any space $\mathscr{F}$ with a Q-metric. They were so introduced (in another terminology in [VIII] (*cf.* [V]).

§5.4. The classes $h^{\pm}$ were introduced (without special names) by Azizov [9] in connection with operators of the class $\mathscr{H}$, about which see below in §5 of Chapter III. More general classes of this same type were considered earlier by Langer [9].

§6.1. The example at the end of this paragraph is Ovchinnikov's [1]. For other examples, see Markin [1].

§6.2, §6.3. A systematic investigation of the Gram operator $G_{\mathscr{L}}$ was first undertaken in [VIII]. There too the term 'Gram operator' was introduced and historical information about it was given. Theorem 6.4 in its complete form was established by Ginzburg [4] who indicated that he was following the ideas of Nevanlinna [4].

§6.4. The law of inertia (Theorem 6.7), although, it is true, not in such a general formulation, we find already in Ginzburg [3]. The rank of indefiniteness $\varkappa_{\mathscr{L}}$ was formally introduced in [V], although essentially it was considered earlier by many authors (mainly in particular cases).

§6.5. Corollary 6.12 is contained in a more general proposition due to Ginzburg in [VIII].

§6.6. An extensive bibliography on W-spaces and G-spaces is contained in [III] and [IV]. Proposition 6.14 see in [III]. The first part of Proposition 6.13 was established essentially by Ginzburg [VIII] (see also Langer [2]); the second part was pointed out by Azizov. The universality of Krein spaces was first established by Ginzburg [VIII]; the proof in the text was given by Azizov.

Apart from the generalization of Krein spaces given in §6, this concept has been generalized in two other directions. The first of them relates to Banach spaces with a form $Q(x, y)$ ($|Q(x, y)| \leqslant (\|x\|\,\|y\|)$) continuous in them. The second considers normed spaces (in particular, Banach spaces) $\mathscr{N} = \mathscr{N}^{+} [\dot{+}] \mathscr{N}^{-}$, where $\mathscr{N}^{\pm}$ are arbitrary closed subspaces, $P^{\pm}$ are the corresponding projectors, and an indefinite metric is given by the functional $J_{\nu}(x) = \|P^{+}x\|^{\nu} - \|P^{-}x\|^{\nu}$ ($1 \leqslant \nu < \infty$). For a short survey of these directions and their bibliography (up to 1970) see [IV].

§7.1. The material of this paragraph is traditional; our exposition follows M. Krein's lectures [XVII] and Ando [II].

§7.2. Lemma 7.7 is due to Ginzburg [4], but its proof in the text and also that of Proposition 7.6 was taken by us from [XVII].

§7.3, §7.4. The concept of projective completeness of a lineal was introduced independently by Scheibe [1] and in a particular case by I. Iokhvidov [21] (see [VIII]) to whom the term 'projective completeness' is due. Lemma 7.11 can be found in Louhivaara [2], other results in §7.3 in Scheibe [1], and Lemma 7.12 in Nevanlinna [4]. J-orthogonal projections, but only on non-degenerate subspaces of the space $\Pi_\varkappa$ (see §9) were considered by Pontryagin [1], and after him by I. Iokhvidov [21] (see also [VIII]). But for spaces of a more general type (G-spaces and W-spaces) the development of the theory of projection was begun by Nevanlinna [4] and was later continued by Louhivaara [2]. The final results were obtained by Ginzburg [4] (see also [VIII]).

§7.5. Regular subspaces (in a more general situation than ours) were introduced and investigated by Ginzburg [4] (see also [VIII]). Theorem 7.16, summing up many investigations (*cf.* [V], [IV], and their bibliographies), appears in such a general formulation here for the first time, apparently. Separate fragments of it have been proved by various authors in the course of almost 30 years.

§7.6. Theorem 7.19 (there is a particular case of it for $\Pi_\varkappa$ spaces in [XIV]) is essentially contained in a more general result of Scheibe's [1] (*cf.* also Theorem 2.1 in [VIII]. The proof given in the text is due to I. Iokhvidov.

§8.1, §8.2. The concept of an angular operator was introduced and investigated almost at the same time and independently in articles by Philips [1], [2] and by Ginzburg [3], [4]. The results in §8.1 and §8.2 are due to these authors. The term 'angular operator' itself was first adopted in [VIII] at M. Krein's suggestion instead of the original name 'angular coefficient' given by Ginzburg [3] (*cf.* Exercise 1 to §8). We borrowed the elegant version of the proof of Lemma 8.4 from [V].

§8.3, §8.4. The results in these two paragraphs have been widely applied after the publications of Philips and Ginzburg mentioned above; however, Proposition 8.9 with the remarks 8.8 and 8.9 connected with it, and Theorem 8.10 are published in connected form for the first time here, it seems.

§8.5. In establishing Theorems 8.11 and 8.11′ Ginzburg [3] did not find it necessary to have recourse to the definition (8.9) of the adjoint operator, since the angular operator K for $\mathscr{L} \in \mathscr{M}^+$ (respectively Q for $\mathscr{L} \in \mathscr{M}^-$) was treated by him immediately as defined everywhere in $\mathscr{H}$, mapping $\mathscr{H}^+$ into $\mathscr{H}^-$ (resp. $\mathscr{H}^-$ into $\mathscr{H}^+$), and annihilating $\mathscr{H}^-$ (resp. $\mathscr{H}^+$). However, later (*cf.* §§8.6, 8.7) such a treatment turns out to be inconvenient. Bognar [V] for the transition to the adjoint operator replaces, as we do, the angular operator K by the operator $\widetilde{K} = KP^+$ defined everywhere in $\mathscr{H}$, and then considers $\widetilde{K}^*$.

§8.6. Theorem 8.15, which is due to Azizov [3], was established by him for a far more general case; it is referred to in [IV] (but again not in the most general formulation). Corollary 8.16 was proved earlier by M. Krein and Shmul'yan [3]. Similar results are obtained by Khatskevich [1].

§8.7, §8.8. The results in these paragraphs are due to Azizov; in particular Theorem 8.17 was published in an article by Azizov and Kondras [1]. The classes $\mathscr{A}^\pm$ appeared first in M. Krein's lectures [XVII]. They are systematically considered by Ritsner [4]. Later in Lemma 3.14 of Chapter III the reader will encounter a generalization of the classes $\mathscr{A}^\pm$.

§9.1. Historically the spaces $\Pi_\varkappa$ $(0 < \varkappa < \infty)$ were the first infinite-dimensional spaces with an indefinite metric to be studied by mathematicians. In the case $\varkappa = 1$ they were considered by Sobolev [1] (for the first reference to him see Pontryagin [1]), and for an arbitrary integer $\varkappa > 0$, on the example of the 'model' separable space l^2 (see Exercise 1 to §9), they were first studied by Pontryagin [1]. Here essentially all the results in §9.1 can be found, including Lemma 9.5. The proof of the latter in the text is due to I. Iokhvidov.

Thus, although in this paragraph (see Definitions 9.1) the spaces $\Pi_\varkappa$ are regarded as particular cases of Krein spaces, the latter themselves arose in the literature as generalizations of Pontryagin spaces (their axiomatics and the notation $\Pi_\varkappa$ itself are introduced in [XIV]; for more details about this, see [XVI].

§9.2. Theorem 9.6 was established by I. Iokhvidov (see [VIII]); and the new proof given in the text, and Remark 9.8, are also his. Theorem 9.9 is essentially contained already in Pontryagin [1].

§9.3. In Theorem 9.11 the implication b) ⇒ a), which is new compared with Theorem 9.6, is due to Ginzburg [4]. I. Iokhvidov [9] proved another (less effective) characterization of $\Pi_\varkappa$ spaces in the class of Krein spaces $\mathscr{H}$: it is necessary and sufficient that any non-closed lineal $\mathscr{L}$ be complete relative to the intrinsic norm $|\cdot|_{\mathscr{L}}$ (*cf.* Exercises 11 and 12 to §9).

§9.4. The formula (9.2) was estabished by I. Iokhvidov (see [XV], Theorem 4.1). The sharpening of it formulated in the second part of Theorem 9.12 is due to Bognar [V], as is the whole of the proof, as given in the text, of this theorem.

§9.5. Proposition 9.13 is due to I. Iokhvidov.

§9.6. $W^{(\varkappa)}$-spaces were considered in Azizov's article [7]. Proposition 9.15 and the first part of Corollary 9.16 are essentially contained already in [IV]; the second part of Corollary 9.16 are essentially contained already in [IV]; the second part of Corollary 9.16 and also Proposition 9.17 were noticed by Azizov.

Exercises to §9 Bognar's work [2] was the source for the results given in Exercises 9–12.

§10.1, §10.2. Dual pains were introduced and studied by Phillips [3].

The first part of Theorem 10.2 is due to Phillips [3], the second part to Azizov. Other approaches to the description of extensions of dual pairs to maximal dual pairs are found in Phillips [3], Langer (Langer [7]; [V]) and Ritzner [3], [4]. Theorem 10.4 is due to Azizov.

For certain applications of the theory of dual pairs not included in this monograph, see the literature cited in [V], Chapter V.

§10.3. We find the first investigations of Q-orthonormalized systems in Nevanlinna's [3]; the most important of his results are reproduced in the survey [VIII]. J-orthonormalized systems were introduced and first investigated by Ginzburg in [VIII], and later—in great detail and in the more general scheme of G-orthonormalized systems (see §6.6)—by V. A. Shtraus [1], [2], to whom, in particular, Proposition 10.6 is due. All the rest of the exposition in §10.3 is due to I. Iokhvidov. We draw attention to the fact that our terminology differs sharply from that of V. A. Shtraus [2] (more precisely, it is almost the exact opposite of his).

§10.4. Proposition 10.9 is due to V. A. Shtraus [2]; there too is published Proposition 10.10 (with an indication that is is due to Azizov). Remark 10.11 and the argument following it are published here by the authors for the first time.

Propositions 10.12 and 10.13 are due to V. A. Shtraus [2]. The proof of Proposition 10.12 given in the text is due to Azizov.

We shall return to various bases in *J*-spaces in Chapter 4.

2 FUNDAMENTAL CLASSES OF OPERATORS IN SPACES WITH AN INDEFINITE METRIC

In this chapter the main classes of linear operators are defined and their simplest properties are described. As already mentioned, our main purpose is to give an account of operators acting in Krein and Pontryagin spaces. However, in the process of studying the properties of these operators it turns out to be necessary to consider also more general spaces with an indefinite metric, namely, W-spaces, G-spaces, and so on (see 1.§6). Most of the operators acting in these spaces have "definite" analogues. In the indefinite case we shall prefix the names with the letters W-, G-, J-, π-, *etc.*, depending on whether the operators are acting in a W-space, a G-space, a Krein space (J-space), a Pontryagin space, etc. If an operator acts from one indefinite space into another, this will also be indicated by appropriate prefixes, whose meaning will be clear from the context.

We mention, as will be clear from the size of the sections, that we devote most attention to the operators specific for an indefinite metric, namely, to plus-operators (§4).

§1 The adjoint operator T^c

1 Let $\mathscr{H}_1$ and $\mathscr{H}_2$ be a G_1-space and a G_2-space respectively (see 1§6.6) with the indefinite forms $[\cdot,\cdot]_i = (G_i\cdot,\cdot)_i$ $(0 \notin \sigma_p(G_i),\ i = 1, 2)$. The symbol $T: \mathscr{H}_1 \to \mathscr{H}_2$ will in future mean that the linear operator T operates from a domain of definition $\mathscr{D}_T \subset \mathscr{H}_1$ into a range of values $\mathscr{R}_T \subset \mathscr{H}_2$.

Definition 1.1: Let $T: \mathscr{H}_1 \to \mathscr{H}_2$, $\bar{\mathscr{D}}_T = \mathscr{H}_1$. The operator T^c defined on the lineal

$$\mathscr{D}_{T^c} = \{y \in \mathscr{H}_2 \mid \text{there is a } z \in \mathscr{H}_1 \text{ such that } [Tx, y]_2 = [x, z]_1 \text{ for all } x \in \mathscr{D}_T\}$$

by the formula $T^c y = z$ is called the (G_1, G_2)-*adjoint of the operator T*.

The connection of T^c with the ordinary adjoint operator T^* follows immediately from this definition:

$$T^c = G_1^{-1} T^* G_2. \tag{1.1}$$

In particular, if $G_i = I_i$ is the identity operator in $\mathscr{H}_i$ $(i = 1, 2)$, then $T^c = T^*$. But if $\mathscr{H}_1 = \mathscr{H}_2 = \mathscr{H}$ is a J-space, then

$$T^c = JT^*J, \tag{1.2}$$

and therefore T^c and T^* are *unitarily equivalent*. From the relation (1.1) and, in particular, from (1.2) it follows that many properties of T^* are preserved for T^c.

2 We give a geometrical interpretation of the concept of the adjoint operator T^c, which will enable us later to prove easily a number of assertions. In doing this we shall for simplicity and brevity of exposition assume everywhere below in this section that $\mathscr{H}_1 = \mathscr{H}_2 = \mathscr{H}$ is a J-space with the indefinite form $[\cdot, \cdot]$.

Definition 1.2: Every lineal $\mathscr{L}$ in the space of graphs $\mathscr{H}_\Gamma = \mathscr{H} \times \mathscr{H} = \{\langle x, y\rangle \mid x, y \in \mathscr{H}\}$ is called a *linear relation operating in* $\mathscr{H}$, *with the domain of definition*

$$\mathscr{D}_{\mathscr{L}} = \{x \mid \text{there is a } y \text{ such that } \langle x, y\rangle \in \mathscr{L}\},$$

and the range of values

$$\mathscr{R}_{\mathscr{L}} = \{y \mid \text{there is an } x \text{ such that } \langle x, y\rangle \in \mathscr{L}\},$$

with the kernal Ker $\mathscr{L} = \{x \mid \langle x, \theta\rangle \in \mathscr{L}\}$, *and with the indefiniteness* $\operatorname{Ind}\mathscr{L} = \{y \mid \langle \theta, y\rangle \in \mathscr{L}\}$.

In particular, if it follows from $\langle \theta, y\rangle \in \mathscr{L}$ that $y = \theta$, then the relation $T: x \to y(\langle x, y\rangle \in \mathscr{L})$ is a linear operator, and $\mathscr{L}$ is its graph, which we shall denote but the symbol $\Gamma_T = \{\langle x, Tx\rangle \mid x \in \mathscr{D}_T\}$.

In the space $\mathscr{H}_\Gamma$ we introduce the scalar product

$$(\langle x_1, y_1\rangle, \langle x_2, y_2\rangle)_\Gamma = (x_1, x_2) + (y_1, y_2). \tag{1.3}$$

It is well-known that Γ_T is a subspace in $\mathscr{H}_\Gamma$ with the scalar product (1.3) if and only if T is a closed operator, i.e., if it follows from $x_n \to x_o$, $Tx_n \to y_o$ that $x_o \in \mathscr{D}_T$ and $y_o = Tx_o$.

We introduce in $\mathscr{H}_\Gamma$ the indefinite metric

$$[\langle x_1, y_1\rangle, \langle x_2, y_2\rangle]_\Gamma = i([x_1, y_2] - [y_1, x_2]). \tag{1.4}$$

Hence, it follows that $[\cdot,\cdot]_\Gamma = (J_\Gamma\cdot,\cdot)_\Gamma$, where $J_\Gamma\langle x, y\rangle = \langle -iJy, iJx\rangle$, and therefore $J_\Gamma = J_\Gamma^* = J_\Gamma^{-1}$.

Consequently the space $\mathscr{H}_\Gamma$ admits the canonical decomposition

$$\mathscr{H}_\Gamma = \mathscr{H}_\Gamma^+ \oplus \mathscr{H}_\Gamma^-, \tag{1.5}$$

where

$$\mathscr{H}_\Gamma^\pm = P_\Gamma^\pm \mathscr{H}_\Gamma, \qquad P_\Gamma^\pm = \tfrac{1}{2}(I_\Gamma \pm J_\Gamma),$$

and therefore

$$\mathscr{H}_\Gamma^\pm = \{\langle x, \pm iJx\rangle \mid x \in \mathscr{H}\}. \tag{1.6}$$

Let T be a linear operator acting in $\mathscr{H}$, and let Γ_T be its graph. Then

$$P_\Gamma^\pm \Gamma_T = \{\langle (I \mp iJT)x, \pm iJ(I \mp iJT)x\rangle \mid x \in \mathscr{D}_T\}. \tag{1.7}$$

In particular, if Γ_T is a non-negative lineal, then its angular operator K_T can be written symbolically in the form of a pair

$$K_T = \langle V, -JVJ\rangle, \tag{1.8}$$

where

$$V = (I + iJT)(I - iJT)^{-1}; \tag{1.9}$$

here it is borne in mind that the operator $I - iJT$ is necessarily invertible and this means that

$$K_T\langle x, y\rangle = \langle Vx_l - JVJy\rangle \tag{1.10}$$

for $\langle x, y\rangle \in \mathscr{D}_{K_T}$.

□ From Definition 1.1 of the operator T^c and from the Definition 1.1.11 it follows immediately that

1.3. If $T\colon \mathscr{H} \to \mathscr{H}$ is a linear operator with $\bar{\mathscr{D}}_T = \mathscr{H}$, then $\Gamma_{T^c} = \Gamma_T^{[\perp]}$. ■

Corollary 1.4: *In order that a vector x_0 should belong to $\mathscr{D}_T \cap \mathscr{D}_{T^c}$ and that $Tx_0 = T^c x_0$ it is necessary and sufficient that the vector $\langle x_0, Tx_0\rangle$ be isotropic in Γ_T.*

□ The condition imposed on x_0 is equivalent to the fact that $\langle x_0, Tx_0\rangle \in \Gamma_T \cap \Gamma_{T^c} = \Gamma_T \cap \Gamma_T^{[\perp]}$, i.e., $\langle x_0, Tx_0\rangle$ is isotropic in Γ_T by the Definition 1.1.12. ■

3 From Proposition 1.3 it also follows that:

1.5 The operator T^c is closed.

1.6 If for a densely defined operator T there is a densely defined inverse operator, then $(T^{-1})^c = (T^c)^{-1}$.

1.7 If an opeatror T_2 is an extension of an operator T_1 ($T_1 \subset T_2$) and if $\bar{\mathscr{D}}_{T_1} = \mathscr{H}_1$, then $T_2^c \subset T_1^c$.

1.8 In order that the operator $T^{cc} \equiv (T^c)^c$ should exist it is necessary and sufficient that $\bar{\mathscr{D}}_T = \mathscr{H}$ and that the linear operator T admit closure; in addition, $\bar{T}^c = T^{c,}$ $T^{cc} = \bar{T}$, and, in particular, $\bar{T} = T^{**}$.

□ Proposition 1.5 follows from the fact that the J_τ-orthogonal complement is a subspace (see I, Proposition 7.1). Proposition 1.6 follows from Proposition 1.3 if we note that $\Gamma_{\tau^{-1}} = \{\langle Tx, x\rangle \mid x \in \mathscr{D}_T\}$. To prove 1.7 and 1.8 it is necessary to use 1 Formula (1.8) and 1 Proposition 7.4 respectively. ■

It follows from (1.2) that *in a Krein space the operator T^c is bounded only when the operator T is bounded.* We note also a number of other properties of T^c.

1.9 Let $T_1, T_2: \mathscr{H} \to \mathscr{H}$ and $\overline{\mathscr{D}_{T_1} \cap \mathscr{D}_{T_2}} = \mathscr{H}$. Then $(T_1 + T_2)^c \supset T_1^c + T_2^c$, and if at least one of the operators T_1 or T_2 is continuous and defined everywhere, then $(T_1 + T_2)^c = T_1^c + T_2^c$.

1.10 Let $T_1, T_2: \mathscr{H} \to \mathscr{H}$, and $\bar{\mathscr{D}}_{T_2} = \bar{\mathscr{D}}_{T_2T_1} = \mathscr{H}$. Then $(T_2T_1)^c \supset T_1^cT_2^c$, and if either T_2 is bounded, or if the operator T_1^{-1}, defined everywhere and bounded, exists, then $(T_2T_1)^c = T_1^cT_2^c$.

□ We verify only the last assertion, the others are verified analogously. Thus suppose that T_1^{-1} is a bounded operator and $\mathscr{R}_{T_1} = \mathscr{H}$. Since we are regarding the inclusion $T_1^cT_2^c \subset (T_2T_1)^c$ as proved, it remains only to verify that $(T_2T_1)^c \subset T_1^cT_2^c$. Let $x_0 \in \mathscr{D}_{(T_2T_1)^c}$. Then

$$[T_2T_1x, x_0] = [x, (T_2T_1)^cx_0] = [T_1^{-1}T_1x, (T_2T_1)^cx_0]$$
$$= [T_1x, (T_1^{-1})^c(T_2T_1)^cx_0]$$

for all $x \in \mathscr{D}_{T_2T_1}$. Since $\mathscr{R}_{T_1} = \mathscr{H}$, so $\mathscr{D}_{T_2} = T_1\mathscr{D}_{T_2T_1}$, and therefore $x \in \mathscr{D}_{T_2^c}$ and $T_2^cx_0 = (T_1^{-1})^c(T_2T_1)^cx_0$. Using Proposition 1.6, we obtain $T_1^cT_2^cx_0 = (T_2T_1)^cx_0$. ■

4 □ From the Definition 1.1 of a J-adjoint operator T^c it follows immediately that:

1.11 If a subspace $\mathscr{L}$ is invariant relative to the operator T, i.e., *if $\overline{\mathscr{L} \cap \mathscr{D}_T} = \mathscr{L}$ and $T(\mathscr{D}_T \cap \mathscr{L}$, then*

$$T^c(\mathscr{D}_{T^c} \cap \mathscr{L}^{[\perp]}) \subset \mathscr{L}^{[\perp]}.$$ ■

□ Since for a closed operator we always have Ker $T = \mathscr{R}_{T^*}^{\perp}$ and the relations $\mathscr{R}_{T^*} = \bar{\mathscr{R}}_{T^*}$ and $\mathscr{R}_T = \bar{\mathscr{R}}_T$ are equivalent (see, e.g. [XIX]), it follows from the equations $\mathscr{R}_{T^c} = J\mathscr{R}_{T^*}$ and $\mathscr{R}_{T^c}^{[\perp]} = (J\mathscr{R}_{T^*})^{[\perp]} = J.\ J\mathscr{R}_{T^*}^{[\perp]} = \mathscr{R}_{T^*}^{[\perp]}$ (= Ker T), in which 1, Propositions 7.1 and 7.3 have been used, that:

1.12 If T is a closed operator and $\bar{\mathscr{D}}_T=\mathscr{H}$, then always Ker $T=\mathscr{R}_{T^c}^{[\perp]}$, *but* $\mathscr{R}_{T^c}=(\text{Ker } T)^{[\perp]}$ *if and only if* $\mathscr{R}_T=\bar{\mathscr{R}}_T$. ■

Let T be a linear operator and λ one of its eigenvalues. We recall that the set of all vectors x in the space, for which there exists a natural number $p(x)$ such that $(T-\lambda I)^{p(x)}x=\theta$, forms a lineal of *'root vectors corresponding to the point* λ' (briefly, a *root lineal*), and it is denoted by the symbol $\mathscr{L}_\lambda(T)$.

The following theorem characterizes the disposition, in the field $\mathscr{H}$, of the invariant subspaces of the operators T and T^c, and also their root lineals.

Theorem 1.13: *Let $T:\mathscr{H}\to\mathscr{H}$, let $\mathscr{D}_T=\mathscr{H}$, $\mathscr{L}(\subset\mathscr{D}_T)$ be an invariant subspace of the operator T, let $\mathscr{N}(\subset\mathscr{D}_{T^c})$ be on invariant subspace of the operator T^c, and let $\sigma(T\mid\mathscr{L})\cap\sigma^*(T^c\mid\mathscr{N})=\emptyset$. Then $\mathscr{L}[\perp]\mathscr{N}$. This implies, in particular, that $\mathscr{L}_\lambda(T)$ is J-orthogonal to $\mathscr{L}_\mu(T^c)$ when $\lambda\neq\mu$.*

□ Let $x\in\mathscr{L}$, $y\in\mathscr{N}$. We bring into consideration the following functions of the complex variable λ:

$$g_{x,y}^{(1)}(\lambda)=[(T\mid\mathscr{L}-\lambda I)^{-1}x,y] \quad\text{and}\quad g_{x,y}^{(2)}(\lambda)=[x,(T^c\mid\mathscr{N}-\bar{\lambda}I)^{-1}y].$$

The first of these functions is defined and holomorphic (because the resolvent is holomorphic) on the set $\rho(T\mid\mathscr{L})$ of regular points of the operator $T\mid\mathscr{L}$, and the second on the set $\rho^*(T^c\mid\mathscr{N})$. Let $|\lambda|>\max\{\|T\mid\mathscr{L}\|,\|T^c\mathscr{N}\|\}$. Then

$$g_{x,y}^{(1)}(\lambda)=[(T\mid\mathscr{L}-\lambda I)^{-1}x,y]=\left[-\frac{1}{\lambda}\sum_{n=0}^{\infty}\left(-\frac{1}{\lambda}T\mid\mathscr{L}\right)^n x,y\right]$$
$$=-\frac{1}{\lambda}\sum_{n=0}^{\infty}\left[\left(-\frac{1}{\lambda}T\mid\mathscr{L}\right)^n x,y\right].$$

Similarly we obtain that

$$g_{x,y}^{(2)}(\lambda)=-\frac{1}{\lambda}\sum_{n=0}^{\infty}\left[x,\left(-\frac{1}{\bar{\lambda}}T^c\mid\mathscr{N}\right)^n y\right].$$

Since

$$\left[\left(-\frac{1}{\lambda}T\mid\mathscr{L}\right)^n x,y\right]=\left[x,\left(-\frac{1}{\bar{\lambda}}T^c\mid\mathscr{N}\right)^n y\right],$$

we have $g_{x,y}^{(1)}(\lambda)=g_{x,y}^{(2)}(\lambda)$ when $|\lambda|>\max\{\|T\mid\mathscr{L}\|,\|T^c\mid\mathscr{N}\|\}$, i.e. each of these functions admits analytic continuation $g_{x,y}(\lambda)$ on to $\rho(T\mid\mathscr{L})\cup\rho^*(T^c\mid\mathscr{N})$. From the condition $\sigma(T\mid\mathscr{L})\cap\sigma^*(T^c\mid\mathscr{N})=\emptyset$ we conclude that $\rho(T\mid\mathscr{L})\cup\rho^*(T^c\mid\mathscr{N})=\mathbb{C}$ and the function $g_{x,y}(\lambda)$ is bounded on $\mathbb{C}$. Therefore, by the well-known Liouville theorem, $g_{x,y}(\lambda)\equiv$ const. In combination with the fact that $g_{x,y}(\lambda)\to 0$ as $|\lambda|\to\infty$, we obtain that $g_{x,y}(\lambda)\equiv 0$. In particular, for $\lambda\in\rho(T\mid\mathscr{L})$ and arbitrary $x\in\mathscr{L}$ and $y\in\mathscr{N}$ we have $[x,y]=g^{(1)}_{(T\mid\mathscr{L}-\lambda I)x,y}(\lambda)=0$, i.e. $\mathscr{L}[\perp]\mathscr{N}$.

We now prove that $\overline{\mathscr{L}_\lambda(T)}[\perp]\overline{\mathscr{L}_\mu(T^c)}$ when $\lambda\neq\bar{\mu}$. Since the form $[\cdot,\cdot]$ is continuous, it suffices to verify that $\mathscr{L}_\lambda(T)[\perp]\mathscr{L}_\mu(T^c)$.

Let $x\in\mathscr{L}_\lambda(T)$, $y\in\mathscr{L}_\mu(T^c)$. We put

$$\mathscr{L}=\text{Lin}\{(T-\lambda I)^p x\}_0^\infty, \qquad \mathscr{N}=\text{Lin}\{(T^c-\mu I)^p y\}_0^\infty.$$

These are finite-dimensional invariant subspaces of the operators T and T^c respectively, and $\sigma(T|\mathscr{L})=\{\lambda\}$, and $\sigma(T^c|\mathscr{N})=\{\mu\}$. Since $\lambda\neq\bar\mu$ by hypothesis, so $\sigma(T|\mathscr{L})\cap\sigma^*(T^c|\mathscr{N})=\emptyset$, and from what has been proved above $[x,y]=0$, i.e. $\mathscr{L}_\lambda(T)\,[\perp]\,\mathscr{L}_\mu(T^c)$ when $\lambda\neq\bar\mu$. ■

Before going on to the next proposition we recall some definitions and notation.

Definition 1.14: Let T be a closed linear operator in an arbitrary Hilbert space $\mathscr{H}$. A point λ_0 is called a *normal point* of T ($\lambda_0\in\tilde\rho(T)$) if it is either one of its regular points or if it is a *normal eigenvalue* ($\lambda_0\in\tilde\sigma_p(T)$). The latter means that there is a decomposition of the space $\mathscr{H}=\mathscr{L}\dot{+}\mathscr{N}$ into subspaces $\mathscr{L}$ and $\mathscr{N}$ such that $\mathscr{L}\subset\mathscr{D}_T$, $\dim\mathscr{L}<\infty$, $T:\mathscr{L}\to\mathscr{L}$, $T:\mathscr{N}\to\mathscr{N}_1$ $\sigma(T|\mathscr{L})=\{\lambda_0\}$ and $\lambda_0\in\rho(T|\mathscr{N})$.

□ Hence, in particular, it follows that:

1.15 Normal eigenvalues are isolated points of the spectrum, and therefore in every open set they make up no more than a countable subset. ■

The symbols $\sigma_{p,1}(T)$ and $\sigma_{p,2}(T)$ denote those $\lambda\in\sigma_p(T)$ for which $\bar{\mathscr{R}}_{T-\lambda I}\neq\mathscr{H}$ and $\bar{\mathscr{R}}_{T-\lambda I}=\mathscr{H}$ respectively. It is clear that $\sigma_{p,1}(T)\cap\sigma_{p,2}(T)=\emptyset$ and $\sigma_p(T)=\sigma_{p,1}\cup\sigma_{p,2}(T)$.

Therem 1.16: *Let $\mathscr{H}$ be a Krein space and $T:\mathscr{H}\to\mathscr{H}$ a closed operator with $\bar{\mathscr{D}}_T=\mathscr{H}$. Then*

a) $\lambda\in\rho(T)\Leftrightarrow\bar\lambda\in\rho(T^c)$;
b) $\lambda\in\sigma_r(T)\Leftrightarrow\bar\lambda\in\sigma_{p,2}(T^c)$;
c) $\lambda\in\sigma_{p,1}(T)\Leftrightarrow\bar\lambda\in\sigma_{p,1}(T^c)$;
d) $\lambda\in\tilde\sigma_p(T)\Leftrightarrow\bar\lambda\in\tilde\sigma_p(T^c)$;

and, as a consequence of a)–d),

e) $\lambda\in\sigma(T)\Leftrightarrow\bar\lambda\in\sigma(T^c)$;
f) $\lambda\in\sigma_c(T)\Leftrightarrow\bar\lambda\in\sigma_c(T^c)$;

□ The proof of these assertions follows immediately from formula (1.2) and the fact that analogous assertions hold for the spectra of the operators T and T^*. As an example, we examine the proof of the implication d). By 1.8 it suffices to verify that $\lambda\in\tilde\sigma_p(T)\Rightarrow\bar\lambda\in\tilde\sigma_p(T^c)$. Since (see [X]) $\lambda\in\sigma_p(T)\Rightarrow\bar\lambda\in\tilde\sigma_p(T^*)$, so it then follows by Definition 1.14 that there is a decomposition $\mathscr{H}=\mathscr{L}\dot{+}\mathscr{N}$ such that $\mathscr{L}\subset\mathscr{D}_{T^*}$, $\dim\mathscr{L}<\infty$, $T^*:\mathscr{L}\to\mathscr{L}$, $T^*:\mathscr{N}\to\mathscr{N}$, $\sigma(T^*|\mathscr{L})=\{\bar\lambda\}$ and $\bar\lambda\in\rho(T^*|\mathscr{N})$. Hence, we obtain without difficulty that $\mathscr{H}=J\mathscr{L}\dot{+}J\mathscr{N}$, $T^cJ\mathscr{L}\subset J\mathscr{L}$, $T^c:J\mathscr{N}\to J\mathscr{N}$, $\sigma(T^c|J\mathscr{L})=\sigma(T^*|\mathscr{L})(=\{\bar\lambda\})$, $\rho(T^c|J\mathscr{N})=\rho(T^*|\mathscr{N})$, and therefore $\bar\lambda\in\tilde\sigma_p(T^c)$. ■

5 Now let T be a linear operator acting in a W-space $\mathscr{H}$, and let $\mathscr{H}^0$ be an isotropic subspace of $\mathscr{H}$, $\mathscr{H}^0\subset\mathscr{D}_T$ and $T\mathscr{H}^0\subset\mathscr{H}^0$. We form the factor-space $\hat{\mathscr{H}} = \mathscr{H}/\mathscr{H}^0$ (see 1, §1.8) and bring into consideration the operator $\hat{T}$:

$$\mathscr{D}_{\hat{T}} = \{\hat{x} \mid x \in \mathscr{D}_T\}, \qquad \hat{T}\hat{x} = \widehat{Tx} \tag{1.11}$$

It is not difficult to verify (see, e.g., [XIX]) that the operator $\hat{T}$ *is properly defined*. Moreover, closedness (respectively boundedness) of T in the norm $\| x \|$ generated by the scalar product (x, y) implies the closedness (respectively boundedness) of $\hat{T}$ in the norm $\| \hat{x} \|^{\wedge} = \inf\{\| x \| \mid x \in \hat{x}\}$ which, also is generated by the scalar product $(\hat{x}, \hat{y}) = (x, y)$ where $x \in \hat{x}$, $y \in \hat{y}$ and x, y are orthogonal to $\mathscr{L}_0$; and $\bar{\mathscr{D}}_T = \mathscr{H}$ implies $\bar{\mathscr{D}}_{\hat{T}} = \hat{\mathscr{H}}$.

Let $\mathscr{H}$ be a Krein space, T a linear operator, $\bar{\mathscr{D}}_T = \mathscr{H}$; let $\mathscr{L}^0$ be a neutral subspace contained in $\mathscr{D}_T \cap \mathscr{D}_{T^c}$ and let $T\mathscr{L}^0 \subset \mathscr{L}^0$, $T^c\mathscr{L}^0 \subset \mathscr{L}^0$. We form the Krein space (see 1, Corollary 5.8) $\hat{\mathscr{H}} = \mathscr{L}^{0[\perp]}/\mathscr{L}^0$ and consider in it the operators $\hat{T}$ and $\hat{T}^c$ generated by the operators T and T^c respectively in accordance with formula (1.11). The following theorem holds:

Theorem 1.17: $\bar{\mathscr{D}}_{\hat{T}} = \hat{\mathscr{H}}$; $\hat{T}^c = \widehat{T^c}$.

□ The density of $\mathscr{D}_{\hat{T}}$ in $\hat{\mathscr{H}}$ was noted earlier. Therefore the operator $\widehat{T^c}$ is defined (see Definition 1.1). Since $\mathscr{L}_0 \subset \mathscr{D}_T$, so $\hat{x} \in \mathscr{D}_{\hat{T}}$ if and only if $x \in \mathscr{D}_T$. Hence, if $x \in \mathscr{D}_{\hat{T}}$ and $y \in \widehat{\mathscr{D}_{T^c}}$, then

$$[\hat{x}, \hat{T}^c\hat{y}]^{\wedge} = [\hat{T}\hat{x}, \hat{y}]^{\wedge} = [Tx, y] = [x, T^c y] = [\hat{x}, \widehat{T^c}\,\hat{y}]^{\wedge}, \tag{1.12}$$

i.e, $\hat{T}^c\hat{y} = \widehat{T^c}\,\hat{y}$, and therefore it remains to prove that $\mathscr{D}_{\hat{T}^c} \supset \mathscr{D}_{\widehat{T^c}}$. Indeed, if $y \in \mathscr{D}_{\widehat{T^c}}$, then, reading the system of equalities (1.12) from right to left, we obtain that $\hat{y} \in \mathscr{D}_{\hat{T}^c}$. ■

Exercises and problems

1 Verify that Definition 1.1 is correct if the condition $\bar{\mathscr{D}}_T = \mathscr{H}_1$ in it is replaced by the condition of 'G_1-denseness' of $\mathscr{D}_T$: $D_T^{[\perp]} = \{\theta\}$, and prove that a) T^c is a closed operator; b) if $\mathscr{R}_T$ is G_2-dense and Ker $T = \{\theta\}$, then $(T^{-1})^c = (T^c)^{-1}$ [III].

2 Prove that when $\mathscr{D}_{T_1+T_2} = \mathscr{D}_{T_2}$ $(\mathscr{D}_{T_1T_2} = \mathscr{D}_{T_2})$ and $\mathscr{D}_{T_1^c} = \mathscr{H}_2$, the relation $(T_1 + T_2)^c = T_1^c + T_2^c$ $((T_1T_2)^c = T_2^cT_1^c$ holds ([III]).

3 Prove that if, for any T: $\mathscr{H}_1 \to \mathscr{H}_2$, the G_1-denseness of $\mathscr{D}_T$ implies the G_2-denseness of $\mathscr{D}_{T^c}$, then $0 \in \rho(G_1)$ ([III]).

4 Investigate which of the implications in Theorem 1.16 continue to hold in the case of G-spaces with $0 \notin \rho(G)$ ([III]).

5 Let T be a linear operator bounded in $\Pi_\varkappa = \Pi^+ [\dotplus] \Pi^-$. Prove that, if the space $\Pi_\varkappa$ is not separable, then there is a separable subspace $\Pi_\varkappa'(\subset \pi_\varkappa)$, invariant relative to T and T^c, such that $\Pi_k = \Pi_\varkappa' [\dotplus] \Pi_0$, where Π_0 is a Hilbert space relative to the scalar product—$[x, y]$ and that the operator T is represented relative to the indicated decomposition by the matrix

$$T = \begin{Vmatrix} T' & 0 \\ 0 & T_0' \end{Vmatrix}. \qquad ([XVII])$$

Hint: Fix in Π^+ any basis $e_1, e_2, \ldots, e_\varkappa$, and choose as $\Pi'_\varkappa$ the closed linear envelope of all vectors of the form $T^{n_1}(T^c)^{n_2}T^{n_3}\ldots T^{n_{2p-1}}(T^c)^{n_{2p}}e_k$, $k = 1, 2, \ldots, \varkappa$; $0 \leqslant n_p < \infty$, $p = 1, 2, \ldots$.

6 Let H_i be J_i-spaces ($i = 1, 2$), T: $\mathscr{H}_1 \to \mathscr{H}_2$. We define the space $\mathscr{H}_3 = \mathscr{H}_1^+ \oplus \mathscr{H}_2^-$ and the operators $P_1^+ + P_2^- T$: $\mathscr{H}_1 \to \mathscr{H}_3$, $P_1^+ + P_2^- TP_2^-$: $\mathscr{H}_1 \to \mathscr{H}_3$, $P_2^- TP_1^-$: $\mathscr{H}_1 \to \mathscr{H}_2$. Show that the following conditions are equivalent:
a) Ker $P_2^- TP_1^- \mid \mathscr{H}_1^- = \{\theta\}$;
b) Ker $(P_1^+ + P_2^- TP_1^-) = \{\theta\}$;
c) Ker $(P_1^+ + P_2^- T) = \{\theta\}$ (*cf.* I. Iokhvidov [14], [18]).

7 Verify that under the conditions of Exercise 6 the following requirements are equivalent:
a) $\mathscr{R}_{P_2^- TP_2^-} = \mathscr{H}_2^-$;
b) $\mathscr{R}_{P_1^+ + P_2^- TP_1^-} = \mathscr{H}_3$;
c) $\mathscr{R}_{P_1^+ + P_2^- T} = \mathscr{H}_3$ (*cf.* I. Iokhvidov [14], [18]).

8 Independently of Theorem 1.13 verify that $\mathscr{L}^\lambda(T)\,[\perp]\,\mathscr{L}_\mu(T^c)$ when $\lambda \neq \bar{\mu}$ (*cf.* Pontryagin [1]).
Hint: Consider the minimal non-negative integers p and q such that $(T - \lambda I)^p x = \theta$ and $(T^c - \mu I)^q y = \theta$ if $x \in \mathscr{L}_\lambda(T)$ and $y \in \mathscr{L}_\mu(T^c)$, and apply induction over $p + q$.

§2 Dissipative operators

1 *Definition 2.1:* A linear operator A with an arbitrary domain of definition $\mathscr{D}_A$, operating in a W-space $\mathscr{H}$, is said to be *W-dissipative* if $\operatorname{Im}[Ax, x] \geqslant 0$ for all $x \in \mathscr{D}_A$, and to be *maximal W-dissipative* if it is W-dissipative and coincides with any W-dissipative extension of it.

In particular, when $W = I$ this definition brings us back to the (usual) dissipative and maximal dissipative operators in $\mathscr{H}$ (see [XI], [XVIII]).

□ From Zorn's lemma easily follows the Proposition

2.2 *Every W-dissipative operator can be extended into a maximal one.* ■

Now let A be a J-dissipative operator. From Definition 2.1 we have

2.3 *An operator A is J-dissipative if and only if each of the operators JA and AJ is J-dissipative. Moreover, A is a maximal J-dissipative operator if and only if each of the operators JA and AJ is a maximal J-dissipative operator.*
□ The proof of the second part of Proposition 2.3 uses the fact that an operator T is an extension of an operator S if and only if the operators JT and TJ are extensions of the operators JS and SJ respectively.

It is easy to see that *an operator A is J-dissipative if and only if its graph* Γ_A *in* $\mathscr{H} \times \mathscr{H}$ *in the metric* (1.4) *is non-negative.*

2.4 *Let A be a densely defined J-dissipative operator, and let the lineal* $\mathscr{L}$ *be a non-negative extension of its graph in* $\mathscr{H} \times \mathscr{H}$. *Then* $\mathscr{L}$ *is the graph of a J-dissipative operator with is an extension of the operator A.*

□ In the proof here we need only the proposition that $\mathscr{L}$ is the graph of some operator, i.e., that $\langle\theta, y\rangle \in \mathscr{L}$ implies $y=\theta$. Since $\mathscr{L}$ is a non-negative subspace in $[\langle\theta, y\rangle, \langle\theta, y\rangle]_\Gamma = 0$, $\langle\theta, y\rangle$ is an isotropic vector in $\mathscr{L}$ and therefore it is J_Γ-orthogonal to Γ_A, i.e., $[y, x]=0$ for all $x \in \mathscr{D}_A$. Since $\mathscr{D}_A = \mathscr{H}$ it follows that $y=0$. ■

Corollary 2.5: *A densely defined J-disspative operator A admits a closure which is J-dissipative, and therefore, if A is a maximal J-dissipative operator and* $\bar{\mathscr{D}}_A = \mathscr{H}$, *then A is closed.*

□ This follows from 1 Proposition 3.6. ■

Theorem 2.6: *A closed J-disspative operator A admits maximal closed J-dissipative extensions. Moreover the latter are densely defined and their graphs are maximal non-negative subspaces in* $\mathscr{H} \times \mathscr{H}$.

□ Since a non-negative subspace in a Krien space can be extended into a maximal subspace by adding a uniformly positive subspace orthogonal to it (relative to the indefinite metric) (see I (8.8)), we shall prove that such an extension $\tilde{\Gamma}$ of the graph Γ_A in $\mathscr{H} \times \mathscr{H}$ is the graph of a J-dissipative closed extension $\tilde{A}$ of the operator A. For, in the opposite case there would be a vector $\langle\theta, y_0\rangle$ in $\tilde{\Gamma}$ with $y_0 \neq \theta$. Since such a vector is neutral by the form (1.4), it is isotropic in $\tilde{\Gamma}$, but by the construction of $\tilde{\Gamma}$ its isotropic part coincides with the isotropic part of Γ_A; but Γ_A is the graph of the operator A and therefore $y_0 = \theta$—a contradiction.

Thus, every closed J-dissipative operator A admits extension into a closed J-dissipative operator $\tilde{A}$ whose graph is a maximal non-negative subspace in $\mathscr{H} \times \mathscr{H}$. Hence it follows that $\tilde{A}$ is a maximal J-dissipative operator. It remains to verify that it is densely defined. Let y_0 be J-orthogonal to $\mathscr{D}_{\tilde{A}}$. Then the neutral vector $\langle\theta, y_0\rangle$ is J_Γ-orthogonal to $\Gamma_{\tilde{A}}$. Therefore $y_0 = \theta$. ■

We now dwell on other criteria for dissipative operators to be maximal.

2.7 A closed J-dissipative operator A is maximal if and only if $\bar{\mathscr{D}}_A = \mathscr{H}$ *and* $(-A^c)$ *is J-dissipative, and therefore the operators A and* $(-A^c)$ *can be maximal J-dissipative only simultaneously.*

□ Taking Proposition 1.3 and Theorem 2.6 into account, this assertion coincides with Theorems 8.11 and 8.11′ in Chapter 1. ■

Before formulating the next result we recall a definition.

A point $\lambda \in \mathbb{C}$ is called a *point of regular type* of a linear operator $T: \mathscr{H} \to \mathscr{H}$ if the operator $(T-\lambda I)^{-1}$ exists and is continuous. The set of all regular points of an operator T is called its *field of regularity*.

We point out that here neither the operator T nor the operator $(T-\lambda I)^{-1}$ is assumed to be defined on the whole of $\mathscr{H}$.

Lemma 2.8: *Let T be a closed dissipative operator in a Hilbert space* $\mathscr{H}$.

Then the open lower half-plane $\mathbb{C}^-$ *belongs to the set of points of regular type of the operator* T *and* $\|(T-\lambda I)^{-1}\| \leqslant 1/|\operatorname{Im}\lambda|$ $(\lambda \in \mathbb{C}^-)$. *Moreover,* T *is a maximal dissipative operator if and only if* $\mathbb{C}^- \cap \rho(T) \neq \emptyset$ *or, equivalently,* $\mathbb{C}^- \subset \rho(T)$.

□ If $\operatorname{Im}\lambda < 0$, then it can be verified directly that, since T is dissipative $((T-\lambda I)x, \quad (T-\lambda I)x) \geqslant \operatorname{Im}^2 \quad \lambda(x,x)$ and therefore $\|(T-\lambda I)^{-1}\| \leqslant 1/|\operatorname{Im}\lambda|$. Consequently λ is a point of regular type of the operator T. Therefore, if $\mathbb{C}^- \cap \rho(T) \neq \emptyset$, then T is a maximal dissipative operator. In the opposite case $\lambda \in \mathbb{C}^- \cap \rho(T)$ would be an eigenvalue of a dissipative extension $\tilde{T} \supset T$. To complete the proof of the lemma we verify that if T is a maximal dissipative operator, then $\mathbb{C}^- \subset \rho(T)$. For, suppose $x_0 \in \mathscr{R}^{\perp}_{T-\lambda I}$ for some $\lambda \in \mathbb{C}^-$. If $x_0 \neq \theta$, then the vector $\langle x_0, \bar{\lambda} x_0\rangle$ is positive by the form (1.4) and is J_Γ-orthgonal to the maximal (by virtue of Theorem 2.6) non-negative subspace Γ_T, which is impossible, and therefore $x_0 = \theta_1$ i.e., $\bar{\mathscr{R}}_{T-\lambda I} = \mathscr{H}$. Since $(T-\lambda \mathrm{I})^{-1}$ is bounded and T is a closed operator, so $\lambda \in \rho(T)$, i.e., $\mathbb{C}^- \subset \rho(T)$. ■

We now suppose that a maximal uniformly positive subspace $\mathscr{L}$ is contained in the domain of definition of the operator T. By Lemma 9.5 of Chapter 1 this supposition is always fulfilled if $\mathscr{H}$ is a Pontryagin space Π_x. Taking account of §7.6 of Chapter 1 we shall, without loss of generality, suppose that $\mathscr{L} = \mathscr{H}^+$. Then, relative to the canonical decomposition $\mathscr{H} = \mathscr{H}'' \oplus \mathscr{H}^-$, the operator T is expressible in matrix form

$$T = \| T_{ij} \|_{i,j=1}^{2} \tag{2.1}$$

where

$$T_{11} = P^+TP^+ \mid \mathscr{H}^+,\ T_{12} = P^+TP^- \mid \mathscr{H}^-,\ T_{21} = P^-TP^+ \mid \mathscr{H}^+,$$
$$T_{22} = P^-TP^- \mid \mathscr{H}^-.$$

For example,

$$J = \| J_{ij} \|_{i,j=1}^{2},\ J_{11} = I^+,\ J_{12} = 0,\ J_{21} = 0,\ J_{22} = -\mathrm{I}^-.$$

Theorem 2.9: *Let* A *be a closed* J*-dissipative operator,* $\mathscr{H}^+ \subset \mathscr{D}_A$. *Then the following conditions are equivalent*:

a) *A is a maximal J-dissipative operator;*
b) $-A_{22}$ *is a maximal dissipative operator in* $\mathscr{H}^-$;
c) $\{\lambda \mid \operatorname{Im}\lambda > 2\|AP^+\|\} \cap \rho(\mathrm{A}) \neq \emptyset$;
d) $\{\lambda \mid \operatorname{Im}\lambda > 2\|AP^+\|\} \subset \rho(A)$.

If these conditions are satisfied, then $\|(A-\lambda I)^{-1}\| = O(1/\operatorname{Im}\lambda)$ $(\operatorname{Im}\lambda \to \infty)$.

□ Let A be a maximal J-dissipative operator. By Proposition 2.3, JA is a maximal dissipative operator. It is easily verified that, since JA is dissipative, $(JA)_{22} = -A_{22}$ is also dissipative. We prove that it is maximal, by checking that $\mathbb{C}^- \cap \rho(-A_{22}) \neq \emptyset$ (see Lemma 2.8), or, what comes to the same thing, that $\mathbb{C}^+ \cap \rho(A_{22}) \neq \emptyset$. Since $JAP^- + \lambda I = (I - JAP^+(JA + \lambda I)^{-1})(JA + \lambda I)$ when $\lambda \in \mathbb{C}^+$, so, when $\mathrm{Im}\,\lambda > \| AP^+ \|$ we have (see Lemma 2.8) $-\lambda \in \rho(JAP^-)$, and since $(-A_{22} + \lambda I^-)^{-1} = P^-(JAP^- + \lambda I)^{-1}P^- \mid \mathscr{H}^-$, so $\lambda \in \rho(A_{22})$.

Now let $(-A_{22})$ be a maximal dissipative operator in $\mathscr{H}^-$. Since $\mathscr{H}^+ \subset \mathscr{D}_A$, every non-trivial J-dissipative extension of the operator A would imply a non-trivial dissipative extension of the operator $(-A_{22})$, and therefore A is a maximal J-dissipative operator and AJ is a maximal dissipative operator. Therefore, from the equality

$$A - \lambda I = -(I - 2AP^+(AJ + \lambda I)^{-1})(AJ + \lambda I)$$

(which is proper because $J\mathscr{D}_A = \mathscr{D}_A$), we obtain, taking account of Lemma 2.8, that, if $\mathrm{Im}\ \lambda > 2\| AP^+ \|$, then $\lambda \in \rho(A)$, and moreover $\| (A - \lambda I)^{-1} \| = 0(1/\mathrm{Im}\ \lambda)$ when $\mathrm{Im}\ \lambda \to \infty$.

Thus we have verified the following implications: a) $\Leftrightarrow$ b) $\Rightarrow$ d). The implication d) $\Leftrightarrow$ c) is trivial. We verify that c) $\Rightarrow$ a). If $\mathrm{Im}\ \lambda > 2\| AP^+ \|$ and $\lambda \in \rho(A)$, then A cannot admit closed J-dissipative extensions $\widetilde{A}_1$ for otherwise we would have $\lambda \in \sigma_p(\widetilde{A})$, which contradicts the implication a) $\Rightarrow$ d), which has already been proved. ■

2 *Definition 2.10:* Let T: $\mathscr{H}_1 \to \mathscr{H}_2$ and S:$\mathscr{H}_1 \to \mathscr{H}_1$ be two linear operators. The operator T is said to be S-bounded (or *S-continuous*) if even for one point $\lambda_0 \in p(S)$ (but then, as follows from Hilbert's identity for the resolvent, also for all points) the operator $T(S - \lambda_0 I)^{-1}$ is bounded and defined everywhere. If, in addition, $T(S - \lambda_0 I)^{-1}$ is a completely continuous operator ($T(S - \lambda_0 I)^{-1} \in \mathscr{S}_\infty$), then the operator T is said to be *S-completely continuous*.

In particular, as is easily seen, *bounded* (*respectively, completely continuous*) *operators A with* $\mathscr{D}_A = \mathscr{H}_1$ *are S-bounded* (*respectively, S-completely continuous*) *for any operator S with* $\rho(S) \neq \emptyset$.

A closed operator T: $\mathscr{H}_1 \to \mathscr{H}_2$ is called a Φ-*operator* if $\mathscr{R}_T = \overline{\mathscr{R}}_T$, $\dim \mathrm{Ker}\ T < \infty$ and $\dim \mathscr{R}_T^\perp < \infty$; and the number $\mathrm{ind}\ T \equiv \dim \mathscr{R}_T^\perp - \dim \mathrm{Ker}\ T$ is called its *index*. In particular, if $\mathrm{ind}\ T = 0$, then such an operator is called a Φ_0-*operator*. Let A: $\mathscr{H} \to \mathscr{H}$, $\lambda_0 \in \mathbb{C}$, and let $T = A - \lambda_0 I$ be a Φ_0-operator; then the point λ_0 will be called a Φ_0-*point* of the operator A.

We observe that, since $\mathscr{R}_T^\perp = \mathrm{Ker}\ T^*$ when $\overline{\mathscr{D}}_T = \mathscr{H}_1$, the definition of a Φ-operator can in this case be given formally in a different way: namely, by replacing $\mathscr{R}_T^\perp$ in the definition by $\mathrm{Ker}\ T^*$. We shall use both these definitions in future.

We note also that if A: $\mathscr{H} \to \mathscr{H}$ and if λ_0 is a normal point of the operator A, then λ_0 is a Φ_0-point of this operator. In particular, the Φ_0-points of any completely continuous operator fill the set $\mathbb{C} \setminus \{\theta\}$.

□ We state without proof a theorem made up of several results form the survey [X] which are applicable to our present interests.

Theorem 2.11: *Let $T_1: \mathscr{H}_1 \to \mathscr{H}_2$ and $T_2: \mathscr{H}_2 \to \mathscr{H}_3$ be Φ-operators, and $\bar{\mathscr{D}} T_2 = \mathscr{H}_2$. Then $T_2 T_1: \mathscr{H}_1 \to \mathscr{H}_3$ is a Φ-operator and* ind $T_2 T_1 =$ ind T_1 + ind T_2. *If also $T_3: \mathscr{H}_1 \to \mathscr{H}_2$ is a completely continuous operator, then $T_1 + T_3: \mathscr{H}_1 \to \mathscr{H}_2$ is a Φ-operator and* ind$(T_1 + T_3) =$ ind T_1.

Let $T: \mathscr{H} \to \mathscr{H}$ be a Φ_0-operator. If T is an S-completely continuous operator, then $T + S$ is also a Φ_0-operator. The set of Φ_0-points of the operator T is open. If Ω is a connected component of this set and if $\Omega \cap \rho(T) \neq \emptyset$, then $\Omega \subset \tilde{\rho}(T)$. ■

Corollary 2.12: *If A is a closed maximal J-dissipative operator, $\mathscr{H}^+ \subset \mathscr{D}_A$, then A_{12} is on A_{22}-continuous operator. If also $A_{11} - A_{11}^* \in \mathscr{S}_\infty$ and A_{12} is an A_{22}-continuous operator, then $\mathbb{C}^+ \subset \tilde{\rho}(A)$.*

□ Let $\lambda \in \rho(A) \cap \mathbb{C}^+$. Then $AP^- - \lambda I = [I - AP^+(A - \lambda I)^{-1}](A - \lambda I)$, and since by Theorem 2.9 $\|(A - \lambda I)^{-1}\| = 0(1/\text{Im } \lambda)$ when Im $\lambda \to \infty$, so $\mathbb{C}^+ \cap \rho(AP^-) \neq \emptyset$. This implies, by Lemma 2.8, the inclusion $\mathbb{C}^+ \subset \rho(AP^-)$. Since $[(AP^- - \lambda I)^{-1}]_{12} = (1/\lambda) A_{12}(A_{22} - \lambda I^-)^{-1}$, we have, by Definition 2.10, that A_{12} is an A_{22}-continuous operator.

Let $\lambda_0 \in \mathbb{C}^+$. Since $A_{11} - \lambda_0 I^+ = (\frac{1}{2}(A_{11} + A_{11}^*) - \lambda_0 I^+) + \frac{1}{2}(A_{11} - A_{11}^*)$ and $\lambda_0 \in \rho(\frac{1}{2}(A_{11} + A_{11}^*))$ but $A_{11} - A_{11}^* \in \mathscr{S}_\infty$, so, by Theorem 8.11, λ_0 is a Φ_0-point of the operator A_{11}. Moreover, since A_{11} is a bounded operator, $\mathbb{C}^+ \cap \rho(A_{11}) \neq \emptyset$, and therefore $\mathbb{C}^+ \subseteq \tilde{\rho}(A_{11})$. We now represent the operator $A - \lambda_0 I$ by the expression

$$A - l_0 I = \left(\begin{Vmatrix} A_{11} - \lambda_0 I^+ & 0 \\ A_{21} & I^- \end{Vmatrix} + \begin{Vmatrix} 0 & A_{12}(A_{22} - \lambda_0 I^-)^{-1} \\ 0 & 0 \end{Vmatrix} \right) \times \begin{Vmatrix} I^+ & 0 \\ 0 & A_{22} - \lambda_0 I^- \end{Vmatrix}.$$

The first term in the round brackets and the second factor are Φ_0-operators, and the second term in the round brackets is a completely continuous operator. It follows from Theorem 2.11 that $A - \lambda_0 I$ is a Φ_0-operator, and from Theorem 2.9 it follows that $\mathbb{C}^+ \cap \rho(A) \neq \emptyset$, and therefore (see Theorem 2.11) $\mathbb{C}^+ \subset \tilde{\rho}(A)$. ■

Corollary 2.13: *A closed π-dissipative operator A in $\Pi_\varkappa = \Pi_+ \oplus \Pi^-$ is maximal if and only if $\mathbb{C}^+ \cap \rho(A) \# \emptyset$, and in that case $\mathbb{C}^+ \subset \tilde{\rho}(A)$.*

□ Let A be a closed maximal π-dissipative operator. It follows from Proposition 2.7 that $\bar{\mathscr{D}}_A = \Pi_\varkappa$ and therefore (see 1, Lemma 9.5) we can assume without loss of generality that $\Pi_+ \subset \mathscr{D}_A$. By Corollary 2.12 we have that A_{12} is an A_{22}-continuous operator. Since the operators A_{11} and $A_{12}(A_{22} - \lambda_0 I^-)^{-1}(\lambda_0 \subset \rho(A_{22}))$ are finite-dimensional and continuous, they are completely continuous, and therefore $\mathbb{C}^+ \in \rho(A)$.

Conversely, let $\lambda_0 \in \mathbb{C}^+ \cap \rho(A)$. Then from the equality

$$AP^- - \lambda_0 I = (I - AP^+(A - \lambda_0 I)^{-1})(A - \lambda_0 I)$$

and the facts that $AP^+(A-\lambda_0 I)^{-1}$ is finite-dimensional and that $\mathbb{C}^- \cap \sigma_p(AP^-) = \emptyset$ it follows that $\lambda_0 \in \rho(AP^-)$, and therefore $\lambda_0 \in \rho(A_{22})$. It remains to apply Lemma 2.8 and Theorem 2.9, from which it follows that A is a maximal π-dissipative operator. ■

3 In this paragraph our main purpose is to study the structure of root lineals of dissipative operators in indefinite spaces.

With every densely defined linear operator $T: \mathcal{H} \to \mathcal{H}$ we associate the pair of operators

$$T_R = \tfrac{1}{2}(T + T^c) \quad \text{and} \quad T_I = \frac{1}{2i}(T - T^c)$$

which we shall call the *J-real part* and the *J-imaginary part* of this operator respectively. We note that $\mathcal{D}_{T_R} = \mathcal{D}_{T_I} = \mathcal{D}_T \cap \mathcal{D}_{T^c}$, and if $x \in \mathcal{D}_T \cap \mathcal{D}_{T^c}$, then

$$\operatorname{Im}[Tx, x] = [T_I x, x] \tag{2.2}$$

□ Consequently the following holds:

2.14 If $\mathcal{D}_A \subset \mathcal{D}_{A^c}$, then the operator A is J-dissipative if and only if $[A_I x, x] \geqslant 0$ $(x \in \mathcal{D}_A)$. In particular, this is true if A is a continuous operator defined everywhere in $\mathcal{H}$. ■

Theorem 2.15: *Let A be a J-dissipative operator, $\overline{\mathcal{D}}_A = \mathcal{H}$. Then the relations $\operatorname{Im}[Ax_0, x_0] = 0$ and $x_0 \in \operatorname{Ker} A_I$ (i.e., $Ax_0 = A^c x_0$) are equivalent.*

□ If $x_0 \in \operatorname{Ker} A_I$, then $\operatorname{Im}[Ax_0, x_0] = 0$ by 12.2).

Now suppose $\operatorname{Im}[Ax_0, x_0] = 0$. Then the vector $\langle x_0, Ax_0 \rangle$ of the non-negative subspace Γ_A is neutral relative to the form (1.4), and therefore it is isotropic in Γ_A (see 1, Proposition 1.17). It only remains to use Corollary 1.4. ■

Theorem 2.15 gives a basis for introducing the following notation in the case of an arbitrary W-dissipative operator A:

$$\operatorname{Ker} A_I = \{x \mid \operatorname{Im}[Ax, x] = 0\}, \tag{2.3}$$

although the symbol A_I has no meaning in such a general case.

From Theorem 2.15 we obtain

Corollary 2.16: *Let A be a J-dissipative operator, $\overline{\mathcal{D}}_A = \mathcal{H}$. Then $\operatorname{Ker}(A - \lambda I) \subset \operatorname{Ker}(A^c - \lambda I) \cap \operatorname{Ker} A_I$ for all $\lambda = \bar{\lambda}$.*

□ If $x \in \operatorname{Ker}(A - \lambda I)$, then $\operatorname{Im}[Ax, x] = \operatorname{Im} \lambda[x, x] = 0$, and therefore our assertion follows from Theorem 2.15 and the definition of the operator A_I. ■

Corollary 2.17: *If A is a maximal closed J-dissipative operator, then* $\operatorname{Ker}(A-\lambda I)=\operatorname{Ker}(A^c-\lambda I)$ *when* $\lambda=\bar{\lambda}$ *and therefore the residual spectrum of the operator A contains no real points.*

□ By Proposition 2.7, A and $(-A^c)$ are maximal J-dissipative operators simultaneously. Since $A^{cc}=A$ (see Proposition 1.8), it follows from Corollary 2.16 that the kernals of $(A-\lambda I)$ and $(A^c-\lambda I)$ coincide: $\operatorname{Ker}(A-\lambda)=\operatorname{Ker}(A^c-\lambda I)$ when $\lambda=\bar{\lambda}$, and from Theorem 1.16 b) that the operator A has no real residual spectrum. ■

□ We observe that the the set introduced in formula (2.3) coincides, as may easily be seen, with the set $\{x\in\mathscr{D}_A \mid [Ax,y]=[x,Ay] \text{ for all } y\in\mathscr{D}_A\}$. Therefore, carrying out an argument analogous to that used in Exercise 8 to §1 we obtain

Corollary 2.18: *Let A be a W-dissipative operator, let $\mathscr{L}\subset\mathscr{L}_\lambda(A)\cap\operatorname{Ker}A_I$, and $A\mathscr{L}\subset\mathscr{L}$. Then the lineal $\mathscr{L}$ is W-orthogonal to all $\mathscr{L}_\mu(A)$ with $\mu\neq\bar{\lambda}$.* ■

Lemma 2.19: *Let $\mathscr{H}^0$ be an isotropic subspace of a W-space $\mathscr{H}$, let A be a W-dissipative operator, $\bar{\mathscr{D}}_A=\mathscr{H}$, and $\mathscr{H}^0\subset\mathscr{D}_A$. Then $A\mathscr{H}^0\subset\mathscr{H}^0$.*

□ It is easy to see that $\mathscr{H}^0\subset\operatorname{Ker}A_I$. Therefore $[Ax,y]=[x,Ay]$ for any $x\in\mathscr{H}^0$ and $y\in\mathscr{D}_a$. Since $\bar{\mathscr{D}}_A=\mathscr{H}$, $Ax\in\mathscr{H}^0$. ■

4 Before proceeding with the exposition of the properties of dissipative operators in spaces with an indefinite metric we recall the definition of a Riesz projector and some of its properties.

Let T be a linear operator in an arbitrary Hilbert space $\mathscr{H}$, let σ be its *bounded spectral set* (with the meaning given in [VI]) i.e., the isolated part of the spectrum (or, in particular, the whole spectrum) of the operator T, and let Γ_σ be a Jordan closed rectifiable contour lying in $\rho(T)$ and containing σ, and suppose $\sigma(T)\backslash\sigma$ lies outside this contour. Then the operator

$$P_\sigma=-\frac{1}{2\pi i}\oint_{\Gamma_\sigma}(\mathrm{T}-\lambda I)^{-}\,\mathrm{d}\lambda_1 \tag{2.4}$$

where the integral is understood to be the strong limit of the integral sums, is called a *Riesz projector*.

We enumerate the basic properties of the integral (22).

Theorem 2.20: *The following assertions hold:*

a) *the operator P_σ does not depend on the choice of the contour Γ_σ isolating the set σ, and it is a projector;*
b) $P_\sigma\mathscr{H}\subset\mathscr{D}_T$;

c) *the subspaces* $P_\sigma\mathcal{H}$ *and* $(I-P_\sigma)\mathcal{H}$ *are invariant relative to the operator* T, *and also* $\sigma(T\mid P_\sigma\mathcal{H})=\sigma$, $\sigma(T\mid(I-P_\sigma)\mathcal{H})=\sigma(T)\backslash\sigma$. *In particular, if* $\sigma=\{\lambda_0\}$ *is a normal ergenvalue of the operator* T, *then* $P_\sigma\mathcal{H}=\mathcal{L}_{\lambda_0}(T)$;

d) *if* T *is a bounded operator and* $\sigma=\sigma(T)$, *then* $P_\sigma=I$;

e) *if* σ_1 *and* σ_2 *are bounded spectral sets of the operator* T *and* $\sigma_1\cap\sigma_2=\emptyset$, *then* $P_{\sigma_1\cup\sigma}=P_{\sigma_1}+P_{\sigma_2}$ *and* $P_{\sigma_1}P_{\sigma_2}=P_{\sigma_2}P_{\sigma_1}=0$;

f) *if* σ *is a bounded spectral set of the operator* T, *and* $\mathcal{L}$ *is an invariant subspace of this operator, and* $\sigma(T\mid\mathcal{L})\subset\sigma$, *then* $\mathcal{L}\subset P_\sigma\mathcal{H}$.

□ Proofs of these assertions can be found, eg, in [VI]. ■

Theorem 2.21: *Let* $\mathcal{H}$ *be a W-space, A a closed W-dissipative operator,* $\sigma(\subset\mathbb{C}^+(\mathbb{C}^-))$ *be a bounded spectral set of A, and let* P_σ *be the corresponding Riesz projector. Then* $P_\sigma\mathcal{H}$ *is a non-negative (non-positive) subspace which is invariant relative to A.*

□ The proof will be carried out for the case $\sigma\subset\mathbb{C}^+$, since the case $\sigma\subset\mathbb{C}^-$ is treated exactly analogously. Without loss of generality we can assume, by virtue of assertions b) and c) in Theorem 2.20, that $A:P_\sigma\mathcal{H}\to P_\sigma\mathcal{H}$ is a bounded operator and $\sigma(A)=\sigma$. Then from assertion d) of the same theorem we have $P_\sigma=I$. Therefore, $[x,x]=[P_\sigma x,x]$ and it suffices for us to verify the inequality $\operatorname{Re}[P_\sigma x,x]\geqslant 0$. To do this we choose as Γ_σ the contour consisting of a segment $[-a,a]$ of the real axis and a semicircle $\{ae^{i\varphi}\}_{\varphi=0}^{\pi}$ with a sufficiently large $a>0$. Then

$$P_\sigma=-\frac{1}{2\pi i}\int_{-a}^{a}(A-\alpha I)^{-1}\,d\alpha+\frac{1}{2\pi}\int_0^{\pi}\left(I-\frac{e^{-i\varphi}}{a}A\right)^{-1}d\varphi.$$

Since P_σ does not depend on the contour Γ_σ separating σ, the limit $\lim_{a\to\infty}P_\sigma=P_\sigma(=I)$ exists. Since

$$\lim_{a\to\infty}\frac{1}{2\pi}\int_0^{\pi}\left(I-\frac{e^{-i\varphi}}{a}A\right)^{-1}d\varphi=\tfrac{1}{2}I,$$

so there exists also

$$\lim_{a\to\infty}-\frac{1}{2\pi i}\int_{-a}^{a}(A-\alpha I)^{-1}\,d\alpha=\frac{i}{2\pi}\int_{-\infty}^{\infty}(A-\alpha I)^{-1}\,d\alpha=\tfrac{1}{2}I.$$

Therefore

$$\operatorname{Re}[P_\sigma x,x]=$$

$$2\operatorname{Re}\left[\frac{i}{2\pi}\int_{-\infty}^{\infty}(A-\alpha I)^{-1}\,d\alpha x,x\right]=-\frac{1}{\pi}\int_{-\infty}^{\infty}\operatorname{in}[(A-\alpha I)^{-1}x,x]\,d\alpha\geqslant 0.$$

The last inequality follows because A is a W-dissipative operator. ■

Corollary 2.22: *Let $\Lambda \subset \mathbb{C}^+ (\mathbb{C}^-)$ be a certain set of eigenvalues of a W-dissipative operator A. Then C Lin$\{\mathscr{L}^\lambda(A)\}_{\lambda \subseteq \Lambda}$ is a non-negative (non-positive) subspace.*

□ As in the proof of Theorem 2.21 we assume for definiteness that $\Lambda \subset \mathbb{C}^+$, and by virtue of I 3.6 it suffices to verify that Lin$\{\mathscr{L}_\lambda(A)\}_{\lambda \in \Lambda}$ is a non-negative lineal. Let $x \in$ Lin$\{\mathscr{L}_\lambda(A)\}_{\lambda \in \Lambda}$. Then there are $\lambda_1, \lambda_2, \ldots, \lambda_n \in \Lambda$, and positive integers $p_1, p_2, \ldots, p_n$, and vectors $x_1, x_2, \ldots, x_n$ such that $x = \Sigma_{i=1}^n x_i$, $(A - \lambda_i I)^{p_i} x_i = \theta$. The subspace $\mathscr{L} = \text{Lin}\{x_i, (A - \lambda_i)x_i, \ldots, (A - \lambda_i \text{I})^{p_i - 1} x_i\}_1^n$ is finite-dimensional, is contained in $\mathscr{D}_A$ and is invariant relative to the operator A; moreover, $\sigma(A \mid \mathscr{L}) = \{\lambda_i\}_1^n \subset \mathbb{C}^+$.

By Theorem 2.21 the subspace $\mathscr{L}$ is non-negative, and therefore the vector $x \in \mathscr{L}$ is also non-negative. ■

Corollary 2.23: *Let A be a maximal π-dissipative operator in* $\Pi_\varkappa$, with $\bar{\mathscr{D}}_A = \Pi_\varkappa$. *The $\sigma(A) \cap \mathbb{C}^+$ consists of not more than $\varkappa$ (taking algebriac multiplicity into account) normal eigenvalues.*

□ The proof of this assertion follows directly from a comparison of Corollaries 2.13 and 2.22 with Proposition 9.2 in Chapter I. ■

Theorem 2.24: *If A is a closed J-dissipative operator, if $\sigma^\pm (\subset \mathbb{C}^\pm)$ are its bounded spectral sets, and if $\Gamma_{\sigma^\pm} \subset \mathbb{C}^\pm$ and $\Gamma_{\sigma^+} = \Gamma_{\sigma^-}^* (\equiv \{\lambda \mid \bar{\lambda} \in \Gamma_{\sigma^-}\})$, then the subspace $\text{P}_{\sigma^- \cup \sigma^-}\mathscr{H}$ is non-degenerate.*

□ Let $\mathscr{H}^0$ be an isotropic subspace of the subspace $P_{\sigma^+ \cup \sigma^-}\mathscr{H}$. By Theorem 2.20 $P_{\sigma^+ \cup \sigma^-}\mathscr{H} \subset \mathscr{D}_A$, and even more so $\mathscr{H}^0 \subset \mathscr{D}_A$, and therefore (see Lemma 2.19) $\text{A}\mathscr{H}^0 \subset \mathscr{H}^0$. Since $\mathscr{H}^0 \subset \text{Ker A}_\text{I}$ and, as is easy to see, $\Gamma_{\sigma^+ \cup \sigma^-} \subset \rho(A \mid \mathscr{H}^0)$, so $(A - \lambda I)^{-1} \mid \mathscr{H}^0 = [(A - \bar{\lambda}\text{I})^{-1}]^c \mid \mathscr{H}^0$ when $\lambda \in \Gamma_{\sigma^- \cup \sigma^-}$.

(For, $[Ax, y] = [x, Ay]$ for any $x \in \mathscr{H}^0$ and $y \in \mathscr{D}_A$. Therefore

$$[(A - \lambda I)x, y] = [x, (A - \bar{\lambda} I)y].$$

Now let

$$x = (A - \lambda I)^{-1} z \; (z \in \mathscr{H}^0) \quad \text{and} \quad y = (A - \bar{\lambda} I)^{-1} w \; (\text{w} \in \mathscr{H}).$$

Then

$$[z, (\text{A} - \bar{\lambda} I)^{-1} w] = [(A - \lambda I)^{-1} z, w].$$

Since this is true for any $w \in \mathscr{H}$, we have $[(A - \bar{\lambda} I)^{-1}]^c z = (A - \lambda I)^{-1} z$ for any $z \in \mathscr{H}^0$.)

Since

$$P^c_{\sigma^+ \cup \sigma^-} = -\frac{1}{2\pi i} \oint_{\Gamma_{\sigma^- \cup \sigma^-}} [(A - \bar{\lambda} I)^{-1}]^c \, \text{d}\lambda,$$

so

$$P^c_{\sigma^+ \cup \sigma^-} \mid \mathscr{H}^0 = P_{\sigma^+ \cup \sigma^-} \mid \mathscr{H}^0 = I \mid \mathscr{H}^0.$$

Hence $[x, x_0] = [x, P^c_{\sigma^+ \cup \sigma^-} x_0] = [P_{\sigma^+ \cup \sigma^-} x, x_0] = 0$ for any $x \in \mathscr{H}$ and $x_0 \in \mathscr{H}^0$, i.e., $\mathscr{H}^0$ is isotropic in the whole of $\mathscr{H}$. Therefore $\mathscr{H}^0 = \{\theta\}$. ■

Corollary 2.25: *Let A be a J-dissipative operator, and let the non-real λ and $\bar{\lambda}$ be its normal points. Then $\mathscr{L} = \text{Lin}\{\mathscr{L}_\lambda(A), \mathscr{L}_{\bar{\lambda}}(A)\}$ is a projectionally complete subspace.*

□ The proof follows immediately from the facts that $\mathscr{L}$ is finite-dimensional (by the definition of a normal point) and is non-degenerate (by Theorems 2.20 and 2.24) taken in conjunction with 1, Corollary 7.18. We note that in the conditions of the corollary it is not excluded, for example, that $\lambda \in \rho(A)$, and in that case $\mathscr{L}_\lambda(A) = \{\theta\}$.

5 Now let $\mathscr{H} = \Pi_\varkappa$ be a Pontryagin space. We investigate the structure of root lineals of π-dissipative operators.

Theorem 2.26: *If A is a closed π-dissipative operator and $\alpha = \bar{\alpha} \in \sigma_p(A)$, then the root lineal $\mathscr{L}_\alpha(A)$ is expressible in the form $\mathscr{L}_\alpha(A) = \mathscr{N}_\alpha [\dot{+}] \mathscr{M}_\alpha$, where $\dim \mathscr{N}_\alpha < \infty$, $A\mathscr{N}_\alpha \subset \mathscr{N}_\alpha$, $\mathscr{N}_\alpha \subset \text{Ker}\,(A - \alpha I)$ and $\mathscr{M}_\alpha$ is a non-degenerate subspace or, in particular, $\mathscr{M}_\alpha = \{\theta\}$. Moreover, if $d_1(\alpha), d_2(\alpha), \ldots, d_{r_\alpha}(\alpha)$ are the orders of the elementary divisors of the operator $A \mid \mathscr{N}_\alpha$, then*

$$\sum_{\alpha = \bar{\alpha} \in \sigma_p(A)}^{r_\alpha} \sum_{i=1}^{r_\alpha} \left[\frac{1}{2} d_i(\alpha)\right] + \sum_{\text{Im}\,\lambda > 0} \dim \mathscr{L}_\lambda(A) \leqslant \varkappa. \tag{2.5}$$

□ First of all we note that, by virtue of Theorem 2.6, A can be assumed, without loss of generality, to be a maximal closed π-dissipative operator, and therefore $\bar{\mathscr{D}}_A = \Pi_\varkappa$.

Since every non-negative subspace in $\Pi_\varkappa$ has a dimension not exceeding $\varkappa$ (see I.9.2) it suffices in proving our theorem to verify that the indicated decomposition of $\mathscr{L}_\alpha(A)$ exists, and that in every $\mathscr{N}_\alpha$ there is a neutral subspace, invariant relative to A, of dimension $\sum_{i=1}^{r_\alpha} [\frac{1}{2} d_i(\alpha)]$, and then to use Corollaries 2.18 and 2.22.

We consider the decomposition of $\text{Ker}(A - \alpha I)$ into its isotropic subspace $\mathscr{L}^1_\alpha$ and a non-degenerate subspace $\mathscr{M}_\alpha$: $\text{Ker}(A - \alpha I) = \mathscr{L}^1_\alpha [\dot{+}] \mathscr{M}_\alpha$. From Theorem 1.9.9. it follows that $\mathscr{M}_\alpha$ is a projectionally complete subspace, and therefore $\Pi_k = \mathscr{M}_\alpha [\dot{+}] \Pi'_{\varkappa_1}$, where $\Pi'_{\varkappa_1} = \mathscr{M}^{[\perp]}_\alpha$ is again a Pontryagin space with $\varkappa_1 \leqslant \varkappa$. Using Corollary 2.16 and Proposition 1.11 we obtain that $\Pi'_{\varkappa_1}$ is invariant relative to the operator A, and therefore it sufficies for us to prove the theorem under the assumption that $\text{Ker}(A - \alpha I) = \mathscr{L}^1_\alpha$, i.e., $\text{Ker}(A - \alpha I)$ is a neutral (and therefore a finite-dimensional) subspace. We introduce the notation $\mathscr{L}_\alpha^p = \mathrm{Ker}(A - \alpha I)^p$ $(p = 1, 2, \ldots)$. Since $\mathscr{L}_\alpha^1$ is finite-dimensional, it follows that all the $\mathscr{L}_\alpha^p$ $(p = 1, 2, \ldots)$ are finite-dimensional. We verify that, if $d_1^{(p)}, d_2^{(p)}, \ldots, d_{s_p}^{(p)}$ $(d_i^{(p)} \leqslant d_j^{(p)}$ when $i \leqslant j)$ are the orders of the elementary divisors of the operator $A \mid \mathscr{L}_\alpha^{p}$, then a neutral subspace of dimension $\Sigma_{i=1}\, [\frac{1}{2}d_i^{(p)}]$ exists in $\mathscr{L}_\alpha^p$.

From Theorem 2.15 it follows immediately that, if $(A - \alpha I)^l x = \theta$, then

$$(A - \alpha I)^p x \in \mathscr{D}(A^c - \alpha I)^{l-p}$$

and

$$(A^c - \alpha I)^{l-p}(A - \alpha I)^p x = \theta \ (p = [\tfrac{1}{2}l], \ldots, l-1).$$

Let the vectors x_i be such that

$$(A - \alpha I)^{d_i^{(p)}} x_i = \theta \ (i = 1, 2, \ldots, s_p),$$
$$\mathscr{L}_\alpha^p = \mathrm{Lin}\{x_i, (A - \alpha I)x_i, \ldots, (A - \alpha I)^{d_i^{(p)}-1} x_i\}_1^{s_p}$$

then

$$\mathscr{L} = \mathrm{Lin}\{(A - \alpha I)^{[(d_i^{(p)}+1)/2]} x_i, \ldots, (A - \alpha I)^{d_i^{(p)}-1} x_i\}_1^{s_p}$$

is a neutral subspace. For, suppose that

$$\left[\frac{d_i^{(p)}+1}{2}\right] \leqslant q_i \leqslant d_i^{(p)} - 1 \quad \text{and} \quad \left[\frac{d_i^{(p)}+1}{2}\right] \leqslant q_j \leqslant d_j^{(p)} - 1 \qquad (i \leqslant j)$$

Then $q_j \geqslant d_i^{(p)} - q_i$, and therefore

$$[(A - \alpha I)^{q_i} x_i, (A - \alpha I)^{q_i} x_i] =$$
$$[(A - \alpha I)^{q_i} x, (A - \alpha I)^{d_i^{(p)} - q_i}(A - \alpha I)^{q_j - (d_i^{(p)} - q_i)} z_i]$$
$$= [(A^c - \alpha I)^{d_i^{(p)} - q_i}(A - \alpha I)^{q_i} x_i, (A - \alpha I)^{q_j - (d_i^{(p)} - q_i)} x_j] = 0 \quad ■$$

Remark 2.27: The decomposition $\mathscr{L}_\alpha(A) = \mathscr{N}_\alpha [\dot{+}] \mathscr{M}_\alpha$ can be chosen in various ways, but the number of non-prime elementary divisors of a given order r of the operator $A \mid \mathscr{N}_\alpha$ is an invariant of the operator A and the number α for any choice of $\mathscr{N}_\alpha$, because it is the same as the number of elementary divisors of order $r - 1$ of the operator generated by the operator A in the factor-space $\mathscr{L}_\alpha(A)/\mathrm{Ker}(A - \alpha I)$, and the latter is finite-dimensional and does not depend on the choice of $\mathscr{N}_\alpha$. ■

Corollary 2.28: *For a closed π-dissipative operator A in $\Pi_\varkappa$ all the root subspaces corresponding to the real eigenvalues are, except for not more than $\varkappa$ of them, negative eigen-subspaces.*

□ Let $\{\alpha\}$ be the set of those real eigenvalues for which there is in $\mathrm{Ker}(A - \alpha I)$ at least one non-negative vector. It follows from Theorem 2.26 that, in particular, all those α for which $\mathscr{L}_\alpha(A) \neq \mathrm{Ker}(A - \alpha I)$ enter into this set, and from Corollary 2.18 it follows that the set $\{\alpha\}$ is finite and consists of not more than $\varkappa$ elements. ■

Corollary 2.29: *Let $\mathscr{H}$ be a $W^{(\varkappa)}$-space (i.e., in accordance with Proposition I.9.15, it is assumed that the set $\sigma|w^{(\varkappa)} \cap [0, \infty)$ contains precisely $\varkappa(<\infty)$ (taking algebriac multiplicity into account) eigenvalues, let $\mathscr{R}_{W^{(\varkappa)}} = \overline{\mathscr{R}}_{W^{(\varkappa)}}$, and let A be a $W^{(\varkappa)}$-dissipative operator defined on $\mathscr{H}$. Then the assertion of Theorem 2.26 holds for the operator A.*

□ Let dim Ker $W^{(\varkappa)} = \varkappa_0 (\leqslant \varkappa)$. By Lemma 2.19 A Ker $W^{(\;)}_{\varkappa} \subset$ Ker $W^{(\;)}_{\varkappa}$, and therefore in the Pontryagin $\Pi_{\varkappa - \varkappa_0} = \mathscr{H}/\text{Ker } W^{(\varkappa)}$ (see Corollary I.9.16) the π-dissipative operator $A^\wedge$ generated by the operator A is properly defined by the formula (1.11). Theorem 2.26 holds for the operator $A^\wedge$. It is not difficult to see that the left-hand side of the inequality (2.5), which was set up for the operator A exceeds the corresponding sum for the operator $A^\wedge$ by not more than $\varkappa_0$ units, and therefore the assertion of Theorem 2.26 is true also for the operator A. ■

6 *Definition 2.30:* An operator A is said to be *strictly J-dissipative* if $\text{Im}[Ax, x] > 0$ for $(\theta \neq)x \in \mathscr{D}_A$, and, in particular, to be *uniformly J-dissipative* if there is a constant $\gamma_A > 0$ such that $\text{Im}[Ax, x] \geqslant \gamma_A \|x\|^2$ $(x \in \mathscr{D}_A)$.

Theorem 2.31: *Every closed strictly or uniformly J-dissipative operator A can be extended into a maximal closed J-dissipative operator $\tilde{A}$ which is also respectively strictly or uniformly J-dissipative, and moreover the latter can be chosen so that $\gamma_{\tilde{A}} = \gamma_A$.*

□ We consider the graph Γ_A of the operator A. In both cases, by the form (1.4) it is a positive subspace, and therefore (see Remark 1.8.9) Γ_A admits extension into a maximal non-negative subspace which is positive. Since the latter is non-degenerate it is the graph of a maximal J-dissipative operator $\tilde{A}$ (see Theorem 2.6) and it is, moreover, strictly J-dissipative.

Now let A be a closed uniformly J-dissipative operator. This is equivalent to saying that the closed operator $A - i\gamma_A J$ is J dissipative. By Theorem 2.6 this operator admits extension into a closed maximal J-dissipative operator $\widetilde{A - i\gamma_A J}$, and therefore the operator $\tilde{A} = \widetilde{A - i\gamma_A J} + i\gamma_A J$ is a maximal J-dissipative operator and it is uniformly J-dissipative: $\text{Im}[\tilde{A}x, x] \geqslant \gamma_A \|x\|^2 (x \in \mathscr{D}_{\tilde{A}})$. ■

To conclude this paragraph we note that

2.32 Strictly J-dissipative operators have no real eigenvalues. All real points are points of regular type of closed uniformly J-dissipative operators; they are regular points of these operators if and only if the latter are maximal J-dissipative operators.

□ If A is a strictly J-dissipative operator and $(A - \lambda_0 I)x_0 = \theta$ when $\lambda_0 = \bar{\lambda}_0$ then $\text{Im}[Ax_0, x_0] = 0$, which, by definition, is possible only when $x_0 = \theta$, i.e., $\lambda_0 \notin \sigma_p(A)$.

Now let A be a uniformly J-dissipative closed operator. Since $\|(A-\lambda_0 I)x\| \geqslant \mathrm{Im}[(A-\lambda_0 I)x, x]/\|x\| = \mathrm{Im}[Ax, x]/\|x\| \geqslant \gamma_A\|x\|$ when $(\theta \neq)x \in \mathscr{D}_A$ and $\lambda_0 = \bar{\lambda}_0$, so λ_0 is a point of regular type for the operator A. If also A is a maximal J-dissipative operator, then from Corollary 2.17 we have $\lambda_0 \in \rho(A)$. Conversely, if A is a uniformly J-dissipative operator and $\lambda_0 = \bar{\lambda}_0 \in \rho(A)$, then A is a maximal J-dissipative operator, since otherwise by Theorem 2.31 it would admit uniformly J-dissipative extensions for which the point λ_0 would already be an eigenvalue—we have obtained a contradiction. ■

Exercises and problems

1 Show that in Theorem 2.9 the condition $\mathscr{H}^+ \subset \mathscr{D}_A$ is essential:
a) give an example of a maximal J-dissipative operator with $\rho(A) = \emptyset$;
b) give an example of a non-maximal J-dissipative operator with $\mathbb{C}^+ \subset \rho(A)$.

2 Let $\mathscr{H}$ be a $G^{(\varkappa)}$-space, i.e. (*cf.* Proposition I.9.15) $0 \notin \sigma_p(G^{(\varkappa)})$ and $\sigma(G^{(\varkappa)} \cap (0, \infty)$ consists of $\varkappa < \infty$ eigenvalues (taking multiplicity into account), and let A be a $G^{(\varkappa)}$-dissipative operator. Prove that then $\dim \mathrm{Lin}\{\mathscr{L}_\lambda(A) \mid \mathrm{Im}\,\lambda > 0\} \leqslant \varkappa$.

3 Give an example of a maximal π-dissipative with $\sigma(A) \cap \mathbb{C}^+ \neq \emptyset$ and $\sigma(A) \cap \mathbb{C}^- \neq \emptyset$.

4 Let $\mathscr{H}$ be a $G^{(\varkappa)}$-space and let A be a $G^{(\varkappa)}$-dissipative operator in $\mathscr{H}$. Prove that $\dim\,(A-\alpha I)\mathscr{L}_\alpha(A) < \infty$ when $\alpha = \bar{\alpha}$.

5 Under the conditions of Example 4 determine the orders $\delta_1(\alpha), \delta_2(\alpha), \ldots, \delta_r(\alpha)$ of the elementary divisors of A in $(A-\alpha I)\mathscr{L}_\alpha(A)$; we shall call the numbers $d_j^{(\alpha)} = \delta_j(\alpha) + 1$ $(j = 1, 2, \ldots, r)$ the orders of the non-prime $(d_j \geqslant 2)$ elementary divisors of the operator corresponding to α. Prove that the formula (2.5) holds for A (*cf.* Usvyatsova [1]).

6 Let $W_T = \{[Tx, x] \mid x \in \mathscr{D}_T\}$. Prove that either $\widetilde{W}_T = \mathbb{C}$, or $\widetilde{W} = \mathbb{R}$, or $\widetilde{W}_T$ is a certain angle with its vertex at the origin of coordinates and its aperture $\theta_T \leqslant \pi$.
Verify that if $\theta_T \leqslant \pi$ then there is a number $\varphi \in \mathbb{R}$ such that $e^{i\varphi}T$ is a J-dissipative operator (*cf.* [XI]).

7 Prove that any (maximal) J-dissipative operator can be appoximated in the uniform operator topology by (maximal) uniformly J-dissipative operators.

8 Give an example of a strictly J-dissipative operator which does not admit closure.

9 Prove that every (including those not densely defined and non-closed ones) uniformly J-dissipative operator admits maximal closed uniformly J-dissipative extensions.

10 Prove that if A is a closed maximal π-dissipate operator and $\mathscr{H}^+ \subset \mathscr{D}_A$, then A_{12} is an A_{22}-completely continuous operator.

11 Prove that if A is a W-dissipative operator and $\lambda \in \sigma_p(A)$ and $Ax_0 = \lambda x_0$, $Ax_1 = \lambda x_1 + x_0$, then x_0 is an isotropic vector in $\mathrm{Ker}(A - \lambda I)$.
Hint: Use the fact that the form $\mathrm{Im}[Ax, x]$ is non-negative on vectors $x \in \mathscr{D}_A$.

12 Prove that in Corollary 2.28 the requirement that the operator A be closed can be dropped.
Hint: Use instead of Theorem 2.26 the result of Exercise 11.

13 Let $\mathscr{H} = \mathscr{H}^+ \oplus \mathscr{H}^-$ be a J-space, let $\mathscr{L}^+ \in \mathscr{M}^+$, and let $K_{\mathscr{L}^+}$ be its angular operator. Prove that if $\mathscr{L}^+$ and $\mathscr{L}^{+[\perp]}$ have no infinite-dimensional uniformly definite subspaces, then $K_{\mathscr{L}^+}$ is a Φ-operator, and if $K_{\mathscr{L}^+} = T|K_{\mathscr{L}^+}|$ is its polar representation, then T is also a Φ-operator, and ind $T =$ ind $K_{\mathscr{L}^+}$ (Azizov). *Hint:* Use the result of Exercise 17 to §8 in Chapter 1, the definition of a Φ-operator in §2.2, and Theorem 2.11.

14 Suppose $\mathscr{L}^+ \in \mathscr{M}^+$ and $\mathscr{L}^{+[\perp]}$ contain no infinite-dimensional uniformly definite subspaces. Let $K_{\mathscr{L}^+} = T|K_{\mathscr{L}^+}|$ be the polar representation of the angular operator $K_{\mathscr{L}^+}$ of the subspace $\mathscr{L}^+$, let $|K_{\mathscr{L}^+}| = I^+ + S$, $S \in \mathscr{S}_\infty$ (see Exercise 17 to §8, Chapter 1), and let K_{M^+} be the angular operator of the uniformly positive subspace $M^+ \in \mathscr{M}^+$. Then $K_{\mathscr{L}^+} - K_{M^+}$ is a Φ-operator, and ind$(K_{\mathscr{L}^+} - K_{M^+}) =$ ind $K_{\mathscr{L}^+}$ (Azizov).
Hint: Verify that $K_{\mathscr{L}^+} - K_{M^+} = TV + S_1$, where $V: \mathscr{H}^+ \to \mathscr{H}^+$ is a linear homeomorphism, and $S_1 \in \mathscr{S}_\infty$; use the result of Exercise 13 and Theorem 2.11.

§3 Hermitian, symmetric, and self-adjoint operators

1 *Definition 3.1:* An operator A operating in a W-space $\mathscr{H}$ is said to be *W-Hermitian* if Im$[Ax, x] = 0$ for all $x \in \mathscr{D}_A$. In particular, an operator A is *W-symmetric* if also $\overline{\mathscr{D}}_A = \mathscr{H}$, and it is a *maximal W-symmetric* operator if it does not admit W-symmetric extensions $\tilde{A} \supset A$, $\tilde{A} \neq A$.

From Exercise 1 to §1, Chapter I it follows that *an operator A is W-Hermitian if and only if $[Ax, y] = [x, Ay]$ for all $x, y \in \mathscr{D}_A$*. Therefore, if A is an operator in a G-space (we recall that $0 \notin \sigma_p(G)$), then its *G-symmetry is equivalent to the inclusion $A \subset A^c$*. Since the operator A^c is closed (see 1.5), *a G-symmetric operator admits closure.*

Definition 3.2: A G-symmetric operator is said to be *G-self adjoint* if $A = A^c$.

We now return to the study of the properties of W-Hermitian operators. It follows from Zorn's lemma that

□ *3.3 Every W-symmetric operator admits extension into a maximal W-symmetric operator.* ■

□ From definitions 2.1 and 3.1 we obtain immediately

3.4 An operator A is W-Hermitian if and only if A and $(-A)$ are simultaneously W-dissipative. ■

From now on to simplify the discussion we shall again, as in §1 and §2, speak mainly about operators acting in a J-space $\mathscr{H}$, although many of the results are also valid in the case of more general spaces.

□ It follows from Proposition 3.4 that

3.5 An operator A is J-Hermitian if and only if its graph is neutral in the metric (1.4) ■

□ Moreover, the following holds:

3.6 A J-symmetric operator A is maximal if and only if Γ_A is a maximal semi-definite subspace in $\mathscr{H}_\Gamma = \mathscr{H} \times \mathscr{H}$. In addition, J-self-adjointness of the operator A is equivalent to the neutral subspace Γ_A being hyper-maximal.

This goemetrical proposition is a consequence of Propositions 3.5, 2.4, and Proposition 4.1 in Chapter 1; it can be rephrased in terms of operators thus:

3.7 A J-symmetric operator A is maximal if and only if at least one of the operators A or $(-A)$ is a maximal J-dissipative operator. Moreover, J-selfadjointness of the operator A is equivalent to A and $(-A)$ being simultaneously maximal J-dissipative operators. ■

□ Propositions 3.4–3.7 enable a number of the assertions in §2 to be made more precise for J-symmetric operators. Thus, for example, we have

Corollary 3.8 (*cf* 2.3): *J-symmetry (maximal J-symmetry, J-self-adjointness) of an operator A is equivalent to each of the operators JA and AJ being symmetric (maximal symmetric, self-adjoint).*

Corollary 3.9 (*cf.* Theorem 2.9): *Let A be a J-symmetric operator and $\mathscr{H}^+ \subset \mathscr{D}_A$. Then the following assertions are equivalent:*

a) *A is a maximal J-symmetric operator;*
b) *A_{22} is a maximal symmetric operator in $\mathscr{H}^-$;*
c) $\{\lambda \mid |\operatorname{Im} \lambda| > 2\| AP^+ \|\} \cap \rho(A) \neq \emptyset$;
d) $(\{\lambda \mid \operatorname{Im} \lambda > \| AP^+ \|\} \subset \rho(A)) \vee (\{\lambda \mid -\operatorname{Im} \lambda > 2\| AP^+ \|\} \subset \rho(A))$.

Corollary 3.10 (*cf.* Corollary 2.12): *If A is a maximal J-symmetric J-selfadjoint operator, if $\mathscr{H}^+ \subset \mathscr{D}_A$, and if A_{12} is an A_{22}-completely continuous operator, then $\mathbb{C}^+ \subset \tilde{\rho}(A)$ or $\mathbb{C}^- \subset \tilde{\rho}(A)$ $(\mathbb{C}^+ \cup \mathbb{C}^- \subset \rho(A))$. Moreover, if $\mathbb{C}^+ \subset \tilde{\rho}(A)$ $(\mathbb{C}^- \subset \tilde{\rho}(A))$, then $\mathbb{C}^-$ $(\mathbb{C}^+)$ consists of points of regular type and not more than a countable set of eigenvalues of finite algebraic multiplicity of the operator A.*

□ In view of Corollary 2.12 explanation is needed only for the last assertion. But this follows from the J-symmetry of the operator A and Theorem 1.16. ■

□ **Corollary 3.11** (*cf.* Theorem 2.21): *Suppose $\mathscr{H}$ is a W-space, A a W-symmetric operator, and σ is its bounded spectral set:* $\sigma \subset \mathbb{C}^+$ or $\sigma \subset \mathbb{C}^-$. *Then $P_\sigma \mathscr{H}$ is a neutral subspace.* ■

Corollary 3.12 (*cf.* Theorem 2.24): *Suppose A is a J-self-adjoint operator, and σ^+ is a bounded spectral set, $\sigma^+ \subset \mathbb{C}^+$. Then $\sigma^- = \sigma^{+*}(\subset \mathbb{C}^-)$ is also a spectral set of the operator A, the projector $P_{\sigma^+ \cup \sigma^-}$ is J-self adjoint (we shall also call such a projector J-orthogonal), and therefore $P_{\sigma^+ \cup \sigma^-}\mathscr{H}$ is a projectionally complete subspace. Moreover, if σ_1^+ $(\subset \mathbb{C}^+)$ is another spectral set of the operator A and $\sigma^+ \cap \sigma_1^+ = \emptyset_1$ and $\sigma_1^- = \sigma_1^- = \sigma_1^{+*}$, then $P_{\sigma^- \cup \sigma^-}\mathscr{H}$ $[\perp]$ $P_{\sigma_1^+ \cup \sigma_1^-}\mathscr{H}$.*

□ By Theorem 1.16 *the spectrum of a J-self-adjoint operator is symmetric about the real axis*, and therefore $\sigma^- = \sigma^{+*}$ is a spectral set of the operator A together with σ^+. Let $\Gamma_{\sigma^+\cup\sigma^-} = \Gamma_{\sigma^+} \cup \Gamma_{\sigma^-}$, where Γ_{σ^-} is a contour consisting of regular point of the operator A and surrounding σ^+, $\Gamma_{\sigma^+} \subset \mathbb{C}^+$, and $\Gamma_{\sigma^-} = \Gamma^*_{\sigma^+}$. Then

$$P^c_{\sigma^+\cup\sigma^-} = \left(-\frac{1}{2\pi i}\oint_{\Gamma_{\sigma^+\cup\sigma^-}} (A-\lambda I)^{-1}\,d\lambda\right)^c = \frac{1}{2\pi i}\oint_{\Gamma_{\sigma^+\cup\sigma^-}} (A-\bar{\lambda} I)^{-1}\,d\bar{\lambda}$$

$$= -\frac{1}{2\pi i}\oint_{\Gamma_{\sigma^+\cup\sigma^-}} (A-\lambda I)^{-1}\,d\lambda = P_{\sigma^+\cup\sigma^-}.$$

As for the projectional completeness of the subspace $P_{\sigma^+\cup\sigma^-}\mathscr{H}$, it follows from the fact that the whole space splits up into the sum of the subspaces $P_{\sigma^+\cup\sigma^-}\mathscr{H}$ and $(I - P_{\sigma^+\cup\sigma^-})\mathscr{H}$ which are J-orthogonal to one another:

$$[P_{\sigma^+\cup\sigma^-}x, (I - P_{\sigma^+\cup\sigma^-})y] = [x, P_{\sigma^+\cup\sigma^-}(I - P_{\sigma^+\cup\sigma^-})y] = 0.$$

The J-orthogonality of $P_{\sigma^+\cup\sigma^-}\mathscr{H}$ and $P_{\sigma^+\cup\sigma^-}\mathscr{H}$ follows from Theorem 2.20. ■

□ *Remark 3.13:* In Corollary 3.12 we could, formally, weaken the conditions on the operator A by premising, not that it is J-self-adjoint, but only that it is J-symmetric and σ^+, σ^- satisfy the conditions of Theorem 2.24. But this relaxation is only formal, since from the conditions on the operator A it follows that it has at least one pair of non-real regular points λ_0 and $\bar{\lambda}_0$ symmetrically situated relative to the real axis. If $A \neq A^c$, then $\lambda_0, \bar{\lambda}_0 \in \sigma_p(A^c)$; but this contradicts assertion b) of Theorem 1.16, and therefore A is a J-selfadjoint operator. ■

Corollary 3.14 (*cf.* Corollary 2.22): *Suppose that Λ is the set of eigenvalues of a J-symmetric operator A and $\Lambda \cap \Lambda^* = \emptyset$. Then C* $\mathrm{Lin}\{\mathscr{L}_\lambda(A)\}_{\lambda\in\Lambda}$ *is a neutral subspace.*

□ Corollary 2.18 has to be used in proving this assertion. ■

Corollary 3.15 (*cf.* Corollary 2.23): *Let A be a π-selfadjoint operator in $\Pi_\varkappa$. Then its non-real spectrum consists of not more than $2\varkappa$ (taking multiplicity into account) normal eigenvalues situated symmetrically about the real axis.*

□ The symmetry of the spectrum of a J-self-adjoint, and, in particular, of a π-self-adjoint, operator follows from Theorem 1.16 (*cf.* Exercise 1 below). It remains only to use Proposition 3.7 and Corollary 2.23. ■

Corollary 3.16: *Let A be a π-self-adjoint operator. Then $\sigma_r(A) = \emptyset$.*

□ This assertion follows from Corollaries 2.17 and 3.15. ■

2 *Definition 3.17:* We shall say that a J-symmetric operator A is *semi-bounded below* if $\gamma_A = \inf\{[Ax, x]/(x, x) \mid (\theta \neq) x \in \mathcal{D}_A\} > -\infty$. In particular, an operator A is said to be *J-non-negative* ($A \overset{J}{\geqslant} 0$), *J-positive* ($A \overset{J}{>} 0$), or *uniformly J-positive* ($A \overset{J}{\gg} 0$) if, respectively, $[Ax, x] \geqslant 0$ when $x \in \mathcal{D}_A$, $[Ax, x] > 0$ when $\theta \neq x \in \mathcal{D}_A$ (in both cases $\gamma_A \geqslant 0$), or $[Ax, x] \geqslant \gamma(x, x)$ when $x \in \mathcal{D}_A$ and for some $\gamma > 0$ (i.e, $\gamma_A > 0$).

In particular, if $J = I$ then Definition 3.17 repeats the corresponding definitions for symmetric operators and in this case we shall omit the symbol 'I' above the signs $\geqslant$, $>$, $\gg$.

Definition 3.18: Let A and B be J-symmetric operators and let $\mathcal{D}_A \subset \mathcal{D}_B$. We shall say that $A \overset{J}{\geqslant} B$ if $A - B \overset{J}{\geqslant} 0$.

Theorem 3.19: *Every J-non-negative operator can be extended into a J-self-adjoint J-non-negative operator.*

Before proving this theorem we shall deal with some auxiliary propositions and we introduce the concept of a 'quasi-inverse' operator. Let T be a densely defined operator. Then we shall say that the operator

$$T^{(-1)} = Q(T \mid \bar{\mathcal{R}}_{T^*})^{-1} P \tag{3.1}$$

where P is the orthoprojector on to $\bar{\mathcal{R}}_T$, and Q is the orthoprojector on to $\bar{\mathcal{R}}_{T^*}$, is *quasi-inverse to T.*

Lemma 3.20: *If A and B are bounded operators with $\mathcal{D}_A = \mathcal{D}_B = \mathcal{H}$ and $A^*A \leqslant B^*B$, then $A = KB$, where $K = \overline{AB^{(-1)}}$ and therefore $\| K \| \leqslant 1$.*

□ $KB = \overline{AB^{(-1)}}B = AB^{(-1)}B = A.$ ■

Lemma 3.21: *Let $\mathcal{H}$ be a Hilbert space, $\mathcal{L}$ a subspace of it, $\mathcal{H} = \mathcal{L} \oplus \mathcal{L}^\perp$, and relative to this decomposition let a bounded operator T be expressed in matrix form: $T = \| T_{ij} \|_{i,j=1}^2$. For the operator T to be non-negative (respectively, positive, uniformly positive) it is necessary and sufficient that the operators T_{11} and T_{22} be non-negative (respectively, positive, uniformly positive), that $T_{21} = T_{12}^*$, and that there be an operator $S_{12}: \mathcal{L}^\perp \to \mathcal{L}$, $\mathcal{D}_{S_{12}} = \mathcal{L}^\perp$, $\| S_{12} \| \leqslant 1$ (respectively, $\| S_{12}x \| < \| x \|$ when $(\theta \neq) x \in \mathcal{R}_{T_{22}^{1/2}}$; $\| S_{12} \| < 1$), Ker $T_{22} \subset$ Ker S_{12}, $\mathcal{R}_{S_{12}} \subset \mathcal{R}_{T_{11}}$, such that $T_{12} = T_{11}^{1/2} S_{12} T_{22}^{1/2}$.*

□ *Sufficiency*: Let T_{11}, T_{22}, and S_{12} be the operators indicated in the theorem. Then

$$T = \begin{Vmatrix} T_{11} & T_{12} \\ T_{12}^* & T_{22} \end{Vmatrix} = \begin{Vmatrix} T_{11} & T_{11}^{1/2} S_{12} T_{22}^{1/2} \\ T_{22}^{1/2} S_{12}^* T_{11}^{1/2} & T_{22} \end{Vmatrix}$$

$$= \begin{Vmatrix} T_{11}^{1/2} & 0 \\ 0 & T_{22}^{1/2} \end{Vmatrix} \begin{Vmatrix} I & S_{12} \\ S_{12}^* & I \end{Vmatrix} \begin{Vmatrix} T_{11}^{1/2} & 0 \\ 0 & T_{22}^{1/2} \end{Vmatrix}. \tag{3.2}$$

It can be verified immediately that the middle factor is a non-negative operator if $\| S_{12} \| \leqslant 1$, and is uniformly positive if $\| S_{12} \| < 1$. This remark enables the non-negativeness and uniform positiveness of T in the cases indicated in the theorem to be proved easily.

Now let $T_{11} > 0$, $T_{22} > 0$, and $\| S_{12} T_{22}^{1/2} x \| < \| T_{22}^{1/2} x \|$ for all $(\theta \neq) x \in \mathscr{L}^{\perp}$. Since $\bar{\mathscr{R}}_{T_{22}^{1/2}} = \mathscr{L}^{\perp}$, so $\| S_{12} \| \leqslant 1$, i.e., $T \geqslant 0$. But if we had $x \in \operatorname{Ker} T$, $x = x_1 + x_2$ where $x_1 \in \mathscr{L}$, $x_2 \in \mathscr{L}^{\perp}$, then this, taking (3.2) into account, would imply the system of equalities

$$\begin{cases} T_{11}^{1/2} x_1 + S_{12} T_{22}^{1/2} x_2 = \theta, \\ S_{12}^{*} T_{11}^{1/2} x_1 + T_{22}^{1/2} x_2 \quad = \theta. \end{cases} \tag{3.3}$$

Hence we obtain

$$\| S_{12} T_{22}^{1/2} x_2 \| = \| T_{11}^{1/2} x_1 \| \geqslant \| S_{12}^{*} T_{11}^{1/2} x_1 \| = \| T_{22}^{1/2} x_2 \|,$$

which is possible only when $x_2 = \theta$. But then we conclude from the first inequality in (3.3) that $T_{11}^{1/2} x_1 = \theta$, and since T_{11} is positive we have $x_1 = \theta$, i.e., $x = \theta$. Therefore $T > 0$.

Necessity: The equality $T_{21} = T_{12}^{*}$ follows from the self-adjointness of T. Since $(T_{ii}, x_i) = (Tx_i, x_i)$, $i = 1, 2$, $x_1 \in \mathscr{L}_1 x_2 \in \mathscr{L}^{\perp}$, the conclusion regarding T_{11} and T_{12} is correct.

Now let $T \gg 0$. Then it is clear that the operator $S_{12} = T_{11}^{-1/2} T_{12} T_{22}^{-1/2}$ satisfies our requirements. But if T is an arbitrary non-negative operator, then $T + (1/n) I \gg 0$, $n = 1, 2, \ldots$, and therefore it follows from what has been proved above that there are operators $S_{12}^{(n)}$ with

$$\| S_{12}^{n} \| < 1, \qquad T_{12} = \left(T_{11} + \frac{1}{n} I \right)^{1/2} S_{12}^{(n)} \left(T_{22} + \frac{1}{n} I \right)^{1/2}.$$

Without less of generality (see 1.8.20) we may suppose that, as $n \to \infty$, the sequence $S_{12}^{(n)}$ converges in the weak operator topology to an operator S_{12}' with $\| S_{12}' \| < 1$, and therefore $T_{12} = T_{11}^{1/2} S_{12}' T_{22}^{1/2}$. We take as the operator S_{12} $S_{12} = Q S_{12}' P = \overline{(T_{11}^{1/2})^{(-1)} T_{12} (T_{22}^{1/2})^{(-1)}}$, where P is the ortho-projector on to $\bar{\mathscr{R}}_{T_{22}}$ and Q is the orthoprojector on to $\bar{\mathscr{R}}_{T_{11}}$. The further verification of the properties of the operator S_{12} presents no difficulties. ■

□ Suppose that A is a Hermitian operator defined on a subspace $\mathscr{D}_A$ of a Hilbert space $\mathscr{H} = \mathscr{D}_A^{\perp} \oplus \mathscr{D}_A$ and that relative to this decomposition A can be expressed in the form of the 'vector'

$$A = \begin{Vmatrix} A_{12} \\ A_{22} \end{Vmatrix}.$$

This operator will be a contraction if and only if A_{22} is a contraction and $A_{12}^{*} A_{12} \leqslant I_2 - A_{22}^{2}$ (here $I_2 = I \mid \mathscr{D}_A$), or, what is equivalent by Lemma 3.20,

$$A_{12} = K(I_2 - A_{22}^{2})^{1/2}, \quad \text{where } K = \overline{A_{12}[(I_2 - A_{22}^{2})^{1/2}]^{(-1)}} \quad \text{and} \quad \| K \| \leqslant 1.$$

A Hermitian operator A always admits selfadjoint extensions $\tilde{A}$ on to the whole space $\mathscr{H}$, and these extensions can be expressed in matrix form

$$\tilde{A} = \begin{Vmatrix} X_{11} & A_{12} \\ A_{12}^* & A_{22} \end{Vmatrix}, \qquad X_{11} = X_{11}^*. \tag{3.4}$$

If A is a contraction then it is easily seen that, if $I_1 = I \mid \mathscr{D}_A^{\perp}$,

$$\begin{aligned} A_{11}^{-} &\equiv -I_1 + \overline{A_{12}(I_2 + A_{22})^{(-1)}A_{12}^*} = -I_1 + K(I_2 - A_{22})K^* \\ &\leqslant I_1 - K(I_2 + A_{22})K^* = I_1 - \overline{A_{12}(I_{22} - A_{22})^{(-1)}A_{22}^*} \equiv A_{11}^{+} \end{aligned} \tag{3.5}$$

Also, $\tilde{A}$ is a selfadjoint contraction if and only if $-I \leqslant \tilde{A} \leqslant I$. By Lemma 3.21 this is equivalent to the system

$$\begin{cases} -I_1 \leqslant X_{11} \leqslant I_1, \\ A_{12} = (I_1 - X_{11})^{1/2} S_{12}'(I_2 - A_{22})^{1/2}, \\ A_{12} = (I_1 + X_{11})^{1/2} S_{12}''(I_2 + A_{22})^{1/2}, \end{cases}$$

where S_{12}' and S_{12}'' have the properties indicated in Lemma 3.21. Since $A_{12} = K(I_2 - A_{22}^2)^{1/2}$, this system can be rewritten in the equivalent form

$$\begin{cases} -I_1 \leqslant X_{11} \leqslant I_1, \\ K(I_2 + A_{22})^{1/2} = (I_2 - X_{11})^{1/2} S_{12}', \\ K(I_2 - A_{22})^{1/2} = (I_1 + X_{11})^{1/2} S_{12}'', \end{cases}$$

which in turn is equivalent to the inequalities

$$-I_1 + K(I_2 - A_{22})K^* \leqslant X_{11} \leqslant I_1 - K(I_2 + A_{22})K^*.$$

Taking (3.5) into account we have proved

Lemma 3.22: *If the operator*

$$A = \begin{Vmatrix} A_{12} \\ A_{22} \end{Vmatrix}$$

is a Hermitian contraction then it admits extension into a self-adjoint operator $\tilde{A}$ *of the form* (3.4), *and the latter will be a contraction if and only if* $A_{11}^{-} \leqslant X_{11} \leqslant A_{11}^{+}$. ■

□ *Proof of Theorem 3.19*: By Corollary 2.5 we can suppose without loss of generality that A is a closed J-non-negative operator. Since A being a J-non-negative J-symmetric operator is equivalent to iA being a J-dissipative operator, the graph Γ_{iA} of the latter is a non-negative subspace in the space $\mathscr{H}_\Gamma$ (see §1.2), and its angular operator $K_{iA} = \langle V, \; -JVJ\rangle$, where $V = (J - JA)(I + JA)^{-1}$ (see (1.8), (1.9)) is a contraction since $JA \geqslant 0$, so V is a Hermitian contraction. In accordance with Lemma 3.22 V has a self-adjoint extension $\tilde{V}$ with $\| \tilde{V} \| \leqslant 1$ which generates the angular operator $\{\tilde{V}, \; -J\tilde{V}J\rangle$ of a certain maximal non-negative subspace in $\mathscr{H}_\Gamma$ that is an extension of Γ_{iA} and which is, by Proposition 2.4, the graph of a certain

J-dissipative operator $i\tilde{A}$. It follows once again from (1.9) that $\tilde{V} = (I - J\tilde{A})(I + J\tilde{A})^{-1}$, or what is equivalent,

$$\tilde{A} = J(I - \tilde{V})(I + \tilde{V})^{-1} = J - 2J(I + V)^{-1}.$$

From Propositions 1.6, 1.9 and Corollary 3.8 it follows at once that $\tilde{A}$ is a J-selfadjoint operator, and this combined with the fact that iA is a J-dissipative operator proves the theorem. ■

Corollary 3.23: *Let A be a J-symmetric operator semi-bounded below: $[Ax, x] \geqslant \gamma_A(x, x)$ $(x \in \mathscr{D}_A)$. Then it has J-selfadjoint extensions $\tilde{A}$ semi-bounded below such that $\gamma_{\tilde{A}} = \gamma_A$. For any such extension $0 \in \sigma(\tilde{A} - \gamma_{\tilde{A}} J)$.*

□ The existence of the required $\tilde{A}$ is equivalent to asserting the existence of J-selfadjoint J-non-negative extensions $\widetilde{A - \gamma_A J}$ of the J-non-negative operator $A - \gamma_A J$. Such as extension exists by virtue of Theorem 3.19 and therefore $\tilde{A} = \widetilde{A - \gamma_A J} + \gamma_A J$ is the required extension of the operator A. For, $[\tilde{A}x, x] = [\widetilde{A - \gamma_A J}x + \gamma_A Jx, x] = [\widetilde{A - \gamma_A J}x, x] + \gamma_A(x, x) \geqslant \gamma_A(x, x)$, and therefore $\gamma_{\tilde{A}} \geqslant \gamma_A$. On the other hand, since $\tilde{A} \supset A$, so $\gamma_A \geqslant \gamma_{\tilde{A}}$, i.e., $\gamma_{\tilde{A}} = \gamma_A$.

We now verify that $0 \in \sigma(\tilde{A} - \gamma_A J)$. Suppose the contrary: $0 \in \rho(\tilde{A} - \gamma_A J)$. But then the J-non-negative operator $\tilde{A} - \gamma_A J$ would be uniformly J-positive, and therefore we would have $\gamma_{\tilde{A}} > \gamma_A$, which is impossible. ■

Corollary 3.24: *Let A be a J-symmetric operator. Then $A \overset{J}{\geqslant} 0$ $(A \overset{J}{>} 0,\ A \overset{J}{\gg} 0)$ if and only if iA is a J-dissipative (strictly J-dissipative, uniformly J-dissipative) operator. Moreover, iA is a maximal J-dissipative operator if and only if $A = A^c \overset{J}{\geqslant} 0$.*

□ The first part of the assertion is trivial, and we have already used it in proving Theorem 3.19. If $A = A^c \overset{J}{\geqslant} 0$, then (see the proof of Theorem 3.19) Γ_{iA} is a maximal non-negative subspace in $\mathscr{H}_\tau$, and therefore by 2.4 iA is a maximal J-dissipative operator. Conversely, if iA is a maximal J-dissipative operator, then A has no non-trivial J-non-negative extensions, and therefore it follows by Theorem 3.19 that $A = A^c \overset{J}{\geqslant} 0$ ■

Corollary 3.25: *Let $A \overset{J}{\geqslant} 0$. Then $\sigma_p(A) \cap (\mathbb{C}^+ \cup \mathbb{C}^-) = \emptyset$. If $\Lambda = \{\lambda \in \sigma_p(A) \mid \lambda > 0(<0)\}$, then $\mathscr{L}_\lambda(A) = \mathrm{Ker}(\mathrm{A} - \lambda I)$ and $\mathrm{Lin}\{\mathrm{Ker}\ (A - \lambda I) \mid \lambda \in \Lambda\} \subset \mathscr{P}^{++} \cup \{\theta\}$ $(P^{--} \cup \{\theta\})$. If $\lambda = 0$, then, generally speaking, $\mathscr{L}_0(A) \neq \mathrm{Ker}\ A$, but the lengths of the Jordan chains of the operator $A \mid \mathscr{L}_0(A)$ do not exceed 2.*

□ By virtue of Theorem 3.19 we can suppose that $A = A^c \overset{J}{\geqslant} 0$. If $\lambda \in \sigma_p(A) \cap (\mathbb{C}^+ \cup \mathbb{C}^-)$ and $Ax_0 = \lambda_0 x_0$ $(x_0 \neq \theta)$, then from Corollary 3.14 we obtain that $[x_0, x_0] = 0$, and therefore $[Ax_0, x_0] = 0$. Using the Cauchy–Bunyakovski inequality, we obtain $Ax_0 = \theta$—a contradiction. Essentially we have proved that the condition $x_0 \in \text{Ker}(A - \lambda_0 I) \cap \mathscr{P}^0$ implies $\lambda_0 = 0$. Therefore, if $(0 \neq)\ \lambda \in \sigma_p(A)$ and $Ax = \lambda x$, then $[x, x] = [Ax, x]/\lambda$, hence $\text{Ker}(A - \lambda I) \subset \mathscr{P}^{++} \cup \{\theta\} (\mathscr{P}^{--} \cup \{\theta\})$ if $\lambda > 0$ (<0), and therefore (see §2, Problem 11) $\mathscr{L}_\lambda(A) = \text{Ker}(A - \lambda I)$ when $\lambda \neq 0$. The assertion about $\text{Lin}\{\text{Ker}(A - \lambda I)\}_{\lambda \in \Lambda}$ follows immediately from Corollary 2.18. Now let $Ax_2 = x_1$, $Ax_1 = x_0$, $Ax_0 = \theta$, i.e., $A^3 x_2 = \theta$. Then $0 = [A^3 x_2, x_2] = [AAx_2, Ax_2]$, and therefore $x_1 = Ax_2 \in \text{Ker}\ A$, i.e., $x_0 = \theta$. ■

Corollary 3.26: *If* $A = AA^c \overset{J}{\geqslant} 0$, *then* $\sigma_r(A) = \emptyset$.

□ By Corollary 2.17 in combination with Proposition 3.7 $A (= A^c)$ has no real points in the residual spectrum. If $\bar{\lambda} \neq \lambda$ belonged to $\sigma_r(A)$, then by Theorem 1.16 $\lambda \in \sigma_p(A)$, contrary to Corollary 3.25. ■

We now investigate the spectrum of a *J*-self-adjoint *J*-non-negative operator.

Theorem 3.27: *Let* $A = A^c \overset{J}{\geqslant} 0$ *and* $\rho(A) \neq \emptyset$. *Then* $\mathbb{C}^+ \cup \mathbb{C}^- \subset \rho(A)$.

□ We assume at first that A is a continuous operator. Let $\bar{\lambda}_0 \neq \lambda_0 \in \sigma(A)$. From Corollaries 3.25 and 3.26 we conclude that $\lambda_0 \in \sigma_c(A)$, and therefore there is a sequence $\{x_n\}_1^\infty$ of normalized vectors ($\| x_n \| = 1$, $n = 1, 2, \ldots$) such that

$$(A - \lambda_0 I)x_n \to \theta \quad \text{when } n \to \infty. \tag{3.6}$$

Since

$$[(A - \lambda_0 I)x_n, x_n] = [(A - \text{Re}\ \lambda_0 I)x_n, x_n] - i\ \text{Im}\ \lambda_0 [x_n, x_n]$$

and $\text{Im}\ \lambda_0 \neq 0$, so $[x_n, x_n] \to 0$ as $n \to \infty$, which implies that

$$((JA)^{1/2} x_n, (JA)^{1/2} x_n) = (JAx_n, x_n) = [Ax_n, x_n] \to 0 \qquad (n \to \infty).$$

From the continuity of the operator A and with it also that of $(JA)^{1/2}$ it follows that

$$Ax_n = J(JA)^{1/2}(JA)^{1/2} x_n \to \theta \qquad (n \to \infty).$$

Taking account of (3.6) we obtain $x_n \to \theta$ $(n \to \infty)$—a contradiction.

Now let A be an arbitrary *J*-self-adjoint *J*-negative operator, and $(\bar{\mu}_0 \neq)\mu_0 \in \rho(A)$. It follows from Corollary 3.12 that $\bar{\mu}_0 \in \rho(A)$, and so the operator $B = (A - \bar{\mu}_0 I)^{-1} A (A - \mu_0 I)^{-1}$ is bounded, $B = B^c \overset{J}{\geqslant} 0$. By what has been proved above, $\sigma(B) \cap (\mathbb{C}^+ \cup \mathbb{C}^-) = \emptyset$. A well-known theorem of Dunford (see [VII]) about the mapping of the spectrum asserts that $\sigma(B) = \{\mu \mid \mu = \lambda/(\lambda - \bar{\mu}_0)(\lambda - \mu_0),\ \lambda \in \sigma(A)\}$. It follows from $\bar{\lambda}_0 \neq \lambda_0$ and $|\lambda_0| \neq |\mu_0|$ that the number $\lambda_0/(\lambda - \bar{\mu}_0)(\lambda - \mu_0)$ is not real, and so $\lambda_0 \in \rho(A)$, i.e.,

$$(\mathbb{C}^+ \cup \mathbb{C}^-) \backslash \{\lambda \mid |\lambda| = |\mu_0|\} \subset \rho(A). \tag{3.7}$$

We now take $(\bar{\mu}_1 \neq)\mu_1 \in \rho(A)$ with $|\mu_1| \neq |\mu_0|$, and carrying out a similar argument we obtain

$$(\mathbb{C}^+ \cup \mathbb{C}^-)\backslash\{\lambda \mid |\lambda| = |\mu_1|\} \subset \rho(A). \tag{3.8}$$

It follows from (3.7) and (3.8) that $\mathbb{C}^+ \cup \mathbb{C}^- \subset \rho(A)$. ■

Corollary 3.28: *Let A be a π-self-adjoint π-non-negative operator in $\Pi_\varkappa$. Then $\mathbb{C}^+ \cup \mathbb{C}^- \subset \rho(A)$, and the set $\sigma(A) \cap (0, \infty)$ consists of not more than $\varkappa$ normal eigenvalues (taking multiplicating into account).*

□ From Corollary 3.15 and Theorem 3.27 we conclude that $\mathbb{C}^+ \cup \mathbb{C}^- \subset \rho(A)$. The second part follows from Corollaries 3.24 and 2.23. ■

Corollary 3.29: *Let $A = A^c \overset{J}{\gg} 0$. Then $\mathbb{C}^+ \cup \mathbb{C}^- \cup \{0\} \subset \rho(A)$. If σ^+ (respectively σ^-) is a bounded spectral set of the operator A, $\sigma^+ \subset (0, \infty)$ (resp. $\sigma^- \subset (-\infty, 0)$), and P_{σ^+} (resp. P_{σ^-}) is the corresponding Riesz projector* (2.4), *then $P_{\sigma^+}\mathscr{H}$ (resp. $P_{\sigma^-}\mathscr{H}$) is a uniformly positive (resp. uniformly negative) subspace. In particular if $A = A^c \overset{J}{\gg} 0$ is a bounded operator, then with $\sigma^+ = \sigma(A) \cap (0, \infty)$, $\sigma^- = \sigma(A) \cap (-\infty, 0)$ the subspaces $P_{\sigma^+}\mathscr{H}$ and $P_{\sigma^-}\mathscr{H}$ are J-orthogonal to one another, are maximal uniformly definite, and $\mathscr{H} = P_{\sigma^+}\mathscr{H} [\dot{+}] P_{\sigma^-}\mathscr{H}$.*

□ From Corollary 3.24 and Proposition 2.32 we conclude that $0 \in \rho(A)$. We use Theorem 3.27 and obtain the inclusion $\mathbb{C}^+ \cup \mathbb{C}^- \cup \{0\} \subset \rho(A)$.

From Corollary 3.12 follows the projectional completeness of the subspaces $P_{\sigma^+}\mathscr{H}$ and $P_{\sigma^-}\mathscr{H}$, and from Theorem 2.21 and Corollary 3.24, their uniform definiteness with the proper signs. If A is a bounded operator, then, since $\sigma(A) = \sigma^+ \cup \sigma^-$, $\sigma^+ \cap \sigma^- = \emptyset$, we have by Theorem 3.20 $P_{\sigma^+} + P_{\mu^-} = I$, i.e., using Corollary 3.12, $\mathscr{H} = \mathrm{P}_{\sigma^+}\mathscr{H} [\dot{+}] \mathrm{P}_{\sigma^-}\mathscr{H}$. The maximal uniform definiteness of $P_{\sigma^+}\mathscr{H}$ and $P_{\sigma^-}\mathscr{H}$ now follows from Proposition 1.25 in Chapter 1. ■

3 Here we present some examples which illustrate the preciseness of the conditions imposed on the operators in the theorems in §§2–3, and the essential difference of the indefinite case from the finite one.

Example 3.30: A Hermitian operator which does not admit closed dissipative extensions (*cf.* Theorems 2.6, 2.31, 3.19).

□ Let $\mathscr{H} = \mathscr{H}^+ \oplus \mathscr{H}^-$ be an infinite-dimensional space with $\mathscr{H}^+ = \mathrm{Lin}\{e^+\}$, $\|e^+\| = 1$, and let φ be a discontinuous linear functional defined on $\mathscr{H}^-$. Then the operator $A: x^- \to \varphi(x^-)e^+$ with $\mathscr{D}_A = \mathscr{H}^-$ is Hermitian, since $(Ax^-, x^-) = 0$. If it admitted a dissipative extension $\tilde{A}$, then $\mathscr{D}_{\tilde{A}} = \mathscr{H}$, and therefore (see Corollary 2.5) $\tilde{A}$ would be a closed operator. But then by Banach's Theorem $\tilde{A}$ would be a continuous operator, which contradicts the discontinuity of φ.

The operator constructed also serves as an example of a Hermitian operator which does not admit closure and therefore the closure $\bar{\Gamma}_A$ in $\mathscr{H}_\Gamma$ of its graph Γ_A would no longer be the graph of an operator. ■

Example 3.31: An operator $A = A^c \overset{J}{>} 0$ *with* $\sigma(A) = \mathbb{C}$ (*cf.* Lemma 2.8, and Theorems 2.9 and 3.27).

□ Let $\mathscr{H}$ be a J-space and $(\mathscr{L}_1, \mathscr{L}_2)$ a maximal dual pair (see 1. §10.1) of definite (but not uniformly definite) subspaces $\mathscr{L}_1$ and $\mathscr{L}_2$. We define an operator A: $\mathscr{D}_A = \mathscr{L}_1 [\dot{+}] \mathscr{L}_2$, $A(x_1 + x_2) = x_1 - x_2$, $x_1 \in \mathscr{L}_1$, $x_2 \in \mathscr{L}_2$. By Lemma 7.7 and Corollary 7.17 in Chapter 1 we have $\bar{\mathscr{D}}_A = \mathscr{H}_1$, $\mathscr{D}_A \neq \mathscr{H}$, and since $A\mathscr{D}_A \subset \mathscr{D}_A$, so $\sigma(A) = \bar{\mathbb{C}}$. It remains to verify that $A = A^c \overset{J}{>} 0$. Since $[A(x_1 + x_2), x_1 + x_2] = [x_1, x_1] - [x_2, x_2] > 0$ when $x_1 + x_2 \neq \theta$, so $A \overset{J}{>} 0$. Now let $y \in \mathscr{D}_{A^c}$ and $([A(x_1 + x_2), y] =)[x_1 - x_2, y] = [x_1 + x_2, z]$ for all $x_1 + x_2 \in \mathscr{D}_A$ and, in particular, $[x_1, y] = [x_1, z]$ and $[-x_2, y] = [x_2, z]$. Therefore $y + z \in \mathscr{L}_1$ and $y - z \in \mathscr{L}_2$, and therefore $y \in \mathscr{D}_A (= \mathscr{L}_1 [\dot{+}] \mathscr{L}_2)$, i.e., $\mathscr{D}_{A^c} \subset \mathscr{D}_A$, which proves that A is J-selfadjoint. We note also that $A^2 = I | \mathscr{D}_A$, and $P_1 = \frac{1}{2}(I + A)$ and $P_2 = \frac{1}{2}(I - A)$ are the J-orthogonal projectors from $\mathscr{D}_A$ on to $\mathscr{L}_1$ and $\mathscr{L}_2$ respectively, and also $P_1 \overset{J}{\geqslant} 0$, $P_2 \overset{J}{\geqslant} 0$, and $\sigma(P_1) = \sigma(P_2) = \mathbb{C}$. ■

Example 3.32: A π-self-adjoint operator, in a finite-dimensional Pontryagin space $\Pi_\varkappa$, *for which the inequality* (2.5) *becomes an equality* (*cf* 2.26).

□ Let $\mathscr{H} = \text{Lin}\{e_0, e_1, e_2\}$ be a unitary three-dimensional space with an orthonormalized basis $\{e_0, e_1, e_2\}$. Each vector $x \in \mathscr{H}$ is expressed in the form $x = \xi_0 e_0 + \xi_1 e_1 + \xi_2 e_2$. Let $y = \eta_0 e_0 + \eta_1 e_1 + \eta_2 e_2$. We consider the indefinite form $[x, y] = -\xi_0 \bar{\eta}_2 - \xi_1 \bar{\eta}_1 - \xi_2 \bar{\eta}_0$. It is generated by the operator J: $Je_0 = -e_2$, $Je_1 = -e_1$, $Je_2 = -e_0$, the eigenvalues of which are $\lambda = -1$ and $\lambda = 1$ with multiplicities 2 and 1 respectively, i.e., $\mathscr{H} = \Pi_\varkappa$ with $\varkappa = 1$. We define a linear operator A; $Ae_0 = \theta$, $Ae_1 = e_0$, $Ae_2 = e_1$. Since $[Ax, x] = [\xi_1 e_0 + \xi_2 e_1, \xi_0 e_0 + \xi_1 e_1 + \xi_2 e_2] = \xi_1 \bar{\xi}_2 + \bar{\xi}_1 \xi_2$ is a real number we have $A = A^c$. If follows from the definition of the operator A that it has a single eigenvalue $\lambda = 0$ and a single Jordan chain of length $d = 2\varkappa + 1$ $(= 3)$. Therefore $[d/2] = \varkappa (= 1)$. ■

Example 3.33: A closed positive operator A which does not admit positive selfadjoint extensions..

□ We presuppose that A is a closed, non-selfadjoint, positive operator with finite-dimensional deficiency numbers (i.e., dim $\mathscr{R}^{\perp}_{A - \lambda I} < \infty$ and dim $\mathscr{R}^{\perp}_{A - \bar{\lambda} I} < \infty$ when $\lambda \neq \bar{\lambda}$), then $\lambda = 0$ is a point of regular type for A, and that $\gamma_A = 0$. By Corollary 3.23 for any selfadjoint non-negative extension $\tilde{A}$ of this operator $0 \in \sigma(\tilde{A})$. From [I] (see also Azizov, I. Iokhvidov, and V. Shtrauss [1]) it follows that $\mathscr{R}_{\tilde{A}} = \mathscr{R}_A$ and therefore, whenever $0 \notin \rho(\tilde{A})$, we have $0 \in \sigma_p(\tilde{A})$, i.e., no non-negative selfadjoint extension of the operator A with the properties indicated above will be positive.

We give an actual example of such an operator A.

Let $\mathcal{H} = \text{Lin}\{e\} \oplus L_2(0,1)$ with $\|e\| = 1$. We consider in $L_2(0,1)$ the selfadjoint contraction V_{22}: $V_{22}x(t) = tx(t)$ $(x(t) \in L_2(0,1))$ and the contraction K: $L_2(0,1) \to \text{Lin}\{e\}$, $Kx(t) = \int_0^1 x(t)\,\mathrm{d}t \cdot e$. Then (see the proof of Lemma 3.22) the operator

$$V = \left\| \begin{matrix} K(I_{22} - V_{22}^2)^{1/2} \\ V_2 \end{matrix} \right\| \quad \text{will be a Hermitian contraction.}$$

Since it follows from $(I - V)x(t) = \theta$ that $x(t) = \theta$, the operator

$$A\colon A(I - V)x(t) = (I + V)x(t)$$

will be properly defined on $\mathcal{D}_A = (I - V)L_2(0,1)$. Since

$$(A(I - V)x(t), (I - V)x(t) = ((I + V)x(t), (I - V)x(t))$$
$$= \int_0^1 (1 - t^2)|x(t)|^2\,\mathrm{d}t - \left| \int_0^1 (1 - t^2)^{1/2} x(t)\,\mathrm{d}t \right|^2$$

is a positive number when $x(t) \neq \theta$, it follows that A is a Hermitian operator. If we suppose that the vector $z(t) = \lambda e + z_1(t)$ $(z_1(t) \in L_2(0,1))$ is orthogonal to $\mathcal{D}_A$, then we obtain that

$$\int_0^1 (1 - t)\overline{x(t)} z_1(t)\,\mathrm{d}t - \lambda \int_0^1 (1 - t^2)^{1/2}\overline{x(t)}\,\mathrm{d}t = 0,$$

and therefore

$$z_1(t) = \frac{\lambda(1 + t)^{1/2}}{(1 - t)^{1/2}} \in L_2(0,1),$$

which is possible if $\lambda = 0$, i.e., $z(t) = 0$. Thus, $\bar{\mathcal{D}}_A = \mathcal{H}$ and A is a positive operator. That A is closed follows from the facts that V is closed and $A = 2(I - V)^{-1} - I$.

From Corollary 3.24 and taking Lemma 2.8 into account, we obtain that $\lambda = -1$ is a point of regular type for the operator V, i.e., $\mathcal{R}_A = (I + V)L_2(0,1)$ is a subspace and therefore $\lambda = 0$ is a point of regular type for the operator A. The finiteness of the deficiency numbers of the operator follows from the fact that $\mathcal{R}_A^\perp = \text{Lin}\{e + [(1 - t)/(1 + t)]^{1/2}\}$, and therefore $\dim \mathcal{R}_A^\perp = 1$. It remains to verify that $\gamma_A = 0$. To do this we observe that, if $x_n(t) \in L_2(0,1)$ are chosen so that the sequence $\{(1 - t^2)^{1/2} x_n(t)\}$ converges in $L_2(0,1)$ to $x_0(t) = 1$, then $\lim_{n\to\infty}(A(I - V)x_n(t), (I - V)x_n(t)) = 0$, but $\lim_{n\to\infty}((I - V)x_n(t), (I - V)x_n(t)) = 2$, and therefore

$$\gamma_A = \inf\left\{ \frac{(A(I - V)x(t), (I - V)x(t))}{((I - V)x(t), (I - V)x(t))} \,\middle|\, x(t) \in L_2(0,1) \right\} = 0. \qquad \blacksquare$$

Example 3.34: A J-selfadjoint differential operator.

□ Let $\mathcal{H} = L_2(-1,1)$. We introduce a J-metric into it by means of the operator J: $Jx(t) = x(-t)$ (*cf.* Remark 1.2.3 and Exercise 1 to 1.§3). We consider the differential operator $A = (d^2/dt^2) - 2(d/dt) + 1$ with the boundary conditions $x(-1) = x(1)$, $x'(-1) = x'(1)$ and the maximal domain of definition in $L_2(-1, 1)$. Since we have for twice-differentiable functions x(t) and y(t) with the given boundary conditions

$$[Ax, y] = \int_{-1}^{1} Ax(t)\overline{y(-t)}\,dt = \int_{-1}^{1} (x''(t) - 2x'(t) + x(t))\overline{y(-t)}\,dt$$

$$= x'(t)\overline{y(-t)} - 2x(t)\overline{y(-t)}\Big|_{-1}^{1} + \int_{-1}^{1} x(t)\overline{y(-t)}\,dt$$

$$+ \int_{-1}^{1} x'(t)\overline{y'(-t)}\,dt - 2\int_{-1}^{1} x(t)\overline{y'(-t)}\,dt$$

$$= x(t)\overline{y'(-t)}\Big|_{-1}^{1} + \int_{-1}^{1} x(t)(\overline{y''1-t)} - \overline{2y'(-t)} + \overline{y(-t)})\,dt$$

$$= \int_{-1}^{1} x(-t)\overline{Ay(t)}\,dt - [x, Ay],$$

we see that A is a J-symmetric operator. The J-selfadjointness of this operator follows from the independence of the boundary conditions. It can be verified immediately that the eigenvalues of this operator are the numbers (non-real if $n \neq 0$) $\lambda_n = -\pi^2 n^2 + 1 - 2\pi ni$ for $n = 0, \pm 1, \pm 2, \ldots$, and $\mathscr{L}_{\lambda_n}(A) = \text{Ker}(A - \lambda_n I) = \text{Lin}\{e^{i\pi nt}\}$. ■

Example 3.35: *A completely continuous J-selfadjoint integral operator.*
□ Let σ be a real function of bounded variation on [a, b], let $\omega = \text{Var}\,\sigma$, and let $K(s, t)$ be a Hermitian non-negative kernel (see IV, §3.3) which is continuous with respect to each of the variables and is bounded with respect to the set of variables. Then the operator A: $(Af)(s) = \int_a^b K(s, t)f(t)\,d\sigma(t)$ is completely continuous (see, e.g., I. Iokhvidov [2], I. Iokhvidov and Ektov [1]), and that it is J-non-negative in the Krein space $L^2_{\overline{\infty}}(a, b)$ follows from the fact that $[Af, f] = \int_a^b \int_a^b K(s, t)f(s)\overline{f(t)}\,d\sigma(s)\,d\sigma(t) \geqslant 0$. ■
□ In conclusion we indicate one way of constructing J-selfadjoint operators having a set of properties prescribed in advance.

Example 3.36: Let $\tilde{\mathscr{H}}$ be a $\tilde{J}$-space, where $\tilde{\mathscr{H}}$ and $\tilde{J}$ are the same as in Example 1.3.9, let $G = 0$, and V be a unitary operator. Now let A_1, be a closed operator densely defined in $\mathscr{H}_1$. Then the operator

$$A\colon A(x_1 + x_2) = A_1 x_1 + V^* A_1^* V x_2, \quad \text{where } x_1 \in \mathscr{D}_A,\ x_2 \in V^* \mathscr{D}_{A_1^*},$$

is a J-selfadjoint operator. For,

$$A^c = \tilde{J}A^*\tilde{J} = \begin{Vmatrix} 0 & V \\ V^* & 0 \end{Vmatrix} \begin{Vmatrix} A_1 & 0 \\ 0 & V^*A_1^*V \end{Vmatrix} \begin{Vmatrix} 0 & V \\ V^* & 0 \end{Vmatrix}$$

$$= \begin{Vmatrix} 0 & V \\ V^* & 0 \end{Vmatrix} \begin{Vmatrix} A_1^* & 0 \\ 0 & V^*A_1^*V \end{Vmatrix} \begin{Vmatrix} 0 & V \\ V^* & 0 \end{Vmatrix} = \begin{Vmatrix} A_1 & 0 \\ & V^*A_1V \end{Vmatrix} = A.$$

In this example, by construction, $A(\mathscr{D}_A \cap \mathscr{H}_i \subset \mathscr{H}_i$, $i = 1, 2$. We note also that the operator B given by the matrix

$$B = \begin{Vmatrix} 0 & B_1 \\ B_2 & 0 \end{Vmatrix}, \qquad \text{where } B_1 = B_1^*,\ B_2 = B_2^*$$

will also be $\tilde{J}$-selfadjoint. Moreover, the conditions $B_i \geqslant 0$ (resp. $B_i > 0$, $B_i \gg 0$) are equivalent to the conditions $B \overset{J}{\geqslant} 0$ (resp. $B \overset{J}{>} 0$, $B \overset{J}{\gg} 0$). We leave the verification of these statements to the reader. ■

Exercises and problems

1 Reformulate Theorem 1.16 for a J-selfadjoint operator and by so doing obtain the 'theorem on the symmetry of the spectrum of a J-selfadjoint operator' (Langer [2]).

2 Prove that if A is a W-Hermitian operator and λ, $\mu \in \sigma_p(A)$ with $\lambda \neq \bar{\mu}$, then $\mathscr{L}_\lambda(A)\,[\perp]\,\mathscr{L}_\mu(A)$.
Hint: Carry out an argument similar to that used in solving Exercise 8 to §1.

3 Give examples of closed, densely defined operators A and B in the Krein space $\tilde{\mathscr{H}}$ of Example 3.36 such that:
a) the operator AA^c does not admit closure:
b) $\mathscr{D}_{BB^c} = \{\theta\}$.

4 Prove that if A is a closed operator in a Pontryagin space and $\mathscr{D}_A = \Pi_\varkappa$, then AA^c is a π-selfadjoint operator (Kholevo [1]).

5 Let A be a closed operator in a J-space $\mathscr{H}$ with $\bar{\mathscr{D}}_A = \mathscr{H}$; suppose at least one of the following conditions holds: a) $\mathscr{H}^+ \subset \mathscr{D}_A$; b) $\mathscr{H}^- \subset \mathscr{D}_A$; c) A is a Φ-operator. Then $(AA^c)^c = AA^c$ (Azizov).
Hint: In cases a) and b) $A = AP^+A^c + AP^-A^c$, where one of the terms is bounded and the other is J-selfadjoint: in case c) prove that $\bar{\mathscr{D}}_{AA^c} = \mathscr{H}$ and ind $AA^c = 0$.

6 Construct an example of a bounded G-self-adjoint operator with a spectrum not symmetrical about the real axis (*cf.* Exercise 1) (Azizov [5]).

7 Prove that if A is a G-self-adjoint operator, and $A = BG$ with $B = B^*$ and GB a closed operator, then a 'theorem on the symmetry of the spectrum', similar to the theorem in Exercise 1, holds for the operator A in $\mathbb{C}\backslash\{0\}$ (*cf.* with Exercise 6) (Azizov [2]).

8 Give an example of a π-non-negative completely continuous operator which has not only eigenvalues but also principal vectors.
Hint: Use the method indicated in Example 3.36 (*cf.* Example 3.32).

9 Use the method in Example 3.36 to construct J-self-adjoint operators A_1, A_2, A, A_4 such that:
a) $\sigma_r(A_1) \neq \emptyset$, $\sigma_p(A_1) \cap \mathbb{C}^+ \neq \emptyset$ (we recall that $\sigma_r(A) = \emptyset$ always when $A = A^*$);
b) $\sigma(A_2) = \mathbb{C}$ (*cf.* Example 3.3); we recall that $\mathbb{C}^+ \cap \mathbb{C}^- \subset \rho(A)$ always when $A = A^*$);
c) $\sigma_c(A_3) \cap \mathbb{C}^+ \neq \emptyset$;
d) the operator A_4 has an infinite chain of eigenvectors and principal vectors.

10 Let $\mathscr{H}$ be a J-space, and P a J-orthogonal projector on to a uniformly positive sub-space. Then there is an $\varepsilon = \varepsilon(P) > 0$ such that all J-orthogonal projector P' such that $\| P' - P \| < \varepsilon$ will map $\mathscr{H}$ on to the uniformly positive subspace $P'\mathscr{H}$ [XVIII].
Hint: First verify that for sufficiently small ε the operator P' maps $P\mathscr{H}$ homeomorphically on to $P'\mathscr{H}$, and then prove that $[P'Px, P'Px] \geqslant \alpha \| Px \|^2$ $(\geqslant \alpha / \| (P' \mid P\mathscr{H})^{-1} \|^2 \| P'Px \|^2$ for some $\alpha > 0)$.

11 Let $\mathscr{F} = \{P\}$ be a commutative family of J-orthoprojectors, with $\mathscr{R}_P \subset \mathscr{P}^+$ for all $P \in \mathscr{F}$, and the subspace $\mathscr{L}_+ \subset \mathscr{P}^+$ and $P\mathscr{L}_+ \subset \mathscr{L}_+$ $(P \in \mathscr{F})$. Prove that C Lin $\{\mathscr{R}_P, \mathscr{L}_+\}_{P \in \mathscr{F}} \subset \mathscr{P}^+$ (Langer [9]).
Hint: Verify that the operator

$$P_{1,2,\ldots,n} = \sum_{0 < i_1 + i_2 + \cdots + i_n \leqslant n} (-1)^{i_1 + i_2 + \cdots + i_n} P_1^{i_1} P_2^{i_2} \ldots P_n^{i_n} \qquad (i_k = 0,1)$$

is a J-orthoprojector on to $\text{Lin}\{\mathscr{R}_{\mathscr{P}_k}\}_{k=1}^n$ $(P_k \in \mathscr{F},\ k = 1, 2, \ldots, n)$, that $\mathscr{R}_{P_{1,2,\ldots,n}} \subset \mathscr{P}^+$ and $P_{1,2,\ldots,n}\mathscr{L}_+ \subset \mathscr{L}_+$.
Then use the relation

$$[P_{1,2,\ldots,n}x + y,\ P_{1,2,\ldots,n}x + y] = [P_{1,2,\ldots,n}(x + y),\ P_{1,2,\ldots,n}(x + y)] + [(I - P_{1,2,\ldots,n})y,\ (I - P_{1,2,\ldots,n})y],$$

which holds for all $x, y \in \mathscr{H}$.

§4 Plus-operators. J-non-contractive and J-bi-non-contractive operators

1 In this section the object of the investigation will be 'plus-operators', that is, operators whose existence depends on the indefiniteness of the spaces in which they operate, and also some sub-classes of plus-operators.

Definition 4.1: Let $\mathscr{H}_1$ and $\mathscr{H}_2$ be a W_1-space and a W_2-space with the metrics $[\cdot,\cdot]_1$ and $[\cdot,\cdot]_2$ respectively. A linear operator V: $\mathscr{H}_1 \to \mathscr{H}_2$ is called a *plus-operator* if its dormain of definition contains positive vectors, and non-negative vectors are carried by it into non-negative vectors $(\mathscr{D}_V \cap \mathscr{P}^{++}(\mathscr{H}_1) \neq \emptyset;\quad V(\mathscr{P}^+(\mathscr{H}_1) \cap \mathscr{D}_V) \subset \mathscr{P}^+(\mathscr{H}_2))$.

Definition 4.2: An operator V: $\mathscr{H}_1 \to \mathscr{H}_2$ is said to be (W_1, W_2)-*non-contractive if* $[Vx, Vx]_2 \geqslant [x, x]_1$ for all $x \in \mathscr{D}_V$.

If follows from Definitions 4.1 and 4.2 that we may take as an example of a plus-operator either an operator V having a non-negative range of values $\mathscr{R}_V$ or a V which is 'collinear' with a (W_1, W_2)-non-contractive operator U, i.e., $V = \lambda U$ with $\lambda \neq 0$. It turns out that, in fact, *these examples exhaust the whole class of plus-operators.*

Theorem 4.3: *Let* V: $\mathscr{H}_1 \to \mathscr{H}_2$ *be a plus-operator. Then there is a number* $\mu \geqslant 0$ *such that* $[Vx, Vx]_2 \geqslant \mu[x, x]_1$ $(x \in \mathscr{D}_V)$.

□ If $\mathscr{D}_V$ is indefinite, then we are in the conditions of lemma I 1.35 and corollary I 1.36 ($\mathscr{D}_V = \mathscr{F}_1$, $V =$) which it follows that

$$[Vx, Vx]_2 \geqslant \mu[x, x], \quad \text{when } \mu(V) \leqslant \mu \leqslant \mu_+(V).$$

Since it follows from the Definition 4.1 of a plus-operator that

$$\mu_+(V) = \lim_{[x, x]_1 > 0} \left\{ \frac{[Vx, Vx]_2}{[x, x]_1} \right\} \geqslant 0,$$

it is sufficient to put $\mu = \mu_+(V) \geqslant 0$.

But if $\mathscr{D}_V$ is semi-definite, i.e., in our case non-negative (since by definition of a plus-operator $\mathscr{D}_V \cap \mathscr{P}^{++}(\mathscr{H}_1) \neq \emptyset$), then we may again take as μ $\mu_+(V) = \inf\{[Vx, Vx]_2 \mid x \in \mathscr{D}_V, [x, x]_1 = 1\} \geqslant 0$. ■

Definition 4.4: A plus-operator V is said to be *strict* if $\mu_+(V) > 0$, and *non-strict* if $\mu_+(V) = 0$.

We note at once that the range of values of a non-strict plus-operator is a non-negative lineal. However, this property is not characteristic for the given class of operators. Indeed, let $V = P^+$ be the orthoprojector on to the subspace $\mathscr{H}^+$ of the canonical decomposition $\mathscr{H} = \mathscr{H}^+ \oplus \mathscr{H}^-$ of a Krein space $\mathscr{H}$. It is clear that $\mathscr{R}_{P^+} = \mathscr{H}^+ \subset \mathscr{P}^+$, but none the less $\mu_+(P^+) = 1 > 0$.

Corollary 4.5: *A plus-operator V is strict if and only if it is collinear with some (W_1, W_2)-non-contractive operator T: $V = \lambda T$, and in this case $0 < |\lambda|^2 \leqslant \mu_+(V)$.*

□ If V is a strict plus-operator, then by Theorem 4.3 the operator $T = (1/\lambda)V$ $(0 < |\lambda|^2 \leqslant \mu_+(V))$ is a (W_1, W_2)-non-contractive operator. Conversely, if $V = \lambda T$ with $\lambda \neq 0$ and T is a (W_1, W_2)-non-contractive operator, then $[Vx, Vx]_2 \geqslant \lambda^2[x, x]_1$. Therefore $\mu_+(V) \geqslant |\lambda|^2$ (>0), i.e., V is a strict plus operator. ■

It should be noted that *plus-operators may be unbounded, and, what is more, they may not admit closure.* As an example of such an operator it suffices to consider an operator $V = \varphi(\cdot)x_0$, where φ is a linear discontinuous functional on $\mathscr{H}_1$, $(\theta \neq)x_0 \in \mathscr{P}^+(\mathscr{H}_2)$.

In this connection, criteria for the discontinuity of plus-operator acting in Krein spaces are of interest. We recall that the symbols P_i^+ $(i = 1, 2)$ denote the canonical projectors (see 1.§3.1).

Theorem 4.6: *In order that a plus-operator V acting from a J_1-space $\mathscr{H}_1$ into a J_2-space $\mathscr{H}_2$ should be bounded (should admit closure) it is necessary and sufficient that the operator $P_2^+ V$ should be bounded (should admit closure).*

□ It follows from Theorem 4.3 that the inequality

$$\| P_2^+ Vx \|_2^2 + \mu \| x \|_1^2 \geqslant \| P_2^- Vx \|_2^2 \tag{4.1}$$

holds for arbitrary $x \in \mathscr{D}_V$.

Let $P_2^+ V$ be a bounded operator. Then

$$\| Vx \|_2 \leqslant \| P_2^+ Vx \|_2 + \| P_2^- Vx \|_2 \leqslant (\| P_2^+ V \| + \sqrt{\| P_2^+ V \|^2 + \mu}) \| x \|_1,$$

and therefore the plus-operator V is also bounded. Conversely, if V is a bounded operator, then it is obvious that $P_2^+ V$ is also a bounded operator.

Now suppose that the plus-operator V admits closures, *i.e.*, if follows from $Vx_n \to y$ and $x_n \to \theta$ that $y = \theta$. We verify that $P_2^+ V$ also admits closure. Let $P_2^+ Vx_n \to y^+$ and $x_n \to \theta$. If the sequences $\{P_2^+ V\}$ and $\{x_n\}$ are fundamental, then by (4.1) $\{Vx_n\}$ is fundamental. Since the operator V admits closure, $Vx_n \to \theta$, and therefore $P_2^+ Vx_n \to \theta$, i.e., $y^+ = \theta$, and the operator $P_2^+ V$ admits closure.

Conversely, suppose $P_2^+ V$ admits closure. Then V has the same property. For, if $Vx_n \to y$ and $x_n \to \theta$, then $P_2^\pm Vx_n \to P_2^\pm y$, and the fact that $P_2^+ V$ can be closed implies that $P_2^+ y \to \theta$; and then it follows from (4.1) that $P_2^- Vx_n \to \theta$, i.e., $y = \theta$. ■

Corollary 4.7: *Under the conditions of Theorem* 4.6 *let* $\mathscr{H}_2 = \Pi_\varkappa$ *be a Pontryagin space. Then the following assertions are equivalent:*

a) *V is a continuous operator;*
b) *the operator V admits closure.*

□ The eigenvalence of assertions a) and b) follows directly from Theorem 4.6 when we take into account that a finite-dimensional operator (in our case, $P_2^+ V$) admits closure if and only if it is bounded. ■

Corollary 4.8: *Under the conditions of Theorem* 4.6 *let* $\mathscr{H}_1 = \Pi_\varkappa$ *and* $\mathscr{H}_2 = \Pi'_\varkappa$ *be Pontryagin spaces, let* $\bar{\mathscr{D}}_V = \mathscr{H}_1$ *and* $\mathscr{P}^{++}(\mathscr{H}_1) \cap \mathrm{Ker}\ V = \emptyset$. *Then V is a continuous operator.*

□ By Theorem 4.6 if suffices to verify that $P_2^+ V$ is a bounded operator. Let $\{e_i\}_1^\varkappa$ be a basis in $\mathscr{H}_2^+$. Then the operator $P_2^+ V$ can be expressed in the form of a sum

$$P_2^+ Vx = \sum_{i=1}^{\varkappa} \varphi_i(x) e_i, \tag{4.2}$$

where $\varphi_1, \varphi_2, \ldots, \varphi_\varkappa$ are linear functionals defined on $\mathscr{D}_V$. Continuity of $P_2^+ V$ is equivalent to the continuity of the whole set of functionals $\{\varphi_i\}_1^\varkappa$. Therefore if, under the conditions of the Corollary, the operator $P_2^+ V$ were unbounded, then at least one of the functionals $\varphi_1, \varphi_2, \ldots, \varphi_\varkappa$ would be unbounded. Suppose for definiteness that φ_1 is unbounded. As is well-known ([XIII]) $\overline{\mathrm{Ker}\ \varphi_1} \supset \mathscr{D}_V$, and since $\bar{\mathscr{D}}_V = \mathscr{H}_1$, so $\overline{\mathrm{Ker}\ \varphi_1} = \mathscr{H}_1$. By Lemma 1.9.5 there is at least one positive K-dimensional subspace $\mathscr{L}^+$ in $\mathrm{Ker}\ \varphi_1 (\subset \mathscr{D}_V)$. From (4.2) we have, on the one hand, $\dim P_2^+ V\mathscr{L}^+ < \varkappa$; but on the other hand, since $\mathscr{P}^{++}(\mathscr{H}_1) \cap \mathrm{Ker}\ V = \emptyset$, we have $\dim P_2^+ V\mathscr{L}^+ = \dim V\mathscr{L}^+ = \dim \mathscr{L}^+ = \varkappa$; so we have a contradiction. ■

2 In this paragraph we shall assume, if nothing to the contrary is stated, that V is a continuous strict plus-operator defined on the whole of a J_1-space $\mathscr{H}_1 = \mathscr{H}_1^+ \oplus \mathscr{H}_1^-$ and acting into a J_2-space $\mathscr{H}_2 = \mathscr{H}_2^+ \oplus \mathscr{H}_2^-$, where, as before, $J_i = P_i^+ - P_i^-$, and $P_i^\pm$ are the ortho-projectors from $\mathscr{H}_i$ on to $\mathscr{H}_i^\pm$ $(i = 1, 2)$.

Definition 4.9: A strict plus-operator V is said to be *doubly strict* if V^c is also a strict plus-operator.

We point out that, in general, V^c will not be a plus-operator for every plus-operator V (not even if V is a strict plus-operator) (see Exercise 32 below).

4.10: A strict plus-operator V will be doubly strict if and only if V^* is a strict plus operator.

□ It is sufficient to use the Definition 4.9 and formula (1.1). ■

Theorem 4.11: *If V is a strict plus-operator, then there is a $\delta > 0$ such that the inequality*

$$\| Vx \|_2 \geqslant \delta \| x \|_1 \tag{4.3}$$

holds for all $x \in \mathscr{P}^+(\mathscr{H}_1)$. The number

$$\frac{\mu_+(V)}{(\| V \|^2 + \mu_+(V))^{1/2} + \| V \|}$$

can be taken as δ.

□ Let $x \in \mathscr{P}^+(\mathscr{H}_1)$, $\| x \|_1 = 1$. From the strictness of the plus-operator V and Corollary 4.5 we conclude that the form $[(V^c V - \mu_+(V)I_1)x, y]_1$ is non-negative in $\mathscr{H}_1$, and therefore we obtain from the Cauchy–Bunyakovski inequality for any $y \in \mathscr{H}_1$ with $\| y \|_1 = 1$ that

$$| [(V^c V - \mu_+(V)I_1)x, y]_1 |^2 \leqslant [(V^c V - \mu_+(V)I_1 x, x]_1] [(V^c V - \mu_+(V)I_1)y, y]_1$$
$$\leqslant (\| V \|^2 + \mu_+(V)) \| Vx \|_2^2.$$

Hence, if follows that

$$\| (V^c V - \mu_+(V)I_1)x \|_1^2 \leqslant (\| V \|^2 + \mu_+(V)) \| Vx \|_2^2$$

(for, if $(V^c V - \mu_+(V)I_1)x = \theta$, this is obvious, and in the opposite case we put $y = J_1(V^c V - \mu_+(V)I_1)x / \| (V^c V - \mu_+(V)I_1)x \|_1$ in the above inequality). On the other hand,

$$\| (V^c V - \mu_+(V)I_1)x \|_1 \geqslant \mu_+(V) - \| V \| \, \| Vx \|_2$$

and therefore

$$\mu_+(V) - \| V \| \, \| Vx \|_2 \leqslant (\| V \|^2 + \mu_+(V))^{1/2} \| Vx \|_2,$$

which is equivalent to the inequality

$$\| Vx \|_2 \geqslant \frac{\mu_+(V)}{(\| V \|^2 + \mu_+(V))^{1/2} + \| V \|}$$ ■

Corollary 4.12: *Let V be a strict plus-operator, with $\mathscr{D}_V = \mathscr{H}_1$. Then*

$$\| P_2^+ Vx \|_2 \geqslant (\delta/\sqrt{2} \| x \|_1 \quad \text{for any } x \in \mathscr{P}^+(\mathscr{H}_1).$$

□ It suffices to note that, because the vector Vx is non-negative, we have

$$\| Vx \|_2^2 = \| P_2^+ Vx \|_2^2 + \| P_2^- Vx \|_2^2 \leqslant 2 \| P_2^+ Vx \|_2^2,$$

and we then use the inequality (4.3). ■

□ From Corollary 4.5 and Theorem 4.11 we obtain directly

Corollary 4.13: *A continuous strict plus-operator V maps every non-negative (positive, uniformly positive) lineal $\mathscr{L}$ $(\subset \mathscr{H}_1)$ homeomorphically on to a non-negative (positive, uniformly positive) lineal $V\mathscr{L}(\subset \mathscr{H}_2)$.* ■

One of the trivial consequences of Theorem 4.11 is the fact that Ker V is a negative subspace. It turns out that a more exact assertion is also true:

4.14 If V is a strict plus-operator and $\mathscr{D}_V = \mathscr{H}_1$, then the subspace Ker V *is uniformly negative.*

□ Since the form $[(V^c V - \mu_+(V)I_1)x, y]$ is non-negative, we obtain from the Cauchy–Bunyakovski inequality that, when $x \in \text{Ker } V$ and $y = J_1 x$,

$$|[(V^c V - \mu_+(V)I_1)x, y]_1|^2 = \mu_+^2(V)(x, x)_1^2$$
$$\leqslant -\mu_+(V)[x, x]_1 \|(V^c V - \mu_+(V)I_1)\| (x, x)_1.$$

Hence, if $V^c V = \mu_+(V)I_1$, then Ker $V = \{\theta\}$, and in the opposite case

$$-[x, x]_1 \geqslant \frac{\mu_+(V)}{\| V^c V - \mu_+(V)I \|} \| x \|_1^2,$$

which proves our assertion. ■

Let T be an arbitrary linear operator acting from a J_1-space $\mathscr{H}_1$ into a J_2-space $\mathscr{H}_2$ and let $P_1^+ \mathscr{D}_T \subset \mathscr{D}_T$, or, what is equivalent, $P_1^- \mathscr{D}_T \subset \mathscr{D}_T$. Then the operator T can be represented in the form of a matrix

$$T = \| T_{ij} \|_{i,j=1}, \tag{4.4}$$

where

$$T_{11} = P_2^+ T P_1^+ \mid \mathscr{H}_1^+, \qquad T_{12} = P_2^+ T P_1^- \mid \mathscr{H}_1^-,$$
$$T_{21} = P_2^- T P_1^+ \mid \mathscr{H}_1^+, \qquad T_{22} = P_2^- T P_1^- \mid \mathscr{H}_1^-.$$

We note that in the case when the J_1-space $\mathscr{H}_1$ coincides with the J_2-space $\mathscr{H}_2$, i.e., $\mathscr{H}_1 = \mathscr{H}_2$, $J_1 = J_2$, the matrix (4.4) coincides with the matrix (2.1).

Theorem 4.15: *Let V be a strict plus-operator and $\mathscr{D}_V = \mathscr{H}_1$. Then for all $\mathscr{L} \in \mathscr{M}^+ \mid \mathscr{H}_1)$ the deficiency* def $V\mathscr{L} \equiv \dim \mathscr{H}_2^+ / P_2^+ V\mathscr{L}$ *is the same.*

□ Let $\mathscr{L} \in \mathscr{M}^+$ and let K be the angular operator of the subspace $\mathscr{L}$. By formula (4.4) the matrix $\| V_{ij} \|_{i,j=1}^{2}$ corresponds to the operator V. Therefore

$$V\mathscr{L} = \{(V_{11} + V_{12}K)x_1^+ + (V_{21} + V_{22}K)x_1^+ \mid x_1^+ \in \mathscr{H}_1^+\}$$

and

$$P_2^+ V\mathscr{L} = \{(V_{11} + V_{12}K)x_1^+ \mid x_1^+ \in \mathscr{H}_1^+\},$$

i.e.,

$$P_2^+ V\mathscr{L} = (V_{11} + V_{12}K)\mathscr{H}_1^+. \tag{4.5}$$

By Corollary 4.12 we have

$$\| (V_{11} + V_{12}K)x_1^+ \|_2 \geqslant \delta/\sqrt{2} \, \| x_1^+ + Kx_1^+ \|_1 \qquad (\geqslant \delta/\sqrt{2} \, \| x_1^+ \|_1)$$

and therefore

$$\operatorname{Ker}(V_{11} + V_{12}K) = \{\theta\}, \qquad \mathscr{R}_{V_{11} + V_{12}K} = \bar{\mathscr{R}}_{V_{11} + V_{12}K}, \tag{4.6}$$

and moreover the norms of the operators $(V_{11} + V_{12}K)^{-1}$ (see (3.1)) are uniformly bounded with respect to $K \in \mathscr{H}_1^+ (\equiv \mathscr{H}_1^+ (\mathscr{H}_1^+, \mathscr{H}_1^-))$:

$$\| (V_{11} + V_{12}K)^{(-1)} \| \leqslant \sqrt{2}/\delta. \tag{4.7}$$

It follows from formula (4.5) that the deficiency of $V\mathscr{L}$ coincides with the co-dimension of the subspace $(V_{11} + V_{12}K)\mathscr{H}_1^+$ in $\mathscr{H}_2^+$. Let K_0 be the angular operator of the subspace $\mathscr{L}_0 \in \mathscr{M}^+ (\mathscr{H}_1)$, and let K be an angular operator such that $\| K - K_0 \| < (\delta/\sqrt{2}) \| V_{12} \|$. Then, from the from the fact that

$$(V_{11} + V_{12}K)\mathscr{H}_1^+ = (I_2^+ + V_{12}(K - K_0)(V_{11} + V_{12}K_0)^{(-1)})(V_{11} + V_{12}K_0)\mathscr{H}_1^+$$

and

$$\| V_{12}(K - K_0)(V_{11} + V_{12}K_0)^{(-1)} \| < 1,$$

we obtain that the dimensions of the subspaces

$$((V_{11} + V_{12}\mathrm{K})\mathscr{H}_1^+)^\perp \cap \mathscr{H}_1^+ \quad \text{and} \quad ((V_{11} + V_{12}K_0)\mathscr{H}_1^+)^\perp \cap \mathscr{H}_2^+$$

coincide, i.e., the deficiencies of the subspaces $V\mathscr{L}_0$ and $V\mathscr{L}$ coincide. By virtue of the linear connectivity of the operator ball the validity of the theorem follows. ■

The theorem just proved justifies the introduction of the following

Definition 4.16: The number $\delta_+(V) = \dim \mathscr{H}_2^+ / P_2 V\mathscr{L}$ $(\mathscr{L} \in \mathscr{M}^+ (\mathscr{H}_1))$ is called the *plus-deficiency* of a strict continuous plus-operator V with $\mathscr{D}_V = \mathscr{H}_1$.

We now introduce a criterion for the double strictness of a strict plus-operator.

Theorem 4.17: *The following assertions for a strict-operator are equivalent:*

a) *V is doubly strict;*
b) *the subspace $V\mathscr{L}_0 \in \mathscr{M}^+ (\mathscr{H}_2)$ for at least one $\mathscr{L}_0 \in \mathscr{M}^+ (\mathscr{H}_1)$;*

c) *$V\mathscr{L} \in \mathscr{M}^+(\mathscr{H}_2)$ for any $\mathscr{L} \in \mathscr{M}^+(\mathscr{H}_1)$;*

d) *for at least one angular operator $K_0 \in \mathscr{K}_1^+$ the operator $(V_{11} + V_{12}K_0)^{-1}$ exists and is defined on the whole of $\mathscr{H}_2^+$;*

e) *for any angular operators $K \in \mathscr{K}_1^+$ the operator $(V_{11} + V_{12}K)^{-1}$ exists and is defined on the whole of $\mathscr{H}_2^+$.*

□ We notice at once that from (4.5) and (4.6) the implications b) ⇔ d) and c) ⇔ e) follow immediately. Further, c) ⇔ b) since, by Theorem 4.15, the plus-deficiency $\delta_+(V)$ does not depend on the choice of $\mathscr{L} \in \mathscr{M}^+$. It is therefore sufficient for us to prove a) ⇒ d) and c) ⇒ a).

a) ⇒ d). It follows from Proposition 4.10 that

$$V^* = \| (V^*)_{ij} \|_{i,j=1}^2 \qquad ((V^*)_{ij} = V_{ji}^*, i, j = 1, 2)$$

is a strict plus-operator whenever V is, and therefore from formula (4.6) written for V^* with $K = 0$, (i.e., $\mathscr{L} = \mathscr{H}_1^+$) we have

$$\text{Ker } V_{11}^* \mid \mathscr{H}_2^+ = \{\theta\} \quad \text{and} \quad \mathscr{R}_{V_{11}^*} = \bar{\mathscr{R}}_{V_{11}^*}.$$

Comparing these relations with (4.6) we conclude that the operator V_{11} is continuously invertible on the whole of $\mathscr{H}_2^+$, i.e., d) holds when $K_0 = 0$. It can be verified immediately that in this case

$$\mu_+(V) \| V_{11}^{-1} \|^2 \leqslant 1 - \inf\left\{ \frac{\| V_{21} V_{11}^{-1} x_2^+ \|_2^2}{\| x_2^+ \|_2^2} \,\middle|\, x_2^+ \neq \theta \right\}, \tag{4.8}$$

and therefore $\mu_+^{1/2}(V) V_{11}^{-1}$ is a contraction.

c) ⇒ a). First of all we prove that V^c is a plus-operator. If this were not so, there would be a vector $(\theta \neq)\ x_0 \in \mathscr{P}^+(\mathscr{H}_2)$ such that $V^x x_0 \in \mathscr{P}^{--}(\mathscr{H}_1)$. Let Q be the angular operator of any maximal negative subspace containing $V^c x_0$. Then $K(= Q^*)$ is the angular operator of a maximal positive subspace J_1-orthogonal to $V^c x_0$ (see Theorem 1.8.11). Consequently $V^c x_0 [\perp] (P_1^+ + K)\mathscr{H}_1{}^+$ and therefore $x_0 [\perp] V(P_1^+ + K)\mathscr{H}_1^+$, i.e., $x_0 (\in \mathscr{P}^+(\mathscr{H}_2))$ is J_2-orthogonal to the maximal positive subspace $V(P_1^+ + K)\mathscr{H}_1^+$ (see Corollary 4.13)—we have obtained a contradiction; so V^c is a plus-operator.

We verify that V^c is a strict plus-operator. In the opposite case $\mathscr{R}_{V^c} \subset \mathscr{P}^+(\mathscr{H}_1)$, and since Ker $V = \mathscr{R}_{V^c}^{[\perp]}$ (see 1.12) and since by 4.14 Ker V is a uniformly negative subspace, it follows from Proposition 1.7.4, Corollary 1.7.17, and Proposition 1.1.25 that $\bar{\mathscr{R}}_{V^c}$ is a maximal uniformly positive subspace (with an angular operator $K' \in \mathscr{K}_2^+ \equiv \mathscr{K}^+(\mathscr{H}_2^+, \mathscr{H}_2^-)$ with $\| K' \| < 1$). The operator V^c acts from $\mathscr{H}_2 = \mathscr{H}_2^+ \oplus \mathscr{H}_2^-$ into $\mathscr{H}_1 = \mathscr{H}_1^+ \oplus \mathscr{H}_1^-$ and has, as is easily verified, the matrix representation

$$V^c = \| (V^c)_{ij} \|_{i,j=1}^2, \quad \text{where } (V^c)_{ii} = V_{11}^* \ (i = 1, 2),$$
$$\text{and} \quad (V^c)_{ij} = -V_{ji}^* (i \neq j).$$

Since

$$V^c \mathscr{H}_2^+ \subset \mathscr{R}_{V^c}, \quad \text{so } -V_{12}^* = K' V_{11}^* \quad \text{and} \quad V_{22}^* = -K' V_{21}^*.$$

Hence, we have

$$\inf\left\{\frac{[V^c x, V^c x]_1}{[x, x]_2}\,\middle|\, x \in \mathcal{P}^{++}(\mathcal{H}_2)\right\} \geqslant \frac{1 - \|K'\|^2}{2} \inf\left\{\frac{\|V^c x\|_1^2}{\|x\|_2^2}\,\middle|\, x \in \mathcal{P}^{++}(\mathcal{H}_2)\right\}.$$

To show that the operator V^c is strict it suffices to show that the quantity on the right-hand side is positive. To do this we note that it follows from the strictness of the operator V and from Corollary 4.13 that the angular operator of the subspace $V\mathcal{H}_1^{\pm}$, which coincides with $V_{21}V_{11}^{-1}$, has a norm less than 1, and therefore $\|V_{11}^{*-1}V_{21}^{*}\|$ is also <1. Therefore (see Proposition 1.8.7)

$$\inf\left\{\frac{\|V^c x\|_1^2}{\|x\|_2^2}\,\middle|\, x \in \mathcal{P}^{++}(\mathcal{H}_2)\right\} \geqslant \inf\left\{\frac{\|V^c(x_2^+ + Kx_2^+)\|_1^2}{\|x_2^+ + Kx_2^+\|_2^2}\,\middle|\, x_2^+ \in \mathcal{H}_2^+,\ K \in \mathcal{K}_2^+\right\}$$

$$\geqslant \inf\left\{\frac{\|P_1^+ V^c(x_2^+ + Kx_2)\|_1^2}{2\|x_2^+\|_2^2}\,\middle|\, x_2 \in \mathcal{H}_2^+,\ K \in \mathcal{K}_2^+\right\}$$

$$= \inf\left\{\frac{\|(V_{11}^* - V_{21}^*K)x_2^+\|_1^2}{\|x_2^+\|_2^2}\,\middle|\, x_2 \in \mathcal{H}_2^+, K \in \mathcal{K}_2^+\right\}$$

$$\geqslant \frac{1}{2\|V^{*-1}\|^2(\sum_{j=0}^{\infty}\|V^{*-1}V_{21}^*\|^j)} > 0. \qquad \blacksquare$$

Definition 4.18: A continuous (J_1, J_2)-non-contractive operator V with $\mathcal{D}_V = \mathcal{H}_1$ is said to be a *(J_1, J_2)-bi-non-contractive operator* if V^c is a (J_1, J_2)-non-contractive operator.

It follows directly from this definition that a (J_1, J_2)-bi-non-contractive operator is a doubly strict plus-operator. However, *not every doubly strict plus-operator is a (J_1, J_2)-bi-non-contractive operator*; as an example we may take the operator $\frac{1}{2}P_1^+$, which is a doubly strict operator in the J_1-space $\mathcal{H}_1$. The following theorem describes all (J_1, J_2)-bi-non-contractive operators.

Theorem 4.19: *For a continuous (J_1, J_2)-non-contractive operator V with $\mathcal{D}_V = \mathcal{H}_1$ to be a (J_1, J_2)-bi-non-contractive operator it is necessary and sufficient that it be a doubly strict plus-operator.*

Necessity: As already noted, this is trivial.

Sufficiency: We consider the space

$$\mathcal{H}_{\hat{\Gamma}} = \mathcal{H}_2 \oplus \mathcal{H}_1 \tag{4.9}$$

and we introduce into it a Hilbert metric and an indefinite $J_{\hat{\Gamma}}$-metric given respectively by the formulae

$$(\langle x_2, x_1\rangle, \quad \langle y_2, y_1\rangle)_{\hat{\Gamma}} = (x_2, y_2)_2 + (x_1, y_1)_1, \tag{4.10}$$

$$[\langle x_2, x_1\rangle, \quad \langle y_2, y_1\rangle]_{\hat{\Gamma}} = [x_2, y_2]_2 - [x_1, y_1]_1, \tag{4.11}$$

where $x_i, y_i \in \mathcal{H}_i$ $(i = 1, 2)$.

It can be verified directly that the $J_{\hat\Gamma}$-metric is generated by the operator

$$J_{\hat\Gamma} = J_2 \oplus -J_1 \tag{4.12}$$

and that this operator is a selfadjoint involution, and the canonical decomposition of $\mathscr{H}_{\hat\Gamma}$ is defined be the formulae

$$\mathscr{H}_{\hat\Gamma} = \mathscr{H}_{\hat\Gamma}^{+} \oplus \mathscr{H}_{\hat\Gamma}^{-}, \qquad \mathscr{H}_{\hat\Gamma}^{\pm} = \mathscr{H}_2^{\pm} \oplus \mathscr{H}_1^{\mp}. \tag{4.13}$$

Moreover, it is easy to see from formula (4.11) that if T is an operator densely defined in $\mathscr{H}_1$ and mapping into $\mathscr{H}_2$ then the $J_{\hat\Gamma}$-orthogonal complement to its graph $\Gamma_T = \{\langle Tx, x\rangle \mid x \in \mathscr{D}_T\}$ in $\mathscr{H}_{\hat\Gamma}$ coincides with the graph of the operator T^c:

$$\Gamma_T^{[\perp]} = \Gamma_{T^c} = \{\langle y, T^c y\rangle \mid y \in \mathscr{D}_{T^c}\}. \tag{4.14}$$

It follows from the same formula (4.11) that the graph Γ_V of the operator V is a non-negative subspace. Moreover, it is a maximal non-negative subspace. For if it were not, then by Theorem 1.4.5 there would be a vector $\langle x_2^+, x_2^-\rangle$ in $\mathscr{H}_{\hat\Gamma}^{+}$ which was $J_{\hat\Gamma}$-orthogonal to Γ_V, and therefore it would follow from (4.14) that $x_1^- = V^c x_2^+$. But since V is a doubly strict plus-operator, $x_1^- = \theta$. We now use Proposition 4.14 and obtain $x_2^+ = \theta$. Therefore $\Gamma_V \in \mathscr{M}^+(\mathscr{H}_{\hat\Gamma})$, and so $\Gamma_{V^c} \in \mathscr{M}^-(\mathscr{H}_{\hat\Gamma})$ (see (4.14) and Theorem 1.8.11),
i.e., $[y, y]_2 - [V^c y, V^c y]_1 \leqslant 0$ for all $y \in \mathscr{H}_2$. ■

Corollary 4.20: *If V is a doubly strict plus-operator with $\mathscr{D}_V = \mathscr{H}_1$ and if*

$$[Vx, Vx]_2 \geqslant \mu[x, x]_1 \quad \textit{for } \mu > 0 \textit{ and all } x \in \mathscr{H}_1,$$

then

$$[V^c y, V^c y]_1 \geqslant \mu[y, y]_2 \quad \textit{for all } y \in \mathscr{H}_2,$$

(or (what is the same thing) $V^c V - \mu I_1 \geqslant 0$ implies $VV^c - \mu I_2 \overset{J_2}{\geqslant} 0$), and therefore

$$\mu_+(V^c) = \mu_+(V) \textit{ and } \max\{0; \mu_-(V^c)\} = \max\{0; \mu_-(V)\}.$$

□ It suffices to note that $(1/\sqrt{\mu})V$ is a (J_1, J_2)-non-contractive operator and then to use Theorem 4.19. ■

Remark 4.21: We leave the reader to compare Theorems 4.17 and 4.19, and to compile a set of criteria for a (J_1, J_2)-non-contractive operator to be a (J_1, J_2)-bi-non-contractive operator.

3 Here we investigate the interconnection between certain classes of plus-operators V. As before, we shall assume that V is continuous and that $\mathscr{D}_V = \mathscr{H}_1$.

Definition 4.22: A plus-operator V is said to be *focusing* if there is a constant $\gamma > 0$ such that $[Vx, Vx]_2 \geqslant \gamma \| Vx \|_2^2$ for $x \in \mathscr{P}^+(\mathscr{H}_1)$.

Definition 4.23: An operator V is said to be *uniformly (J_1, J_2)-expansive* if there is a constant $\delta > 0$ such that $[Vx, Vx]_2 \geqslant [x, x]_1 + \delta \| x \|_1^2$, i.e. $V^c V - I_1 \overset{J_1}{\gg} 0$, and to be *uniformly (J_1, J_2)-bi-expansive* if the operator V^c has the same property.

Theorem 4.24: *For a plus-operator V the following assertions are equivalent:*

a) *V is collinear with a certain uniformly (J_1, J_2)-expansive operator;*
b) *V is a focusing strict plus-operator;*
c) $\mu_+(V) > \max\{0; \mu_-(V)\}$;
d) *there is a $\nu_0 > 0$ such that $V^c V - \nu_0 I_1 \overset{J_2}{\gg} 0$.*

☐ a) ⇒ b). If V is collinear with a uniformly (J_{11}, J_2)-expansive operator, then V is a strict plus-operator (see Corollary 4.5) and there are numbers $\alpha > 0$ and $\delta > 0$ such that $[Vx, Vx]_2 \geqslant \alpha [x, x]_1 + \delta \| x_1 \|^2$. Therefore when $x \in \mathscr{P}^+(\mathscr{H}_1)$ we have

$$[Vx, Vx]_2 \geqslant \delta \| x \|_1^2 \geqslant \frac{\delta}{\| V \|^2} \| Vx \|_2^2,$$

i.e., V is a focusing plus-operator.

b) ⇒ c). Let V be a focusing strict plus-operator:

$$[VX, Vx]_2 \geqslant \gamma \| Vx \|_2^2 \qquad (x \in \mathscr{P}^+(\mathscr{H}_1)).$$

If $\mu_-(V) \leqslant 0$, then the implication b) ⇒ c) follows from the definition of the strictness of a plus-operator V: $\mu_+(V) > 0$.

Now suppose $\mu_-(V) > 0$. We consider operators $V_\varepsilon = S_\varepsilon^{(2)} V$, where $S_\varepsilon^{(2)} = (1 - \varepsilon) P_2^+ + P_2^-$. For sufficiently small $\varepsilon > 0$ and $x \in \mathscr{P}^+(\mathscr{H}_1)$ we have, taking Definition 4.22 into account,

$$\begin{aligned}[V_\varepsilon x, V_\varepsilon x]_2 &= [(V - \varepsilon P_2^+ V)x, (V - \varepsilon P_2^+ V)x]_2 \\ &= [Vx, Vx]_2 + (\varepsilon^2 - 2\varepsilon)[P_2^+ Vx, P_2^+ Vx]_2 \\ &\geqslant (\gamma + \varepsilon^2 - 2\varepsilon) \| Vx \|_2^2 \geqslant (\gamma + \varepsilon^2 - 2\varepsilon) \| V_\varepsilon x \|_2^2,\end{aligned}$$

and therefore V_ε is a focusing plus-operator. Since $[V_\varepsilon x, V_\varepsilon x] \leqslant [Vx, Vx]_2$ we have $\mu_-(V) \leqslant \mu_-(V_\varepsilon) \leqslant \mu_+(V_\varepsilon) \leqslant \mu_+(V)$.

Now to complete the proof of the implication b) ⇒ c) it sufficies to verify that $\mu_+(V_\varepsilon) < \mu_+(V)$. Let $[x, x]_1 = 1$. Then

$$\begin{aligned}[V_\varepsilon x, V_\varepsilon x]_2 &= [Vx, Vx]_2 + (\varepsilon^2 - 2\varepsilon)[P_2^+ Vx, P_2^+ Vx]_2 \\ &\leqslant (1 - \varepsilon)^2 [Vx, Vx]_2\end{aligned}$$

(since $\varepsilon^2 - 2\varepsilon < 0$ for sufficiently small $\varepsilon > 0$), and therefore $\mu_+(V_\varepsilon) \leqslant (1 - \varepsilon)^2 \mu_+(V) < \mu_+(V)$.

c) ⇒ d). Suppose the contrary: $\mu_+(V) > \max\{0; \mu_-(V)\}$ but nevertheless there is no $\nu > 0$ such that $V^c V - \nu I_1 \overset{J_1}{\gg} 0$. Let the positive numbers ν_1, ν_2, ν_3 satisfy the following inequalities: $\mu_-(V) < \nu_1 < \nu_2 < \nu_3 < \mu_+(V)$. Then (see Corollary 1.1.36) $[Vx, Vx]_2 \geqslant \nu_i[x, x]_1$ $(i = 1, 2, 3)$. From the premise about V we have that there is a sequence of vectors $\{x_n\} \subset \mathscr{H}_1$ with $\| x_n \| = 1$ such that

$$[Ix_n, Vx_n]_2 - \nu_2[x_n, x_n]_1 = [(V^c V - \nu_2 I_1)x_n, x_n]_1 \to 0 \text{ as } n \to \infty.$$

Since the operator $V^* H_2 V - \nu_2 J_1$ is non-negative, we have

$$(V^* J_2 V - \nu_2 J_1)x_n \to \theta. \tag{4.15}$$

Let $\{x_n\}$ be a sub-sequence of the sequence $\{x_n\}$ consisting of semi-definite vectors, for definiteness, let us say of non-negative vectors. Then

$$[(V^c V - \nu_2 I_1)x_n', x_n']_1 \geqslant [(V^c V - \nu_3 I_1)x_n', x_n']_1,$$

and therefore

$$(V^* J_2 V - \nu_3 I_1)x_n' \to \theta. \tag{4.16}$$

If $\{x_n'\} \subset \mathscr{P}^-$, then similarly we obtain

$$(V^* J_2 V - \nu_1 J_1)x_n' \to \theta. \tag{4.17}$$

Comparing the relation (4.15) with the relation (4.16) or (4.17) we obtain $x_n' \to \theta$—a contradiction.

The implication d) ⇒ a) follows immediately from the fact that $1/\sqrt{\nu_0}\, V$ is a uniformly (J_1, J_2)-expansive operator. ■

□ *Remark 4.25*: In proving the implication c) ⇒ d) it was established, essentially, that if V is a strict plus-operator and $\mu_+(V) > \nu > \max\{0; \mu_-(V)\}$, then there is a $\delta_\nu > 0$ such that $[Vx, Vx]_2 \geqslant \nu[x, x]_1 + \delta_\nu \| x \|_1^2$. Moreover, the converse proposition is true: if $[Vx, Vx]_2 \geqslant \nu[x, x_{+1} + \delta_\nu \| x \|_1^2$ for all $x \in \mathscr{H}_1$, $\nu > 0$, then $\mu_+(V) > \nu > \max\{0; \mu_-(V)\}$. For, $\mu_+(v) \geqslant \nu \geqslant \mu_-(V)$ by Corollary I.1.36. We verify that ν cannot coincide with $\mu_\pm(V)$. Suppose, for example, that $\mu_+(V) = \nu$. Then, by the definition of the number $\mu_+(V)$, there would be a sequence $\{x_n\} \subset \mathscr{P}^+(\mathscr{H}_1)$ such that $[x_n, x_n]_1 = 1$ and $[Vx_n, Vx_n]_2 \to \mu_+(V)$, i.e., $[Vx_n, Vx_n]_2 - \mu_+(V)[x_n, x_n]_1 \to 0$, which contradicts the inequality $[Vx_n, Vx_n]_2 - \mu_+(V)[x_n, x_n]_1 \geqslant \delta_\nu \| x \|_1^2 \geqslant \delta_\nu [x_n, x_n]_1 = \delta_\nu > 0$. Similarly it can be verified that $\mu_-(\mathrm{V}) \neq \nu$.

Thus for any plus-operator V (including even non-strict ones)

$$V^c V - \nu I_1 \overset{J_1}{\geqslant} 0 \ (\max\{0; \mu_-(V)\} \leqslant \nu \leqslant \mu_+(V)), \tag{4.18}$$

and when $\max\{0; \mu_-(V)\} \neq \mu_+(V)$

$$V^c V - \nu I_1 \overset{J_1}{\gg} 0 \ (\max\{0; \mu_-(V)\} < \nu < \mu_+(V)). \quad ■ \tag{4.19}$$

This remark enables us to prove the Proposition.

4.26: For a uniformly (J_1, J_2)-expansive operator V to be a uniformly (J_1, J_2)-bi-expansive operator it is necessary and sufficient that V^c be a strict plus-operator.

☐ The necessity is trivial. The sufficiency follows from the fact that, by virtue of Remark 4.25, $\mu_+(V) > 1 > \mu_-(V)$, and therefore we have, from Corollary 4.20, $\mu_+(V^c) > 1 > \mu_-(V^c)$. It remains only to apply Remark 4.25 again. ■

We formulate next a theorem which characterizes the 'power' of the set of uniformly (J_1, J_2)-expansive operators.

Theorem 4.27: *The set of uniformly (J_1, J_2)-expansive operators ((J_1, J_2)-bi-expansive operators) is open in the uniform operator topology in the set of all continuous operators acting from $\mathscr{H}_1$ into $\mathscr{H}_2$, and its closure in this topology coincides with the set of all (J_1, J_2)-non-contractive operators ((J_1, J_2)-bi-non-contractive operators).*

☐ Let V be a uniformly (J_1, J_2)-expansive operator $[Vx, Vx]_2 \geqslant [x, x]_1 + \delta \| x \|_1^2 \; (\delta > 0)_1$ and let V' be such that $\delta - 2 \| V' - V\| \, \| V\| - \| V' - V\|^2 > 0$. Then

$$
\begin{aligned}
[V'x, V'x]_2 &= [(V + V' - V)x, (V + V' - V)x]_2 \\
&\geqslant [Vx, Vx]_2 - (2\| V' - V\| \, \| V\| + \| V' - V\|^2)\| x\|_1^2 \\
&\geqslant [x, x]_1 + (\delta - 2\| V' - V\| \, \| V\| + \| V' - V\|^2)\| x\|_1^2,
\end{aligned}
$$

i.e., V' is a uniformly (J_1, J_2)-expansive operator.

If, in addition, V were a uniformly (J_1, J_2)-bi-expansive operator, then by Theorem 4.17 $0 \in \rho(V_{11})$, and therefore for V' from a sufficiently small neighbourhood of V we have $0 \in \rho(V'_{11})$. It then follows from Theorem 4.17 and Proposition 4.26 that V' is a uniformly (J_1, J_2)-bi-expansive operator.

Since the uniform limit of (J_1, J_2)-non-contractive operators is a (J_1, J_2)-non-contractive operator, the closure of the set of uniformly (J_1, J_2)-expansive operators ((J_1, J_2)-bi-expansive operators) is embedded in the set of (J_1, J_2)-non-contractive operator (J_1, J_2)-bi-non-contractive operators). It remains to verify that if V is a (J_1, J_2)-non-contractive operator ((J_1, J_2)-bi-non-contractive operator), then it can be approximated in norm by uniformly (J_1, J_2)-expansive operators ((J_1, J_2)-bi-expansive operators). To do this we bring into consideration the operators

$$I_\varepsilon^{(1)} = \sqrt{1 + \varepsilon P_1^+} + \sqrt{1 - \varepsilon P_1^-}. \tag{4.20}$$

It can be verified immediately that when $0 < \varepsilon \leqslant 1$ the operators $I_\varepsilon^{(1)}$ are uniformly J_1-bi-expansive and when $\varepsilon \to 0$ they converge in norm to the unit operator in $\mathscr{H}_1$. Since

$$[VI_\varepsilon^{(1)}x, VI_\varepsilon^{(1)}x]_2 \geqslant [I_\varepsilon^{(1)}x, I_\varepsilon^{(1)}x] = [x, x]_1 + \varepsilon \| x \|_1^2,$$

$VI_\varepsilon^{(1)}$ are uniformly (J_1, J_2)-expansive operators ((J_1, J_2)-bi-expansive operators) which as $\varepsilon \to 0$ approximate in norm the (J_1, J_2)-non-contractive ((J_1, J_2)-bi-non-contractive) operator V. ■

□ From Theorems 4.24 and 4.27 follows immediately

Corollary 4.28: *The set of focusing strict plus-operators (focusing doubly strict plus-operators) is open in the set of all continuous operators and its closure coincides with set of all strict plus-operators (doubly strict plus-operators).*

We close this paragraph with the

Remark 4.29: Plus-operator and sub-classes of plus-operators acting from the anti-space to $\mathscr{H}_1$ into the anti-space to $\mathscr{H}_2$ will be called (as operators acting from $\mathscr{H}_1$ into $\mathscr{H}_2$) *minus-operators*, and the names of sub-classes will be changed correspondingly (for example, $(-J_1, -J_2)$-non-contractive operators will be called (J_1, J_2)*-non-expansive*). All the propositions given above for plus-operators can be reformulated in a natural way for minus-operators. We leave the reader to do this, and later, in using such propositions, we shall refer back to this Remark 4.29.

4 If V is a continuous plus-operator acting from a J_1-space $\mathscr{H}_1$ into a J_2-space $\mathscr{H}_2$, with $\mathscr{D}_V = \mathscr{H}_1$, then the operator $V^c V$ acts in $\mathscr{H}_1$, and therefore it is proper to ask about the description of its spectrum.

Theorem 4.30: *Let V and V^c be continuous plus-operators. Then* $\sigma(V^c V) \geqslant 0$.

□ From (4.18) and Theorem 3.24 it follows that $\sigma(V^c V) \subset \mathbb{R}$. Let $-\alpha \in \sigma(V^c V)$, $\alpha > 0$. Then there is a sequence $\{x_n\}_1^\infty \subset \mathscr{H}_1$, $\|x_n\| = 1$, such that

$$(V^c V + \alpha I_1)x_n \to \theta \qquad (n \to \infty). \tag{4.21}$$

Without loss of generality we may suppose that the sequences $\{[x_n, x_n]_1\}_1^\infty$, $\{[Vx_n, Vx_n]_2\}_1^\infty$ and $\{[V^c Vx_n, V^c Vx_n]_1\}_1^\infty$ converge and have the limits β, γ, and δ respectively. From (4.21) we obtain (by mulitplying $(V^c V_+ \alpha I_1)x_n$ scalarly by $J_1 x_n$ and taking the limit as $n \to \infty$) $\gamma + \alpha\beta = 0$ and (multiplying $(V^c V + \alpha I_1)x_n$ scalarly by $J_1 V^c Vx_n$ and taking the limit as $n \to \infty$) $\delta + \alpha\gamma = 0$. The first equality implies $\gamma \geqslant 0$; for, if $\beta \geqslant 0$, then $\gamma \geqslant 0$ because V is a plus-operator, and if $\beta < 0$, then $\gamma \geqslant 0$ because $\alpha \geqslant 0$ by hypothesis and $\gamma = -\alpha\beta$. Now from $\delta + \alpha\gamma = 0$ and the fact that V is a plus-operator we conclude that $\delta \geqslant 0$, i.e., $\delta = \gamma = 0$ and therefore $\beta = 0$ also. Hence,

$$\lim_{n \to \infty} [(V^c V - \mu_+(V)I_1)x_n, x_n]_1 = \gamma - \mu_+(V)\beta = 0,$$

and therefore (*cf.* proof of Theorem 3.24) $(V^cV-\mu_+(V)I_1)x_n \to \theta$ as $n\to\infty$. Comparing the last relation with (4.21), we obtain $x_n\to\theta$—a contradiction. ■

We now suppose that V is a continuous operator acting in a J-space $\mathscr{H}$ with $\mathscr{D}_V=\mathscr{H}$.

Theorem 4.31: *If V is a uniformly J-expansive operator, then its field of regularity contains the unit circle $\mathbb{T}=\{\xi \mid |\xi|=1\}$. Moreover, V is a uniformly J-bi-expansive operator if and only if $\mathbb{T}\subset\rho(V)$.*

□ Suppose $\xi\in\mathbb{T}$ and that ξ is not a point of regular type of the operator V. Then there is a sequence $\{x_n\}\subset\mathscr{H}$ with $\|x_n\|=1$ such that $(V-\xi I)x_n\to\theta$ when $n\to\infty$. Since V is a uniformly J-expansive operator, we have, for some $\delta>0$,

$$([Vx_n,Vx_n]-[x_n,x_n]=\xi[x_n,(V-\xi I)x_n]+\bar{\xi}[(V-\xi I)x_n,x_n]$$
$$+[(V-\xi I)x_n,(V-\xi I)x_n]\geqslant\delta\|x_n\|_1^2,$$

and therefore $x_n\to\theta$ when $n\to\infty$; a contradiction.

Now let V be a uniformly J-bi-expansive operator. From what has just been proved and Theorem 1.16, we conclude that $\mathbb{T}\subset\rho(V)$.

Conversely, let V be a uniformly J-expansive operator and let $\mathbb{T}\subset\rho(V)$. Since

$$VV^c-I=(V-I)(V^c-I)^{-1}(V^cV-I)(V-I)^{-1}(V^c-I),$$

so

$$V^cV-I\overset{J}{\gg}0 \quad\text{implies}\quad VV^c-I\overset{J}{\gg}0. \qquad ■$$

Corollary 4.32: *Let V be a focusing strict plus-operator in a J-space $\mathscr{H}$. Then all ξ such that $\sqrt{\max\{0,\mu_-(V)\}}<|\xi|<\sqrt{\mu_+(V)}$ are points of regular type of the operator V, and they are regular if and only if V is a doubly strict focusing plus-operator.*

□ This assertion follows directly from Theorem 4.24, (4.19), and Theorem 4.31. ■

Exercises and problems

1 Prove that for every strict plus-operator V there is on $\varepsilon=\varepsilon(V)>0$ such that for all plus-operators from an ε-neighbourhood (in norm) of the operator V the plus-deficiency is the same (M. Krein and Shmul'yan [2]).

2 Prove that if V_1, V_2 are strict plus-operators and $V_1-V_2\in\mathscr{S}_\infty$, then $\delta_+(V_1)=\delta_+(V_2)$ (M. Krein and Shmul'yan [2]).

3 Verify that if V is a strict plus-operator acting from a J_1-space $\mathcal{H}_1$ into a J_2-space $\mathcal{H}_2$, then $P_2^+ V$ also is a strict plus-operator.
Hint: Use the fact $[P_2^+ Vx, P_2^+ Vx]_2 \geqslant [Vx, Vx]_2$.

4 Prove that if V is a J-non-contractive continuous operator acting in a Krein space $\mathcal{H}$ and $\mathcal{D}_V = \mathcal{H}$, then the disc $\mathbb{D} = \{\lambda \,\|\, \lambda \,\| < 1\}$ belongs to the field of regularity of the operator V_{11}, $\mathbb{D} \subset \rho(V_{11})$, if and only if V is a J-bi-non-contractive operator (M. Krein and Shmul'yan [2]).
Hint: Use (4.8) and Theorem 4.17.

5 Prove that the conditions a)-e) in Theorem 4.17 are equivalent to the condition f) $0 \notin \sigma_p(V_{11}^*)$.
*Hint:*Use Exercise 4.

6 Let $\mathcal{H} = \mathcal{H}^+ \oplus \mathcal{H}^-$ be a J-space, $\dim \mathcal{H} = \dim \mathcal{H}^+$, and let V be a semi-unitary operator mapping $\mathcal{H}$ into $\mathcal{H}^+$. Verify that V is a strict, but not a doubly strict, plus-operator.
Hint: Use Theorem 4.17.

7 Prove that a continuous (J_1, J_2)-non-contractive operator V is (J_1, J_2)-bi-non-contractive if and only if its graph Γ_V is a maximal non-negative subspace in the $J_{\tilde{\Gamma}}$-space $\mathcal{H}_{\tilde{\Gamma}}$ (4.5) (*cf.* Rintsner [4]).
Hint: Compare (4.14) with Theorem 1.8.11.

8 A plus-operator V is said to be *stable* if all operators from a certain neighbourhood of it are plus-operators. Prove that assertions a)–d) of Theorem 4.24 are equivalent to the stability of the plus-operator V (M. Krein and Shmul'yan [5]).

9 Let $V = \| V_{ij} \|_{i,j=1}^2$ be a continuous uniformly (J_1, J_2)-expansive operator, with $\mathcal{D}_V = \mathcal{H}_1$. Prove that the operator will be uniformly (J_1, J_2)-bi-expansive if and only if $0 \notin \sigma_p(V_{11}^*)$.
Hint. cf. Exercise 5 and Proposition 4.26.

10 Prove that the set of all continuous strict plus-operators acting in a J-space $\mathcal{H}$ forms a subgroup and that $\mu_+(T, T_2) \geqslant \mu_+(T_1)\mu_+(T_2) \geqslant \mu_-(T_1 T_2)$ (M. Krein and Shmul'yan [5]).

11 Prove that if T_1 and T_2 are strict plus-operators in a J-space $\mathcal{H}$, then $\delta_+(T_1 T_2) = \delta_+(T_1) + \delta_+(T_2)$, where, we recall, $\delta_+(T)$ is the plus-deficiency of the operator T (see Theorem 4.15) (M. Krein and Shmul'yan [2]).

12 Prove that every strict plus-operator in the space $\Pi_\varkappa$ is doubly strict (Ginzburg [2]).
Hint: Cf. Exercises 4 and 5, using the fact that $\Pi_\varkappa$ is finite-dimensional.

13 Let V be a (J_1, J_2)-non-contractive operator acting from a Pontryagin space $\Pi_\varkappa$ into a Pontryagin space $\Pi'_{\varkappa'}$, with $\varkappa = \varkappa'$, and $\bar{\mathcal{D}}_V = \Pi_\varkappa$. Then V is a continuous operator (*cf.* I. Iokhvidov [17]).
Hint: Verify that the operator V satisfies the conditions of Corollary 4.8.

14 Give an example showing that the condition $\varkappa = \varkappa'$ in Exercise 13 is essential (Azizov).

15 Let V be a J-non-contractive operator, let $\mathcal{L}_+ \subset \mathcal{P}^+$, $\mathcal{L}_+^0 = \mathcal{L}_+ \cap \mathcal{L}^{[\perp]}$, and $\dim \mathcal{L}_+^0 < \infty$. Then $V\mathcal{L}_+ = \mathcal{L}_+$ implies $V\mathcal{L}_+^0 = \mathcal{L}_+^0$ (Azizov).
Hint: First verify that $\mathcal{L}_+^0 \subset V\mathcal{L}_+^0$, and then use the fact that $\mathcal{L}_+^0$ is finite-dimensional.

16 Let $\mathcal{H} = \mathcal{H}^+ \oplus \mathcal{H}^-$ be a J-space, $\mathcal{H}^\pm$ being separable infinite-dimensional spaces

with orthonormalized bases $\{e_i^+\}_{-\infty}^{\infty}$ $\{e_i^-\}_{-\infty}^{\infty}$ respectively. Verify that the linear operator V defined on the basis as follows

$$Ve_i^+ = e_{i+1}^+,\ i = 0, \pm 1, \ldots i \qquad Ve_i^- = e_{i+1}^+, \qquad i = -1, -2, \ldots i \qquad Ve_0^- = \theta$$

is a J-bi-non-contractive operator and

$$\mathscr{L}_+ = \mathrm{C\ Lin}\{\{e_i^+ + e_i^-\}_{-\infty}^{0}, \{e_i^+\}_1^{\infty}\} \in \mathscr{M}^+ \qquad V\mathscr{L}_+ + \mathscr{L}_+,$$

but nevertheless $V\mathscr{L}_+^0 \neq \mathscr{L}_+^0$, where $\mathscr{L}_+^0 = \mathscr{L}_+ \cap \mathscr{L}_+^{[\perp]}$ (*cf.* Problem 15) (Azizov).

17 Prove that if V is a (J_1, J_2)-non-contractive operator with $\bar{\mathscr{D}}_V = \mathscr{H}_1$ and if $[Vx_0, Vx_0] = [x_0, x_0]$ for some x_0, then $[Vx_0, Vy] = [x_0, y]$ for all $y \in \mathscr{D}_V$, and therefore $Vx_0 \in \mathscr{D}_{V^c}$ and $V^c Vx_0 = x_0$, i.e., $x_0 \in \mathrm{Ker}(V^c V - I_1)$ (Azizov).
Hint: Use the fact that the graph $\Gamma_V = \{\langle Vx, x\rangle \mid x \in \mathscr{D}_V\}$ is non-negative in the $J_{\hat{\Gamma}}$-metric (4.11) and that the vector $\langle Vx_0, x_0\rangle$ $(\in \Gamma_V)$ is isotropic in it.

18 If V is a J-bi-contractive operator, then $\sigma_r(V) \cap \mathbb{T} = \emptyset$ (*cf.* Corollary 2.17) (E. Iokhvidov [1]).
Hint: Use the result of Exercise 17 applied to the operator V^c.

19 Let V be a J-non-contractive operator acting in a J-space $\mathscr{H}$ with $\bar{\mathscr{D}}_V = \mathscr{H}$. Prove that Ker $(V - \xi I) \subset \mathrm{Ker}(V^c - \bar{\xi} I)$ when $\xi \in \mathbb{T}$; in particular, if V is a J-bi-non-contractive operator and $\xi \in \mathbb{T}$, then Ker $(V - \xi I) = \mathrm{Ker}(V^c - \bar{\xi} I)$, and therefore $\bar{\mathscr{R}}_{V-\xi I} = \mathscr{H}$ when $\xi \notin \sigma_p(V)$ (Azizov).
Hint: Use Exercise 17.

20 Let V be a J-bi-non-contractive operator, $\mathscr{L}_0$ the neutral subspace, and $V\mathscr{L}_0 = \mathscr{L}_0$. Prove that the operator $\hat{V}$ induced by the operator V in the $\hat{J}$-space $\hat{\mathscr{H}} = \mathscr{L}_0^{[\perp]}/\mathscr{L}_0$ is $\hat{J}$-bi-non-contractive (Azizov).
Hint: Use Exercise 17 and Theorem 1.17.

21 Let V be a J-non-contractive operator with $\bar{\mathscr{D}}_V = \mathscr{H}$, and let $\mathscr{L} \subset \mathscr{L}_\lambda(V) \cap \mathrm{Ker}\ (V^c V - I)$ and $V\mathscr{L} \subset \mathscr{L}$. Prove that $\mathscr{L}\,[\perp]\,\mathscr{L}_\mu(V)$ when $\lambda\bar{\mu} \neq 1$ (Azizov).
Hint: As in Exercise 8 on §1, use induction with respect to the parameter $p + q$, where p, q are the least non-negative integers for which the equalities $(V - \lambda I)^p x = \theta$, $(V - \mu I)^q y = \theta$ hold for $x \in \mathscr{L}$, $y \in \mathscr{L}_\mu(V)$.

22 Prove that if for an arbitrary J-non-contractive operator V we have $Vx = \lambda x$, $Vy = \mu y$ and $|\lambda| = |\mu| = 1$, $\lambda \neq \mu$, then $[y, x] = 0$.
Hint: Use the result of Exercise 17.

23 Prove that if V is a W-non-contractive operator (W-non-expansive operator) and $Vx_0 = \lambda x_0$, $Vx_1 = \lambda x_1 + x_0$ when $|\lambda| = 1$, then $x_0\,[\perp]\,\mathrm{Ker}(V - \lambda I)$ (*cf.* Exercise 11 on §2).
Hint: Use the fact that $[Vx, Vx] - [x, x]$ is non-negative (non-positive) on $\mathscr{D}_V$.

24 Prove that for a π-non-contractive operator V all the root subspaces $\mathscr{L}_\lambda(V)$ $(|\lambda| = 1)$, with the exception of not more than $\varkappa$ of them, are negative eigen-subspaces (*cf.* Corollary 2.28) (Azizov).
Hint: Use the results of Exercise 22 and 23.

25 A π-non-contractive operator V in a separable space $\Pi_\varkappa$ can have no more than a countable set of different eigenvalues on the circle $\mathbb{T}$ (Azizov, *cf.* I. Iokhvidov [1]).
Hint: Use the results of Exercises 22 and 24.

26 Let $\mathscr{H}_i$ $(i = 1, 2)$ be J_i-spaces. A continuous operator $V: \mathscr{H}_1 \to \mathscr{H}_2$ with $\mathscr{D}_V = \mathscr{H}_1$ is called a *B-plus-operator* if $[x, x]_1 \geqslant 0$, $x \neq \theta \Rightarrow [Vx, Vx] > 0$. It is clear that V is a plus-operator. Prove that in the case when $\mathscr{H}_1 = \Pi_\varkappa$ every B-plus-operator is a strict plus-operator, but the converse assertion if false ([XVI]).

27 Verify that in the first assertion of Exercise 26 the condition $\mathscr{H}_1 = \Pi_x$ is essential. *Hint:* In a J-space which is not a Pontryagin space consider an orthoprojector on to an improper maximal positive subspace and use Proposition 4.14 and Theorem 1.8.11.

28 It is clear that every uniformly (J_1, J_2)-expansive operator is a B-plus operator. Construct an example showing that even if $\mathscr{H}_1 = \mathscr{H}_2 = \Pi_x$ the converse of this assertion is false.

29 Suppose that V is a B-plus-operator and V^c a plus-operator. Show that V^c is also a B-plus-operator. In particular (*cf.* Exercise 12), when $\mathscr{H}_1 = \mathscr{H}_2 = \Pi_x$, for every B-plus-operator V the operator V^c is also a B-plus-operator.
Hint: Consider an arbitrary $\mathscr{L} \in \mathscr{M}^+(\mathscr{H}_2)$; prove that the subspace $(V^c\mathscr{L})^{[\perp]}$ is negative, and use Theorem I 1.19 (*cf.* [XVI]).

30 Prove that for a plus-operator V with $V(\mathscr{P}^{++}(\mathscr{H}_1) \cap \mathscr{D}_V) \cap \mathscr{P}^0(\mathscr{H}_2) \neq \emptyset$ it is always true that $\mathscr{R}_V \subset \mathscr{P}^+(\mathscr{H}_2)$ (Brodskiy [1]).

31 Prove that under the conditions of Theorems 4.6 it follows from $\mathscr{P}^{++}(\mathscr{H}_1) \cap \operatorname{Ker} V = \emptyset$, $\mathscr{D}_V \supset \mathscr{H}_1^+$ and the fact that the operator V_{11} (see (4.4)) is bounded and $V_{11}\mathscr{H}_1^+ = \mathscr{H}_2^+$ that V is bounded (Brodskiy [1], *cf.* I. Iokhvidov [17]).

32 Verify that in a J-space $\mathscr{H} = \mathscr{H}^+ \oplus \mathscr{H}^-$ with $\dim \mathscr{H}^+ = \infty$ the operator

$$V = \begin{Vmatrix} V_{11} & 0 \\ 0 & I^- \end{Vmatrix},$$

where V_{11} is a semi-unitary operator with $V_{11}\mathscr{H}^+ \neq \mathscr{H}^+$, is a strict plus-operator, but V^c is not even a plus-operator (I. Iokhvidov [XVII]).

33 Let $V: \mathscr{H}_1 \to \mathscr{H}_2$ be a strict minus-operator. Then $\operatorname{Ker} P_2^- VP_1^- \mid \mathscr{H}_1^- = \{\theta\}$, and when $\mathscr{D}_V = \mathscr{H}_1$ the equality $\mathscr{R}_{P_2^- VP_1^-} = \mathscr{H}_2^-$ is equivalent to the minus-operator V being doubly strict (*cf.* Ginzburg [2]).
Hint: In proving the first assertion, use the results of Exercise 3 and the equality (4.3), and in proving the second, use Exercise 7 on §1 and Theorem 4.17. The Remark 4.29 has also to be taken into account.

34 Let V: $\mathscr{H}_1 \to \mathscr{H}_2$ be a (J_1, J_2)-bi-non-expansive operator, let $\mathscr{L}_i (\subset \mathscr{H}_i)$ be uniformly positive subspaces, W_i be the Gram operators of the subspaces $\mathscr{L}_i^{[\perp]}$, and let P_i be the J_i-orthogonal projectors on to $\mathscr{L}_i^{[\perp]}$ $(i = 1,2)$. Then $P_2 VP_1 \mid \mathscr{L}_1^{[\perp]}$ is a (W_1, W_2)-bi-non-expansive operator (Azizov).
Hint: Without loss of generality assume that $\mathscr{L}_i \subset \mathscr{H}_i^+$ $(i = 1, 2)$. Write V in matrix form relative to the decompositions $\mathscr{H}_1 = \mathscr{L}_1 [\dot{+}] \mathscr{L}_1^{[\perp]}$, $\mathscr{H}_2 = \mathscr{L}_2 [\dot{+}] \mathscr{L}_2^{[\perp]}$ and calculate the matrices $V^*J_2V (\leqslant J_1)$, $VJ_1V^* (\leqslant J_2)$.

§5. Isometric, semi-unitary, and unitary operators

1 *Definition 5.1*: A linear operator U acting from a W_1-space $\mathscr{H}_1$ into a W_2-space $\mathscr{H}_2$ is said to be

1) (W_1, W_2)-*isometric* if $[Ux, Ux]_2 = [x, x]_1$ when $x \in \mathscr{D}_U$;
2) (W_1, W_2)-*semi-unitary* if it is (W_1, W_2)-isometric and $\mathscr{D}_U = \mathscr{H}_1$;
3) (W_1, W_2)-*unitary* if it is (W_1, W_2)-semi-unitary and $\mathscr{R}_U = \mathscr{H}_2$.

In particular, if $W_1 = I_1$ and $W_2 = I_2$, then 5.1 is the definition of (ordinary) isometric, semi-unitary, and unitary operators.

From Definition 5.1 it can also be seen that an arbitrary operator U mapping a neutral lineal $\mathscr{D}_U$ into a neutral lineal $\mathscr{R}_U$ can serve as an example of a (W_1, W_2)-isometric operator. Therefore, in contrast to (Hilbert) isometric operators, (W_1, W_2)-isometric operators in infinite-dimensional spaces can be unbounded and can have a non-trivial kernel. It is clear that U is a (W_1, W_2)-isometric operator (respectively a (J_1, J_2)-unitary operator) if and only if it is simultaneously a (W_1, W_2)-non-contractive operator and a (W_1, W_2)-non-expansive operator (respectively a (J_1, J_2)-bi-non-contractive operator and a (J_1, J_2)-bi-non-expansive operator), and for such operators the corresponding propositions in §4 hold.

5.2 *In order that a linear operator V with an indefinite domain of definition should be collinear with some (W_1, W_2)-isometric operator U, i.e., $V = \lambda U$* ($\lambda \neq 0$), *it is necessary and sufficient that V should be a plus-operator and a strict minus-operator* (*or a minus-operator and a strict-plus operator*) *simultaneously.*

□ The necessity is trivial. Now suppose, for example, that V is a plus-operator and a strict minus-operator. Then (see Theorem 4.3) there are constants $\alpha \geqslant 0$ and $\beta > 0$ such that $[Vx, Vx]_2 \geqslant \alpha [x, x]_1$ and $-[Vx, Vx]_2 \geqslant -\beta [x, x]_1$ when $x \in \mathscr{D}_V$, and therefore $(\alpha - \beta)[x, x]_1 \leqslant 0$. Since $\mathscr{D}_V$ is indefinite (by hypothesis), $\alpha = \beta > 0$, and $U = (1/\sqrt{\beta})V$ is a (W_1, W_2)-isometric operator. ■

Later we shall use the following simple proposition, which follows immediately from the polarization formula (see Exercise 1 on I §1).

5.3 *For an operator U to be (W_1, W_2)-isometric it is necessary and sufficient that $[Ux, Uy]_2 = [x, y]_1$ for all $w, y \in \mathscr{D}_U$.* ■

□ *Remark 5.4:* It follows form Proposition 5.3 that *if $x, y \in \mathscr{D}_U$, then $\{x [\perp] y\} \Leftrightarrow \{Ux [\perp] Uy\}$, and therefore a lineal $\mathscr{L} (\subset \mathscr{D}_U)$ is isotropic in $\mathscr{D}_U$ if and only if $U\mathscr{L}$ is isotropic in $\mathscr{R}_U$.* ■

The possibility was mentioned above of a (W_1, W_2)-isometric operator with a neutral domain of definition being unbounded. It turns out that the last condition is not essential; there are even (J_1, J_2)-*semi-unitary* unbounded operators.

□ *Example 5.5:* Let $\mathscr{H} = \mathscr{H}^+ \oplus \mathscr{H}^-$ be an infinite-dimensional J-space with infinite-dimensional $\mathscr{H}^\pm$, let $(\tilde{\mathscr{L}}_+, \tilde{\mathscr{L}}_-)$ be a maximal dual pair of semi-definite subspaces of the classes $h^\pm$ respectively (see Exercise 4 on §1.10), let $\mathscr{L}_0 = \tilde{\mathscr{L}}_+ \cap \tilde{\mathscr{L}}_-$ with dim $\mathscr{L}_0 = 1$, and let $\tilde{\mathscr{L}}_\pm = \mathscr{L}_0 + \mathscr{L}_\pm$, where $\mathscr{L}_\pm$ are definite lineals dense in $\tilde{\mathscr{L}}_\pm$. Then relative to the scalar product $\pm [x, y]$ $(x, y \in \mathscr{L}_\pm)$ the lineals $\mathscr{L}_\pm$ are Hilbert spaces (see Definition 1.5.9 and Propositions 1.1.23). Therefore there are isometric operators $U^\pm$ mapping $\mathscr{H}^\pm$ on to $\mathscr{L}_\pm$. We now define an operator U on elements $x = x^+ + x^-$, $x^\pm \in \mathscr{H}^\pm$, by the formula $U(x^+ + x^-) = U^+x^+ + U^-x^-$. It is easy to see that this is a J-semi-unitary operator, and that the image of the subspace $\mathscr{H}^+$ is the lineal $\mathscr{L}_+$ (not closed in $\mathscr{H}$). From Proposition 5.2 and Corollary 4.13 we conclude that U is an unbounded operator. ■

In this example a *sufficient condition* for the operator U to be unbounded turned out to be that $U\mathscr{H}^+$ *was not closed*. It turns out that a condition of this sort is also *necessary*.

Theorem 5.6: *Let U be a (J_1, J_2)-semi-unitary operator. Then the following conditions are equivalent.*

a) *U is a continuous operator;*
b) U$\mathscr{H}_1^{\pm}$ *are subspaces;*
c) *$\mathscr{R}_U$ is a subspace.*

□ a) ⇒ b) follows immediately from Proposition 5.2 and Corollary 4.13.

b) ⇒ c) we verify that $U\mathscr{H}_1^{\pm}$ are uniformly definite subspaces. To do this it is sufficient (see Proposition 1.5.6) to prove that the subspace $U\mathscr{H}_1^{\pm}$ are complete relative to the intrinsic norm $|[Ux, Ux]_2|^{1/2}$ $(x \in \mathscr{H}_1^{\pm})$. But the latter follows immediately from the facts that $\mathscr{H}_1^{\pm}$ are Hilbert spaces with the scalar products $\pm[x, y]_1$, and $\pm[Ux, Uy]_2$ in $\mathscr{H}_1^{\pm}$, and $U|\mathscr{H}_1^{\pm}$ are isometric operators. Thus $U\mathscr{H}_1^{\pm}$ are uniformly definite subspaces. It only remains to apply Theorem 1.5.7.

c) ⇒ a). Since $[Ux, Uy]_2 = [x, y]_1$ for all $x, y \in \mathscr{H}_1$ and $\mathscr{H}_1$ is not degenerate, Ker $U = \{\theta\}$. Therefore the operator U^{-1} exists. It follows from the same relation that the fact that U is a (J_1, J_2)-semi-unitary operator is equivalent to the inclusion.

$$U^{-1} \subset U^c \tag{5.1}$$

Therefore U^{-1} admits closure and it is defined on the subspace $\mathscr{R}_U$, i.e., U^{-1} is a continuous operator. Hence if follows by Banach's theorem that U is continuous. ■

□ *Remark 5.7:* Essentially it is proved in the implications a) ⇒ b) ⇒ c) that $U\mathscr{H}_1^{\pm}$ and $U\mathscr{H}$ are regular subspaces. ■

□ **Corollary 5.8:** *Every (J_1, J_2)-unitary operator is bounded.* ■

Remark 5.9: If $\mathscr{H}_i$ are G_i-spaces, i.e. $0 \notin \sigma_p(G_i)$ $(i = 1, 2)$, then the relation (5.1) also holds for (G_1, G_2)-semi-unitary operators, and therefore if $\mathscr{R}_U$ is a subspace, then U^{-1} and U are bounded operators. In particular, all (G_1, G_2)-unitary operators are bounded, and the condition that U be a (G_1, G_2)-unitary operator can be rewritten in the form

$$U^{-1} = U^c \;(= G_1^{-1}U^*G_2), \qquad \mathscr{D}_U = \mathscr{H}_1, \qquad \mathscr{R}_U = \mathscr{H}_2. \tag{5.2}$$

2□ Now let $\mathscr{H}_i$ be J_i-spaces, $i = 1, 2$, and let $U = \|U_{ij}\|_{i,j=1}^2$ be the matrix representation of a (J_1, J_2)-semi-unitary (J_1, J_2)-bi-non-contractive operator.

By Remark 4.21 and Theorem 4.17 the operator U_{11} is continuously invertible on the whole of $\mathscr{H}_2^+$, and $U_{21}U_{11}^{-1}$ $(\equiv \Gamma)$ is the angular operator of the uniformly positive subspace $U\mathscr{H}_1^+$, and therefore $\| \Gamma \| < 1$. We bring into consideration the operator

$$U(\Gamma) = \| U(\Gamma)_{ij} \|_{i,j=1}^2 = \begin{Vmatrix} (I_2^+ - \Gamma^*\Gamma)^{-1/2} & \Gamma^*(I_2^+ - \Gamma\Gamma^*)^{-1/2} \\ \Gamma(I_2^+ - \Gamma^*\Gamma)^{-1/2} & (I_2 - \Gamma\Gamma^*)^{-1/2} \end{Vmatrix} \tag{5.3}$$

operating in the space $\mathscr{H}_2 = \mathscr{H}_2^+ \oplus \mathscr{H}_2^-$.

Using Formula (5.2) it can be verified immediately that $U(\Gamma)$ is a J_2-unitary operator, and it follows from 3.21 after a straightforward verification that $U(\Gamma)$ is a uniformly positive operator. By the construction of $U(\Gamma)$ we have that $V = U(\Gamma)^{-1}U$ is a (J_1, J_2)-semi-unitary (J_1, J_2)-bi-non-contractive operator, mapping $\mathscr{H}_1^+$ into $\mathscr{H}_2^+$, and therefore (see Remark 5.4) also $\mathscr{H}_1^-$ into $\mathscr{H}_2^-$. So V has the matrix representation $V = \| V_{ij} \|_{i,j=1}^2$, where V_{11}: $\mathscr{H}_1^+ \to \mathscr{H}_2^+$ is unitary, V_{22}: $\mathscr{H}_1^- \to \mathscr{H}_2^-$ is a semi-unitary operator, and $V_{12} = 0$, $V_{21} = 0$.

Conversely, let $\mathscr{L}_+$ be an arbitrary maximal uniformly positive subspace in $\mathscr{H}_2$ and let Γ be its angular operator. We introduce the J_2-unitary operator $U(\Gamma)$ in accordance with formula (5.3). Let $\dim \mathscr{H}_1^+ = \dim \mathscr{H}_2^+$, $\dim \mathscr{H}_1^- \leqslant \dim \mathscr{H}_2^-$, and V_{11}: $\mathscr{H}_1^+ \to \mathscr{H}_2^+$ be unitary, V_{22}: $\mathscr{H}_1^- \to \mathscr{H}_2^-$ a semi-unitary operator, and $V_{21} = 0$, $V_{12} = 0$. Then $V = \| V_{ij} \|_{i,j=1}^2$ is simultaneously a semi-unitary operator and a (J_1, J_2)-semi-unitary operator, and $U = U(\Gamma)V$ is a (J_1, J_2)-semi-unitary (J_1, J_2)-bi-non-contractive operator, mapping $\mathscr{H}_1^+$ into the given space $\mathscr{L}_+ \subset \mathscr{H}_2$.

We summarize the above argument:

Theorem 5.10: *A one-to-one correspondence has been established between all triples of operators* (Γ, V_{11}, V_{22}), *where* Γ: $\mathscr{H}_2^+ \to \mathscr{H}_2^-$ *is a uniform contraction,* V_{11}: $\mathscr{H}_1^+ \to \mathscr{H}_2^+$ *is a unitary operator,* V_{22}: $\mathscr{H}_1^- \to \mathscr{H}_2^-$ *a semi-unitary operator, and all* (J_1, J_2)*-semi-unitary* (J_1, J_2)*-bi-non-contractive operators*:

$$(\Gamma, V_{11}, V_{22}) \to U = U(\Gamma)V,$$

where $U(\Gamma)$ *is constructed according to formula* (5.3), *and*

$$V = \| V_{ij} \|_{i,j=1}^2, \qquad V_{21} = 0, \; V_{12} = 0;$$

$$U \to (\Gamma; V_{11}, V_{22}),$$

where $\Gamma = U_{21}U_{11}^{-1}$, $V_{11} = P_2^+ U(\Gamma)^{-1}UP_1^+ \mid \mathscr{H}_1^+$, $V_{22} = P_2^{-1}U(\Gamma)^{-1}UP_1^- \mid \mathscr{H}_1^-$. ■

□ **Corollary 5.11:** *Under the conditions of Theorems* 5.10 *the operator* U *is* (J_1, J_2)*-unitary if and only if* V_{22} *is a unitary operator.* ■

Corollary 5.12: *If* $U = \| U_{ij} \|_{i,j=1}^2$ *is a* (J_1, J_2)*-semi-unitary* (J_1, J_2)*-bi-non-contractive operator, then* $U_{21} \in \mathscr{S}_\infty \Rightarrow U_{12} \in \mathscr{S}_\infty$

□ This follows from the implication

$$U_{21} \in \mathcal{S}_\infty \Leftrightarrow \Gamma \in \mathcal{S}_\infty \Leftrightarrow \Gamma^* \in \mathcal{S}_\infty \Rightarrow U_{21} \in \mathcal{S}_\infty.$$ ■

Corollary 5.13: *If $U = U(\Gamma)V$ is a J-unitary operator and if $\Gamma \in \mathcal{S}_\infty$, then $\mathbb{C}\backslash\mathbb{T} \subset \tilde{\rho}(U)$.*

□ Since

$$\Gamma \in \mathcal{S}_\infty, \qquad I_2^+ - (I_2^+ - \Gamma^*\Gamma)^{-1/2}, \qquad I_2^- - (I_2^- - \Gamma\Gamma^*)^{-1/2} \in \mathcal{S}_\infty,$$

and therefore $U - V \in \mathcal{S}_\infty$. Since V is a unitary operator, it only remains to use Theorem 2.11, taking into account that the spectrum of a unitary operator lies on the unit circle. ■

Remark 5.14: In §§4 and 5 we have not so far investigated the spectral properties of J-non-contractive and J-isometric operators, since later, in §6, we shall establish, by means of the Cayley–Nayman transformation, a connection between the classes of operators and the classes of J-dissipative and J-symmetric operators respectively, and, in so doing, a connection between their spectral properties.

3 In this paragraph we introduce and describe a special class of J-unitary operators, namely, stable J-unitary operators. This description is obtained as a consequence of a more general result.

Definition 5.15: A J-unitary operator U acting in a J-space $\mathcal{H}$ is said to be *stable* if $\| U^n \| \leqslant c < \infty$ for $n = 0, 1, 2, \ldots$.

We notice at once that if follows from Formulae (5.2) and (1.2) that $\| U^{-n} \| = \| U^{*n} \| = \| U^n \|$, and therefore for a stable U we also have $\| U^{-n} \| \leqslant c$ for $n = 0, 1, 2, \ldots$.

Definition 5.16: A group $\mathcal{U} = \{U\}$ is said to be *amenable* if on the space $B(\mathcal{U})$ of bounded complex-valued functions on U there is an *invariant mean* $f_\mathcal{U}$, i.e., a linear continuous function $f_\mathcal{U}$ having the properties

a) $f_\mathcal{U}(1) = 1$;
b) $f_\mathcal{U}(\varphi) \geqslant 0$ when $\varphi \in B(\mathcal{U})$ and $\varphi(U) \geqslant 0$ for arbitrary $U \in \mathcal{U}$;
c) $f_\mathcal{U}(\varphi(U\,\cdot\,)) = f_\mathcal{U}(\varphi(\,\cdot\,U)) = f_\mathcal{U}(\varphi)$ for any $U \in \mathcal{U}$.

It follows at once from properties a) and b) that for real functions $\varphi \in B(\mathcal{U})$ the inequalities

$$\inf\{\varphi(U) \mid U \in \mathcal{U}\} \leqslant f_\mathcal{U}(\varphi) \leqslant \sup\{\varphi(U) \mid U \in \mathcal{U}\} \tag{5.4}$$

hold.

Remark 5.17: Examples of amenable groups are (see, e.g., [XII]):

a) soluble groups and, in particular, commutative groups
b) compact groups and, in particular, bounded groups of operators in a finite-dimensional space, etc.

Theorem 5.18: *An amenable group* $\mathscr{U} = \{U\}$ *of J-unitary operators is bounded (in norm) if and only if there is a dual pair* $(\mathscr{L}^+, \mathscr{L}^-)$ *of maximal uniformly definite subspaces invariant relative to* $\mathscr{U}$, i.e., $U\mathscr{L}^{\pm} \subset \mathscr{L}^{\pm}$ *for all* $U \in \mathscr{U}$.

□ Let $\mathscr{U}$ be a bounded amenable group, $\|U\| \leqslant c < \infty$ when $U \in \mathscr{U}$. We consider a function $\varphi_{x,y}$: $U \to (Ux, Uy)$ $(x, U \in \mathscr{H},\ U \in \mathscr{U})$. This function is bounded on $\mathscr{U}$: $|\varphi_{x,y}(U)| = |(Ux, Uy)| \leqslant c^2 \|x\|\ \|y\|$, and therefore it goes into the domain of definition of the invariant mean $f_{\mathscr{U}}$. It is easy to see that the function $\langle x, y\rangle \to f_{\mathscr{U}}(\varphi_{x,y}) \equiv (x, y)_1$ is a non-negative sesquilinear form on $\mathscr{H}$. From formula (5.4) we obtain

$$\frac{1}{c^2}\|x\|^2 \leqslant \inf\{\varphi_{x,x}(U) \mid U \in \mathscr{U}\} \leqslant (x, x)_1 \leqslant \sup\{\varphi_{x,x}(U) \mid U \in \mathscr{U}\} \leqslant c^2\|x\|_1^2$$

i.e., $(x, y)_1$ is a scalar product equivalent to the original scalar product in the sense of the norms generated by them. By property c) of the invariant mean we have $(Ux, Uy)_1 = f_{\mathscr{U}}(\varphi_{Ux,Uy}) = f_{\mathscr{U}}(\varphi_{x,y}(\cdot\, U)) = f_{\mathscr{U}}(\varphi_{x,y}) = (x, y)_1$, $(x, y)_1$, i.e. the group U is a group of unitary operators in the new scalar product $(x, y)_1$. Since $(x, y)_1 = (Wx, y)$, where W is uniformly positive, $\{W^{1/2}UW^{-1/2}\}$ is a group of unitary operators (in the original scalar product), and moreover $W^{1(2}UW^{-1/2}$ comments for any $U \in \mathscr{U}$ with the bounded operator $G = W^{-1/2}JW^{-1/2}$, which is self-adjoint and continuously invertible $(0 \in \rho(G))$.

Consequently the spectral subspaces $\tilde{\mathscr{L}}_+, \tilde{\mathscr{L}}_-$ of the operator G corresponding to its positive and negative spectrum respectively are invariant relative to the operators $W^{1/2}UW^{-1/2}$, and therefore the subspaces $\mathscr{L}^{\pm} = W^{-1/2}\tilde{\mathscr{L}}_{\bullet}$ are invariant relative to all $U \in \mathscr{U}$. Since $\tilde{\mathscr{L}}_+$ is orthogonal to $\tilde{\mathscr{L}}_-$ and $\mathscr{L}_+ \oplus \tilde{\mathscr{L}}_- = \mathscr{H}$, it is immediately verifiable that $\mathscr{L}^+$ is J-orthogonal to $\mathscr{L}^-$, and from the continuous invertibility of $W^{-1/2}$ we have $\mathscr{H} = \mathscr{L}^+ [+] \mathscr{L}^-$. Moreover, since $\tilde{\mathscr{L}}_{\pm}$ are definite subspaces in the G-metric, $\mathscr{L}^{\pm}$ are definite subspaces in the J-metric. From Corollary 1.7.17 it follows that $\mathscr{L}^{\pm}$ are uniformly definite subspaces. Thus the first part of the theorem has been proved.

Now suppose that a set $\mathscr{U} = \{U\}$ of J-unitary operators (it need not even be assumed that $\mathscr{U}$ is a group) has a general invariant dual pair of maximal uniformly definite subspaces $(\mathscr{L}^+, \mathscr{L}^-)$, i.e., $U\mathscr{L}^{\pm} \subset \mathscr{L}^{\pm}$ $(U \in \mathscr{U})$. Then $\mathscr{H} = \mathscr{L}^+ [+] \mathscr{L}^-$. We introduce in $\mathscr{H}$ a new scalar product, equivalent to the original one (see Theorem 1.7.19)

$$(x, y)_1 = [x_+, y_+] - [x_-, y_-], \text{ where } x = x_+ + x_-,\ y = y_+ + y_-,\ x_{\pm}, y_{\pm} \in \mathscr{L}^{\pm}.$$

Relative to this scalar product every operator $U \in \mathscr{U}$ is unitary. Therefore, if $m\|x\| \leqslant \|x\|_1 \leqslant M\|x\|$, then $\|U\| \leqslant M/m$, i.e., the set $\mathscr{U}$ is bounded. ■

□ *Remark 5.19:* In proving Theorem 5.18 it has been established that *a bounded amenable group $\mathscr{U} = \{U\}$ of operators U* (we did not use the fact that they are J-unitary) *is similar to the group of unitary operators $W^{1/2}UW^{-1/2}$, and therefore the spectrum of each of these operators is unitary, i.e.,* $\sigma(U) \subset \mathbb{T}$. ■

Corollary 5.20: *A J-unitary operator U has an invariant dual pair of uniformly definite subspaces if and only if it is stable.*

□ This follows immediately from Theorem 5.18 taking into account that the group generated by a J-unitary operator is commutative and therefore (see Remark 5.17) amenable. ■

Definition 5.21: A continuous operator T acting in a Krein space $\mathscr{H}$ is said to be *normally decomposable* if its spectrum $\sigma(T)$ can be split into two non-intersecting spectral sets $\sigma(T) = \sigma_1(T) \cup \sigma_2(T)$ such that the subspaces $P_{\sigma_i(T)}\mathscr{H}$ and $P_{\sigma_2(T)}\mathscr{H}_1$ invariant relative to T (see Theorem 2.20), are uniformly definite.

Definition 5.22: A stable J-unitary operator is said to be *strongly stable* if all J-unitary operators from a certain neighbourhood of it are stable.

The following theorem links the concepts of normal decomposability for J-unitary operators with strong stability.

Theorem 5.23: *A J-unitary operator stable if and only if it is normally decomposable.*

□ Suppose a J-unitary operator U is strongly stable. By Theorem 5.18 and Theorem 1.7.19 we may assume without loss of generality that $U\mathscr{H}^{\pm} = \mathscr{H}^{\pm}$, and therefore U is simultaneously also a (Hilbert) unitary operator. We verify that $\sigma(U \mid \mathscr{H}^{+}) \cap \sigma(U \mid \mathscr{H}^{-}) = \emptyset$. We assume the contrary, and let $\xi_0 \in \sigma(U \mid \mathscr{H}^{+}) \cap \sigma(U \mid \mathscr{H}^{-})$, and let $\widehat{(e^{i\varphi_1}, e^{i\varphi_2})}$ be an open arc of the unit circle containing the point $\xi_0 = e^{i\varphi_0}$: $0 \leqslant \varphi_1 < \varphi_0 < \varphi_2 < 2\pi$. Let E_{φ} denote the spectral function of the unitary operator U. We define an operator

$$U_{\varphi_1,\varphi_2}\colon\ U_{\varphi_1,\varphi_2} \mid (E_{\varphi_2} - E_{\varphi_1})\mathscr{H} = \xi_0 I \mid (E_{\varphi_2} - E\varphi_1)\mathscr{H},$$

and

$$U_{\varphi_1,\varphi_2} \mid (I - E_{\varphi_2} + E_{\varphi_1})\mathscr{H} = U \mid (I - E_{\varphi_2} + E_{\varphi_1})\mathscr{H}.$$

It follows from the construction of U_{φ,φ_2} that it is simultaneously unitary and J-unitary, that $(E_{\varphi_2} - E_{\varphi_1})\mathscr{H}$ is invariant relative to U_{φ_1,φ_2}, and that, having chosen the length of the arc $\widehat{(e^{\varphi_1}, e^{\varphi_2})}$ suitably, we can make the operator U_{φ_1,φ_2} arbitrarily close in norm to the operator U. Since $U\mathscr{H}^{\pm} = \mathscr{H}^{\pm}$,

$$(E_{\varphi_2} - E_{\varphi_1})\mathscr{H} = (E_{\varphi_2} - E_{\varphi_1})\mathscr{H}^+ \oplus (E_{\varphi_2} - E_{\varphi_1})\mathscr{H}^-,$$

and therefore $(E_{\varphi_2} - E_{\varphi_1})\mathscr{H}$ is a Krein $\tilde{J}$-space, where $\tilde{J} = J \mid (E_{\varphi_2} - E_{\varphi_1})\mathscr{H}$. To complete the proof of this part of the theorem it suffices to show that in any neighbourhood of the J-unitary operator $\xi_0 I$ we can find unstable J-unitary operators. For example, the operators $\xi_0 U(\Gamma)$, where $U(\Gamma)$ is defined by the formula (5.3) and $\Gamma \neq 0$ will be unstable J-unstable J-unitary operators. Indeed, the operator $U(\Gamma)$ is uniformly positive, $U(\Gamma) \neq 1$, and therefore the spectrum of $U(\Gamma)$ contains at least one point $(0<)\ \lambda_0 \neq 1$, and then $\lambda_0\xi_0 \in \sigma(\xi_0 U(\Gamma))$ and $|\lambda_0\xi_0| \neq 1$. Hence, it follows (see Remark 5.19) that $\xi_0 U(\Gamma)$ is not a stable operator. By choosing the norm of the operator Γ sufficiently small, we can ensure that the operators $\xi_0 U(\Gamma)$ are sufficiently close to $\xi_0 I$.

Now let a J-unitary operator U be normally decomposable, i.e., $\sigma(U) = \sigma_1 U \sigma_2$, $\sigma_1 \cap \sigma_2 = \emptyset$, and the subspaces $\mathscr{L}_+ = P_{\sigma_1}\mathscr{H}$ and $\mathscr{L}_- = P_{\sigma_2}\mathscr{H}$ are uniformly positive and uniformly negative respectively. Since $U\mathscr{L}_+ = \mathscr{L}_+$, so (*cf.* 1.11 with Formula (5.2)) $U\mathscr{L}_+^{[\perp]} = \mathscr{L}_+^{[\perp]}$, and therefore $\{\mathscr{L}_+, \mathscr{L}_+^{[\perp]}\}$ is a dual pair (invariant relative to U) of uniformly definite subspaces. By Corollary 5.20 the operator U is stable. Therefore (see Remark (5.19)) its spectrum is unitary, i.e., $\sigma(U) \subset \mathbb{T}$. Just as in Corollary 3.12 it can be verified that P_{σ_1} and P_{σ_2} are J-selfadjoint projectors. Let Γ_1 and Γ_2 be corresponding non-intersecting contours, symmetric relative to the unit-circle, which surround σ_1 and σ_2. Then (see, e.g., [VI]) for perturbations V_ε, sufficiently small in norm, of the operator U the spectrum of the operator $V = U + V_\varepsilon$ will also be divided into two non-intersecting sets σ_1' and σ_2' surrounding Γ_1 and Γ_2 respectively. Moreover, if V is a J-unitary operator, then $P_{\sigma_i'}$ are J-selfadjoint projectors, sufficiently close in norm (because of the closeness of V and U) to the projectors P_{σ_i} $(i = 1, 2)$, and therefore (see Problem 10 on §3) they are, together with the latter, projectors on to uniformly definite subspaces. It follows from Corollary 5.20 that V is a stable operator, and therefore U is a strongly stable operator. ■

4 In conclusion we introduce one more class of operators closely connected with J-isometric operators.

Definition 5.24: Let $\mathscr{H}_i$ be J_i-spaces, $i = 1, 2$. A bounded operator $U: \mathscr{H}_1 \to \mathscr{H}_2$ with $\mathscr{D}_U = \mathscr{H}_1$ is said to be *partially* (J_1, J_2)*-isometric* if Ker U is a regular subspace, and $U \mid (\text{Ker } U)^{[\perp]}$ is a (J_1, J_2)-isometric operator.

It follows directly from Definition 5.24 that continuous (J_1, J_2)-semi-unitary operators, and, in particular, (J_1, J_2)-unitary operators, are partially (J_1, J_2)-isometric.

5.25: If U is a partially (J_1, J_2)-isometric operator, then the operator U^c will also be partially (J_1, J_2)-isometric.

□ Since $(\text{Ker } U)^{[\perp]}$ is a regular subspace whenever Ker U is, and since the operator $U | (\text{Ker } U)^{[\perp]}$ is continuous, so by Remark 5.7 $\mathscr{R}_U = \mathscr{R}_{U | (\text{Ker } U)^{[\perp]}}$ is a regular subspace, and therefore Ker U^c $(= \mathscr{R}_u{}^{[\perp]})$ is also a regular subspace. Since $U | (\text{Ker } U)^{[\perp]}$ is (J_1, J_2)-isometric, it follows that $U^c | \mathscr{R}_u = (U | (\text{Ker } U)^{[\perp]})^{-1}$ is a (J_1, J_2)-isometric operator. It remains only to use Definition 5.24. ■

□ **Corollary 5.26:** *The operator U^c which is (J_1, J_2)-conjugate to a (J_1, J_2)-semi-unitary operator U is a partially (J_2, J_1)-isometric operator.* ■

□ **Corollary 5.27:** *Let U be a partially (J_1, J_2)-isometric operator. Then $U^c U$ is a J_1-orthogonal projector on to* $(\text{Ker } U)^{[\perp]}$. ■

Exercises and problems

1 Let V be a (J_1, J_2)-semi-unitary operator (not necessarily bounded); then V^{-1} is a continuous operator (*cf*, [III]).
Hint: Since $\mathscr{D}_V = \mathscr{H}_1$, the operator V^c is bounded (*cf.* [XXII]). It remains to use the inclusion (5.1).

2 Prove that if U is a (J_1, J_2)-semi-unitary operator with $\bar{\mathscr{R}}_u = \mathscr{H}_2$, then U is a (J_1, J_2)-unitary operator.
Hint: Use the result of Exercise 1.

3 Prove that if V is a (J_2, J_2)-isometric operator, and $\mathscr{D}_V$ is closed, and $\bar{\mathscr{R}}_V$ is a non-degenerate subspace, then V is a continuous operator (*cf.* I. Iokhvidov [5]).

4 Give examples of (J_1, J_2)-isometric, unbounded, densely defined operators U: $\mathscr{H}_1 \rightarrow \mathscr{H}_2$ for which $U^{-1} = U^c$.
Hint: Use the device set out in Example 3.36.

5 Prove that an arbitrary (J_1, J_2)-unitary operator U can be factorized into a product $U = U_1 U_2$, where U_1 is a J_2-unitary operator with $U_1^2 = I_2$, and U_2 is a (J_1, J_2)-unitary operator iwth $U_2 \mathscr{H}_1^{\pm} = \mathscr{H}_2^{\pm}$.
Hint: Put $U_1 = U(\Gamma) J_2$ and $U_2 = J_2 V$, where $U(\Gamma)$ and V are the same as in Theorem 5.10.

6 Prove that if U is a (J_1, J_2)-unitary operator, then $|U| = (U^*U)^{1/2}$ is a J-unitary operator. Derive, based on this, the polar expansion $U = V|U|$ of a (J_1, J_2)-unitary operator and prove that V is a (J_1, J_2)-unitary operator and $V\mathscr{H}_1^{\pm} = \mathscr{H}_2^{\pm}$.
Hint: Use Theorem 5.10.

7 Prove that a bounded group of π-unitary operators in a finite-dimensional space has an invariant dual pair of definite subspaces (Azizov, Shul'man).
Hint: Use Theorem 5.18 taking Remark 5.17b) into account.

8 Give an example of a J-unitary operator with a single invariant dual pair $(\mathscr{H}^+, \mathscr{H}^-)$ which is not strongly stable.

Hint: Take as U, for example, the operator U such that $U|\mathscr{H}^+ = I^+$ and $U|\mathscr{H}^-$ is the shift operator along the orthonormalized basis $\{e_i^-\}_{-\infty}^{\infty}$ in $\mathscr{H}^-$.

9 Let $U = \| U_{ij} \|_{i,j=1}^{2}$ be the matrix representation of a J-unitary operator relative to the canonical decomposition $\mathscr{H} = \mathscr{H}^+ \oplus \mathscr{H}^-$, and suppose there are open sets Ω_1 and Ω_2 such that

$$\sigma_1 = \cup \{\sigma(U_{11} + U_{12}K) \mid K \in \mathscr{K}^+\} \subset \Omega_1, \qquad \sigma_2 = \cup \{\sigma(U_{22} + QU_{12}) \mid Q \in \mathscr{K}^-\} \subset \Omega_2,$$
$$\Omega_1 \cap \Omega_2 = \emptyset, \qquad \varphi(U) \subset \Omega_1 \cup \Omega_2.$$

Prove that if $\rho(U) \cap \{\xi\colon\ |\xi| = 1\} \neq \emptyset$, then U is a strongly stable J-unitary operator (Langer [11]).

10 Let $\mathscr{U} = \{U\}$ be a group of J-unitary operators acting in a J-space $\mathscr{H}$, let $\mathscr{L}$ be its invariant subspace, and let $\mathscr{L}^0 = \mathscr{L} \cap \mathscr{L}^{[\perp]}$. Then $U\mathscr{L}^0 = \mathscr{L}^0$ $(U \in \mathscr{U})$.
Hint: Use Remark 5.4.

11 Let V be a continuous J-semi-unitary operator, let $\mathscr{L}_0$ be a neutral subspace, and $V\mathscr{L}_0 = \mathscr{L}_0$. Prove that $\mathscr{L}_0^{[\perp]}$ is also invariant relative to V and that the operator $\hat{V}$ induced by V in the $\hat{J}$-space $\hat{\mathscr{H}} = \mathscr{L}_0^{[\perp]}/\mathscr{L}_0$ (see Corollary 1.5.8) is $\hat{J}$-semi-unitary.
Hint: Use Proposition 1.11.

12 Under the conditions of Exercise 11, let V be a J-unitary operator. Prove that then $\hat{V}$ is a $\hat{J}$-unitary operator (*cf.* [XV]).
Hint: Use Exercise 20 on §4 and Exercise 11.

13 Prove that if V is a continuous J-semi-unitary operator, then the uniform definiteness of a subspace $\mathscr{L}$ is equivalent to the uniform definiteness of the subspace $V\mathscr{L}$.
Hint: Use the definition of a J-isometric operator and the continuity of the operators V and V^{-1} (see Exercise 1).

14 The spectrum of a J-unitary operator U is symmetric about the unit circle, i.e., $\{\lambda \in \sigma(U)\} \Leftrightarrow \{\bar{\lambda}^{-1} \in \sigma(U)\}$.
Hint: Use the relation (5.2) and Theorem 1.16).

15 Let U be a J-unitary operator in $\mathscr{H}$, and let σ_1 be its bounded spectral set, $|\sigma_1| > 1$. Prove that then $\sigma_2 = (\sigma_1^*)^{-1}$ is also a spectral set for U, that the projector $P_{\sigma_1 \cup \sigma_2}$ is J-self-adjoint, and therefore $\mathscr{L} = P_{\sigma_1 \cup \sigma_2}\mathscr{H}$ is a projectionally complete space. Moreover, if σ_1' $(|\sigma_1'| > 1)$ and $\sigma_2' = (\sigma_1'^*)^{-1}$ is a second pair of such spectral sets for U, and $\sigma_1 \cap \sigma'_1 = \emptyset$, and $\mathscr{L}' = P_{\sigma_1' \cup \sigma_2'}\mathscr{H}$, then $\mathscr{L} [\perp] \mathscr{L}'$.
Hint: Use Exercise 14. For the rest the proof is entirely similar to that of Corollary 3.12.

16 Let V be a J-isometric operator with $\lambda, \mu \in \sigma_p(V)$ and $\lambda\bar{\mu} \neq 1$. Then $\mathscr{L}_\lambda(V) [\perp] \mathscr{L}_\mu(V)$ (*cf*, [XIV]).
Hint: *cf.* Exercises 21 and 22 on §4.

§6 The Cayley–Neyman transformation

1 We shall find it convenient to introduce this transformation not merely for linear operators but also in a more general setting—for linear relations (see Definition 1.2)

Let $\mathscr{H}$ be a given Hilbert space. We consider the set $\mathscr{L}$ of all linear relations T, i.e., of all possible lineals of the space $\mathscr{H} \times \mathscr{H}$. Thus $T = \{\langle x, y\rangle\}$ is the linear set of ordered pairs $\langle x, y\rangle$ $(x, y \in \mathscr{H})$, and linear operations in $\mathscr{H} \times \mathscr{H}$ (and therefore also in T) are defined in the natural way (component-wise). We recall (see §1.1) that a linear relation T is the graph Γ_A of some linear operator $A: \mathscr{H} \to \mathscr{H}$. If and only if it follows from $\langle w, y\rangle \in T$ and $x = \theta$ that $y = \theta$; so that $T = \Gamma_A = \{\langle x, Ax\rangle\}_{x \in \mathscr{D}_A}$ $(\mathscr{D}_A \subset \mathscr{H})$. Such linear relations T are called *single-valued linear relations*. The sets

$$\left.\begin{aligned} \mathscr{D}_T &= \{x \in \mathscr{H} \mid \text{there is an } y \in \mathscr{H} \text{ such that } \langle x, y\rangle \in T\} \\ \mathscr{R}_T &= \{y \in \mathscr{H} \mid \text{there is an } x \in \mathscr{H} \text{ such that } \langle x, y\rangle \in T\} \end{aligned}\right\} \tag{6.1}$$

are called respectively the *domain of definition* and the *range of values* of the linear relation T. For all $T \in \mathscr{L}$ we also introduce the following definitions and notation:

$$\operatorname{Ker} T = \{x \in \mathscr{H} \mid \langle x, \theta\rangle \in T\}, \qquad \operatorname{Ind} T = \{y \in \mathscr{H} \mid \langle \theta, y\rangle \in T\}, \tag{6.2}$$

$$\lambda T = \{\langle x, \lambda y\rangle \mid \langle x, y\rangle \in T\} (\lambda \in \mathbb{C}), \qquad -T = (-1)T, \tag{6.3}$$

$$T^{-1} = \{\langle y, x\rangle \mid \langle x, y\rangle \in T\}. \tag{6.4}$$

Ker T and Ind T are called respectively (see Definition 1.1) the *kernel* and the *indefiniteness* of the linar relation T. It is clear that $\lambda T, T^{-1} \in \mathscr{L}$. We also introduce the *identical linear relation* $I = \{\langle x, x\rangle\}_{x \in \mathscr{H}}$ and the *sum* of two linear relations $T_1, T_2 \in \mathscr{L}$:

$$T_1 + T_2 = \{\langle x, y_1 + y_2\rangle \mid \langle x, y_k\rangle \in T_k,\ k = 1, 2\}. \tag{6.5}$$

It is clear that $T_1 + T_2 \in \mathscr{L}$ and $\mathscr{D}_{T_1 + T_2} = \mathscr{D}_{T_1} \cap \mathscr{D}_{T_2}$. In particular, we consider the linear relation

$$T_\lambda = T - \lambda I \qquad (T \in \mathscr{L}, \lambda \in \mathbb{C}) \tag{6.6}$$

and by means of it we introduce for a linear relation T its *resolvent set*

$$\rho(T) = \{\lambda \in \mathbb{C} \mid \operatorname{Ker} T_\lambda = \{\theta\},\ \mathscr{R}_{T_\lambda} = \mathscr{H}\}, \tag{6.7}$$

the *spectrum*

$$\sigma(T) = \mathbb{C} \backslash \rho(T) \tag{6.8}$$

and, in particular, the *point spectrum*

$$\sigma_p(T) = \{\lambda \in \mathbb{C} \mid \operatorname{Ker} T_\lambda \neq \{\theta\}\}. \tag{6.9}$$

Points $\lambda \in \rho(T)$ are called *regular points*, but $\mathscr{L} \in \sigma_p(T)$ are *eigenvalues* of the linear relation T. For eigenvalues λ the lineal Ker $T_\lambda = \{x \in \mathscr{H} \mid \langle x, \lambda x\rangle \in T\}$ is called the *eigenlineal*, and its non-zero vectors are called *eigenvectors* of the linear relation T corresponding to the eigenvalues λ)[1].

We can introduce a further classification of points $\lambda(\in \mathbb{C})$ of the point spectrum $\sigma_p(T)$ of an arbitrary linear relation T by dividing these points into

---

[1] It will be convenient later, when Ind $T \neq \{\theta\}$, also to suppose that, by definition, $\infty \in \sigma_p(T)$ $(\subset \sigma(T))$.

two sub-classes $\sigma_{p1}(T)$ and $\sigma_{p,2}(T)$ (*cf.* Theorem 1.16)

$$\sigma_{p,1}(T) = \{\lambda \in \sigma_p(T) \mid \bar{\mathcal{R}}_{T_\lambda} \neq \mathcal{H}\}, \tag{6.10}$$

$$\sigma_{p,2}(T) = \{\lambda \in \sigma_p(T) \mid \bar{\mathcal{R}}_{T_\lambda} = \mathcal{H}\}, \tag{6.11}$$

The *continuous spectrum* $\sigma_c(T)$ and the *residual spectrum* $\sigma_r(T)$ of a linear relation T are defined by the formulae

$$\sigma_c(T) = \{\lambda \in \sigma(T)\backslash\sigma_p(T) \mid \mathcal{R}_{T_\lambda} \neq \bar{\mathcal{R}}_{T_\lambda} = \mathcal{H}\}, \tag{6.12}$$

$$\sigma_r(T) = \{\lambda \in \sigma(T)\backslash\sigma_p(T) \mid \bar{\mathcal{R}}_{T_\lambda} \neq \mathcal{H}\} \tag{6.13}$$

respectively

From (6.7)–(6.9), and (6.12), (6.13) it now follows that $\sigma(T)$ is the union of four mutually non-intersecting parts:

$$\sigma(T) = \sigma_{p,1}(T) \cup \sigma_{p,2}(T) \cup \sigma_c(T) \cup \sigma_r(T).$$

The *field of regularity* $r(T)$ of a linear relation T is the set

$$r(T) = \{\lambda \in \mathbb{C}\backslash\sigma_p(T) \mid \mathcal{R}_{T_\lambda} = \bar{\mathcal{R}}_{T_\lambda}\}. \tag{6.14}$$

It is clear that $\rho(T) \subset r(T)$ (since $\bar{\mathcal{H}} = \mathcal{H}$), and that $r(T)\backslash\rho(T) \subset \sigma_r w(T)$.

It is easy to convince onself that the definitions introduced of the sets $\rho(T)$, $\sigma(T)$, $\sigma_{p,1}(T)$, $\sigma_{p,2}(T)$, $\sigma_c(T)$ and $\sigma_r(T)$ for a linear relation T generalize the definitions of the corresponding sets for *closed* linear operators A. When $T = \Gamma_A$ the two sets of definitions are simply identical.

2 For a linear relation $T \in \mathcal{L}$ we now introduce the *direct* ($\mathbf{K}_\zeta$) and the *inverse* ($\mathbf{K}_\zeta^{-1}$) *Cayley–Neyman transformations* defined for a non-real parameter $\zeta (\neq \bar{\zeta})$.

To do this we first introduce the corresponding transformations *element-wise*, i.e., for all pairs $\langle x, y\rangle$, $\langle u, v\rangle \in \mathcal{H} \times \mathcal{H}$,

$$k_\zeta\langle x, y\rangle = \langle y - \zeta z, y - \bar{\zeta} x\rangle, \tag{6.15}$$

$$k_\zeta^{-1}\langle u, v\rangle = \frac{1}{\zeta - \bar{\zeta}}\langle v - u, \zeta v - \bar{\zeta} u\rangle. \tag{6.16}$$

□ A direct calculation will verify the proposition.

6.1 The transformations k_ζ and k_ζ^{-1} are mutually inverse bijections of $\mathcal{H} \times \mathcal{H}$ on to $\mathcal{H} \times \mathcal{H}$. ■

Now for any $(\bar{\zeta} \neq)\ \zeta \in \mathbb{C}$ and $S, T \in \mathcal{L}$ we put

$$\mathbf{K}_\zeta(T) = \{k_\zeta\langle x, y\rangle\}_{\langle x, y\rangle \in T}, \tag{6.17}$$

$$\mathbf{K}_\zeta^{-1}(T) = \{k_\zeta^{-1}\langle u, v\rangle\}_{\langle u, v\rangle \in S}. \tag{6.18}$$

6.2 The transformation $\mathbf{K}_\zeta (\zeta \neq \bar{\zeta})$ maps $\mathcal{L}$ on to $\mathcal{L}$ bijectively, and $\mathbf{K}_\zeta^{-1}$ carries out the inverse transformation.

□ This assertion follows directly from 6.1. ■

6.3 *Let* $T \in \mathscr{L}$ *and* $S = \mathsf{K}_\zeta(T)$ $(\zeta \neq \bar{\zeta})$ *or, equivalently,* $T = \mathsf{K}_\zeta^{-1}(S)$. *Then*

$$\text{Ker } S = \text{Ker } T_{\bar{\zeta}}, \qquad \text{Ind } S = \text{Ker } T_\zeta, \tag{6.19}$$

$$\text{Ker } T = \text{Ker } (\zeta S - \bar{\zeta} I), \qquad \text{Ind } T = \text{Ker } S_1, \tag{6.20}$$

$$\mathscr{D}_S = \mathscr{R}_{T_\zeta}, \qquad \mathscr{R}_S = \mathscr{R}_{T_1}, \qquad \mathscr{D}_T = \mathscr{R}_{S_1}. \tag{6.21}$$

□ The formulae (6.19) and (6.20) follow directly from a comparison of the definitions (6.2), (6.3), (6.5) with (6.17). The same applies to (6.21) for the verification of which the definitions (6.1) must be brought in again. ■

We now investigate how the spectrum of a linear relation $T(\in \mathscr{L})$ and its components are transformed on transition from the linear relation T to another linear relation $S = \mathsf{K}_\zeta(T)$. To do this we introduce, for $\zeta \neq \bar{\zeta}$, a mapping Φ_ζ: $\tilde{\mathbb{C}} \to \tilde{\mathbb{C}}$ in the extended complex plane $\tilde{\mathbb{C}} = \mathbb{C} \cup \{\infty\}$:

$$\mu \equiv \Phi_\zeta(\lambda) = \begin{cases} (\lambda - \bar{\zeta})(\lambda - \zeta)^{-1}, & \lambda \neq \zeta, \lambda \neq \infty, \\ \infty, & \lambda = \zeta, \\ 1, & \lambda = \infty. \end{cases} \tag{6.22}$$

Theorem 6.4: *The function* $\mu = \Phi_\zeta(\lambda)$ *maps*

$$\sigma_{p,1}(T), \sigma_{p,2}(T), \sigma_c(T), \sigma_c(T), r(S), \textit{ and } \rho(T)$$

bijectively on to

$$\sigma_{p,1}(S), \sigma_{p,2}(S), \sigma_c(S), \sigma_r(S), r(S), \textit{ and } \rho(S)$$

respectively.

□ When $\lambda \neq \zeta$, $\lambda \neq \infty$, we have by (6.17), (6.15) and (6.22)

$$\begin{aligned} S_\mu &= S - \mu I \\ &= \{\langle y - \zeta x, y - \bar{\zeta} x - \mu(y - \zeta x) \mid \langle x, y \rangle \in T\} \\ &= \left\{\left\langle y - \zeta x, y - \bar{\zeta} x - \frac{\lambda - \bar{\zeta}}{\lambda - \zeta}(y - \zeta x)\right\rangle \middle| \langle x, y \rangle \in T\right\} \\ &= \left\{\left\langle y - \zeta x, \frac{\bar{\zeta} - \zeta}{\lambda - \zeta}(y - \lambda x)\right\rangle \middle| \langle x, y \rangle \in T\right\}, \end{aligned}$$

from which it follows that when $(\zeta \neq)\lambda \in \mathbb{C}$

$$\text{Ker } S_\mu = \text{Ker } T_\lambda, \qquad \mathscr{R}_{S_\mu} = \mathscr{R}_{T_\lambda}. \tag{6.23}$$

We note that, taking (6.22) and Proposition 6.3 into account, the relations (6.23) remain valid for all $\lambda \in \tilde{\mathbb{C}}$ if, for any linear relation T, we put $\text{Ker } T_\infty = \text{Ind } T$ by definition.[1]

[1] We point that, on this understanding, the Definitions (6.10) and (6.11) make sense also for $\lambda = \infty \in \sigma_p(T)$ (see the Footnote on p. 143), and so the assertions of Theorem 6.4 remain vald in this case. This is also true when $\lambda = \zeta \in \sigma_p(T)$.

In accordance with this we shall also suppose (*cf.* (6.7))[1] that

$$\lambda = \infty \in \rho(\mathrm{T}) \Leftrightarrow \{\mathscr{D}_T = \mathscr{H},\ \mathrm{Ind}\ T = \{\theta\}\}.$$

All the assertions of the Theorem now follow directly from a comparison of the formulae (6.23) with the definitions (6.7)–(6.14). ∎

Corollary 6.5: *The function*

$$\lambda = \Psi_\zeta(\mu) = \begin{cases} (\zeta\mu - \bar{\zeta})(\mu - 1)^{-1}, & \mu \neq 1, \mu \neq \infty, \\ \infty, & \mu = 1, \\ \zeta, & \mu = \infty \end{cases} \tag{6.24}$$

effects for the linear relations S and $T = \mathbf{K}_\zeta^{-1}(S)$ *the mapping inverse to the mapping* Φ_ζ (see Theorem 6.4).

3 We return to the most important case for us when the Hilbert space $\mathscr{H}$ generating the set $\mathscr{L}$ ($\subset \mathscr{H} \times \mathscr{H}$) of linear relations is a W-space, i.e., it is equipped with an indefinite W-metric $[x, y] = (Wx, y)(x, y \in \mathscr{H})$ (see 1.§6.6).

Definition 6.6: A linear relation $T(\in \mathscr{L})$ is said to be *W-non-expansive* (respectively, *W-non-contractive*, *W-isometric*) in $\mathscr{H}$ if for all $\langle x, y\rangle \in T$ we have $[y, y] \leqslant [x, x]$ (respectively, $[y, y] \geqslant [x, x]$, $[y, y] = [x, x]$).

It is clear that Definition 6.6 generalizes the corresponding definitions for linear operators V (see Definitions 4.2 and 5.1). The latter definitions are obtained from Definition 6.6 when $T = \Gamma_V$ is the graph of the operator V.

In the case when $\mathscr{H}$ is a G-space, i.e. $[x, y] = (Gx, y)$ $(0 \notin \sigma_p(G)$ we introduce

Definition 6.7: For a $T \in \mathscr{L}$ the linear relation

$$T^c = \{\langle u, v\rangle \in \mathscr{H} \times \mathscr{H} \mid [y, u] = [x, v] \text{ for all } \langle x, y\rangle \in T\}$$

is called its *G-adjoint*.

Remark 6.8: It can be seen from this definition that $T^c(\in \mathscr{L})$ exists for any linear relation T. However, even in the case when $T = G_A$ is the graph of a linear operator A $(\mathscr{D}_A, \mathscr{R}_A \subset \mathscr{H})$ the linear relation T^c is not necessarily a graph (i.e., a single-valued linear relation)—see Exercise 1 below. However, comparison of the Definitions 6.7 and 1.1 shows that, when $T = \Gamma_A$ and $\bar{\mathscr{D}}_A = \mathscr{H}$, these definitions are equivalent, i.e., $T^c = \Gamma_{A^c}$

Definition 6.9: A linear relation T is said to be *G-symmetric* if $T \subset T^c$, and to be *G-selfadjoint* if $T^c = T$. We note that the condition $T \subset T^c$ is equivalent

[1] We recall that the condition Ind $T = \{\theta\}$ is equivalent to T being the graph of a linear operator. But then by Banach's theorem, in the case when this operator is closed, the condition $\mathscr{D}_T = \mathscr{H}$ implies that the operator is bounded, and this gives meaning to the notation, $\infty \in p(T)$ in this case.

in an obvious way to the requirement that $[y, x] = [x, y]$ for all $\langle x, y\rangle \in T$, and in this form we generalize it to an arbitrary W-space, and call such T *W-symmetric*.

When $T = \Gamma_A$ with $\bar{\mathscr{D}}_A = \mathscr{H}$ the Definition 6.9 goes over into the definitions (of the graph) of a G-symmetric and a G-selfadjoint operator A respectively (*cf*, with the Definitions 3.1 and 3.2 respectively).

Returning to the general case of a W-space we introduce

Definition 6.10: A linear relation is said to be *W-dissipative* if $\operatorname{Im}[y, x] \geqslant 0$ for all $\langle x, y\rangle \in T$. When $T = \Gamma_A$, this condition is equivalent to the operator A being *W-dissipative* (*cf.* Definition 2.1).

We examine how the Cayley–Neyman transformation affects one or other of the properties of a linear relation T given in the formulae (6.6)–(6.9). The identity

$$[y - \bar{\zeta}x, y - \bar{\zeta}x] - [y - \zeta x, y - \zeta x] = 4 \operatorname{Im} \zeta \operatorname{Im}[y, x] \tag{6.25}$$

is established by a simple calculation. From it follows immediately the proposition.

6.11 *Let the linear relations S, $T(\in\mathscr{L})$ be connected by the mutually inverse transformations* (*cf.* (6.15)–(6.18))

$$\left.\begin{aligned} S = \mathbf{K}_\zeta(T) &= \{\langle y - \zeta x, y - \bar{\zeta}x\rangle \mid \langle x, y\rangle \in T\}, \\ T = \mathbf{K}_\zeta^{-1}(S) &= \left\{\frac{1}{\zeta - \bar{\zeta}}\langle u - v, \bar{\zeta}u - \zeta v\rangle \mid \{u, v\rangle \in S\right\}. \end{aligned}\right\} \tag{6.26}$$

Then the following assertions are equivalent:

a) *T is a W-dissipative* (*respectively, W-symmetric*) *linear relation*;
b) *S is a W-non-contractive linear relation when $\operatorname{Im} \zeta > 0$ and is W-non-expansive when $\operatorname{Im} \zeta < 0$* (*respectively, a W-isometric linear relation for any $\zeta \neq \bar{\zeta}$*).

□ It is sufficient to compare the Definitions 6.6, 6.8 and 6.9 with the formulae (6.25)–(6.26). ■

Remark 6.12: It is sometimes useful to consider the so-called *W-accumulative linear relations T* distinguished by the condition $\operatorname{Im}[y, x] \leqslant 0$ for all $\langle x, y\rangle \in T$, and the corresponding linear operators. It is clear that only the factor (-1) makes them different from dissipative linear relations and operators respectively. Proposition 6.11 can easily be reformulated in terms of W-accumulative linear relations; it is clear from (6.25) that in the reformulation the rôles of the conditions $\operatorname{Im} \zeta > 0$ and $\operatorname{Im} \zeta < 0$ are interchanged.

We add further that *maximal W-accumulative operators*, and, in particular, *J-accumulative operators*, are defined in the natural way, and the theory of them is entirely analogous to the theory of maximal W-dissipative (and J-dissipative) operators developed in §2.

4 In this and the following paragraphs, returning to the main purpose of our book, we shall specialize our examination still further, restricting it to a single-valued linear relations $T\ (\in \mathscr{L})$, i.e., to graphs of linear operators. Applying the transformation $\mathbf{K}_\zeta$ to T we shall see to it that $S = \mathbf{K}_\zeta(T)$ *is also a single-valued linear relation*, i.e., a graph. To do this *it is necessary and sufficient*, in view of the first of the formulae (6.26), *to ensure that* $\zeta \notin \sigma_p(T)$ (see (6.6) and (6.9)). For, if $T = \Gamma_A = \{\langle x, Ax\rangle\}_{x \in \mathscr{D}_A}$ then destruction of the single-valuedness of the linear relation $S = \mathbf{K}_\zeta(T)$ is equivalent (see (6.26)) to the simultaneous satisfaction for some $x \in \mathscr{D}_A$ of the conditions $y - \zeta x = \theta$ and $y - \bar{\zeta} x \neq \theta$, where $y = Ax$, i.e., $x \neq \theta$ and $\zeta \in \sigma_p(A)\ (\equiv \sigma_p(T))$.

Thus, when $(\bar{\zeta} \neq)\ \zeta \in \sigma_p(A)$ we have $S = \mathbf{K}_\zeta(T) = \Gamma_V$, where $V = V_\zeta$ is a linear operator for which

$$\mathscr{D}_V = \{Ax - \zeta x\}_{x \in \mathscr{D}_A}; \qquad \mathscr{R}_V = \{Ax - \bar{\zeta} x\}_{x \in \mathscr{D}_A}, \tag{6.27}$$

which can be written shortly in the form

$$V = (A - \bar{\zeta} I)(A - \zeta I)^{-1} \tag{6.28}$$

or equivalently

$$V = I + (\zeta - \bar{\zeta})(A - \zeta I)^{-1}. \tag{6.29}$$

Moreover, it is clear that $1 \notin \sigma_p(V)$. A simple calculation (*cf.* (6.26)) shows that here

$$A = (\zeta V - \bar{\zeta} I)(V - I)^{-1} \tag{6.30}$$

or equivalently

$$A = \zeta I + (\zeta - \bar{\zeta})(V - I)^{-1}. \tag{6.31}$$

It is clear that we can also work in the reverse order, starting with a linear operator V for which $1 \notin \sigma_p(V)$, and for any $\zeta (\neq \bar{\zeta})$ specify the operator A by the formula (6.30) or (6.31).

It is also clear that

$$\mathscr{D}_A = \{Vy - y\}_{y \in \mathscr{D}_V}, \qquad \mathscr{R}_A = \{\zeta Vy - \bar{\zeta} y\}_{y \in \mathscr{D}_V}, \qquad \zeta \notin \sigma_p(A),$$

and that the formula (6.28) (or (6.29)) is the inverse of the transformation (6.30) (or (6.31)). Traditionally it is precisely in this way that the direct and inverse Cayley–Neyman transformations of operators are defined, and we, allowing a certain freedom, will keep for them the notations $V = \mathbf{K}_\zeta(A)$, $A = \mathbf{K}_\zeta^{-1}(V)$.

The reader will without difficulty reformulate for this particular case Proposition 6.11 and Remark 6.12 which remain valid, of course, on passing from the operator A to its Cayley–Neyman transform, as we shall call, for brevity, the operator $V = \mathbf{K}_\zeta(A)$, and reversely.

This proposition can be developed further in several directions in the case when it is possible to choose the parameter $(\bar{\zeta} \neq)\zeta$ so that $\zeta \in \rho(A)$. In this case it follows at once from (6.29) that $V = \mathbf{K}_\zeta(A)$ is a bounded operator with $\mathscr{D}_V = \mathscr{H}$.

Theorem 6.13: *Let A be a maximal closed J-dissipative operator in a Krein space and let $(\bar{\zeta} \neq)\zeta \in \rho(A)$. Then when* $\operatorname{Im} \zeta > 0$ $(\operatorname{Im} \zeta < 0)$ *the operator $V = \mathbf{K}_\zeta(A)$ is a J-bi-non-contractive (J-bi-non-expansive) operator and $(V - I)\mathscr{H} = \mathscr{H}$.*

Conversely, if V is a J-bi-non-contractive (J-bi-non-expansive) operator and $1 \notin \sigma^p(V)$, then the operator $A = \mathbf{K}_\zeta^{-1}(V)$ is when $\operatorname{Im} \zeta > 0$ $(\operatorname{Im} \zeta < 0)$ *a closed maximal J-dissipative operator and $\zeta \in \rho(A)$.*

□ If A is a closed maximal J-dissipative operator, then (see Proposition 2.7) $\overline{(V - I)\mathscr{H}} = \bar{\mathscr{D}}_A = \mathscr{H}$ and $(-A)^c$ is a J-dissipative operator. Further, by 6.11, V is when $\operatorname{Im} \zeta > 0$ $(\operatorname{Im} \zeta < 0)$ a J-non-contractive (J-non-expansive) operator. From (6.29) taking account of 1.6 and 1.9 we have $V^c = I + (\bar{\zeta} - \zeta)(A^c - \bar{\zeta} I)^{-1} = \mathbf{K}_{\bar{\zeta}}(A^c)$, and the first pair of Theorem 6.13 now follows from Remark 6.12.

Conversely, when $1 \notin \sigma_p(V)$ we have, by Exercise 18 and 19 on §4, $\overline{(V - I)\mathscr{H}} = \mathscr{H}$, and therefore the operator $A = \mathbf{K}_\zeta^{-1}(V)$ (see (6.31)) is densely defined and (exactly the same as the operator $-A^c = \mathbf{K}_{\bar{\zeta}}^{-1}(V^c)$) it is under the conditions of Theorem 6.13 a J-dissipative operator. Moreover (see (6.31) and (6.29) $\zeta \in \rho(A)$, and, by Proposition 2.7, A is a closed maximal J-dissipative operator. ■

Remark 6.14: It is easy to understand that Theorem 6.13 can be reformulated and proved if the term '*J*-dissipative' is replaced everywhere by the term '*J*-accumulative' (see Remark 6.12) and the conditions $\operatorname{Im} \zeta > 0$ and $\operatorname{Im} \zeta < 0$ change places.

Corollary 6.15: *If A is a J-selfadjoint operator ($A = A^c$) and $(\bar{\zeta} \neq) \zeta \in \rho(A)$, then $U = \mathbf{K}_\zeta(A)$ is a J-unitary operator and $1 \in \sigma_p(U)$.*

Conversely, if U is a J-unitary operator, $1 \notin \sigma_p(U)$, and $\zeta \neq \bar{\zeta}$, then $A = \mathbf{K}_\zeta^{-1}(U)$ is a J-selfadjoint operator.

□ The first assertion follows from the fact that a J-selfadjoint operator A is a maximal J-dissipative and a maximal J-accumulative operator simultaneoulsy and therefore by Theorem 6.13 and Remark 6.14, for any $\zeta \neq \bar{\zeta}$, $U = \mathbf{K}_\zeta(A)$ is a bounded J-bi-non-expansive and a J-bi-non-contractive operator simultaneously, i.e. (see §5.1) it is a J-unitary operator.

The converse argument proceeds entirely analogously. ■

It is not difficult to obtain the result contained in Corollary 6.15 directly without using Theorem 6.13 (*cf.* Exercise 2 below).

6.16 Under the conditions of Theorem 6.13, the operators corresponding to bounded uniformly J-dissipative operators A ($\mathscr{D}_A = \mathscr{H}$) are when $\operatorname{Im} \zeta > 0$ $(\operatorname{Im} \zeta < 0)$ *bounded uniformly J-bi-expansive (uniformly J-bi-contractive) operators $V = \mathbf{K}_\zeta(A)$ and conversely.*

□ Suppose, for example, that $\operatorname{Im}\zeta > 0$. Since $\operatorname{Im}[Ax, x] \geqslant \gamma_A \|x\|^2$, it follows from (6.25) (when $y = Ax$) and from (6.28) when $g = Ax - \zeta x$, $Vg = Ax - \bar{\zeta}x$ that

$$[Vg, Vg] \geqslant [g, g] + 4 \operatorname{Im}\zeta \,.\, \gamma_A \|A - \zeta I\|^{-2} \|g\|^2$$

and similarly for V^c (taking account of the fact that for bounded A we have $\operatorname{Im}[-A^c x, x] = \operatorname{Im}[Ax, x]$ for all $x \in \mathcal{H}$), i.e., V is a uniformly J-bi-expansive operator. In the converse of this assertion it has to be taken into account that for a uniformly J-bi-expansive operator V we have $\mathbb{T} \subset \rho(V)$ (see Theorem 4.31) and, in particular, $1 \in \rho(V)$; otherwise the arguments are analogous. ■

5 Returning to the general theory of Cayley–Neyman transformations $V = \mathbf{K}_\zeta(A)$ and $A = \mathbf{K}_\zeta^{-1}(V)$ for linear operators A and V $(\zeta \neq \bar{\zeta})$, we continue first of all the study started earlier (see 6.2) of spectral questions connected with these transformations. With this purpose we shall explain the connection between invariant subspaces of the operators A and V, and also that between the root lineals of these operators.

6.17 Let A be a linear operator in $\mathcal{H}$, let $(\bar{\zeta} \neq)\zeta \notin \sigma_p(A)$, and let $V = \mathbf{K}_\zeta(A)$.

If the subspace $\mathcal{L} \subset \mathcal{D}_A$, if $A\mathcal{L} \subset \mathcal{L}$ and $\zeta \in \rho\ (A \mid \mathcal{L})$, then $\mathcal{L} \subset \mathcal{D}_V$, $V\mathcal{L} \subset \mathcal{L}$ and $1 \notin \sigma(V \mid \mathcal{L})$. Conversely, if $\mathcal{L} \subset \mathcal{D}_V$, $V\mathcal{L} \subset \mathcal{L}$, and $1 \notin \sigma(V \mid \mathcal{L} \cup \sigma_p(V)$, then $\mathcal{L} \subset \mathcal{D}_A$, $A\mathcal{L} \subset \mathcal{L}$ and $\zeta \in \rho(A \mid \mathcal{L})$.

□ Since (see (6.29)) $V = I + (\zeta - \bar{\zeta})(A - \zeta I)^{-1}$, and by hypothesis $\zeta \in \rho(A \mid \mathcal{L})$, i.e., $(A - \zeta I)\mathcal{L} = \mathcal{L}$ and $\mathcal{L} = (A - \zeta I)^{-1}\mathcal{L}$, so $\mathcal{L} \subset \mathcal{D}_V$ and $V\mathcal{L} \subset \mathcal{L}$. Moreover, $V \mid \mathcal{L} = \mathbf{K}_\zeta(A \mid \mathcal{L})$, $A \mid \mathcal{L}$ and $\mathrm{V} \mid \mathcal{L}$ are closed (and bounded) operators, and from the condition $\zeta \in \rho(A \mid \mathcal{L})$ it follows (see Theorem 6.6) that $1 \in \rho(V \mid \mathcal{L})$.

The converse implication is verified similarly by means of formula (6.31) and Corollary 6.5. ■

Corollary 6.18: *Let A and V be the same as in* 6.17, $\zeta \notin \sigma_p(A)(\Leftrightarrow 1 \notin \sigma_p(V))$. *Then $\lambda \in \sigma_p(A) \Leftrightarrow \{\nu = (\lambda - \bar{\zeta})(\lambda - \zeta)^{-1} \in \sigma_p(V)\}$ and the root lineals $\mathcal{L}_\nu(V)$ and $\mathcal{L}_\lambda(A)$ coincide.*

□ The first assertion follows directly from Theorem 6.4. Now let $x \in \mathcal{L}_\lambda(A)$, i.e., there is $p \in \mathbb{N}$ such that $(A - \lambda I)^p x = \theta$. Then $\mathcal{L} \equiv \operatorname{Lin}\{x, Ax, \ldots, A^p x\}$ is a finite dimensional invariant subspace of the operator A: $\mathcal{L} \subset \mathcal{D}_A$, $A\mathcal{L} \subset \mathcal{L}$ and $\sigma(A \mid \mathcal{L}) = \{\lambda\}$, i.e., $\zeta \in \rho(A \mid \mathcal{L})$ (because $\zeta \neq \lambda$). Therefore (see 6.17) $\mathcal{L} \subset \mathcal{D}_V$, $V\mathcal{L} \subset \mathcal{L}$ and $(V \mid \mathcal{L}) = \mathbf{K}_\zeta\ (A \mid \mathcal{L})$. By Theorem 6.4 the number $\nu = (\lambda - \bar{\zeta})(\lambda - \zeta)^{-1}$ is the only point of the spectrum of the finite-dimensional operator $V \mid \mathcal{L}$, and so $x \in \mathcal{L} \subset \mathcal{L}_\nu(V \mid \mathcal{L}) \subset \mathcal{L}_\nu(V)$, i.e., $\mathcal{L}_\lambda(A) \subset \mathcal{L}_\nu(V)$. The converse is proved in a similar way. ■

Remark 6.19: It is clear that in Corollary 6.18 the first assertion can be reformulated equivalently thus:

$$(\nu \in \sigma_p(V)\} \Leftrightarrow \{\lambda = (\zeta\nu - \bar{\zeta})(\nu - 1)^{-1} \in \sigma_p(A)\}. \qquad \blacksquare$$

Unfortunately, the requirement that $\zeta \in p(A \mid \mathscr{L})$ (respectively, $1 \in p(V \mid \mathscr{L})$ imposed in Proposition 6.17 turns out to be extremely stringent. Later, in Chapter 3, we shall see that in particular cases the 'preservation' of certain invariant subspaces under the Cayley-Neyman transformation can be ensured under less burdensome conditions.

Remark 6.20: In all the discussions in §6.4 and §6.5 it might have appeared that, in contrast to the arbitrary number ζ $(\neq \bar{\zeta})$ on which some requirement such as $\zeta \notin \sigma_p(A)$ or $\zeta \in p(A)$ was imposed, the number 1, of which it was always required that $1 \notin \sigma_p(V)$, played some special rôle. But in fact this rôle can be played by any number ε with $|\varepsilon| = 1$, if $\varepsilon \notin \sigma_p(V)$. This leads only to an insignificant modification of the formulae for the mutually inverse Cayley transformations themselves:

$$V = \varepsilon(A - \bar{\zeta}I)(A - \zeta I)^{-1} = \varepsilon(\zeta - \bar{\zeta})(A - \zeta I)^{-1} \tag{6.32}$$

$$A = (\zeta V - \bar{\zeta}\varepsilon I)(V - \varepsilon I)^{-1} = \zeta I + \varepsilon(\zeta - \bar{\zeta})(V - \varepsilon I)^{-1}, \tag{6.33}$$

and also to corresponding modifications in the formulation of the theorems and propositions 6.13–6.19, and in particular to the formulae for the transformation of the point spectra in Corollary 6.18 and Remark 6.19:

$$\{\lambda \in \sigma_p(A)\} \Leftrightarrow \{\nu = \varepsilon(\lambda - \bar{\zeta})(\lambda - \zeta)^{-1} \in \sigma_p(V)\},$$
$$\{\nu \in \sigma_p(V)\rangle \Leftrightarrow \{\lambda = (\zeta\nu - \bar{\zeta}\varepsilon)(\nu - \varepsilon)^{-1} \in \sigma_p(A)\}.$$

6 In conclusion we consider the Cayley–Neyman transformation $V = \mathbf{K}_\zeta(A)$ $(\bar{\zeta} \neq \zeta \notin \sigma_p(A))$ linking the operators A and V acting in a Hilbert space $\mathscr{H} = \mathscr{H}_1 \oplus \mathscr{H}_2$ (a typical situation for a Krein space). We are interested in the case when the operators A and V admit matrix representations (*cf.* (4.4)) $A = \| A_{ij} \|_{i,j=1}^2$, $V = \| V_{ij} \|_{i,j=1}^2$, generated by the corresponding ortho-projectors P_1 and P_2 $(P_k\mathscr{H} = \mathscr{H}_k, k = 1, 2)$ (for such a representation if necessary that, for example, $\mathscr{H}_1 \subset \mathscr{D}_A$ and $\zeta \in p(A)$). In this case the relation $V(A - \zeta I) = A - \bar{\zeta}I$, which follows from (6.28), can be rewritten in matrix form as

$$\begin{Vmatrix} V_{11} & V_{12} \\ V_{21} & V_{22} \end{Vmatrix} \begin{Vmatrix} A_{11} - \zeta I_1 & A_{12} \\ A_{21} & A_{22} - \zeta I_2 \end{Vmatrix} = \begin{Vmatrix} A_{11} - \bar{\zeta} I_1 & A_{21} \\ A_{21} & A_{22} - \bar{\zeta} I_2 \end{Vmatrix}.$$

From this we have at once the relations

$$V_{11}(A_{11} - \zeta I_1) + V_{12}A_{21} = A_{11} - \bar{\zeta}I_1 \tag{6.34}$$

$$V_{11}A_{12} + V_{12}(A_{22} - \zeta I_2) = A_{12} \tag{6.35}$$

$$V_{21}(A_{11} + \zeta I_1) + V_{22}A_{21} = A_{21}$$

$$V_{21}A_{12} + V_{22}(A_{22} - \zeta I_2) = A_{22} - \bar{\zeta}I_2.$$

In particular, in the case important for us later, when $\zeta \in p(A_{22})$, it follows from (6.35) that

$$V_{12} = -(V_{11} - I_1)A_{12}(A_{22} - \zeta I)^{-1} \tag{6.36}$$

and if additionally $\zeta \in p(A_{11})$, the elimination of the operator V_{12} from the system (6.34), (6.35) gives

$$V_{11}[A_{11} - \zeta I_1 - A_{12}(A_{22} - \zeta I_2)^{-1}A_{21}] = A_{11} - \bar{\zeta} I_1 \models A_{12}(A_{22} - \zeta I_2)^{-1}A_{21}.$$

or

$$(V_{11} - I_1)[I_1 - A_{12}(A_{22} - \zeta I_2)^{-1}A_{21}(A_{11} - \zeta I_1)^{-1}] = (\zeta - \bar{\zeta})(A_{11} - \zeta I_1)^{-1}. \tag{6.37}$$

It is precisely in this form that the relations (6.36) and (6.37) will be used later in Chapter 3 (Theorem 1.13).

§Exercises and problems

1 If T is a single-valued linear mapping into $\mathscr{H}$, then T^c is a single-valued linear mapping into $\mathscr{H}$ if and only if $\bar{\mathscr{D}}_T = \mathscr{H}$.

2 Derive Corollary 6.15 directly from the formulae (6.29) and (6.31).

3 Prove that Corollary 6.15 can be generalized as follows. Let A be a maximal J-dissipative operator and a maximal J-symmetric operator simultaneously, and let $(\bar{\zeta} \neq)\zeta \in p(A)$. Then $V = \mathbf{K}_\zeta(A)$ is, when $\operatorname{Im} \zeta > 0$ ($\operatorname{Im} \zeta < 0$), a bounded J-bi-non-contractive (J-bi-non-expansive) operator and a J-semi-unitary operator simultaneously and $1 \notin \sigma_p(V)$. The converse proposition is true, and so are the analogues of both the direct and converse propositions when the word 'J-dissipative' is replaced by 'J-accumulative' and the conditions $\operatorname{Im} \zeta > 0$ ($\operatorname{Im} \zeta < 0$) are replaced by $\operatorname{Im} \zeta < 0$ ($\operatorname{Im} \zeta > 0$).
Hint: cf. Theorem 6.13 and Proposition 6.11.

4 Let V be a W-non-contractive operator in $\mathscr{H}$, let σ be its bounded spectral set, and let P_σ be the corresponding Riesz projector. Then when $|\sigma| > 1$ (respectively $|\sigma| < 1$) the sub space $\mathscr{L} = P_\sigma \mathscr{H}$ invariant relative to V is non-negative (non-positive). In particular, for a W-isometric operator the subspace $\mathscr{L} = P_\sigma \mathscr{H}$ is neutral (*cf.* Theorem 2.21).
Hint: To the graph of the restriction $V | \mathscr{L}$ (for any $\zeta \neq \bar{\zeta}$) apply Proposition 6.11, Corollary 6.5 and formula (6.31), and then Proposition 6.17 and Theorem 2.21.

5 Let the operator V be the same as in Exercise 4, and let Λ with $|\Lambda| > 1$ ($|\Lambda| < 1$) be a certain set of its eigenvalues. Then $C \operatorname{Lin}\{\mathscr{L}_\lambda(V)\}_{\lambda \in \Lambda}$ is a non-negative (non-positive) subspace. In the case of a W-isometric V both these subspaces are neutral.
Hint: Prove this by analogy with Corollary 2.22, basing the proof on the result of Exercise 4.

6 Let V be a bounded π-non-contractive operator with $\mathscr{D}_V = \Pi_\varkappa$. Then $\sigma(V) \cap \{\lambda \mid |\lambda| > 1\}$ consists of not more than $\varkappa$ (taking algebraic multiplicity into account) normal eigenvalues (*cf.* Corollary 2.23) (Brodskiy [1]).
Hint: Use the results of Exercise 5 on §1, Exercise 22 on §4, then Remark 6.19, Proposition 6.17 and Corollary 2.23.

7 Let $\wedge$ be the set of eigenvalues of a J-isometric operator V and let $\Lambda \cap (A^*)^{-1}[\emptyset$. Then $C\,\mathrm{Lin}\{\mathscr{L}_\lambda(V)\}_{\lambda\in\Lambda}$ is a neutral subspace (*cf.* Corollary 3.14) ([XIV]).
Hint: Use the results of Exercise 5, and Exercise 16 on §5.

8 Let U be a π-unitary operator in $\Pi_\varkappa$. Then its non-unitary spectrum consists of not more than $2\varkappa$ (taking multiplicity into account) normal eigenvalues situated symmetrically relative to the unit circle $\mathbb{T}$ (*cf.* Corollary 3.15) ([XIV]).
Hint: Use the results of Exercise 5 on §1, Exercise 22 on §4, Remark 6.20, and Corollaries 6.18 and 3.15.

9 Let U be a π-unitory operator; then $\sigma_r(U) = \emptyset$ (*cf.* Corollary 3.16).
Hint: Use the results of Exercise 8 on §4.

10 If V is a π-non-contractive operator ($\mathscr{D}_V = \Pi_\varkappa$), if $\lambda \in \sigma_p(V)$ and $|\lambda| = 1$, then the root lineal $\mathscr{L}_\lambda(V)$ can be represented in the form $\mathscr{L}_\lambda(V) = \mathscr{N}_\lambda [\dotplus] \mathscr{M}_\lambda$, where $\dim \mathscr{N}_\lambda < \infty$, $V\mathscr{N}_\lambda \subset \mathscr{N}_\lambda$, $\mathscr{M}_\lambda \subset \mathrm{Ker}(V - \lambda I)$, and $\mathscr{M}_\lambda$ is a non-degenerate subspace (in particular, it may happen that $\mathscr{M}_\lambda = \{\theta\}$).

If $d_1(\lambda), d_2(\lambda), \ldots, d_{r_\lambda}(\lambda)$ are the orders of the elementary divisors of the operator $V \mid \mathscr{N}_\lambda$, then

$$\sum_{\lambda\in\sigma_p(V), |\lambda|=1} \sum_{j=1}^{r_\lambda} \left[\frac{d_j(\lambda)}{2}\right] + \sum_{|\lambda|>1} \dim \mathscr{L}_\lambda(V) \leqslant \varkappa$$

(Azizov [8]; *cf.* [XIV]).
Hint: Use the results of Exercise 5 on §1, Remark 6.20, Corollary 6.18 and Theorem 2.26.

11 Prove that a J-symmetric operator A is J-non-negative if and only if the J-isometric operator $V = \mathbf{K}_\zeta(A)$ $(\zeta \neq \bar{\zeta})$ satisfies the condition $\mathrm{Re}[\zeta(J - V)x, x] \geqslant 0$ for all $x \in \mathscr{D}_V$ (Azizov, L. I. Sukhocheva).

12 Prove that if V is a J-unitary operator and $\mathrm{Re}[\zeta(I - V)x, x] \geqslant 0$ for some $\zeta \in \mathbb{C}$ and for all $x \in \mathscr{H}$, then $\sigma(V) \subset \mathbb{T}$ (Azizov, Sukhocheva).

13 Let A and V be the same as in Proposition 6.17, and let $\mathscr{L}$ be a finite-dimensional subspace. Then $\{\mathscr{L} \subset \mathscr{D}_A, A\mathscr{L} \subset \mathscr{L}\} \Leftrightarrow (V \subset \mathscr{D}_V, V\mathscr{L} \subset \mathscr{L})$ (*cf.* [III]).
Hint: Use Corollary 6.18.

14 Let V be a W-non-contractive operator, and let $Vx_0 = \varepsilon x_0$, $Vx_1 = \varepsilon x_1 + x_0$ with $|\varepsilon| = 1$. Then x_0 is an isotropic vector in $\mathrm{Ker}(V - \varepsilon I)$.
Hint: Apply the transformation $\mathbf{K}_\zeta^{-1}$ $(\zeta \neq \bar{\zeta})$ to $V \mid \mathscr{L}_\varepsilon(VV)$ and use the results of Exercise 11 on §4.

15 Prove that if $A \subset B$, where $A, B \in \mathscr{L}$, then $\mathbf{K}_\zeta(A) \subset \mathbf{K}_\zeta(B)$ and $\mathbf{K}_\zeta^{-1}(A) \subset \mathbf{K}_\zeta^{-1}(B)$ when $\zeta \neq \bar{\zeta}$.

Remarks and bibliographical indications on chapter 2

In §§1,4,5 the exposition is carried out at first in the most general form—for operators acting from one space $\mathscr{H}_1$ into another $\mathscr{H}_2$ ($T:\mathscr{H}_1 \to \mathscr{H}_2$), and only later is it made concrete for the case of operators acting in a single space. This is the first time, apparently, that this has been done (at any rate, so systematically; *cf.* the monograph [V] and our survey [IV]) and mainly in the interests of the theory of extensions of operators (see Chapter 5). In this connection it should be borne in mind when reading these notes and especially the bibliographical indications that in them, *with rare exceptions, no account is taken of generalizations (as compared with the primary sources) made in the text to the case of operators* $T: \mathscr{H}_1 \to \mathscr{H}_2$.

§1.1. Adjoint operators (relative to an indefinite metric) were first considered in the space $\Pi_\varkappa$ by Pontryagin [1], in J-spaces by I. Iokhvidov [5], [6] and by Langer [2], in G-spaces by I. Iokhvidov [12] and next by Azizov and I. Iokhvidov [1].

§1.2. The idea of using the indefinite metric (1.4) applied to graphs of linear operators for discovering the properties of the operators themselves was first put forward and used by Phillips [1]. Later Shmul'yan [4], [5] and others developed it. Our exposition here follows Ritsner's monograph [4].

§1.3. The formula (1.10) was established by Azizov. For all the rest of the material see [III].

§1.4. The Propositions 1.11 and 1.12 are 'folk-lore'. Sources of Theorem 1.13 can be found already in Pontryagin's article [1]. The formulation and proof given in the text are due to Azizov who used an idea of Langer's in [XVI]. Normal points are studied in [X]. Theorem 1.16 represents a certain development of Langer's results [2].

§1.5. The device presented here of passing to the factor-space $\mathscr{H}/\mathscr{H}^0$ was first applied in [XV]; Theorem 1.17 we find essentially in Langer [9].

§2.1. W-dissipative operators were introduced in the book [VI]. π-dissipative operators were first introduced in a particular case by Kuzhel' [5], [6]; they were studied in detail independently by Azizov [1], [4], [5], [8], and also by M. Krein and Langer [3]; J-dissipative operators by Azizov [5], [8] E. Iokhvidov [1], Azizov and E. Iokhvidov [1] in which the main theorem of this paragraph was established. A geoemtrical proof of the well-known Lemma 2.8 (see [VII]) is due to Azizov.

§§2.2, 2.3. The Definition 2.10 and Theorem 2.11 are taken from [X]. Corollaries 2.12 and 2.13 are due to Azizov, as is all the material in §2.3 (see Azizov [8]).

§2.4. Theorem 2.20 has been borrowed from [VI] and [XXII]. The remaining material of this paragraph was obtained by Azizov.

§2.5, §2.6. Theorem 2.26 due to Azizov generalizes the corresponding result of Pontryagin for π-Hermitian operators (see also [XIV], [XVI]). The other results of these paragraphs were also established by Azizov.

§3.1. In an abstract formulation (but in a different terminology) π-Hermitian and π-self-adjoint operators were first studied by Pontryagin [1], who mentions that his attention was drawn to them by S. L. Sobolev (see the remark on Chapter 3 below). After Pontryagin they were studied by I. Iokhvidov [1], [2] (in the 'finite-dimensional' case *cf.* Potapov [1]), and in more detail see [XIV]. J-self-adjoint operators were considered by Ginzburg [2], I. Iokhvidov [6] and later (in great detail) by Langer [1]–[3]. G-symmetric and G-self-adjoint operators are encountered partially even in Langer [2], and they are considered in detail in [III], to which we refer the reader for details.

Proposition 3.7 and its Corollaries 3.8, 3.9 are found in an article by Azizov and E. Iokhvidov [1]. As regards Corollary 3.12 see I. Iokhvidov [6], Langer [2]. Remark 3.13 is due to Azizov, Corollaries 3.14 and 3.15 go back to Pontryagin [1] (*cf.* [XIV]).

§3.2. In proving Theorem 3.19 it would have been possible to use Proposition 3.7 and to refer to a well-known 'definite' result. However, we decided to bring in what we think is a simple proof, due to Azizov. Lemmas 3.20 and 3.21 used in it have such a tangled pre-history that we are inclined to attribute them to 'folk-lore'. Lemma 3.22 is due to Azizov.

Corollary 3.25 is found, essentially, in Potapov [1].

Theorem 3.27 in the case of a continuous operator A is due to Ginzburg [2], and in the form in which it is formulated—to Langer [8].

§3.3. All the examples, except Example 3.33, are due to Azizov.

§4.1. Plus-operators V in a real space Π, ($\mathscr{D}_V = \Pi_1$, V bounded) were first considered by M. Krein (see M. Krein and Rutman [1], and later in an arbitrary (complex) $\Pi_\varkappa$ by Brodskiy [1]). The latter, in contrast to our Definition 4.1, imposed the condition $\bar{\mathscr{D}}_V = \Pi_\varkappa$ (we point out that from this condition $\mathscr{D}_V \cap \mathscr{P}^{++} = \emptyset$ already follows—see Lemma 1.9.5). The definition of a plus-operator V in a general J-space $\mathscr{H}$ and the name 'plus-operator' itself were introduced in the articles of M. Krein and Shmul'yan [1], [2]. In contrast to Brodskiy, with these authors $\mathscr{D}_V = \mathscr{H}$ and the operator V is bounded *a priori*. For such operators they formulated Theorem 4.3 and presented for it a proof which remains valid in the more general case (see [XVI] for the previous history of this theorem), and they established a classification (which had appeared earlier in a particular case in Brodskiy's [1]) of plus-operators into strict and non-strict (this terminology itself is due to them).

J-non-contractive (J-non-expansive) operators in a Krein space were introduced and studied by Ginzburg [1], [2], generalizing the corresponding considerations in Potapov's [1] for finite-dimensional spaces. Earlier M. Krein [4] (see also [XIV]) had considered in $\Pi_\varkappa$ the so-called *non-decreasing* linear operators $V: \mathscr{D}_V = \Pi_\varkappa$, V bounded and $[Vx, Vx] \geqslant [x, x]$ for $x \in \mathscr{P}^+$. We point out that this inequality means, even in the general conditions of Definition 4.1, that V is a strict plus-operator with $\mu_+(V) \geqslant 1$ (*cf.* [XVI]).

For plus-operators V in $\Pi_\varkappa$ with $\bar{\mathscr{D}}_V = \Pi_\varkappa$ Brodskiy [1] discovered that they are either finite-dimensional (and then, possibly, unbounded), or they are continuous (*cf.* with our Corollary 4.8). As I. Iokhvidov [17] pointed out, the same Brodskiy article contains essentially all the ideas used in the proof of Theorem 4.6 and its corollaries. For these results in a rather fuller and explicit form, see I. Iokhvidov [17]. T. Ya. Azizov pointed out that the requirement that $\mathscr{D}_V \cap \mathscr{H}^+ \neq \emptyset$ imposed in these papers can be omitted in Theorem 4.6. Regarding Corollary 4.7 see I. Iokhvidov [17].

A curious generalization of plus-operators V was recently proposed and investigated in [XVI]: $[Vx, Vx] \geqslant \mu[x, x]$ for some $\mu \in \mathbb{R}$ and all $x \in \Pi_\varkappa$. For such V finite $\mu_\pm(V)$ again exist and $\mu_-(V) \leqslant \mu \leqslant \mu_+(V)$, and a number of facts were established, many of which probably remain true in the more general situation when $V: \mathcal{H}_1 \to \mathcal{H}_2$ (in the spirit of our Definitions 4.1 and 4.2).

§4.2. The concept of a doubly strict plus-operator and the basic facts about such operators in *J*-spaces were established in the articles of M. Krein and Shmul'yan [2], [3]; there is a later bibliography of this topic in [IV]. The proof of Proposition 4.10 in the text was given by I. Iokhvidov [17] (there the priority of Yu. P. Ginzburg on this question is pointed out).

J-bi-non-contractive (*J*-bi-non-expansive) operators were introduced and studied by Ginzburg [2]. Theorem 4.19 and its Corollary 4.20 are due to M. Krein and Shmul'yan [2], but the proof in the text is Azizov's (*cf.* Ritsner [4]). It embodies most of the earlier known criteria for *J*-non-contractive operators *V* to be *J*-bi-non-contractive (see Ginzburg [1], [2], I. Iokhvidov [14], [17], M. Krein and Shmul'yan [1], [2]; in connection with the rejection of the *a priori* requirement for *V* to be continuous, see I. Iokhvidov [17].

See Ritzer [4], E. Iokhvidov [8] for the generalization of the concept of a *J*-bi-non-expansive operator (*cf.* Shmul'yan [4]).

§4.3 Focusing plus-operators were first examined by Krasnosel'skry and A. Sobolev [1], later by A. Sobolev and Khatsevich [1], [2], and in detail by Khatskevich [6], [7], [10], [11], [15]; uniformly *J*-expansive operators—in the book [VI]. Theorem 4.24 was established by Azizov [10] (see Azizov and Khoroshavin [1]). Some of its assertions were obtained independently by M. Krein and Shmul'yan [5]. Theorem 4.27 is contained essentially in [VI].

§4.4. Theorem 4.30 in the 'finite-dimensional' case was established by Potapov [1], in the general case—see Ginzburg [2], M. Krein and Shmul'yan [3]; the proof in the text is due to Azizov. Theorem 4.31 has been borrowed from [VI]; Corollary 4.32 is due to A. Sobolev and Khatskevich [1], [2]

§4, Exercise 12. In connection with Remark 4.29 a warning must be given against mechanical transfer of results of the type in Exercise 12 from plus-operators to minus-operators (respectively, from (J_1, J_2)-non-contractive to (J_1, J_2)-non-expansive operators); it must be remembered that the rôles of the subspaces $\mathcal{H}_i^+$ and $\mathcal{H}_i^-$ (respectively, of the projectors P_i^+ and P_i^-), $i = 1, 2$, are interchanged.

§4, Exercise 26. *B*-plus-operators in space $\Pi_\varkappa$ were introduced by Brodskiy [1] (see [XVI]), together with the name '*B*-plus-operator' itself.

In conclusion we point out to the reader that much more information about the operators acting in $\Pi_\varkappa$ spaces that are mentioned in §4 can be found in the monograph [XVI].

§5.1. Isometric (in particular, unitary) operators in infinite-dimensional spaces with an indefinite metric were first considered by M. Krein (see M. Krein and Rutman [1]), I. Iokhvidov [1], [2], I. Iokhvidov and M. Krein [XIV], [XV]. Proposition 5.2, see M. Krein and Shmul'yan [2]. Example 5.5 in the text was given by Azizov; for the other examples, see I. Iokhvidov [12]. Theorem 5.6 is due to I. Iokhvidov [12].

§5.2. In connexion with Theorem 5.10 see [XIV], M. Krein and Shmul'yan [3], and also *cf.* Azizov [11]. Corollary 5.13 is found in M. Krein [5].

§5.3. Theorem 5.18 (and its corollaries) in the case of a commutative group is due to Phillips [3]. In the general case of amenable groups we are inclined to attribute it to 'folk-lore', since the proof given in the text in no way corresponds to that given in [VI], for example, for a single operator. This fact was formally noted by Azizov and Shmul'yan; see also Exercise 7 on this section. For groups in Π_1 this result was proved by Shmul'yan without the requirement of amenability.

Definitions 5.21 and 5.22 were given by M. Krein [XVII]. Theorem 5.23 is due to M. Krein [XVII] (for details, see [VI]), the proof given here was somewhat modified by Azizov. For similar results for stable and strongly stable J-self-adjoint operators, see Langer [1], McEnnis [1].

§5.4. All the results of this paragraph are in M. Krein and Shmul'yan [3].

§5, Exercises. The operators in Exercise 4 are called J-unitary operators by Shul'man. The condition in Exercise 9 was first considered by Masuda [1].

§6. I. Iokhvidov [1] was the first to apply Cayley–Neumann transformations to operators in spaces with an indefinite matric. Then these transformations were considered in detail in [XIV], [III], and were widely applied by many authors. In §§6.1–6.3 we mainly follow Ritsner [4].

3 INVARIANT SEMI-DEFINITE SUBSPACES

In this chapter we shall set out results on one of the central problems in the theory of operators in Krein spaces and, in particular, in Pontryagin spaces—the problem of the existence of maximal semi-definite invariant subspaces for operators and sets of operators acting in these spaces. It will be assumed that the J-non-contractive operators appearing here are bounded and defined on the whole space.

§1 Statement of the problems

1 We have already encountered the concept of an invariant subspace in Proposition 2.1.11. We now go into it in detail.

Definition 1.1: Let $T: \mathscr{H} \to \mathscr{H}$ be an operator densely defined in a Hilbert space $\mathscr{H}$. We shall say that *the subspace $\mathscr{L}$ is invariant relative to T* if $\overline{\mathscr{D}_T \cap \mathscr{L}} = \mathscr{L}$ and $T: \mathscr{L} \to \mathscr{L}$. In particular, the subspace $\mathscr{L} = \{\theta\}$ is invariant relative to any operator T; in this case we put by definition $\rho(T \mid \{\theta\}) = \mathbb{C}$.

We note that if T is an operator defined everywhere in $\mathscr{H}$, then the condition $\overline{\mathscr{D}_T \cap \mathscr{L}} = \mathscr{L}$ is always satisfied and moreover $\mathscr{L} \subset \mathscr{D}_T$ $(= \mathscr{H})$. It would be possible in Definition 1.1 to require, instead of the condition that $\mathscr{D}_T \cap \mathscr{L}$ be dense in $\mathscr{L}$, the inclusion $\mathscr{L} \subset \mathscr{D}_T$, but then the set of invariant subspaces would be impoverished. On the other hand, it would be possible to extend this class by dropping the condition that $\overline{\mathscr{D}_T \cap \mathscr{L}} = \mathscr{L}$ and leaving only the condition that $\overline{\mathscr{D}_T \cap \mathscr{L}} \subset \mathscr{L}$, but then this would lead to the situation, unnatural in our opinion, when any subspace which intersected $\mathscr{D}_T$ only along the vector θ would be an invariant subspace for T. We therefore stay with the Definition 1.1.

In Pontryagin's foundation-laying work [1] it is proved that (in the terminology we have adopted) every π-selfadjoint operator A in $\Pi_\varkappa$ has a $\varkappa$-dimensional non-negative invariant subspace $\mathscr{L}^+$, which can be chosen so that $\operatorname{Im}\sigma(A \mid \mathscr{L}^+) \geqslant 0$ ($\operatorname{Im}\sigma(A \mid \mathscr{L}^+) \leqslant 0$). Further development of this result led to the following problems.

Problem 1.2: *Does every closed J-dissipative (and, in particular, every J-selfadjoint) operator A have an invariant subspace $\mathscr{L}^+ \in \mathscr{M}^+$? If it does, is there an $\mathscr{L}^+$ such that* $\operatorname{Im}\sigma(A \mid \mathscr{L}^+) \geqslant 0$ *(and for a J-selfadjoint A is there also one such that* $\operatorname{Im}\sigma(A \mid \mathscr{L}^+) \leqslant 0$)?

Problem 1.3: *Let $\mathscr{A} = \{A\}$ be the set of maximal J-dissipative operators A with $\rho(A) \cap \mathbb{C}^+ \neq \emptyset$ whose resolvents commute in pairs, and let $\mathscr{L}_+$ ($\subset \mathscr{P}^+$) be their common invariant subspace, and let $\rho(A \mid \mathscr{L}_+) \cap \mathbb{C}^+ \neq \emptyset$ ($A \in \mathscr{A}$). Does the family $\mathscr{A}$ have a common invariant subspace $\widetilde{\mathscr{L}}_+ \in \mathscr{M}^+$ which contains $\mathscr{L}_+$ and is such that $\rho(A \mid \widetilde{\mathscr{L}}_+) \cap \mathbb{C}^+ \neq \emptyset$ ($A \in \mathscr{A}$)?*

We notice at once that, in such a general formulation, Problem 1.2 has a *negative* answer even for J-selfadjoint J-positive operators (see Theorem 4.1.10 below).

□ *Example 1.4:* Let $\mathscr{H}_1 = \operatorname{Lin}\{e\} \oplus \mathscr{H}_1'$ be a Hilbert space, with $\| e \| = 1$, $\dim \mathscr{H}_1' = \infty$, and $e \perp \mathscr{H}_1'$. We define relative to this decomposition a completely continuous selfadjoint operator G in $\mathscr{H}_1$ by means of the matrix $G = \| G_{ij} \|_{i,j=1}^2$, where $G_{11} = 0$, $G_{12} = \alpha(\cdot, f)e$, $G_{21} = \alpha(\cdot, e)f$, $G_{22} = \alpha G'_{22}$, G'_{22} is a negative completely continuous operator in $\mathscr{H}_1'$, $f \in \mathscr{H}_1' \backslash \mathscr{R}_{G'_{22}}$, $\| f \| = 1$, and $\alpha > 0$ is such that $\| G \| < 1$, and we introduce in $\mathscr{H}_1$ the G-metric $[x, y] = (Gx, y)$. Since $\mathscr{H}_1$ is decomposed into the sum of the neutral one-dimensional subspace $\operatorname{Lin}\{e\}$ and the negative $\mathscr{H}_1'$, and since it follows from the inclusion $f \in \mathscr{H}_1' \backslash \mathscr{R}_{G'_{22}}$ that $0 \notin \sigma_p(G)$, so $\mathscr{H}_1$ is a $G^{(\varkappa)}$-space with $\varkappa = 1$ (see 1.§9.6). Let P be the orthoprojector from $\mathscr{H}_1$ on to $\mathscr{H}_1'$. Then PG is a completely continuous G-selfadjoint operator, which has, as is easily verified, not one non-negative eigenvector.

In the space

$$\mathscr{H} = \mathscr{H}_1 \oplus \mathscr{H}_2, \qquad \mathscr{H}_2 = \mathscr{H}_1, \tag{1.1}$$

we define a J-metric $[\langle x_1, x_2 \rangle, \langle y_1, y_2 \rangle] = [J\langle x_1, x_2 \rangle, \langle y_1, y_2 \rangle]$ by means of the operator J (see 1.(3.9))

$$J = \| J_{ij} \|_{i,j=1}^2, \qquad J_{11} = G,\ J_{12} = (I - G^2)^{1/2}, \qquad J_{21} = J_{12}^*,\ J_{22} = -G.$$

Let A_1 be the orthoprojector from $\mathscr{H}$ on to $\mathscr{H}_1$ ($\subset \mathscr{H}$), and let the linear operator A_2 be such that $\operatorname{Ker} \mathscr{A}_2 = \mathscr{H}_1$, $A_2 \mid \mathscr{H}_2 = | G |^{-1}$. Then $A_1 = A_1^* \geqslant 0$, and $A_2 = A_2^* \geqslant 0$. Since the operator A_1 is bounded, it follows (see 2.Proposition 1.9 and 2.Corollary 3.8) that $A \equiv A_1 J + JA_2 = A^c \overset{J}{\geqslant} 0$. Since $\mathscr{H}_1 \subset \mathscr{D}_A$, we can express A in matrix form relative to the decomposition (1.1): $A = \| A_{ij} \|_{i,j=1}^2$. It can be verified immediately that

$$A_{11} = PG, \qquad A_{12} = P(I - G^2)^{1/2} + (I - G^2)^{1/2} | G |^{-1}, \qquad A_{21} = 0,$$
$$A_{22} = -G | G |^{-1}.$$

It is well-known that the selfadjointness of G implies $\mathscr{R}_{|G|} = \mathscr{R}_G$, and therefore $\mathscr{D}_A = \mathscr{H}_1 \overset{J}{\oplus} \mathscr{R}_G$. From the matrix representation of the operator A it follows that $0 \notin \sigma_P(A)$, i.e., $A \gg 0$.

Since $A_{21} = 0$ and $G|G|^{-1}\mathscr{R}_G = \mathscr{R}_G$, so $A\mathscr{D}_A \subset \mathscr{D}_A$. From this the equality $\mathscr{D}_{A^n} = \mathscr{D}_A$ $(n = 1, 2, 3, \ldots)$ follows. Therefore if $\mathscr{L}^+$ $(\in \mathscr{M}^+)$ were an invariant subspace relative to A, it would also be invariant relative to $A^2 = \|(A^2)_{ij}\|_{ij=1}^2$:

$$(A^2)_{11} = (PG)^2, \qquad (A^2)_{22} = PGP(I - G^2)^{1/2} - (I - G^2)^{1/2}G^{-1},$$
$$(A^2)_{21} = 0, \qquad (A^2)_{22} = I_2.$$

Therefore

$$\text{if } x = \langle x_1, x_2 \rangle \in \mathscr{L}^+ \ (x_i \in \mathscr{H}_i,\ i = 1, 2),$$

then also

$$A^2 x = \langle (PG)^2 x_1 + PGP(I - G^2)^{1/2} x_2 - (I - G^2)^{1/2} G^{-1} x_2, x_2 \rangle \in \mathscr{L}^+.$$

Since

$$\operatorname{Ker}(A^2 - I) = \{\langle (I - G^2)^{1/2} x_2, -Gx_2 \rangle \mid x_2 \in \mathscr{H}_2\} = J\mathscr{H}_2,$$

$\operatorname{Ker}(A^2 - I)$ does not contain maximal non-negative subspaces. Therefore $\mathscr{L}^+ \not\subset \operatorname{Ker}(A^2 - I)$ also. Since $(A^2 - I)\mathscr{D}_A \subset \mathscr{H}_1$, so $(A^2 - I)(\mathscr{L}^+ \cap \mathscr{D}_A) \subset \mathscr{H}_1$ also. From the relations $\overline{\mathscr{L}^+ \cap \mathscr{D}_A} = \mathscr{L}^+$ and $\mathscr{L}^+ \not\subset \operatorname{Ker}(A^2 - I)$ we conclude that $(A_2 - I)(\mathscr{L}^+ \cap \mathscr{D}_A) \neq \{\theta\}$. Since $(A^2 - I)(\mathscr{L}^+ \cap \mathscr{D}_A) \subset \mathscr{L}^+$ and all the non-negative subspaces in $\mathscr{H}_1$ are one-dimensional, there is in $\mathscr{H}_1$ a non-negative eigenvector of the operator A and this will also be an eigenvector for PG; we have obtained a contradiction, because PG has no non-negative eigenvectors. ∎

We note that for the operator A in the above example $\rho(A) = \emptyset$ and therefore (see 2.Theorem 2.9) there are no maximal uniformly positive subspaces in $\mathscr{D}_A$. At the same time, in Pontryagin's theorem mentioned above, since $\overline{\mathscr{D}}_A = \Pi_\varkappa$, the domain $\mathscr{D}_A$ does, by I.Lemma9.5, contain such a subspace. It is therefore expedient to introduce the following

Definition 1.5: We shall say that a set of operators $\mathscr{A} = \{A\}$ in a Krein space $\mathscr{H} = \mathscr{H}^+ \oplus \mathscr{H}^-$ *satisfies condition* (L) and we shall write $\mathscr{A} \in (L)$ if, for any $A \in \mathscr{A}$, there is contained in $\mathscr{D}_A$ at least one maximal uniformly positive subspace $\mathscr{L}^+$ (in the general case, different $\mathscr{L}^+$ for different $A \in \mathscr{A}$); in particular, if $\mathscr{A}$ should consist of a single operator A, then we would write $A \in (L)$ and would assume that $\mathscr{H}^+ \subset \mathscr{D}_A$.

If in Problems 1.2 and 1.3 the operators A satisfy condition (L), then naturally, as we shall see later in Theorem 1.13, the non-negative invariant subspaces are to be sought in the lineal $\mathscr{D}_A$ itself. In such a formulation the Problems 1.2 and 1.3 have not yet been solved. Later we shall introduce additional conditions under which they have affirmative solutions.

2 Another group of problems—for J-non-contractive operators V with $\mathscr{D}_V = \mathscr{H}$—finds its origin in M. Krein's works [4], [5].

Problem 1.6: *Does every J-non-contractive operator V have an invariant subspace* $\mathscr{L}^+ \in \mathscr{M}^+$? *If yes, then is there an* $\mathscr{L}^+$ *such that* $|\sigma(V|\mathscr{L}^+)| \geqslant 1$?

Problem 1.7: *Let* $\mathscr{U} = \{U\}$ *be a commutative family of J-non-contractive operators, and let* $\mathscr{L}_+$ *be their common invariant non-negative subspace (in brief,* $\mathscr{U}\mathscr{L}_+ \subset \mathscr{L}_+$. *Is there an* $\tilde{\mathscr{L}}_+ \in \mathscr{M}^+$ *such that* $\mathscr{L}_+ \subset \tilde{\mathscr{L}}_+$ *and* $\mathscr{U}\tilde{\mathscr{L}}_+ \subset \tilde{\mathscr{L}}_+$?

We shall show that Problem 1.7 is such a general formulation has a *negative* solution even for a family $\mathscr{U}$ consisting of a single operator U.

□ *Example 1.8:* Let $\mathscr{H} = \mathscr{H}^+ \oplus \mathscr{H}^-$ be a J-space, and let $\{e_i^+\}_0^\infty$, $\{e_i^-\}_1^\infty$ be orthonormalized bases in $\mathscr{H}^+$ and $\mathscr{H}^-$ respectively. We define on the basis $\{e_i^+\}_0^\infty \cup \{e_i^-\}_1^\infty$ a J-semi-unitary operator U:

$$Ue_0^+ = e_0^+ + e_1^+ + e_1^-, \qquad Ue_i^\pm = e_{i+1}^\pm \qquad (i = 1, 2, \ldots).$$

The subspace $\mathscr{L}_+ = C\,\mathrm{Lin}\{e_i^+\}_1^\infty$ $(\subset \mathscr{H}^+)$ is uniformly positive and invariant relative to U; moreover it does not admit extension into a maximal non-negative subspace invariant relative to U. For, suppose $\mathscr{L}_+ \subset \tilde{\mathscr{L}}_+$ $(\in \mathscr{M}^+)$ and $U\tilde{\mathscr{L}}_+ \subset \tilde{\mathscr{L}}_+$. Since $P^+\tilde{\mathscr{L}}_+ = \mathscr{H}^+$ (see 1.Theorem 4.5), there must be in $\tilde{\mathscr{L}}_+$ a vector of the form $e_0^+ + x^-$ with $x^- \neq \theta$, $x^- \in \mathscr{H}^-$, and together with it the vector $U(e_0^+ + x^-) = e_0^+ + e_1^+ + e_1^- + Ux^-$. Since $e_1^+ \in \mathscr{L}_+$, so $e_0^+ + x_1^- + Ux^- \in \tilde{\mathscr{L}}_+$ and $(e_0^+ + x^-) - (e_0^+ + e_1^- + Ux^-) \in \tilde{\mathscr{L}}_+$. Since $\tilde{\mathscr{L}}_+$ is non-negative we have $e_1^- = (I - u)x^-$, which, as is easily seen, is impossible; we have obtained a contradiction. We note (*cf.* Theorem 2.8 and Remark 2.4 below) that in this example the 'corner' U_{12} $(= P^+UP^- | \mathscr{H}^-) = 0$. ■

Definition 1.9: A subspace $\mathscr{L}$ is said to be *completely invariant relative to a family of operators* $\mathscr{T} \in \{T\}$ if $\mathscr{L} \subset \mathscr{L}_T$ and $T\mathscr{L} = \mathscr{L}$ for all $T \in \mathscr{T}$ (in this case we shall also write $\mathscr{T}\mathscr{L} = \mathscr{L}$).

To conclude this paragraph we point out that, if $\mathscr{L}_+$ is assumed to be a completely invariant subspace relative to $\mathscr{U}$, then the solutions of the problems 1.6 and 1.7 have not yet been found. Later, in §§2–5, we shall indicate sufficient conditions under which these problems have an affirmative solution.

3 Finally, the third of the main directions is that generated by Phillips's paper [3].

Problem 1.10: *Let* $\mathscr{A} = \{A\}$ *be an algebra of continuous operators acting in the whole of a Hilbert J-space* $\mathscr{H}$; *let* $\mathscr{A}$ *be closed relative to J-conjugation,* *i.e.,* $A \in \mathscr{A} \Rightarrow A^c \in \mathscr{A}$, and let $\mathscr{A}$ *contain I. Under what conditions on* $\mathscr{A}$ *will every dual pair* $(\mathscr{L}^+, \mathscr{L}^-)$ *which is invariant relative to* $\mathscr{A}$ *(i.e.,* $\mathscr{A}\mathscr{L}^\pm \subset \mathscr{L}^\pm$, $\mathscr{L}^\pm \subset \mathscr{P}^\pm$, $\mathscr{L}^+ [\perp] \mathscr{L}^-$*) admit extension into an invariant maximal dual pair? Is commutativity of* $\mathscr{A}$ *sufficient for this condition?*

Problem 1.11: *Under what conditions on a group* $\mathscr{U} = \{U\}$ *of J-unitary operators will every dual pair* $(\mathscr{L}^+, \mathscr{L}^-)$ *which is invariant relative to* $\mathscr{U}$ *admit extension into a maximal dual pair invariant relative to* $\mathscr{U}$*? Is commutativity of* $\mathscr{U}$ *sufficient for this condition?*

Remark 1.12: In Problem 1.11 the invariance of the subspaces $\mathscr{L}^\pm$ implies their complete invariance, because $\mathscr{U}$ is a group and therefore it follows from $U\mathscr{L}^\pm \subset \mathscr{L}^\pm$ and $U^{-1}\mathscr{L}^\pm \subset \mathscr{L}^\pm$ that $U\mathscr{L}^\pm = \mathscr{L}^\pm$. Moreover, if $(\mathscr{L}^+, \mathscr{L}^-)$ is an invariant dual pair of any J-unitary operator U, then it is easy to see that $(C\ \mathrm{Lin}\{U^{-n}\mathscr{L}^+\}_0^\infty,\ C\ \mathrm{Lin}\{U^{-n}\mathscr{L}^-\}_0^\infty)$ is a dual pair which is completely invariant relative to U and which extends $(\mathscr{L}^+, \mathscr{L}^-)$, and even more precisely, it enters into any extension of $(\mathscr{L}^+, \mathscr{L}^-)$ into a completely invariant dual pair.

We point out, regarding their above formulations, that Problems 1.10 and 1.11 have not yet been solved for the case of arbitrary commutative algebras and groups respectively. In §§4 and 5 below we shall give some sufficient conditions for them to be soluble affirmatively (see also 2.Theorem 5.18).

We observe also that, under the conditions of Problem 1.10, the algebra $\mathscr{A}$ coincides with the linear envelope of its J-self-adjoint elements A (A^c) with $\| A \| < 1$, and therefore it is contained in the linear envelope of J-unitary operators

$$U = (I - A^2)^{1/2} + iA, \text{ where } (I - A^2)^{1/2} = -\frac{1}{2\pi i} \oint_{\Gamma_\sigma} \sqrt{\lambda}(I - A^2 - \lambda I)^{-1}\, d\lambda,$$

the contour Γ_σ contains $\sigma(I - A^2)$ within itself and is symmetric about the real axis and lies in the open right-hand half-plane, and $\mathrm{Re}\sqrt{\lambda} > 0$. If now $\mathscr{L}$ is a maximal semi-definite invariant subspace of the operator U, then $U\mathscr{L} = \mathscr{L}$ (see 2.Remark 4.21), and therefore $U^{-1}\mathscr{L} = \mathscr{L}$. Consequently

$$A\mathscr{L} = \frac{1}{2i}(U - U^c)\mathscr{L} = \frac{1}{2i}(U - U^{-1})\mathscr{L} \subset \mathscr{L}.$$

Conversely, if $A\mathscr{L} \subset \mathscr{L}$, then it follows from the definition of the operator $\tilde{I} - A^2)^{1/2}$ that $(I - A^2)^{1/2}\mathscr{L} \subset \mathscr{L}$, and therefore $U\mathscr{L} \subset \mathscr{L}$, and, by virtue of 2.Remark 4.21, $U\mathscr{L} = \mathscr{L}$. ■

Thus, this Remark 1.12 enables us to go into the investigation of Problem 1.11 only, we leave the reader to make the appropriate reformulation for the Problem 1.10 himself.

4 In this paragraph we point out a connection between Problems 1.2, 1.3 and 1.6, 1.7 respectively.

Theorem 1.13: *Let $A \in (L)$ be a maximal J-dissipative operator. Then there is a point $\zeta \in \rho(A) \cap \mathbb{C}^+$ such that, for $\mathscr{L}^+ \in \mathscr{M}^+$, the following conditions are equivalent:*

a) *$\mathscr{L}^+ \subset \mathscr{D}_A$ and $A\mathscr{L}^+ \subset \mathscr{L}^+$;*
b) *$U_\zeta \mathscr{L}^+ \subset \mathscr{L}^+$, where $U_\zeta \equiv \mathbf{K}_\zeta(A) = (A - \bar{\zeta}I)(A - \zeta I)^{-1}$ is the Cagley–Neyman transform of the operator A at the point ζ* (see 2.§6.4).

□ Since $A \in (L)$, we suppose (see Definition 1.5) that $\mathscr{H}^+ \subset \mathscr{D}_A$, and therefore all ζ with Im $\zeta > 2 \| AP^+ \|$ belong to $\rho(A)$ (see 2.Theorem 2.9).

Let $\mathscr{L}^+ \subset \mathscr{D}_A$, $\mathscr{L}^+ \in \mathscr{M}^+$, and $A\mathscr{L}^+ \subset \mathscr{L}^+$. Then all the ζ specified above belong to the field of regularity of the operator $A \mid \mathscr{L}^+$, and since $A \mid \mathscr{L}^+$ is continuous, they are regular for this operator. Therefore

$$(A - \zeta I)^{-1}\mathscr{L}^+ = \mathscr{L}^+ \text{ and } U_\zeta \mathscr{L}^+ = (A - \bar{\zeta}I)(A - \zeta I)^{-1}\mathscr{L}^+ \subset \mathscr{L}^+.$$

Thus, for the implication a) ⇒ b) to hold, it is sufficient to take ζ with Im $\zeta > 2 \| AP^+ \|$.

We shall now prove that among these ζ there are some for which the reverse implication b) ⇒ a) holds. Let $\mathscr{L}^+ \in \mathscr{M}^+$ and $U_\zeta \mathscr{L}^+ \subset \mathscr{L}^+$. It is easy to see (see 2.Proposition 6.16) that a) will follow from b) if and only if $(U_\zeta - I)\mathscr{L}^+ = \mathscr{L}^+$. Let K be the angular operator of the subspace $\mathscr{L}^+$. Since $(U_\zeta - I)\mathscr{L}^+ \subset \mathscr{L}^+$, coincidence of these lineals is equivalent to maximal non-negativity of $(U_\zeta - I)\mathscr{L}^+$, or (see 1.Theorem 4.5) to the equality $\mathscr{H}^+ = P^+(U_\zeta - I)(P^+ + K)\mathscr{H}^+$, which, in turn, is equivalent to $0 \in \rho(U_{\zeta 11} - I^+ + U_{\zeta 12}K)$, where $U_{\zeta ij}$ $(i, j = 1, 2)$ are the blocks of the operator U_ζ (see 2.(2.1)).

Since it follows from 2.Lemma 2.8, 2.Theorem 2.9, and 2.Formula (6.36) that

$$U_{\zeta 12} = -(U_{\zeta 11} - I^+)A_{12}(A_{22} - \zeta I^-)^{-1},$$

so

$$u_{\zeta 11} - I^+ + U_{\zeta 12}K = (U_{\zeta 11} - I^+)(I^+ - A_{12}(A_{22} - \zeta I^-)^{-1}K),$$

and therefore it is sufficient for us to verify that $1 \in \rho(u_{\zeta 11})$ and it is possible to choose ζ with Im $\zeta > 2\| AP^+ \|$ so that $\| A_{12}(A_{22} - \zeta I^-)^{-1} \| < 1$. Using 2.Formula(6.37) and supposing without loss of generality that $\zeta \in \rho(A_{11})$ we have

$$(U_{\zeta 11} - I^+)(I^+ - A_{12})(A_{22} - \zeta I^-)^{-1}A_{21}(A_{11} - \zeta I^+)^{-1}) = (\zeta - \bar{\zeta})(A_{11} - \zeta I^+)^{-1}$$

This equality implies that $\mathscr{R}_{u_{\zeta 11} - I}$ coincides with $\mathscr{H}^+$. Since U_ζ is a J bi-non-contractive operator (see 2.Theorem 6.13), $0 \in \rho(U_{\zeta 11})$, $\| U_{\zeta 11}^{-1} \| \leqslant 1$ (see 2.Remark 4.21), and therefore we have from $I^+ - U_{\zeta 11}^{-1} = U_{\zeta 11}^{-1}(U_{\zeta 11} - I^+)$ that $\mathscr{R}_{I^+ - U_{\zeta 11}^{-1}} = \mathscr{H}^+$, which implies, as is easily seen, the inclusion $1 \in \rho(U_{\zeta 11}^{-1})$, and therefore $1 \in \rho(U_{\zeta 11})$.

We now show that there is a ζ with Im $\zeta > 2 \| AP^+ \|$ such that $\| A_{12}(A_{22} - \zeta I^-)^{-1} \| < 1$.

Since the operator A is J-dissipative

$$0 \leqslant \mathrm{Im}[A_{11}x^+, x^+] + \mathrm{Im}[A_{12}x^-, x^+] + \mathrm{Im}[A_{21}x^+, x^-] + \mathrm{Im}[A_{22}x^-, x^+]$$

for all $x^+\mathscr{H}^+$, $x^- \in \mathscr{H}^- \cap \mathscr{D}_A$. In particular, on substituting instead of x^- the vector $x_\alpha^-(\zeta) = \mathrm{Im}^\alpha\zeta(A_{22} - \zeta I^-)^{-1}x^-, \alpha \in \mathbb{R}$, for all $x^- \in \mathscr{H}^-$ with $\| x^- \| = 1$ and

$$x^+ = i\frac{A_{12}(A_{22} - \zeta I^-)^{-1}x^-}{\| A_{12}(A_{22} - \zeta I^-)^{-1}x^- \|}$$

(here it is assumed that $A_{12}(A_{22} - \zeta I^-)x^- \neq \theta$—for, otherwise, $\| A_{12}(A_{22} - \zeta I^-)^{-1}x^- \| = 0 \leqslant q < 1$ for any $q \geqslant 0$), we obtain

$$\| A_{12}(A_{22} - \zeta I^-)^{-1}x^- \| \leqslant \frac{\mathrm{Im}[A_{11}x^+, x^+]}{\mathrm{Im}^{\alpha}\zeta} + \mathrm{Im}[A_{21}x^+, (A_{22} - \zeta I^-)^{-1}x^-]$$

$$+ \mathrm{Im}^\alpha\zeta\ \mathrm{Im}\ IA_{22}IA_{22} - \zeta I^-)^{-1}x^-, (A_{22} - \zeta I^-)^{-1}x^-]$$

From 2.Theorem 2.9 and 2.Corollary 2.8 we obtain that $\| (A_{22} - \zeta I^-)^{-1} \| 1/\mathrm{Im}\ \zeta$, and when $\mathrm{Im}\ \zeta \geqslant \| \mathrm{Re}\ \zeta \|$ we have $\| A_{22}(A_{22} - \zeta I^-)^{-1} \| \leqslant 1 + \sqrt{2}$, and therefore

$$\| A_{12}(A_{22} - \zeta I^-)^{-1}x^- \| \leqslant \frac{\| A_{11} \|}{\mathrm{Im}^\alpha\zeta} + \frac{\| A_{21} \|}{\mathrm{Im}\ \zeta} + \frac{1 + \sqrt{2}}{\mathrm{Im}^{1-\alpha}\zeta}.$$

Since for $0 < \alpha < 1$ and sufficiently large $\mathrm{Im}\ \zeta$ we have

$$\frac{\| A_{11} \|}{\mathrm{Im}^{\alpha}\zeta} + \frac{\| A_{21} \|}{\mathrm{Im}\ \zeta} + \frac{1 + \sqrt{2}}{\mathrm{Im}^{1-\alpha}\zeta} < 1,$$

and since x^- ($\| x^{-1} \| = 1$) is arbitrary, we obtain

$$\| A_{12}(A_{22} - \zeta I^-)^{-1} \| < 1.$$

■

1.14 If $A \in (L)$, $\bar{\mathscr{D}}_A = \mathscr{H}$, $\mathscr{L}^- \in \mathscr{M}^-$, and A: $\mathscr{L}^- \to \mathscr{L}^-$, then $\mathscr{L}^-$ is an invariant subspace of the operator A.

☐ In accordance with Definition 1.1 it is only necessary to verify that $\overline{\mathscr{D}_A \cap \mathscr{L}^-} = \mathscr{L}^-$. From 1.Corollary 8.16 it follows that $\mathscr{H}^+ + \mathscr{L}^- = \mathscr{H}$. Since $\mathscr{H}^+ \subset \mathscr{D}_A$, we have $\mathscr{D}_A = \mathscr{H}^+ + (\mathscr{L}^- \cap \mathscr{D}_A)$, and therefore $\bar{\mathscr{D}}_A = \mathscr{H}$ implies $\overline{\mathscr{D}_A \cap \mathscr{L}^-} = \mathscr{L}^-$. ■

Theorem 1.13 and Proposition 1.14 show that in future we need investigate only Problems 1.6 and 1.7, and the reader himself will be able to formulate the corresponding consequences for maximal J-dissipative operators $A \in (L)$ or for families of them $\mathscr{A} \in (L)$, and so we shall now dwell on them in the main text. In conclusion we mention that, historically, solution of the Problems 1.6 and 1.7 did not always precede the investigation of Problems 1.2 and 1.3; thus, for example, Pontryagin's theorem mentioned in §1.1 was obtained earlier than Corollary 2.9 set out below in §2, from which we suggest to the reader in Exercise 10 on §2 that he should derive it.

5 In the preceding paragraphs of this §1 we decided that we would investigate the problem of invariant subspaces mainly for J-non-contractive operators and, in particular, for J-semi-unitary and J-unitary operators. For brevity in the further exposition we introduce

Definition 1.15: We shall say that a family $\mathscr{U} = \{U\}$ of J-non-contractive operators has:

a) *property* Φ_+ (respectively, $\Phi_-, \Phi, \Phi^{[\perp]}$) if it has a common invariant subspace $\mathscr{L}^+ \in \mathscr{M}^+$ (respectively, $\mathscr{L}^- \in \mathscr{M}^-$, $\mathscr{L}^\pm \in \mathscr{M}^\pm$, $\mathscr{L}^+ [\perp] \mathscr{L}^-$);

b) *Property* $\tilde{\Phi}_+$ (respectively, $\tilde{\Phi}_-, \tilde{\Phi}$) if each of its common completely invariant non-negative (respectively, invariant non-positive, completely invariant non-negative and non-positive) subspaces (including also $\{\theta\}$) admits extension into a maximal semi-definite subspace (of the corresponding sign) invariant relative to $\mathscr{U}$;

c) *Property* $\tilde{\Phi}^{[\perp]}$ if each of its invariant dual pairs $(\mathscr{L}^+, \mathscr{L}^-)$ with $U\mathscr{L}^+ = \mathscr{L}^+$ $(u \in \mathscr{U})$ admits extension into a maximal dual pair invariant relative to $\mathscr{U}$.

It is clear from Definition 1.15 that if a family $\mathscr{U}$ has the properties $\tilde{\Phi}_\pm$, $\tilde{\Phi}$, $\tilde{\Phi}^{[\perp]}$, then it also has the properties $\Phi_\pm$, Φ, $\Phi^{[\perp]}$ respectively, and the property Φ ($\tilde{\Phi}$) is equivalent to having properties Φ_+ and Φ_- ($\tilde{\Phi}_+$ and $\tilde{\Phi}_-$) simultaneously. We note also that Definition 1.15 is applicable, in particular, to a family $\mathscr{U}$ consisting of a single operator U.

6 To conclude this section we indicate a device which will enable us later to simplify the proof of whether certain J-non-contractive operators (or families of such operators) have the properties $\Phi_\pm$, $\tilde{\Phi}_\pm$, $\Phi^{[\perp]}$ and $\tilde{\Phi}^{[\perp]}$, and to avoid repetitions in the discussions.

Definition 1.16: Let $\mathscr{U} = \{U\}$ be a set of J-bi-non-contractive operators. We shall say that a complex $\mathscr{K}$ of properties of the set $\mathscr{U}$ (the complex may, in particular, consist of just one property) is *Φ-invariant relative to* $\mathscr{U}$ if this same complex of properties is also possesed by every $\hat{\mathscr{U}} = \{\hat{U}\}$ of $\hat{J}$-bi-non-contractive operators acting in the factor-space $\hat{\mathscr{H}} = \mathscr{L}_0^{[\perp]}/\mathscr{L}_0$ (see 2.(1.11)), where $\mathscr{L}_0$ $(\subset \mathscr{P}^0)$ is an arbitrary subspace completely invariant relative to $\mathscr{U}$.

It is well-known that (in our terminology) *properties of families of operators, such as the property of being a group, commutativity, etc., are also Φ-invariant.* Moreover a proposition such as follows also holds:

1.17 Let $\mathscr{U} = \{U\}$ be a family of J-bi-non-contractive operators. The property that $\{\mathscr{L} \subset \mathscr{P}^0$ and $\mathscr{L} \subset U\mathscr{L}$ implies $U\mathscr{L} \subset \mathscr{P}^0\}$ is Φ-invariant, and if $\mathscr{U}$ has this property, if $\mathscr{L}_+$ is a completely invariant subspace relative to $\mathscr{U}$, $\mathscr{L}_+ \subset \mathscr{P}^+$, and $\mathscr{L}_+^0 = \mathscr{L}_+ \cap \mathscr{L}_+^{[\perp]}$, then $U\mathscr{L}_+^0 = \mathscr{L}_+^0$.

□ Let $\mathscr{L}_0$ be an arbitrary neutral subspace completely invariant relative to $\mathscr{U}$, let $\hat{\mathscr{H}} = \mathscr{L}_0^{[\perp]}/\mathscr{L}_0$, and let $\hat{\mathscr{U}} = \{\hat{U}\}$ be a family of $\hat{J}$-bi-non-contractive operators generated by the family $\mathscr{U}$. If $\hat{\mathscr{L}}$ is a neutral subspace in $\hat{\mathscr{H}}$ and if $\hat{\mathscr{L}} \subset \hat{U}\hat{\mathscr{L}}$ for every $\hat{U} \in \hat{\mathscr{U}}$, then $\mathscr{L} = \mathrm{Lin}\{x \in \hat{x} \mid \hat{x} \in \hat{\mathscr{L}}\}$ is a neutral subspace in $\mathscr{H}$ and $\mathscr{L} \subset U\mathscr{L} = \{Ux \in \hat{U}\hat{x} \mid \hat{x} \in \hat{\mathscr{L}}\}$. Then by hypothesis $U\mathscr{L}$ is a neutral subspace, which implies that $\hat{U}\hat{\mathscr{L}}$ $(= \widehat{U\mathscr{L}})$ is neutral.

Now let $\mathscr{L}_+ \subset \mathscr{P}^+$ and $U\mathscr{L}_+ = \mathscr{L}_+$. It follows from this that $\mathscr{L}_+^0 \subset U\mathscr{L}_+^0$ $(\subset \mathscr{L}_+)$. By hypothesis $U\mathscr{L}_+^0$ is a neutral subspace, and since it contains $\mathscr{L}_0^+$, so (see 1 Proposition 1.17 and 1. Definition 1.13) $\mathscr{U}\mathscr{L}_+^0 = \mathscr{L}_+^0$. ■

Definition 1.18: A semi-definite subspace is said to be a *maximal invariant* (respectively a *maximal completely invariant*) *subspace* for the family $\mathscr{U}$ if it does not admit non-trivial semi-definite extensions of the same sign in the set of subspaces invariant (respectively completely invariant) relative to $\mathscr{U}$. A maximal invariant dual pair is defined similarly.

□ It follows directly from Zorn's lemma that

1.19 Every semi-definite invariant (respectively, completely invariant) subspace admits extension into a maximal invariant (respectively, completely invariant) semi-definite subspace of the same sign. A similar proposition also holds for invariant dual pairs[1]). ■

Lemma 1.20: *Let $\{\mathscr{H}_J\}$ be the set of J-spaces, and let $\{\mathscr{U}_{J,\mathscr{K}} = \{U\}\}$ be the set of all families of J-bi-non-contractive operators in which is inherent a certain complex $\mathscr{K}$ of Φ-invariant properties, including the property $\{\mathscr{L} \subset \mathscr{P}^0,\ \mathscr{L} \subset U\mathscr{L} \Rightarrow U\mathscr{L} \subset \mathscr{P}^0\}$. Then either every such family has the property Φ_+, or among them there is a family ($\hat{\mathscr{U}}$) which does not have the property Φ_+ and is such that each of its completely invariant non-negative subspaces is positive.*

□ Suppose a family $\mathscr{U} = \{U\}$ satisfies the condition of the theorem and nevertheless does not have the property Φ_+. We assume that $\mathscr{L}_0$ is a maximal neutral subspace completely invariant relative to $\mathscr{U}$ (see definition 1.18). This enables us to pass to the family $\hat{\mathscr{U}} = \{U\}$ of $\hat{J}$-bi-non-expansive operators acting in the $\hat{J}$-space $\hat{\mathscr{H}} = \mathscr{L}_0^{[\perp]}/\mathscr{L}_0$. In accordance with the condition, this family has the complex of properties $\mathscr{K}$ and, by virtue of Exercise 21 on 1.§8, like $\mathscr{U}$ it does not have the property Φ_+. We now show that all non-negative subspaces $\hat{\mathscr{L}}_+$ which are completely invariant relative to $\mathscr{U}$ are positive. Let $\hat{\mathscr{L}}_+^0$ be the isotropic part of such a subspace $\hat{\mathscr{L}}_+$. Then, by Proposition 1.17, $\hat{\mathscr{L}}_+^0$ is a neutral subspace completely invariant relative to every operator $\hat{U} \in \mathscr{U}$, and this implies that complete invariance of $\mathscr{L} = \mathrm{Lin}\{x \in \hat{x} \mid \hat{x} \in \hat{\mathscr{L}}_+^0\}$ relative to $U \in \mathscr{U}$. Since

[1] We draw the reader's attention to the fact that a maximal invariant semi-definite subspace is not assumed to be, generally speaking, a maximal semi-definite subspace, and the concepts of a 'maximal invariant dual pair' and an 'invariant maximal dual pair' (*cf.* 1.Definition10.1) are, in general, not identical.

$\mathscr{L}_0 \subset \mathscr{L} \subset \mathscr{P}^0$, it follows from the maximality of $\mathscr{L}^0$ as a neutral subspace completely invariant relative to $\mathscr{U}$ that $\mathscr{L}_0$ coincides with $\mathscr{L}$, i.e., $\hat{\mathscr{L}}^0_+ = \{\theta\}$.

Remark 1.21: Corresponding propositions about properties Φ_-, $\Phi^{[\perp]}$, $\tilde{\Phi}_\pm$, $\tilde{\Phi}^{[\perp]}$ can be proved in an entirely similar way (see Exercises 4–6).

Exercises and problems

1 Construct a J-positive completely continuous operator A and verify that if the J-space is not a Pontryagin space, then A^{-1} has no invariant subspaces $\mathscr{L}^\pm \in \mathscr{M}^\pm$ which are contained in $\mathscr{D}_{A^{-1}}$ (Larionov [9]).

2 Let $\mathscr{U} = \{U\}$ be a group of J-unitary operators. Prove that for it the properties: a) amenability; b) solubility; c) commutativity; d) $U \in \mathscr{U} \Rightarrow U^* \in \mathscr{U}$; e) uniform boundedness; f) $U^c = U^* = U^{-1}$ are Φ-invariant in the sense of Definition 1.16.
Hint: In proving d) first verify that under the conditions of Exercise 2 $\hat{U}^* = \hat{U}^*$.

3 Let $\mathscr{U}$ be a family of J-bi-non-contractive operators. Verify that the property {*every neutral invariant subspace of the family $\mathscr{U}$ admits extension into its completely invariant subspace*} is Φ-invariant (Azizov).

4 Let $\mathscr{K}$ *be a certain* Φ-invariant complex of properties inherent in families $\mathscr{U}_{J,\mathscr{K}} = \{U\}$ of J-bi-non-contractive operators and containing the property the Φ-invariance of which was asserted in Exercise 3. Then either every such family has the property Φ_-, or among them there is a family $(\hat{\mathscr{U}})$ which does not have the property Φ_- and is such that each of its invariant non-positive subspaces is negative (Azizov).

5 Prove that under the conditions of Lemma 1.20 (respectively, of Exercise 4) either each family $\mathscr{U}_{j,\mathscr{K}}$ there mentioned has the property $\tilde{\Phi}_+$ (respectively, $\tilde{\Phi}_-$), or among these families there is a family $(\hat{\mathscr{U}})$ which does not have this property and is such that it has a maximal completely invariant non-negative subspace $\mathscr{L}^+$ (respectively, a maximal invariant non-positive subspace $\mathscr{L}^-$) which is definite and $\mathscr{L}^+ \notin \mathscr{M}^+$ (respectively, $\mathscr{L}^- \notin \mathscr{M}^-$) (Azizov).

6 If under the conditions of Lemma 1.20 $\mathscr{U}_{j,\mathscr{K}}$ is a group of J-unitary operators, then either every such group has the property $\Phi^{[\perp]}$ (respectively $\tilde{\Phi}^{[\perp]}$), or there is among them a group $(\hat{\mathscr{U}})$ which does not have this property and is such that all its maximal invariant dual pairs are definite (respectively, which has a maximal invariant definite dual pair $(\mathscr{L}^+, \mathscr{L}^-)$, $\mathscr{L}^\pm \notin \mathscr{M}^\pm$) (Azizov).
Hint on Exercise 4–6. They are proved similarly to Lemma 1.20.

§2 Invariant subspaces of a J-non-contractive operator

1 Let $\mathscr{H} = \mathscr{H}^+ \oplus \mathscr{H}^-$ be the canonical decomposition of a Krein space, and let $V = \| V_{ij} \|_{i,j=1}^2$ be the matrix representation (see 2.(2.1)) of a J-non-contractive operator V relative to this decomposition.

Theorem 2.1: *Every uniformly J-expansive operator V has the property Φ_+. If V is a uniformly J-bi-expansive operator, then it has the property $\tilde{\Phi}$ and it has a single pair $\mathscr{L}^+ \subset \mathscr{P}^+$ and $\mathscr{L}^- \subset \mathscr{P}^-$ of maximal semi-definite invariant subspaces; moreover they are uniformly definite,* $|\sigma(V|\mathscr{L}^+)| \geqslant \sqrt{\mu_+(V)}$ (>1), *and* $|\sigma(V|\mathscr{L}^-)| \leqslant \sqrt{\max\{0;\, \mu_-(V)\}}$ (<1).

□ We begin with the assertion about a uniformly J-bi-expansive operator V. By 2.Theorem 4.13 the unit circle is free from the spectrum of the operator V, and therefore its Cayley transform $A = \mathbf{K}_i^{-1}(V) = i(V+I)(V-I)^{-1}$ will, by virtue of 2.Proposition 6.16, be a bounded uniformly J-dissipative operator and $\sigma(A) \cap (-\infty, \infty) = \emptyset$ (see 2.Proposition 2.32).

We write $\sigma^{\pm} \equiv \sigma(A) \cap \mathbb{C}^{\pm}$, and let $\Gamma_{\pm}$ be Jordan contours lying in $\mathbb{C}^{\pm} \cap \rho(A)$ and surrounding $\sigma^{\pm}$. By 2.Theorem 2.21 the subspaces $\mathscr{L}^+ = P_+\mathscr{H}$ and $\mathscr{L}^- = P_-\mathscr{H}$, where

$$P_{\pm} = -\frac{1}{2\pi i} \oint_{\Gamma_{\pm}} (A - \lambda I)^{-1}\, \mathrm{d}\lambda, \tag{2.1}$$

are respectively non-negative and non-positive invariant subspaces of the operator A. Since $P_+ + P_- = I$ (see 2.Theorem 2.20)

$$\mathscr{L}^+ + \mathscr{L}^- = \mathscr{H} \tag{2.2}$$

and therefore (see 1.Proposition 1.25) $\mathscr{L}^{\pm} \in \mathscr{M}^{\pm}$. From (2.2) and the invariance of $\mathscr{L}^{\pm}$ relative to A it follows immediately that $\mathscr{L}^{\pm}$ are also invariant relative to $V (= \mathbf{K}_i(A) = (A + iI)(A - iI)^{-1})$. Since V is a uniformly J-bi-expansive operator, we have:

a) An arbitrary vector $y \in \mathscr{L}^+$ is, by 2.Theorem 4.17, expressible in the form $y = Vx$ $(x \in \mathscr{L}^+)$, and by 2.Theorem 4.23 there is a $\delta > 0$ such that

$$[y, y] \geqslant [x, x] + \delta \| x \|^2 \geqslant \frac{\delta}{\| V \|^2} \| y \|^2,$$

i.e. $\mathscr{L}^+$ is a uniformly positive subspace;

b) For an arbitrary $x \in \mathscr{L}^-$ the inequality $-[x, x] \geqslant -[Vx, Vx] + \delta \| x \|^2 \geqslant \delta \| x \|^2$ holds (since $Vx \in \mathscr{P}^-$), and therefore $\mathscr{L}^-$ is a uniformly negative subspace.

Therefore $\mathscr{L}^{\pm}$ are Hilbert spaces relative to the scalar products $\pm [x, y] (x, y \in \mathscr{L}^{\pm})$, and so in them $V | \mathscr{L}^+$ is an expansive operator and $V | \mathscr{L}^-$ is contractive, which implies the inequalities $| \sigma(V | \mathscr{L}^+) | \geqslant 1$ and $| \sigma(V | \mathscr{L}^-) | \leqslant 1$. But since (see 2.Remark 4.25) any operator λV with

$$\frac{1}{\sqrt{\mu_+(V)}} < \lambda < \frac{1}{\sqrt{\max\{0; \mu(V)\}}}$$

such that $| \sigma(\lambda V | \mathscr{L}^+) | \geqslant 1$ and $| \sigma(\lambda V | \mathscr{L}^-) | \leqslant 1$ will be uniformly J-bi-expansive the inequalities $| \sigma(V | \mathscr{L}^+) | \geqslant \sqrt{\mu_+(V)}$ and $| \sigma(V | \mathscr{L}^-) | \leqslant \sqrt{\max\{0; \mu_-(V)}$ also follow.

Now let the subspaces $\mathscr{L}_{\pm}$ $(\subset \mathscr{P}^{\pm})$ be such that $V\mathscr{L}_+ = \mathscr{L}_+$ and $V\mathscr{L}_- \subset \mathscr{L}_-$. Just as above we can verify that $\mathscr{L}_{\pm}$ are uniformly definite, and moreover $| \sigma(V | \mathscr{L}_+) | \geqslant \sqrt{\mu_+(V)}$ and $| \sigma(V | \mathscr{L}_-) | \leqslant \sqrt{\max\{0; \mu_-(V)\}}$. From the properties of the Riesz projectors (2.1) (see 2.Theorem 2.20,e)) we obtain $\mathscr{L}_{\pm} \subset \mathscr{L}^{\pm}$, and therefore the operator V has the properties $\tilde{\Phi}_{\pm}$, i.e., the property $\tilde{\Phi}$. At the same time we have verified the uniqueness of the maximal semi-definite invariant subspaces of the operator V.

Now let V be an arbitrary bounded uniformly J-expansive operator. We suppose that we have succeeded in extending the J-space $\mathscr{H}$ into a $\tilde{J}$-space $\tilde{\mathscr{H}} = \mathscr{H}_+ \oplus \mathscr{H}$ $(\tilde{J} = I_+ \oplus J)$, and the operator V into a uniformly J-bi-expansive operator $\tilde{V}$. Then, by what has been proved above, the operator $\tilde{V}$ has a unique maximal uniformly positive invariant subspace $\tilde{\mathscr{L}}^+ = \operatorname{Lin}\{\tilde{x}^+ + K\tilde{x}^+ \mid \tilde{x}^+ \in \mathscr{H}_+ \oplus \mathscr{H}^+\}$ $(K \in \mathscr{K}(\mathscr{H}_+ \oplus \mathscr{H}^+, \mathscr{H}^-), \|K\| < 1)$, and therefore $\mathscr{L}^+$ $(= \tilde{\mathscr{L}}_+ \cap \mathscr{H} = \operatorname{Lin}\{x^+ + Kx^+ \mid x^+ \in \mathscr{H}^+\}) \in \mathscr{M}^+(\mathscr{H})$ and $V\mathscr{H}^+ \subset \mathscr{L}^+$. Now, if $\mathscr{L}_+ \subset \mathscr{P}^+(\mathscr{H})$ and $V\mathscr{L}_+ = \mathscr{L}_+$, then $\tilde{V}\mathscr{L}_+ = \mathscr{L}_+$ also. Hence, $\mathscr{L}_+ \subset \tilde{\mathscr{L}}^+ \cap \mathscr{H} = \mathscr{L}^+$, which proves that the operator V has the property $\tilde{\Phi}_+$. Therefore to complete the proof it remains for us to construct the extensions mentioned above of the space $\mathscr{H}$ and the operator V.

Let V be a uniformly J-expansive operator. Then $A \equiv V^c V - I$ is a uniformly J-positive operator (see 2.Theorem 4.24). By 2.Corollary 3.29 the operator A has uniformly definite invariant subspaces $\mathscr{L}_1^{\pm} \in \mathscr{M}^{\pm}$, and $\mathscr{L}_1^+ [\perp] \mathscr{L}_1^-$, and so without loss of generality (see 1.§7.6) we shall suppose $\mathscr{L}_1^{\pm} = \mathscr{H}^{\pm}$. Since $\{A\mathscr{H}^{\pm} \subset \mathscr{H}^{\pm}\}$ is equivalent to $\{V^c V\mathscr{H}^{\pm} \subset \mathscr{H}^{\pm}\}$, we have $V\mathscr{H}^+ [\perp] V\mathscr{H}^-$. Since V is a J-non-contractive operator, so (see 2.Corollary 4.13) $V\mathscr{H}^+$ is a uniformly positive subspace with the angular operator $V_{22}V_{11}^{-1}$ defined on $\mathscr{R}_{V_{11}}$ and $\|V_{21}V_{11}^{-1}\| < 1$. Let Γ be the extension of the operator $V_{22}V_{11}^{-1}$ on to the whole of $\mathscr{H}^+$ and let $\|\Gamma\| < 1$.

We construct a J-unitary operator $U(\Gamma)$ by the formula 2.(5.3). Then $V = U(\Gamma)W$, where $W = U^{-1}(\Gamma)V$ is a uniformly J-expansive operator, and $W\mathscr{H}^+ \subset \mathscr{H}^+$ and $W\mathscr{H}^+ [\perp] W\mathscr{H}^-$. Now as $\mathscr{H}_+$ we take one more copy of $\mathscr{H}^+$ and we extend the J-unitary operator $U(\Gamma)$ into a J-unitary operator $U(\tilde{\Gamma})$ in $\tilde{\mathscr{H}} = \mathscr{H}_+ \oplus \mathscr{H}$ by putting $\tilde{\Gamma} \supset \Gamma$, $\tilde{\Gamma} \mid \mathscr{H}_+ = 0$, i.e., $U(\tilde{\Gamma}) \mid \mathscr{H}_+ = I_+$. After this it is sufficient now to construct an extension of the operator W into a uniformly $\tilde{J}$-bi-expansive operator $\tilde{W}$ defined on $\tilde{\mathscr{H}}$. Let $[Wx, Wx] \geqslant [x, x] = \delta \|x\|^2$, $\delta > 0$. We define the operator $\tilde{W} \mid \mathscr{H}_+$ by putting $\tilde{W}x_+ = W_{00}x_+ + W_{10}x_+ + W_{20}x_+$, where $x_+ \in \mathscr{H}_+$ and

$$W_{00} = W_{11}^*\colon \mathscr{H}_+ \to \mathscr{H}_+;$$
$$W_{10} = ((1+\alpha)I^+ + W_{12}W_{12}^*)^{1/2}U_{10}\colon \mathscr{H}_+ \to \mathscr{H}^+,$$

U_{10} is a partially isometric operator mapping $\mathscr{H}_+$ on to Ker W_{11}^* $(\subset \mathscr{H}^+)$ and Ker $U_{10} = \mathscr{R}_{W_{00}^*}$;

$$W_{20} = W_{22}W_{12}^*((1+\alpha)I^+ + W_{12}W_{12}^*)^{-1/2}U_{10}\colon \mathscr{H}_+ \to \mathscr{H}^-.$$

It is immediately verifiable that, when $0 < \alpha < \delta/(1 + \|W_{22}\|^2)$, the operator will be uniformly $\tilde{J}$-bi-expansive. ■

2 Let $T = \|T_{ij}\|_{i,j=1}^2$ be the matrix representation of an operator T, defined everywhere, relative to the canonical decomposition $\mathscr{H} = \mathscr{H}^+ \oplus \mathscr{H}^-$ of the

J-space $\mathscr{H}$. We bring into consideration the functions

$$G_T^+(K_+) = K_+T_{11} + K_+T_{12}K_+ - T_{21} - T_{22}K_+, \tag{2.3}$$

$$G_T^-(K_-) = K_-T_{22} + K_-T_{21}K_- - T_{21} - T_{11}K_-, \tag{2.4}$$

whose domains of definition are respectively the operator balls $\mathscr{K}^{\pm}$ (see 1.8.19).

Lemma 2.2: *A subspace $\mathscr{L}^{\pm} \in \mathscr{M}^{\pm}$ with angular operator $K_{\pm}$ is invariant reative to an operator T with $\mathscr{D}_T = \mathscr{H}$ if and only if the operator $K_{\pm}$ is a solution of the equation $G_T^{\pm}(K_{\pm}) = 0$ respectively.*

□ Let $T\mathscr{L}^+ \subset \mathscr{L}^+$ or, what is equivalent, suppose that for every $x^+ \in \mathscr{H}^+$ there is a $y^+ \in \mathscr{H}^+$ which is a solution of the equation

$$T(x^+ + K_+x^+) = y^+ + K_+y^+),$$

which, in its turn, is equivalent to the system

$$\left.\begin{aligned} T_{11}x^+ + T_{12}K_+x^+ &= y^+ \\ T_{21}x^+ + T_{22}K_+x^+ &= K_+y^+ \end{aligned}\right\} \tag{2.5}$$

We now substitute the value of y^+ from the first equation of the system (2.5) into the second and, since x^+ is arbitrary in $\mathscr{H}^+$, we obtain that $G_T^+(K_+) = 0$. Conversely, suppose $G_T^+(K_+) = 0$. Then we take as the required y^+ the vector defined by the first equation of the system (2.5).

The assertion that $T\mathscr{L}^- \subset \mathscr{L}^- \Leftrightarrow G_T^-(K_-) = 0$ is proved similarly.

Definition 2.3: We shall say that an operator T *satisfies the condition* Λ_+ (Λ_-) and we shall write $T \in \Lambda_+$ ($T \in \Lambda_-$) if there is an operator $K_+ \in \mathscr{K}^+$, $\| K_+ \| < 1$ ($K_- \in \mathscr{K}^-$, $\| K_{-1} \| < 1$) such that $G_T^+(K_+)$ (respectively, $G_T^-(K_-)$ is a completely continuous operator.

Remark 2.4: Using 1.Theorem 8.17 we can by a simple calculation satisfy ourselves that $T \in \Lambda_+$ (respectively, $T \in \Lambda_-$) if and only if there is a canonical decomposition $\mathscr{H} = \mathscr{H}_1^+ [+] \mathscr{H}_1^-$ (respectively, $\mathscr{H} = \mathscr{H}_2^+ [+] \mathscr{H}_2^-$) such that the 'corner' $P_1^- TP_1^+ | \mathscr{H}_1^+$ (respectively, $P_2^+ TP_2^- | \mathscr{H}_2^-$), where $P_i^{\pm}$ are the J-orthogonal projectors on to $\mathscr{H}_i^{\pm}$ $(i = 1, 2)$ is completely continuous. Therefore the inclusions $T \in \Lambda_+$ or $T \in \Lambda_-$ do not depend on the actual decompositions of the space as might at first sight appear from the Definition 2.3. It follows from Theorem 2.1 that every uniformly J-bi-expansive operator satisfies the conditions Λ_+ and Λ_-. However, even for such operators there is not always a single decomposition for which both the 'corners' are completely continuous simultaneously, or, what is equivalent, there is not a $K_0 \in \mathscr{K}^+$ with $\| K_0 \| < 1$ such that $G_T^+(K_0) \in \mathscr{S}_\infty$ and $G_T^-(K_0) \in \mathscr{S}_\infty$ simultaneously (see Exercise 2 below).

Nevertheless the following proposition holds:

2.5 If U is a J-bi-non-contractive J-semi-unitary operator, then $\{U \in \Lambda_+\} \Rightarrow \{U \in \Lambda_-\}$. In particular, if U is a J-unitary operator, then $\{U \in \Lambda_+\} \Leftrightarrow \{U \in \Lambda_-\}$.

□ Let $U_{21} \in \mathscr{S}_\infty$. It follows from 2.Corollary 5.12 that $U_{12} \in \mathscr{S}_\infty$, and hence $\{U \in \Lambda_-\}$. The assertion about J-unitary operators is proved similarly. ■

3 Let V be a J-bi-non-contractive operator. We introduce the notation

$$\mathscr{J}_+(V) = \{\mathscr{L}^+ \in \mathscr{M}^+ \mid V\mathscr{L}^+ = \mathscr{L}^+, |\sigma(V|\mathscr{L}^+)| \geqslant \sqrt{\mu_+(V)}\}, \tag{2.6}$$

$$\mathscr{J}_-(V) = \{\mathscr{L}^- \in \mathscr{M}^- \mid V\mathscr{L}^- = \mathscr{L}^-, |\sigma(V|\mathscr{L}^-)| \leqslant \sqrt{\max\{0; \mu_-(V)\}}. \tag{2.7}$$

For the proof of Theorem 2.8 below we need the following simple proposition:

2.6 Let $\mathscr{L}^\pm \in \mathscr{M}^\pm$, and let $K^\pm$ respectively be their angular operators and let $\mathscr{L}^\pm$ be invariant relative to an operator T with $\mathscr{D}_T = \mathscr{H}$. Then $\sigma(T|\mathscr{L}^+) = \sigma(T_{11} + T_{12}K^+)$ and $\sigma(T|\mathscr{L}^-) = \sigma(T_{22} + R_{21}K^-)$.

□ This proposition is a corollary of the fact that the projectors $P^\pm$ map $\mathscr{L}^\pm$ homeomorphically on to $\mathscr{H}^\pm$ (see 1.Theorem 4.5) and

$$T|\mathscr{L}^+ = (P^+ | \mathscr{L}^+)^{-1}(T_{11} + T_{12}K^+)P^+ | \mathscr{L}^+$$

and

$$T|\mathscr{L}^- = (P^- | \mathscr{L}^-)^{-1}(T_{22} + T_{21}K^-)P^- | \mathscr{L}^-. \qquad ■$$

Moreover, we shall repeatedly use the following lemma which follows from, for example [VII], Theorem 4.5.6.

Lemma 2.7: *Let T be a completely continuous operator acting from a Hilbert space $\mathscr{H}_1$ into a Hilbert space $\mathscr{H}_2$. If $\{K_\delta\}$ and $\{F_\delta\}$ are bounded generalized sequences of operators consisting of operators acting from $\mathscr{H}_2$ into $\mathscr{H}_1$ and converging in the weak operator topology to K_0 and F_0 respectively ($K_\delta \to K_0$, $F_\delta \to F_0$), then $\{TK_\delta\}$ converges in the strong operator topology to TK_0 ($TK_s \xrightarrow{(s)} TK_0$) and $F_\delta TK_\delta \to F_0TK_0$.* ■

Theorem 2.8: *If a J-non-contractive operator $V \in \Lambda_-$, then it has the property Φ_+[1]). If V is a J-bi-non-contractive operator, then*

a) $V \in \Lambda_\mp \Rightarrow \mathscr{J}_\pm(C^c) \neq \emptyset$ *and* $\mathscr{J}_\mp(V^c) \neq \emptyset$;

b) $V \in \Lambda_- \Rightarrow \mathscr{J}_-(V^c) = \{\mathscr{N}^- \in \mathscr{M}^- \mid \mathscr{N}^- = \mathscr{L}^{+[\perp]}, \mathscr{L}^+ \in \mathscr{J}_+(V)\}$

and $V \in \Lambda_+ \Rightarrow \mathscr{J}_+(V^c) = \{\mathscr{N}^+ \in \mathscr{M}^+ \mid \mathscr{N}^+ = \mathscr{L}^{-[\perp]}, \mathscr{L}^- \in \mathscr{J}_-(V)\}$;

c) $V \in \Lambda_- \Rightarrow \mathrm{Lin}\{\lambda_\lambda(V) \| \lambda| > 1\} \subset \cap\{\mathscr{L}^+ \mid \mathscr{L}^+ \in \mathscr{J}_+(V)\}$

and $V \in \Lambda_+ \Rightarrow \mathrm{Lin}\{\lambda_\lambda(V) | \lambda| < 1\} \subset \cap\{\mathscr{L}^- \mid \mathscr{L}^- \in \mathscr{J}_-(V)\}$.

[1] Later, in Theorem 3.9, it will be proved by another method that V has the property $\tilde{\Phi}_+$.

□ Let V be a J-non-contractive operator, and let $I_n = (1+\varepsilon_n)P^+ + (1-\varepsilon_n)P^-$, where $0 < \varepsilon_n < 1$. Then $V_n = VI_n$ is a uniformly J-expansive operator:

$$[VI_nx, VI_nx] \geqslant [I_nx, I_nx] \geqslant [x, x] + \varepsilon_n^2 \| x \|^2.$$

Since $V_{n11} = (1+\varepsilon_n)V_{11}$, the operator V_n will be uniformly J-bi-expansive if and only if V is a J-bi-non-contractive operator (see 2.Remark 4.21). It follows from Theorem 2.1 that the operator V_n has an invariant subspace $\mathscr{L}_n^+ \in \mathscr{M}^+$. Let K_n denote its angular operator. By Lemma 2.2 it satisfies the equation $G_{V_n}^+(K_n) = 0$. Since (see 1.Proposition 8.20) the ball $\mathscr{K}^+$ is bicompact in the weak operator topology, we can choose from the sequence $\{K_n\}$ ($\varepsilon_n \to 0$ as $n \to \infty$) a sub-sequence $\{K_{\tilde{n}}\}$ which converges in this topology to a certain operator $K_0 \in \mathscr{K}^+$ as $\tilde{n} \to \infty$. Since $\varepsilon_{\tilde{n}} \to 0$ when $\tilde{n} \to \infty$,

$$K_{\tilde{n}}V_{\tilde{n}11} = (1+\rho_{\tilde{n}})K_{\tilde{n}}V_{11} \rightharpoonup K_0V_{11}, \qquad V_{\tilde{n}21} = (1+\varepsilon_{\tilde{n}})V_{21} \rightharpoonup V_{21},$$
$$V_{\tilde{n}22}K_{\tilde{n}} = (1-\varepsilon_{\tilde{n}})V_{22}K_{\tilde{n}} \rightharpoonup V_{22}K_0.$$

Since it follows from Remark 2.4 that we can without loss of generality assume that $V_{12} \in \mathscr{S}_\infty$, we can use Lemma 2.7 and obtain

$$K_{\tilde{n}}V_{\tilde{n}12}K_{\tilde{n}} = (1-\varepsilon_{\tilde{n}})K_{\tilde{n}}V_{12}K_{\tilde{n}} \rightharpoonup K_0V_{12}K_0.$$

These relations enable us to conclude that K_0 satisfies the equation $G_V^+(K_0) = 0$, i.e., by Lemma 2.2 the subspace $\mathscr{L}^+ = \{x^+ + K_0x^+ \mid x^+ \in \mathscr{H}^+\}$ ($\in \mathscr{M}^+$) is invariant relative to the operator V.

a) Let V be a J-bi-non-contractive operator. We verify that then $|\sigma(V|\mathscr{L}^+)| \geqslant \sqrt{\mu_+(V)}$. It suffices (taking Theorem 2.1 and 2.Remark 4.25 into account) for us to deal with the case when $\mu_+(V) = 1$. By Proposition 2.6 we have $\rho(V|\mathscr{L}^+) = \rho(\tilde{V}_0)$, where $\tilde{V}_0 = V_{11} + V_{12}K_0$. Since (see 2.Remark 4.21) $0 \in \rho(V_{11})$ and $V\mathscr{L}^+ = \mathscr{L}^+$, so $0 \in \rho(V_{11}) \cap \rho(\tilde{V}_0)$. Since (see Exercise 4 on 2.§4, $\mathbb{D} \equiv \{\lambda \mid |\lambda| < 1\} \subset \rho(V_{11})$ and $V_{12} \in \mathscr{S}_\infty$, it follows from 2.Theorem 2.11 that $\mathbb{D} \subset \tilde{\rho}(\tilde{V}_0)$. We shall prove that $\mathscr{L}^+ \in \mathscr{J}_+(V)$, i.e., $\sigma_0 \equiv \mathbb{D} \cap \sigma(V_0) = \emptyset$. Let $\lambda \in \rho(\tilde{V}_0) \cap \mathbb{D}$. The operators $\tilde{V}_n \equiv V_{\tilde{n}11} + V_{\tilde{n}12}K_{\tilde{n}}$ converge strongly, by Lemma 2.7, to the operator $\tilde{V}_0$, and by Theorem 2.1 and Proposition 2.6 we have $\lambda \in \rho(\tilde{V}_{\tilde{n}})$. We now verify that $\sup_{\tilde{n}}(\| (\tilde{V}_{\tilde{n}} - \lambda I^+)^{-1} \| < \infty$. To do this it is, by a well-known Banach–Steinhaus theorem, sufficient to show that the set $\{(\tilde{V}_{\tilde{n}} - \lambda I^+)^{-1}x^+\}$ is bounded for every $x^+ \in \mathscr{H}^+$. Let $(\tilde{V}_{\tilde{n}} - \lambda I^+)^{-1}x_+ = x_{\tilde{n}}^+$. We rewrite this equality in the equivalent form

$$\begin{aligned}[I^+ - \varepsilon_{\tilde{n}}[(\tilde{V}_0 - \lambda I^+)^{-1}V_{12}(K_{\tilde{n}} - K_0) - I^+](\tilde{V}_0 - \lambda I^+)^{-1}(V_{11} - V_{12}K_{\tilde{n}})\\ - [(\tilde{V}_0 - \lambda I^+)^{-1}V_{12}(K_n - K_0)]^2\}x_{\tilde{n}}^+\\ = [I^+ - (\tilde{V}_0 - \lambda I^+)^{-1}V_{12}(K_{\tilde{n}} - K_0)](\tilde{V}_0 - \lambda I^+)^{-1}x^+. \quad (2.8)\end{aligned}$$

Since

$$\begin{aligned}&\| [(\tilde{V}_0 - \lambda I^+)^- V_{12}(K_n - K_0)]^2 \|\\ &\qquad \leqslant 2 \| (\tilde{V}_0 - \lambda I^+)^{-1}V_{12}(K_{\tilde{n}} - K_0)(\tilde{V}_0 - \lambda I^+)^{-1}V_{12} \|,\end{aligned}$$

$K_{\tilde{n}} - K_n \rightharpoonup 0$, and $V_{12} \in \mathcal{S}_\infty$ and by Lemma 2.7

$$(\tilde{V}_0 - \lambda I^+)^{-1} V_{12}(K_{\tilde{n}} - K_0) \xrightarrow{(s)} 0,$$

so (see, for example, [XVIII])

$$\| (\tilde{V}_0 - \lambda I^+)^{-1} V_{12}(K_{\tilde{n}} - K_0)(\tilde{V}_0 - \lambda I^+)^{-1} V_{12} \| \to 0.$$

Therefore (2.8) takes the form $(I^+ + T_{\tilde{n}})x_{\tilde{n}}^+ = S_{\tilde{n}}x^+$, where $\| T_{\tilde{n}} \| \to 0$, and $\{S_{\tilde{n}}\}$ is a uniformly bounded sequence. Hence for sufficiently large values of $\tilde{n}$ we obtain that $x_{\tilde{n}}^+ = \tilde{I}^+ + T_{\tilde{n}})^{-1} S_{\tilde{n}} x^+$ and $\{x_{\tilde{n}}^+\}$ is a bounded sequence, and therefore $\sup_{\tilde{n}} \| (\tilde{V}_{\tilde{n}} - \lambda I^+)^{-1} \| < \infty$.

Now let $\lambda_0 \in \sigma_0$. Since σ_0 is an isolated point of the spectrum of the operator $\tilde{V}_0$, there is an open disc $\mathbb{D}_0$ $(\subset \mathbb{D})$ whose boundary Γ_0 consists of regular points of this operator and is such that $\mathbb{D}_0 \cap \sigma_0 = \{\lambda_0\}$. Since Γ_0 is compact we conclude that the set $\{(\tilde{V}_{\tilde{n}} - \lambda I^+)^{-1} \mid \lambda_0 \in \Gamma_0\}$ is uniformly bounded with respect to λ and $\tilde{n}$, and therefore the sequence of Riesz projectors

$$P_{\tilde{n}} = -\frac{1}{2\pi i} \oint_{\Gamma_0} (\tilde{V}_{\tilde{n}} - \lambda I^+)^{-1} \, d\lambda$$

converges strongly to the projector

$$P_0 = -\frac{1}{2\pi i} \oint_{\Gamma_0} (\tilde{V}_0 - \lambda I^+)^{-1} \, d\lambda$$

on the root space $\mathcal{L}_{\lambda_0}(\tilde{V}_0)$ (see 2.Theorem 2.20). But $P_{\tilde{n}} = 0$ for all $\tilde{n}$, and therefore $P_0 = 0$, which implies that $\mathcal{L}_{\lambda_0}(\tilde{V}_0) = \{\theta\}$. Thus $|\sigma(\tilde{V}_0)| > 1$, i.e., $\mathcal{J}_+(V) \neq \emptyset$.

Now let $V \in \Lambda_+$. We shall suppose that $V_{21} \in \mathcal{S}_\infty$ (see Remark 2.4). We bring into consideration, by 2.Formula (5.3), the operator $U(\Gamma)$ with $\Gamma = V_{22} V_{11}^{-1}$ $(\in \mathcal{S}_\infty)$. The operator $U = U^{-1}(\Gamma) V$ is, together with V, J-binon-contractive and, as is easily verified, $\| U_{22} \| \leqslant 1$. Therefore

$$V_{22} = U_{22} + \Gamma(I^+ - \Gamma^*\Gamma)^{-1/2} U_{12} + ((I^- - \Gamma\Gamma^*)^{-1/2} - I^-) U_{22}$$

is the sum of a compression and a completely continuous operator. This enables us, using the same scheme as we used in proving that $V \in \Lambda_- \Rightarrow \mathcal{J}_+(V) \neq \emptyset$, i.e., again starting from Proposition 2.6 (but this time its second part), to prove that $V \in \Lambda_+ \Rightarrow \mathcal{J}_-(V) \neq \emptyset$. We also notice the following obvious implication:

$$V \in \Lambda_\pm \Leftrightarrow V^c \in \Lambda_\mp. \tag{2.9}$$

Therefore, if $V \in \Lambda_-$ (respectively, Λ_+), then $\mathcal{J}_-(V^c) \neq \emptyset$ (respectively, $\mathcal{J}_+(V) \neq \emptyset$). So assertion a) has been proved.

b) Let $\mathcal{L}^+ \in \mathcal{J}_+(V)$. Then $\mathcal{N}^- \equiv \mathcal{L}^{+[\perp]} \in \mathcal{M}^-$ and $V^c \mathcal{N}^- \subset \mathcal{N}^-$. Since $(V^c)_{22} = V_{22}^*$ and $(V^c)_{21} = -V_{12}^*$, so by Proposition 2.26, $\sigma(V^c \mid \mathcal{N}^-) = \sigma(V_{22}^* - V_{12}^* Q)$, where Q is the angular operator of the subspace $\mathcal{N}^-$. From $V_{12} \in \mathcal{S}_\infty$ we conclude (see 2.Theorem 2.11) that $\{\lambda \mid |\lambda| > 1\} \subset \tilde{\rho}(V^c \mid \mathcal{N}^-)$. Let $|\lambda_0| > 1$ and $V^c x_0 = \lambda_0 x_0$, $\theta \neq x_0 \in \mathcal{N}^-$. Since moreover $x_0 \in \mathcal{P}^+$ (see 2.Exercise 5 on §6), so $x_0 \in \mathcal{P}^0$ and therefore $x_0 \in \mathcal{L}^+ \cap \mathcal{N}^-$, which implies (see

2.Exercise 17 on §4) $VV^c x_0 = x_0$, i.e., $Vx_0 = (1/\lambda_0)x_0$ and $|1/\lambda_0| < 1$—we have obtained a contradiction of the fact that $\mathscr{L}^+ \in \mathscr{J}_+(V)$. Therefore $\mathscr{N}^- \in \mathscr{J}_-(V^c)$. Similarly it can be verified that, if $\mathscr{N}^- \in \mathscr{J}_-(V^c)$, then $\mathscr{L}^+ = \mathscr{N}^{-[\perp]} \in \mathscr{J}_+(V)$, and the second implication in b) holds.

c) Let $\lambda \in \sigma_p(V)$ with $|\lambda| \neq 1$ and let $x \in \mathscr{L}_\lambda(V)$. Then the spectrum of the restriction of the operator V on to the finite-dimensional invariant subspace $\text{Lin}\{(V-\lambda I)^j x\}_0^\infty$ consists of the one point $\{\lambda\}$. By 2.Theorem 1.13 we have $x[\perp]\mathscr{N}$ for all $\mathscr{N} \in \mathscr{J}_+(V^c)$ if $|\lambda| < 1$ or for all $\mathscr{N} \in \mathscr{J}_-(V^c)$ if $|\lambda| > 1$. Using the proposition b) which has already been proved we obtain that $x \in \cap \{\mathscr{L}^- \mid \mathscr{L}^- \in \mathscr{J}_-(V)\}$ $(|\lambda| < 1)$ or, respectively, that $x \in \cap \{\mathscr{L}^+ \mid \mathscr{L}^+ \in \mathscr{J}_+(V)\}$ $(|\lambda| > 1)$. Therefore the inclusions indicated in assertion c) are valid for every $\mathscr{L}_\lambda(V)$ and are therefore also valid for their linear envelope.

Corollary 2.9: *If V is a π-non-contractive operator or a π-bi-non-expansive operator, then $\mathscr{J}_\pm(V) \neq \emptyset$ and the inclusions in assertions* b) *and* c) *of Theorem* 2.8 *hold for V. In particular, π-semi-unitary and π-unitary operators have these properties.*

□ This follows from the fact that V_{21} and V_{12} are finite-dimensional continuous operators and therefore $V \in \Lambda_+ \cap \Lambda_-$. ■

Corollary 2.10: *If $V(\in \Lambda_+)$ is a J-semi-unitary J-bi-non-contractive operator, then $\mathscr{J}_\pm(V) \neq \emptyset$ and assertions* b) *and* c) *in Theorem 2.8 hold for it. In particular, J-unitary operators $U \in \Lambda_+ \cap \Lambda_-$ have these properties.*

□ It is sufficient to compare Proposition 2.5 with Theorem 2.8. ■

Remark 2.11: In accordance with 2.Remark 4.29 all the statements of problems and results in §§1 and 2, as also in §§3–5 later, for *J*-non-contractive and *J*-bi-non-contractive operators can be reformulated without difficulty in terms of *J*-non-expansive and *J*-bi-non-expansive operators, and this we leave the reader to do.

Examples and problems

1 Give an example of a uniformly I-expansive operator which does not have the property Φ_- (and even less, the property $\tilde{\Phi}_-$) (Azizov).
Hint: Consider a J-space $\mathscr{H} = \mathscr{H}^+ \oplus \mathscr{H}^-$, $\dim \mathscr{H} = \dim \mathscr{H}^+$, $\mathscr{H}^- \neq \{\theta\}$ and in it an operator λU where $|\lambda| > 1$ and U is a semi-unitary operator mapping $\mathscr{H}$ into $\mathscr{H}^+$.

2 Let $\mathscr{H} = \mathscr{H}^+ \oplus \mathscr{H}^-$ be a J-space, $\mathscr{H}^+$ and $\mathscr{H}^-$ being two copies of one and the same infinite-dimensional Hilbert space. Verify that the operator $V = \| V_{ij} \|_{i,j=1}^2$, where $V_{11} = \sqrt{3/2}I$, $V_{12} = (1/2)I$, $V_{21} = 0$, $V_{22} = \sqrt{1/2}I$, is a uniformly J-bi-expansive operator. Prove that there is no $K_0 \in \mathscr{K}^+$ with $\| K_0 \| < 1$ such that $G_V^+(K_0) \in \mathscr{S}_\infty$ and $G_V^-(K_0^*) \in \mathscr{S}_\infty$ simultaneously (Azizov).

3 Let V be a J-bi-non-contractive operator with $V \in \Lambda_-(\Lambda_+)$, let σ_0 be the spectral set of the operator V with $\sigma_0 \subset \{\lambda \| \lambda | > 1\}$ (respectively, $\sigma_0 \subset \mathbb{D}$), and let P_{σ_0} be the corresponding Riesz projector. Prove that then $P_{\sigma_0}\mathscr{H} \subset \cap \{\mathscr{L}^+ \mid \mathscr{L}^+ \in \mathscr{J}_+(V)\}$ (respectively, $P_{\sigma_0}\mathscr{H} \subset \cap \{\mathscr{L}^- \mid \mathscr{L}^- \in \mathscr{J}_-(V)\}$ (Azizov).
Hint: Use 2.Theorem 1.13 and Theorem 2.8.

4 Let V be a π-non-contractive operator, let $|\lambda_0| > 1$, and let $\mathscr{L}_0$ be the isotropic part of the lineal $\mathscr{L}_{\bar{\lambda}_0^{-1}}(V)$ (the possibility that $\mathscr{L}_{\lambda_0}(V) = \mathscr{L}_{\bar{\lambda}_0^{-1}}(V) = \mathscr{L}_0 = \{\theta\}$) is not excluded). Verify that then $\dim \mathscr{L}_0 \leqslant \dim \mathscr{L}_{\lambda_0}(V)$ and that the subspace $\mathscr{L}_0 + \mathscr{L}_{\lambda_0}(V)$ is non-degenerate (Azizov).
Hint: Use Theorem 2.8, 2.Theorem 1.13, and 2.Exercise 17 on §4.

5 Let V be a π-non-contractive operator, and let $\mathscr{L}_1, \mathscr{L}_2$ be arbitrary invariant subspaces of it from $\mathscr{M}^+$. Prove that then, if $|\lambda| > 1$, we have

$$\dim(\mathscr{L}_\lambda(V|\mathscr{L}_1) + \mathscr{L}_{\bar{\lambda}^{-1}}(V|\mathscr{L}_1)) = \dim(\mathscr{L}_\lambda(V|\mathscr{L}_2 + \mathscr{L}_{\bar{\lambda}^{-1}}(V|\lambda_2)) = \dim \mathscr{L}_\lambda(V),$$

and if $|\lambda| = 1$, then $\dim \mathscr{L}_\lambda(V|\mathscr{L}_1) = \dim \mathscr{L}_\lambda(V|\mathscr{L}_2)$ (Azizov).
Hint: Use Exercise 4, Theorem 2.8, 2.Theorem 1.13, and 2.Exercise 17 on §4.

6 Let V be a π-semi-unitary operator in $\Pi_\varkappa$, let $\mathscr{L}^+ \in \mathscr{J}_+(V)$ and to $(V|\mathscr{L}^+)| > 1$, and let $\sigma_p(V) \cap \sigma^{*-1}(V|\mathscr{L}^+) = \emptyset$. Then the operator V^c has a single $\varkappa$-dimensional positive, and a single $\varkappa$-dimensional neutral, invariant subspace. When $\varkappa = 1$ the operator V^c has no other invariant subspaces from $\mathscr{M}^+$, and when $\varkappa > 1$ the power of the set of them is either not greater than $2^{\varkappa-2}$ or is not less than that of the continuum (Azizov).
Hint: Use Exercise 6 and the hint for it.

7 Let $\Pi_1 = \Pi_+ \oplus \Pi_-$ with $\Pi_+ = \text{Lin}\{e^+\}$ and let $\{e_i^-\}_1^\infty$ be an orthonormal basis in Π_-. Prove that the linear operator V defined on $e^+ \cup \{e_i^-\}_1^\infty$ as $Ve^+ = \sqrt{2}e^+ + e_1^-$, $Ve_i^- = e_{i+1}^-$ $(i = 1, 2, \ldots)$ is bounded and that its closure is a π-semi-unitary operator which satisfies the conditions of Exercise 6 (Azizov).

8 Let A be a bounded uniformly J-dissipative operator with $\mathscr{D}_A = \mathscr{H}$. Prove that the operator A has a single pair of invariant subspaces $\mathscr{L}^\pm \in \mathscr{M}^\pm$, that $\mathscr{L}^\pm$ are uniformly definite and $\text{Im}\, \sigma(A|\mathscr{L}^+) > 0$, $\text{Im}\, \sigma(A|\mathscr{L}^-) < 0$ ([VI]).
Hint: Use Theorems 1.13 and 2.1.

9 Let A be a maximal J-dissipative operator, let $A \in (L)$, $\bar{\mathscr{D}}_A = \mathscr{H}$, and let A_{12} be an A_{22}-completely continuous operator. Prove that the operator A has at least one invariant subspace $\mathscr{L}^+ \in \mathscr{M}^+$, $\mathscr{L}^+ \subset \mathscr{D}_A$ and $\text{Im}\, \sigma(A|\mathscr{L}^+) \geqslant 0$ (*cf.* Langer [2], M. Krein [5], Azizov and E. Iokhvidov [1]).
Hint: Use Theorems 1.13 and 2.8.

10 Let A be a maximal π-dissipative operator (in particular, a π-self-adjoint operator) in $\Pi_\varkappa$ with $\bar{\mathscr{D}}_A = \Pi_\varkappa$. Prove that it has invariant subspaces $\mathscr{L}^\pm \in \mathscr{M}^\pm$ such that $\text{Im}\, \sigma(A|\mathscr{L}^+) \geqslant 0$, $\text{Im}\, \sigma(A|\mathscr{L}^-) \leqslant 0$. Moreover, $\text{Lin}\{\mathscr{L}_\lambda(A) \mid \text{Im}\, \lambda > 0\} \subset \mathscr{L}^+$ and $C\ \text{Lin}\{\mathscr{L}_\lambda(A) \mid \text{Im}\, \lambda < 0\} \subset \mathscr{L}^{-1}$ (Pontryagin [1], Azizov [4], [8], M. Krein and Langer [3].
Hint: Use Theorem 1.13 and Corollary 2.9.

11 Under the conditions of Exercise 10 let $\theta_A < \pi$ (regarding the symbol θ_A, see 2.Exercise 6 on §2) and let the right-hand (respectively left-hand) boundary ray of the corresponding angle form with the positive (respectively negative) semi-axis an angle $\varphi_1 \geqslant 0$ (respectively $\varphi_2 \geqslant 0$). Prove that the operator A then has no points of the spectrum within the angles $(-\varphi_2, \varphi_1)$ and $(-\pi + \varphi_1, \pi - \sigma_2)$ and if, moreover, Ker A is definite, then the operator A has a single pair of invariant subspaces $\mathscr{L}^\pm \in \mathscr{M}^\pm$, that $\mathscr{L}^\pm$ are definite and $\mathscr{L}^+ = \text{Lin}\{\mathscr{L}_\lambda(A)$, Ker $A \cap \mathscr{P}^+ \mid \text{Im}\, \lambda > 0\}$, and $\mathscr{L}^- = (\text{Lin}\{\mathscr{L}_\mu(A^c)$, Ker $A \cap \mathscr{P}^{-1} \mid \text{Im}\, \mu > 0\})^{[\perp]}$ (Azizov [8]).

§3 Fixed points of linear-fractional transformations and invariant subspaces

1 Let $\mathscr{H}=\mathscr{H}^+\oplus\mathscr{H}^-$ be a J-space and $V=\|V_{ij}\|_{i,j=1}^2$ be a J-bi-non-contractive operator. It follows from 2.Theorem 4.17 that on the operator ball $\mathscr{K}^+\equiv\mathscr{K}^+(\mathscr{H}^+,\mathscr{H}^-)$ (see 1.Proposition 8.19) the *Krein–Shamul'yan linear-fractional transformation*

$$F_V^+ \colon K\to F_V^+(K)=(V_{21}+V_{22}K)(V_{11}+V_{12}K)^{-1}, \tag{3.1}$$

or (in equivalent form)

$$F_V^+(K)V_{11}+F_V^+(K)V_{12}K-V_{21}-V_{22}K=0 \tag{3.2}$$

is properly defined.

3.1 Let K be the angular operator of the subspace $\mathscr{L}^+\in\mathscr{M}^+$. Then $F_V^+(K)$ is the angular operator of the subspace $V\mathscr{L}^+$ ($\in\mathscr{M}^+$), and therefore the function F_V^+ maps the ball $\mathscr{K}^+$ into $\mathscr{K}^+$.

□ If $x=x^+ + Kx^+$, then

$$\begin{aligned} Vx &= V_{11}x^+ + V_{21}x^+ + V_{12}Kx^+ + V_{22}Kx^+ = (V_{11}+V_{12}K)x^+ \\ &\quad + (V_{21}+V_{22}K)x^+ = y^+ + (V_{21}+V_{22}K)(V_{11}+V_{12}K)^{-1}y^+, \end{aligned}$$

where y^+ denotes the vector $(V_{11}+V_{12}K)x^+$. ■

This proposition and the writing of the function F_V^+ in the implicit form (3.2) enables us to introduce the concept of a *generalized linear-fractional transformation* F_V^+ defined by a bounded J-non-contractive operator V ($\mathscr{D}_V=\mathscr{H}$) as a mapping of elements of the ball $\mathscr{K}^+$ into the set of subsets of this ball:

$$F_V^+(K)=\{L'\in\mathscr{K}^+ \mid L'V_{11}+L'V_{12}K-V_{21}-V_{22}K=0\}. \tag{3.3}$$

□ The following proposition is proved in the same way as 3.1 was:

3.2 Let V be a J-non-contractive operator, let K be the angular operator of $\mathscr{L}^+\in\mathscr{M}^+$, and let L be the angular operator of the subspace $V\mathscr{L}^+$. Then $F_V^+(K)=\mathscr{K}^+(L)$, where $\mathscr{K}^+(L)$ is defined by the formula in 1.8.9.

Definition 3.3: A subset $\mathscr{K}\subset\mathscr{K}^+$ is said to be *invariant relative to the generalized linear-fractional transformation* F_V^+ if $F_V^+(K)\cap\mathscr{K}\neq\emptyset$ for any $K\in\mathscr{K}$. Moreover, the restriction $F_V^+\mid\mathscr{K}$ is understood to mean the mapping $F_V^+\mid\mathscr{K}\colon K\ (\in\mathscr{K})\to F_V^+(K)\cap\mathscr{K}$. In particular, if $\mathscr{K}$ consists of a single operator K_0, then K_0 is called a *fixed point of the transformation* F_V^+.

□ From Proposition 3.2 it follows immediately that

3.4 A subspace $\mathscr{L}_0^+\in\mathscr{M}^+$ with the angular operator K_0 is invariant relative to a J-non-contractive operator if and only if $K_0\in F_V^+(K_0)$, i.e., K_0 is a fixed point of F_V^+. ■

We note that this proposition coincides with Lemma 2.2, if in the latter we put $T = V$, a J-non-contractive operator.

2 Proposition 3.4 shows another way of seeking the solution of the problem of invariant subspaces. This way consists in investigating when the function F_V^+ has a fixed point. In order to apply this idea we need some topological concepts and results; we introduce the latter without proof.

Definition 3.5: Let E be a Hausdorff linear topological space, and $\mathscr{K}$ a subset of it. A mapping F which carries points $K \in \mathscr{K}$ into now-empty convex subsets $F(K) \subset E$ is said to be *closed* if the fact that generalized sequences $\{K_\delta\}$ and $\{F_\delta\}$ $(F_\delta \in F(K_\delta))$ converge to K_0 and F_0 respectively implies that $F_0 \in F(K_0)$.

□ **Theorem 3.6:** (Glicksberg [IX]). *Let $\mathscr{K}$ be a non-empty bi-compact convex subset in a locally convex Hausdorff topological space E, and let F be a closed mapping of points $K \in \mathscr{K}$ into non-void convex subsets $F(K) \subset \mathscr{K}$. Then the function F has at least one fixed point in $\mathscr{K}$, i.e., there is a point $K_0 \in \mathscr{K}$ such that $K_0 \in F(K_0)$.* ■

□ It is easy to see that under the conditions of Theorem 3.6 the following proposition holds:

3.7 The set of fixed points of the mapping F is closed ■

In our case the rôle of E will be played by the space of linear continuous operators acting from one Hilbert space into another, and the rôle of $\mathscr{K}$ by the ball $\mathscr{K}^+$ of this space or its closed convex subsets. The topology is the weak operator topology.

We now pass on to the key result of this section.

Theorem 3.8: *Let $\mathscr{H} = \mathscr{H}^+ \oplus \mathscr{H}^-$ be a J-space and $V = \| V_{ij} \|_{i,j=1}^2$ a J-non-contractive operator with $V_{12} \in \mathscr{S}_\infty$; let F_V^+ be a generalized linear-fractional transformation generated by the operator V according to the formula* (3.3), *and let $\mathscr{K}$ $(\subset \mathscr{K}^+)$ be a non-empty convex subset, closed in the weak operator topology,* which is invariant relative *to F_V^+. Then the mapping $F_V^+ \mid \mathscr{K}$ has at least one fixed point. The set of fixed points of the function $F_V^+ \mid \mathscr{K}$ is closed in the same topology.*

□ We verify that we are under the conditions of Theorem 3.6. Let $K \in \mathscr{K}$. Then it follows from a comparison of Proposition 3.2 with 1.Theorem 8.23 that the non-empty set $F_V^+(K)$ is bi-compact and convex, and the same is true of the non-empty set $F_V^+(K) \cap \mathscr{K}$. Thus the function $F_V^+ \mid \mathscr{K}$ carries points from $\mathscr{K}$ into its convex non-empty subsets. It remains to verify that $F_V^+ \mid \mathscr{K}$ is a closed mapping. Let $\{K_\delta\}$ be a generalized sequence of elements from $\mathscr{K}$, and let $F_\delta \in F_V^+(K_\delta) \cap \mathscr{K}$ and $K_\delta \rightharpoonup K_0$, $F_\delta \rightharpoonup F_0$. From the Definition (3.3) we then have

$$F_\delta V_{11} + F_\delta V_{12} K_\delta - V_{21} - V_{22} K_\delta = 0.$$

Since $F_\delta V_{11} \rightharpoonup F_0 V_{11}$, $V_{22}K_\delta \rightharpoonup V_{22}K_0$, and by Lemma 2.7 $F_\delta V_{12}K_\delta \rightharpoonup F_0 V_{12}K_0$, we have $F_0 \in F_V^+(K_0)$. Since $\mathscr{K}$ is closed, $F_0 \in \mathscr{K}$, i.e., $F_0 \in F_V^+(K_0) \cap \mathscr{K}$. It now follows from Theorem 3.6 that $F_V^+ \mid \mathscr{K}$ has at least one fixed point, and from Proposition 3.7 that the set of all such points is closed. ■

3 As a corollary from Theorem 3.8 we obtain

Theorem 3.9: *A J-non-contractive operator V $(\in \Lambda_-)$ has the property $\tilde{\Phi}_+$; a J-bi-non-contractive operator V $(\in \Lambda_+)$ has the property $\tilde{\Phi}_-$; a J-unitary operator U $(\in \Lambda_+ \cap \Lambda_-)$ has the property $\Phi^{[\perp]}$.*

□ Let V be a J-non-contractive operator and $V \in \Lambda_-$. Without loss of generality, by virtue of Remark 2.4, we assume that $V_{12} \in \mathscr{S}_\infty$. If a non-negative subspace $\mathscr{L}_0$ with the angular operator K_0 is completely invariant relative to V, then it follows from Proposition 3.2 that the weakly closed convex set $\mathscr{K}_+(K_0)$ is invariant (see 1.Theorem 8.23) relative to the generalized linear-fractional transformation F_V^+. Therefore by Theorem 3.8 the function $F_V^+ \mid \mathscr{K}_+(K_0)$ has at least one fixed point $\tilde{K}_0$ which, by Proposition 3.2, will be the angular operator of a maximal non-negative invariant subspace $\tilde{\mathscr{L}}_0$ of the operator V, and $\mathscr{L}_0 \in \tilde{\mathscr{L}}_0$.

Now let V be a J-bi-non-contractive operator, $V \in \Lambda_+$. By Remark 2.4 we can suppose that $V_{21} \in \mathscr{S}_\infty$, and therefore $(V^c)_{12} = -V_{21}^* \in \mathscr{S}_\infty$. Consequently we conclude from Theorem 3.8 that the function $F_{V^c}^+$ has a fixed point in any convex weakly closed subset from $\mathscr{K}^+$ which is invariant relative to $F_{V^c}^+$. In particular, if $\mathscr{R}_0 \subset \mathscr{P}^-$, $V\mathscr{R}_0 \subset \mathscr{R}_0$, and Q_0 is the angular operator of the subspace $\mathscr{R}_0$, then it follows from the invariance of $\mathscr{R}_0^{[\perp]}$ relative to V^c (see 2.Proposition 1.11) and from Proposition 3.2 that the weakly closed convex subset $\mathscr{K}_-^*(Q_0)$ (see 1.Theorem 8.23) is invariant relative to $F_{V^c}^+$.

Let $\mathscr{L}_0 = \{x^+ + K_0 x^+ \mid x^+ \in \mathscr{H}^+\}$, where K_0 is a fixed point of the function $F_{V^c}^+ \mid \mathscr{K}^*(Q_0)$. Then $\mathscr{L}_0$ is invariant relative to V^c and is J-orthogonal to $\mathscr{R}_0$ (see 1.Proposition 8.22). Therefore $\mathscr{L}_0^{[\perp]}$ $(\in \mathscr{M}^-)$ is a subspace invariant relative to V and containing $\mathscr{R}_0$.

The last assertion about a J-unitary operator U is proved similarly. Namely, because of Remark 2.4 we can suppose that $U_{12} \in \mathscr{S}_\infty$. If $(\mathscr{L}_+, \mathscr{L}_-)$ is a dual pair with $U\mathscr{L}_\pm = \mathscr{L}_\pm$, and if $K_\pm$ are the angular operators of the subspaces $\mathscr{L}_\pm$ respectively, then, as a convex weakly closed subset $\mathscr{K}$ in $\mathscr{K}^+$ which is invariant relative to F_U^+, we consider the non-empty subset $\mathscr{K} = \mathscr{K}_+(K_+) \cap \mathscr{K}_-^*(K_-)$ (see 1.Theorem 8.23). By Theorem 3.8 $F_U^+ \mid \mathscr{K}$ has a fixed point $\tilde{K}_+$, and therefore $\tilde{\mathscr{L}}_+ = \{x^+ + \tilde{K}_+ x^+ \mid x^+ \in \mathscr{H}^+\}$ is a subspace which is invariant relative to U, contains $\mathscr{L}_+$, and is J-orthogonal to $\mathscr{L}_-$. Therefore $(\tilde{\mathscr{L}}_+, \mathscr{L}_+^{[\perp]})$ is a maximal dual pair which is invariant relative to U and contains $(\mathscr{L}_+, \mathscr{L}_-)$. ■

Corollary 3.10: *Let V be a π-non-contractive or a π-bi-non-contractive operator. Then it has the property $\tilde{\Phi}$. If V is a π-unitary operator, then it has the property $\tilde{\Phi}^{[\perp]}$.*

□ Since V_{12} and V_{21} are finite-dimensional operators, $V \in \Lambda_+ \cap \Lambda_-$, and it only remains to apply Theorem 3.9. ■

Corollary 3.11: *Let $\Pi_\varkappa = \Pi_+ \oplus \Pi_-$ be a Pontryagin space and U be a π-unitary operator. Then U is stable if and only if all its eigen-subspaces $\operatorname{Ker}(U - \lambda I)$ are non-degenerate.*

□ Let U be a stable operator. In accordance with 2.Corollary 5.20 we can without loss of generality suppose that $U\Pi_\pm = \Pi_\pm$, and therefore

$$\operatorname{Ker}(U - \lambda I) = (\operatorname{Ker}(U - \lambda I) \cap \Pi_+)\,[\oplus]\,(\operatorname{Ker}(U - \lambda I) \cap \Pi_-).$$

This equality implies that $\operatorname{Ker}(U - \lambda I)$ are non-degenerate.

Conversely, suppose that all the $\operatorname{Ker}(U - \lambda I)$ are non-degenerate. Then $\lambda \in \sigma_p(U)$ implies $|\lambda| = 1$ (see Exercise 5 on 2.§6). Therefore (see Exercise 22 on 2.§5) $\operatorname{Ker}(U - \lambda I)\,[\perp]\,\operatorname{Ker}(U - \mu I)$ when $\lambda \neq \mu$, and so there are precisely p $(0 < p \leqslant \varkappa)$ different eigenvalues $\lambda_1, \lambda_2, \ldots, \lambda_p$ of the operator U to which correspond non-negative eigenvectors. Since the $\operatorname{Ker}(U - \lambda_i I)$ $(i = 1, 2, \ldots, p)$ are non-degenerate, it follows that

$$\Pi_\varkappa = \operatorname{Ker}(U - \lambda_1 I)\,[+]\,\operatorname{Ker}(U - \lambda_2 I)\,[+] \cdots [+]\,\operatorname{Ker}(U - \lambda_p I)\,[+]\,\mathcal{N},$$

where

$$\mathcal{N} = [\operatorname{Ker}(U - \lambda_1 I)\,[+]\,\operatorname{Ker}(U - \lambda_2 I)\,[+] \cdots [+]\,\operatorname{Ker}(U - \lambda_p I)]^{[\perp]}$$

and $U\mathcal{N} \subset \mathcal{N}$ in accordance with 2.Proposition 1.11, and by construction $U|\mathcal{N}$ has no non-negative eigenvectors. By Corollary 3.10 $\mathcal{N}$ is a negative subspace and therefore $U|\mathcal{N}$ is a unitary operator relative to the scalar product $-[x, y]\,|\mathcal{N}$. Consequently u is a unitary operator relative to the scalar product (which is equivalent to the original one)

$$(x, y)_1 = \sum_{i=1}^{p} (x_i, y_i) - [x_{\mathcal{N}}, y_{\mathcal{N}}],$$

where

$$x = \sum_{i=1}^{p} x_i + x_{\mathcal{N}}, \quad y = \sum_{i=1}^{p} y_i + y_{\mathcal{N}}, \quad x_i, y_i \in \operatorname{Ker}(U - \lambda_i I) \quad \text{for } i = 1, 2, \ldots, p,$$

and

$$x_{\mathcal{N}}, y_{\mathcal{N}} \in \mathcal{N}.$$

From this it follows that the operator U is stable. ■

Theorem 3.9 enables us to make Theorem 2.8 more precise for the case of a J-unitary operator.

Theorem 3.12: *Let U $(\in \Lambda_+ \cup \Lambda_-)$ be a J-unitary operator, let Λ be its non-unitary spectrum, $\Lambda = \Lambda_1 \cup \Lambda_2$, $\Lambda_1 \cap \Lambda_2 = \emptyset$, and let $\Lambda_2^{-1} \equiv \{\lambda^{-1} \mid \lambda \in \Lambda_2\} = \Lambda_1^*$. Then the operator U has invariant subspaces $\tilde{\mathscr{L}}_\pm \in \mathscr{M}^\pm$ such that the non-unitory spectra of $U \mid \tilde{\mathscr{L}}_+$ and $U \mid \tilde{\mathscr{L}}_-$ coincide with Λ_1 and Λ_2 respectively. Moreover, $\Lambda \subset \tilde{\rho}(U)$, and if $\lambda \in \Lambda_1$ (respectively, $\lambda \in \Lambda_2$), then the root subspace $\mathscr{L}_\lambda(U \mid \tilde{\mathscr{L}}_+)$ (respectively, $\mathscr{L}_\lambda(U \mid \tilde{\mathscr{L}}_-)$) coincides with $\mathscr{L}_\lambda(U)$.*

□ $\Lambda \subset \tilde{\rho}(U)$ by virtue of Remark 2.4 and 2.Corollary 5.13. Let

$$\mathscr{L}_+ = C\,\mathrm{Lin}\{\mathscr{L}_\lambda(U) \mid \lambda \in \Lambda_1\}, \qquad \mathscr{L}_- = C\,\mathrm{Lin}\{\mathscr{L}_\lambda(U) \mid \lambda \in \Lambda_2\}.$$

Since $\Lambda_1^{-1} \cap \Lambda_1^* = \emptyset$ by hypothesis, $\mathscr{L}_\pm \subset \mathscr{P}^0$ by virtue of Exercise 7 on 2.§6. By construction $\mathscr{L}_\pm$ are completely invariant subspaces of the operator U and $\lambda_\lambda(U \mid \mathscr{L}_+) = \mathscr{L}_\lambda(U)$ $(\mathscr{L}_\lambda(U \mid \mathscr{L}_-) = \mathscr{L}_\lambda(U))$ when $\lambda \in \Lambda_1$ (respectively, $\lambda \in \Lambda_2$). By Theorem 3.9 there are subspaces $\tilde{\mathscr{L}}_\pm \in \mathscr{M}^\pm$ which are invariant relative to U, which contain $\mathscr{L}_\pm$ respectively. Carrying out an argument similar to that used in proving Theorem 2.8 we realize that the non-unitary spectra of the operators $U \mid \tilde{\mathscr{L}}_+$ and $U \mid \tilde{\mathscr{L}}_-$ consist of normal eigenvalues. By construction $\Lambda_1 \subset \sigma(U \mid \tilde{\mathscr{L}}_+)$ and $\Lambda_2 \subset \sigma(U \mid \tilde{\mathscr{L}}_-)$, and the corresponding root lineals satisfy the requirements of the theorem. It remains only to notice that the skew-connectedness of $\mathscr{L}_\lambda(U)$ and $\mathscr{L}_{\bar{\lambda}^{-1}}(U)$ (see 2.Corollary 3.12) implies that $\sigma(U \mid \tilde{\mathscr{L}}_+) \cap \Lambda_2 = \emptyset$ and $\sigma(U \mid \tilde{\mathscr{L}}_-) \cap \Lambda_1 = \emptyset$. ■

4 In this paragraph we investigate the question of the number of invariant subspaces possessed by J-bi-non-contractive operators which have at least one invariant maximal uniformly positive subspace (from not on *the sets of uniformly definite subspaces from $\mathscr{M}^+$ and $\mathscr{M}^-$ will be denoted by $\mathscr{M}_0^+$ and $\mathscr{M}_0^-$ respectively*). But first we prove the following

Lemma 3.13: *Let $\mathscr{H} = \mathscr{H}^+ \oplus \mathscr{H}^-$ be a J-space and let Q be the angular operator of the subspace $\mathscr{N}^- \in \mathscr{M}^-$. Then $\mathscr{H} = \mathscr{H}^+ \dotplus \mathscr{N}^-$ and the following assertion holds: if the subspace*

$$\mathscr{L} = \{x^+ + Kx^+ \mid x^+ \in \mathscr{H}^+,\ K\colon \mathscr{H}^+ \to \mathscr{H}^-,\ \|K\| < \infty\}$$

and if $1 \in \rho(QK)$, then

$$\mathscr{L} = \{y^+ + Fy^+ \mid y^+ \in \mathscr{H}^+,\ F\colon \mathscr{H}^+ \to \mathscr{N}^-,\ \|F\| < \infty\}$$

and $-1 \in \rho(QP^- F)$, and

$$F = (P^- + Q)K(I^+ - QK)^{-1}.$$

Conversely, if

$$\mathscr{L} = \{y^+ + Fy^+ \mid y^+ \in \mathscr{H}^+,\ F\colon \mathscr{H}^+ \to \mathscr{N}^-,\ \|F\| < \infty\}$$

and

$$-1 \in \rho(QP^- F),$$

then

$$\mathscr{L} = \{x^+ + Kx^+ \mid x^+ \in \mathscr{H}^+,\ K\colon \mathscr{H}^+ \to \mathscr{H}^-,\ \|K\| < \infty\}$$

and

$$1 \in \rho(QK),$$

and

$$K = P^- F(I^+ + QP^- F)^{-1}. \tag{3.4}$$

□ The equality $\mathscr{H} = \mathscr{H}^+ + \mathscr{N}^-$ is obtained from 1.Corollary 8.16. Let $\mathscr{L} = \{x^+ + Kx^+ \mid x^+ \in \mathscr{H}^+,\ K\colon \mathscr{H}^+ \to \mathscr{H}^-,\ \|K\| < \infty\}$ and let $1 \in \rho(QK)$. Then, if $x = x^+ + Kx^+ \in \mathscr{L}$, we have

$$\begin{aligned} x &= x^+ - QKx^+ + Kx^+ + QKx^+ = (I^+ - QK)x^+ + (P^- + Q)Kx^+ \\ &= y^+ + (P^- + Q)K(I^+ - QK)^- y^+ = y^+ + Fy^+, \end{aligned}$$

where

$$y^+ = (I^+ - QK)x^+ \quad \text{and} \quad F = (P^- + Q)K(I^+ - QK)^{-1}.$$

Therefore

$$\mathscr{L} = \{y^+ + Fy^+ \mid y^+ \in \mathscr{H}^+,\ F\colon \mathscr{H}^+ \to \mathscr{N}^-,\ \|F\| < \infty\}.$$

The assertion that $-1 \in \rho(QP^- F)$ follows from the equality

$$\begin{aligned} QP^- F &= QP^-(P^- + Q)K(I^+ - QK)^{-1} = QK(I^+ - QK)^{-1} \\ &= -I^+ + (I^+ - QK)^{-1}. \end{aligned}$$

The other assertions of the lemma are proved similarly. ■

Theorem 3.14: *Let V be a J-bi-non-contractive operator, and let $\mathscr{L}^+$ $(\in \mathscr{M}^+)$ be an invariant subspace of V. Then the operator V has an invariant subspace $\mathscr{N}^- \in \mathscr{M}^-$. If, moreover, V has an invariant subspace $(\mathscr{L}^+ \neq)\mathscr{N}_1^+ \in \mathscr{M}^+$ such that $\mathscr{N}_1^+ + \mathscr{N}^- = \mathscr{H}$, then this operator has not less than a continuum of invariant subspaces in each of the sets $\mathscr{M}_0^+$ and $\mathscr{M}^+ \backslash \mathscr{M}_0^+$.*

□Let K_+ be the angular operator of the subspace $\mathscr{L}^+$. It follows from 1.Lemma 8.4 that $\|K_+\| < 1$, and by Lemma 2.2 $G_V^+(K_+) = 0$. Therefore $V \in \Lambda_+$ and so, by Theorem 2.8, V has an invariant subspace $\mathscr{N}^- \in \mathscr{M}^-$.

We now use results from 1.§7.6 and we shall suppose, without loss of generality, that $\mathscr{L}^+ = \mathscr{H}^+$. Let Q be the angular operator of the subspace $\mathscr{N}^-$ and K_1 the angular operator of a subspace $\mathscr{N}_1^+$ $(\neq \mathscr{H}^+)$. By virtue of 1.Theorem 8.15 assertion c) $1 \in \rho(QK_1)$ and so it follows from Lemma 3.13 that $\mathscr{N}_1^+ = \{y^+ + F_1 y^+\}$, where $y^+ \in \mathscr{H}^+$ and $F_1 = (P^- + Q)K_1(I^+ - QK_1)^{-1}$.

Moreover $F_1 \neq 0$. Since $\mathcal{H}^+$ and $\mathcal{N}^-$ are invariant relative to V, and F_1: $\mathcal{H}^+ \to \mathcal{N}^-$, the subspaces $\mathcal{N}_\alpha^+ = \{y^+ + \alpha F_1 y^+ \mid y^+ \in \mathcal{H}^+\}$ will also be invariant for all $\alpha \in \mathbb{C}$. We note that the $\mathcal{N}_\alpha^+$ are not necessarily semi-definite. But, for α sufficiently small in modulus we have $-1 \in \rho(\alpha QP^- F_1)$ and therefore when, for example, $|\alpha| < \frac{1}{2} \| F_1 \|^{-1}$, it follows by Lemma 3.13 that

$$\mathcal{N}_\alpha^+ = \{x^+ + K_\alpha x^+ \mid x^+ \mathcal{H}^+,\ K_\alpha = \alpha P^- F_1 (I^+ + \alpha QP^- F_1)^{-1}\} \tag{3.5}$$

and since

$$\| K_\alpha \| \leqslant \frac{|\alpha| \, \| F_1 \|}{1 - |\alpha| \, \| F_1 \|} < 1$$

it follows that all such $\mathcal{N}_\alpha^+$ are uniformly positive. Moreover the power of the sets of them is not less than that of the continuum.

Without loss of generality we can now suppose that $\mathcal{N}_1^+ \in \mathcal{M}_0^+$, i.e., $\| K_1 \| < 1$ (the equality $\mathcal{N}_1^+ + \mathcal{N}^- = \mathcal{H}$ is obtained from 1.Corollary 8.16). We consider the set of those $\alpha \in \mathbb{C}$ for which $-1 \in \rho(\alpha QP^- F)$. This is an open set in each connected component of which the function $\alpha \to K_\alpha$ is holomorphic. Let Ξ be the connected component of this set which contains the point $\alpha = 1$ (see Lemma 3.13). We prove that there is a point $\alpha_0 \in \Xi$ such that $\| K_{\alpha_0} \| > 1$. First of all we notice that if $\Xi \neq \mathbb{C}$, then Ξ has a finite boundary point α'. Then, if $\alpha_n \in \Xi$ and $\alpha_n \to \alpha'$, we conclude from the form of the operator K_α (see (3.5)) that $\| K_{\alpha_n} \| \to \infty$ as $n \to \infty$, and the existence of α_0 has been proved. But if $\Xi = \mathbb{C}$ and if, contrary to assumption, $| K_\alpha \| < 1$ ($\alpha \in \mathbb{C}$), then by Liouville's theorem $K_\alpha \equiv$ constant. Therefore $QK_\alpha = \alpha QP^- F_1 (I^+ + \alpha QP^- F_1)^{-1} = I^+ - (I^+ + \alpha QP^- F_1)^{-1} \equiv$ const., which implies the equality $QP^- F_1 = 0$, and therefore $K_\alpha = \alpha P^- F_1 \not\equiv$ const. (because $F_1 \neq 0$)—we have obtained a contradiction.

We now consider continuous functions

$$(\alpha =) f_\tau\colon\ [0, 1] \to \Xi (0 \leqslant \tau < 1) \quad \text{with} \quad f_\tau(0) = 1 \quad \text{and} \quad f_\tau(1) = \alpha_0$$

for all

$$\tau \in [0, 1] \text{ and } f_{\tau_1}(t_1) \neq f_{\tau_2}(t_2) \text{ when } \tau_1 \neq \tau_2 \text{ and } t_1, t_2 \in (0, 1).$$

Since all the functions $f_\tau(t)$ are continuous, the functions $\| K_{f_\tau(t)} \|$ will also be continuous with respect to t for each τ, and moreover

$$\| K_{f_\tau(0)} \| = \| K_1 \| < 1 \quad \text{and} \quad \| K_{f_\tau(1)} \| = \| K_{\alpha_0} \| > 1,$$

and therefore there are $t_\tau \in (0, 1)$ such that $\| K_{f_\tau(t_\tau)} \| = 1$. By hypothesis $f_{\tau_1}(t_{\tau_1}) \neq f_{\tau_2}(t_{\tau_2})$ when $\tau_1 \neq \tau_2$. Therefore the operator V has not less than a continuum of different invariant subspaces $\mathcal{N}^+_{f_\tau(t_\tau)} \in \mathcal{M}^+ \setminus \mathcal{M}_0^+$ with the angular operator $K_{f_\tau(t_\tau)}$. ■

5 In conclusion we shall show how to obtain from results about invariant subspaces for J-non-contractive operators similar results for operators acting in G-spaces.

In 1§6.6 a construction was given for embedding a G-space in a J-space, in which, in particular, a $G^{(\varkappa)}$-space was embedded in $\Pi_\varkappa$. We shall call such an embedding *canonic*, and we shall investigate what happens to certain properties of operators acting in the corresponding spaces. Later we shall use the following general theorem which we introduce here without proof.

□ **Theorem 3.15** (M. Krein [2]): *Suppose a Banach space $\mathscr{B}$ is densely and continuously embedded in a Hilbert space $\mathscr{H}$. If a linear operator $A: \mathscr{B} \to \mathscr{B}$ with $\mathscr{D}_A = \mathscr{B}$ is continuous (respectively, completely continuous) in $\mathscr{B}$ and is also symmetric as an operator in $\mathscr{H}$, then it is continuous (respectively, completely continuous) in $\mathscr{H}$ and it can be extended by continuity on to the whole of $\mathscr{H}$ into a self-adjoint operator $\widetilde{A}$: $\mathscr{H} \to \mathscr{H}$.* ■

As a direct consequence of this theorem we obtain the following proposition which will be important later:

Lemma 3.16: *Let the G-space $\mathscr{H} = \mathscr{H}^+ \oplus \mathscr{H}^-$ with $G\mathscr{H}^\pm \subset \mathscr{H}^\pm$, and let $\mathscr{H}$ be canonically embedded in a $\tilde{J}$-space $\widetilde{\mathscr{H}} = \widetilde{\mathscr{H}}^+ \oplus \widetilde{\mathscr{H}}^-$ with $\mathscr{H}^\pm$ canonically embedded in $\widetilde{\mathscr{H}}^\pm$. If A with $\mathscr{D}_A = \mathscr{H}$ is a G-selfadjoint continuous (respectively, completely continuous) operator, then it extends uniquely into a continuous (respectively, completely continuous) J-selfadjoint operator $\widetilde{A}$:$\widetilde{\mathscr{H}} \to \widetilde{\mathscr{H}}$.*

□ Let $J\,(=|G|^{-1}G)$ be the unitary part of the polar decomposition of the operator G; in the given case J coincides with the difference of the ortho-projectors P^+ and P^- on to $\mathscr{H}^+$ and $\mathscr{H}^-$ respectively: $J = P^+ - P^-$. Then JA is a $|G|$-selfadjoint operator, and $\widetilde{\mathscr{H}}$ by construction (see 1.Proposition 6.14) is the completion of $\mathscr{H}$ relative to the norm $(|G|x, x)^{1/2}$. In accordance with Theorem 3.15 the operator J extends into an operator $\tilde{J} = \widetilde{P}^+ - \widetilde{P}^-$ ($\widetilde{P}^\pm$ are the ortho-projectors in $\widetilde{\mathscr{H}}$ on to $\widetilde{\mathscr{H}}^\pm$), and JA extends into an operator $\widetilde{JA}$ selfadjoint in $\widetilde{\mathscr{H}}$ which will be simultaneously with A continuous or completely continuous. Therefore $\widetilde{A} = \tilde{J}\,\widetilde{JA}$ is the required operator. ■

Corollary 3.17: *If under the conditions of Lemma 3.16 $A = \|A_{ij}\|_{i,j=1}^2$ is a continuous G-selfadjoint operator and $A_{12}, A_{21} \in \mathscr{S}_\infty$, then $\widetilde{A} = \|\widetilde{A}_{ij}\|_{i,j=1}^2$, where the $\widetilde{A}_{ij}$ are the extensions of the operators A_{ij} $(i, j = 1, 2)$ and moreover $\widetilde{A}_{12}, \widetilde{A}_{21} \in \mathscr{S}_\infty$.*

□ Since

$$A = A_1 + A_2, \quad \text{where } A_1 = \begin{Vmatrix} A_{11} & 0 \\ 0 & A_{22} \end{Vmatrix}, \qquad A_2 = \begin{Vmatrix} 0 & A_{12} \\ A_{21} & 0 \end{Vmatrix},$$

A_1 and A_2 are G-selfadjoint operators, and $A_2 \in \mathscr{S}_\infty$, we have only to use Lemma 3.16. ■

Remark 3.18: Suppose $\mathscr{H} + \mathscr{H}^+ \oplus \mathscr{H}^-$ *is a G-space,* $G\mathscr{H}^\pm \subset \mathscr{H}^\pm$, $G_\pm = G\,|\,\mathscr{H}^\pm$, *and* $0 \in \rho(G_-)$. *Then from* $A = A^c$, $\mathscr{D}_A = \mathscr{H}$ *and* $A_{12} \in \mathscr{S}_\infty$ *it follows that* $A_{21} \in \mathscr{S}_\infty$.

□ This follows from the equality $A_{21} = G_-^{-1} A_{12} G_+$ which holds in this case. ■

Remark 3.19: If V^c, *together with* V, *is a continuous operator and* $\mathscr{D}_V = \mathscr{D}_{V^c} = \mathscr{H}$, *then* V *and* V^c *extend into continuous operators* $\tilde{V}$ *and* $\tilde{V}_c$ *in* $\mathscr{H}$, *and* $\widetilde{V^c} = \tilde{V}^c$.

□ This follows immediately from Lemma 3.16 if we use the equality $V = \frac{1}{2}(V + V^c) + i[(V - V^c)/2i]$ and the fact that the operators $\frac{1}{2}(V + V^c)$ and $1/2i(V - V^c)$ are G-selfadjoint. ■

Theorem 3.20: *Suppose* $\mathscr{H} = \mathscr{H}^+ \oplus \mathscr{H}^-$ *is a G-space,* $G\mathscr{H}^\pm \subset \mathscr{H}^\pm$, $G_\pm = G\,|\,\mathscr{H}^\pm$, *and* $0 \in \rho(G_-)$. *If V is a G-bi-non-contractive operator (i.e., V and* V^c *are continuous G-non-contractive operators,* $\mathscr{D}_V = \mathscr{D}_{V^c} = \mathscr{H}$), $V_{12} \in \mathscr{S}_\infty$, $\mathscr{L}_+ \subset \mathscr{P}^\pm$, $\overline{V\mathscr{L}_+} = \mathscr{L}_+$, *then there is an* $\tilde{\mathscr{L}}_+ \in \mathscr{M}^+(\mathscr{H})$ *containing* $\mathscr{L}_+$ *and* such that $V\tilde{\mathscr{L}}_+ \subset \tilde{\mathscr{L}}_+$.

□ Let $\mathscr{H} = \mathscr{H}^+ \oplus \mathscr{H}^-$ be canonically embedded in $\tilde{\mathscr{H}} = \tilde{\mathscr{H}}^+ \oplus \tilde{\mathscr{H}}^-$. It follows from $0 \in \rho(G_-)$ that $\tilde{\mathscr{H}}^- = \mathscr{H}^-$. The operator V satisfies the conditions in Remark 3.19 and it therefore extends into a continuous operator $\tilde{V}$ which is $\tilde{J}$-bi-non-contractive. Since $V_{12} \in \mathscr{S}_\infty$ and $0 \in \rho(G_-)$, so $\tilde{V}_{12} \in \mathscr{S}_\infty$. Therefore, by Theorem 3.9, the operator $\tilde{V}$ has the property $\tilde{\Phi}_+$. The condition $\overline{V\mathscr{L}_+} = \mathscr{L}_+$ implies the complete invariance of the closure of $\mathscr{L}_+$ in $\tilde{\mathscr{H}}$ relative to the operator $\tilde{V}$. Consequently there is an $\tilde{\mathscr{L}} = \{x^+ = \tilde{K}x^+ \,|\, x^+ \in \tilde{\mathscr{K}}^+\}$ $(\in \mathscr{M}^+(\tilde{\mathscr{H}}))$ containing $\mathscr{L}_+$, and $V\tilde{\mathscr{L}} = \tilde{\mathscr{L}}$. It is easy to see that $\tilde{\mathscr{L}}_+ = \tilde{\mathscr{L}} \cap \mathscr{H} (\in \mathscr{M}^+(\mathscr{H}))$ will be the required subspace. ■

Corollary 3.21: *If* $\mathscr{H}$ *is a* $G^{(\varkappa)}$*-space and if V is a* $G^{(\varkappa)}$*-bi-non-contractive operator, then any of its completely invariant non-positive subspaces admits extension into a maximal non-positive invariant subspace.*

□ We are in the conditions of Theorem 3.20 to within a change-over to the anti-space and taking account of the fact that V_{21} is a finite-dimensional operator. So $V_{21} \in \mathscr{S}_\infty$. ■

Exercises and problems

1 Let V be a J-non-contractive operator. We define a generalized linear-fractional transformation F_V^+ on a non-empty subset $\mathscr{K}(\subset \mathscr{K}^+)$: $F_V^+(\mathscr{K}) = U\{F_V^+(K) \,|\, K \in \mathscr{K}\}$.

Prove that if $\mathscr{B} = \{V\}$ is a semi-group of J-non-contractive operators, then $F^{+}_{\mathscr{B}} = \{F^{+}_{V} \mid V \in \mathscr{B}\}$ is also a semi-group in relation to superposition of transformations and that $F^{+}_{V_0} \circ F^{+}_{V_2} = F_{V_1 V_2}$; if $\mathscr{B}$ is a group, then $F^{+}_{\mathscr{B}}$ is a group, and $(F^{+}_{V})^{-1} = F^{+}_{V^{-1}}$.

2 Prove that if U is a J-unitary operator, then $\| K \| < 1 \Leftrightarrow \| F^{+}_{U}(K) \| < 1$ and $\| K \| = 1 \Leftrightarrow \| F_{U}(K) \| = 1$ (M. Krein and Shmul'yan [4], Potapov [1]).
Hint: Use Proposition 3.1.

3 Let V be a J-non-contractive operator, U a J-unitary operator, and $W = U^{-1}VU$. An operator $K_0 \in \mathscr{K}^{+}$ is a fixed point of the transformation F^{+}_{V} if and only if $F^{+}_{U^{-1}}(K_0)$ is a fixed point of the transformation F^{+}_{W}.

4 Prove that the function F^{+}_{V} generated by a J-non-contractive operator $V \in \Lambda_{-}$ has at least one fixed point in $\mathscr{K}^{+}$ (I. Iokhvidov [10]).
Hint: Use Proposition 3.2 and Theorem 3.9.

5 Suppose V is a J-bi-non-contractive operator, and $F^{+}_{V^*}$ has a fixed point K_0 with $\| K_0 \| < 1$. Then the function F^{+}_{V} has either a single fixed point in $\mathscr{K}^{+}$, or it has not less than a continuum of such points, and moreover not less than a continuum on the boundary of the ball $\mathscr{K}^{+}$ (Azizov; *cf.* Khabkevich [5]).
Hint: Use the result of Problem 3 and suppose $K_0 = 0$. Use Theorem 3.14.

6 Suppose $T \in \Lambda_{-}$ is a J-bi-non-contractive operator, $(\mathscr{L}_{+}, \mathscr{L}_{-})$ is a dual pair, $\mathscr{L}_{+}$ is completely invariant relative to T, and $\mathscr{L}_{-}$ is invariant relative to T^{c}. Then there is an extension of this pair into a maximal dual pair with preservation of the properties mentioned (Azizov).
Hint: Suppose $T_{12} \in \mathscr{S}_{\infty}$ and use 1.Corollary 8.24 and Theorem 3.8.

7 Prove that if U is a J-unitary operator, then the function F^{+}_{U} is continuous in the weak operator topology if and only if $U_{12} \in \mathscr{S}_{\infty}$ (Helton [3]).

8 If $V \in \Lambda_{-}$ is a J-semi-unitary J-bi-non-contractive operator, then it has the property $\tilde{\Phi}$ (Azizov).
Hint: Use Proposition 2.5 and Theorem 3.9.

9 Suppose V is a J-non-contractive operator, $\mathscr{L}_{+}$ $(\subset \mathscr{P}^{+})$ is its completely invariant subspace with angular operator K_{+}, and $\operatorname{def} \mathscr{L}_{+} < \infty$. Then V is a J-bi-non-contractive operator and there is a subspace $\tilde{\mathscr{L}}_{+} \in \mathscr{M}^{+}$ which is completely invariant relative to it and which contains $\mathscr{L}_{+}$ (Azizov).
Hint: Verify that F^{+}_{V} is a closed transformation of the convex non-empty bicompact subset $\mathscr{K}_{+}(K_{+})$, and use Theorem 3.6.

10 Suppose $A \in (L)$ is a J-selfadjoint operator, Λ is its non-real spectrum, $\Lambda = \Lambda_1 \cup \Lambda_2$, $\Lambda_1 = \overset{*}{\Lambda_2}$, $\Lambda_1 \cap \Lambda_2 = \emptyset$. Prove that if A_{11} is an A_{22}-completely continuous operator, then the operator A has a maximal non-negative invariant subspace $\mathscr{L}^{+} \subset \mathscr{D}_{A}$ such that the non-real spectrum of the operator $A \mid \mathscr{L}^{+}$ coincides with Λ_1 (or with Λ_2) (Langer [3], M. Krein [5]).
Hint: Use Theorems 1.13 and 3.12.

§4 Invariant subspaces of a family of operators

1 We have already encountered invariant subspaces of groups of J-unitary operators in 2.§5.3. From 2.Theorem 5.18 it follows, in particular, that if $\mathscr{U} = \{U\}$ is a bounded amenable group of J-unitary operators, then without loss of generality these operators can also be supposed to be unitary. So below, instead of the condition of amenability and boundedness, *we shall immediately impose on a group of J-unitary operators the condition that the operators entering into the group are unitary*, i.e., $U\mathcal{H}^{\pm} = \mathcal{H}^{\pm}$. The following theorem makes 2.Theorem 5.18 more precise.

Theorem 4.1: *If $\mathcal{U} = \{U\}$ is a group of operators which are simultaneously J-unitary and unitary, then it has the property $\Phi^{[\perp]}$.*

□ Let $(\mathcal{L}_+, \mathcal{L}_-)$ be a dual pair invariant relative to the group $\mathcal{U}$, and let $(\tilde{\mathcal{L}}_+, \tilde{\mathcal{L}}_-)$ be its extension into a maximal invariant dual pair. Since the properties of being J-unitary and simultaneously unitary are Φ-invariant (see Exercise 2 on §1), we can suppose without loss of generality that $(\tilde{\mathcal{L}}_+, \tilde{\mathcal{L}}_-)$ is a definite pair (see Exercise 6 on §1). Since $\mathcal{U}\tilde{\mathcal{L}}_+ = \tilde{\mathcal{L}}_+$, and all $U \in \mathcal{U}$ are J-unitary, it follows that $\mathcal{U}\tilde{\mathcal{L}}_+^{[\perp]} = \tilde{\mathcal{L}}_+^{[\perp]}$. The duality of the pair $(\tilde{\mathcal{L}}_+, \tilde{\mathcal{L}}_-)$ implies the inclusion $\tilde{\mathcal{L}}_- \subset \tilde{\mathcal{L}}_+^{[\perp]}$. Since (see 1.(10.1)) $\tilde{\mathcal{L}}_+^{[\perp]} = \mathcal{D}^+ [\oplus] \mathcal{N}_1$ (in the present case the $\mathcal{N}_0$ which appears in 1.(10.1) is equal to $\{\theta\}$), and since all $U \in \mathcal{U}$ are simultaneously J-unitary and unitary which implies $\mathcal{U}\mathcal{D}^+ = \mathcal{D}^+$, it follows that $UN_1 = \mathcal{N}_1$. By virtue of 1.Theorem 10.2, in order to establish the inclusion $\tilde{\mathcal{L}}_- \in \mathcal{M}^-$, it is now sufficient to establish that $\tilde{\mathcal{L}}_-$ is the non-positive subspace which is maximal in $\tilde{\mathcal{L}}_+^{[\perp]}$. Let P_1 be the orthoprojector from $\mathcal{L}_+^{[\perp]}$ on to $\mathcal{N}_1$. Since $U\tilde{\mathcal{L}}_- = \tilde{\mathcal{L}}_-$ and $U\mathcal{N}_1 = \mathcal{N}_1$, we have $UP_1\tilde{\mathcal{L}}_- = P_1\tilde{\mathcal{L}}_-$, and therefore $U[(P_1\tilde{\mathcal{L}}_-)^{\perp} \cap \mathcal{N}_1] = (P_1\tilde{\mathcal{L}}_-)^{\perp} \cap \mathcal{N}_1$. The subspace $(P_1\tilde{\mathcal{L}}_-)^{\perp} \cap \mathcal{N}_1 (\subset \tilde{\mathcal{L}}_+^{[\perp]})$ is negative and J-orthogonal to $\tilde{\mathcal{L}}_-$. Consequently, $\mathcal{L} \equiv \operatorname{Lin}\{\tilde{\mathcal{L}}_-, (P_1\tilde{\mathcal{L}}_-)^{\perp} \cap \mathcal{N}_1\}$ is the maximal negative subspace from $\tilde{\mathcal{L}}_+^{[\perp]}$ which is invariant relative to all $U \in \mathcal{U}$. The maximality of $\tilde{\mathcal{L}}_-$ as an invariant negative subspace from $\tilde{\mathcal{L}}_+^{[\perp]}$ relative to the group $\mathcal{U}$ implies the coincidence of $\mathcal{L}_-$ with $\mathcal{L}$. That $\tilde{\mathcal{L}}_+ \in \mathcal{M}^+$ is verified similarly. ■

2 In this paragraph we investigate the question of invariant subspaces for groups of normal J-unitary operators. But first we prove some auxiliary propositions.

Lemma 4.2: *Let $\mathcal{H}$ be a J-space, U be a J-unitary and simultaneously positive operator, and let E_λ be its spectral function. Then*

a) $\mathcal{L}_1 = C\operatorname{Lin}\{E_\lambda\mathcal{H} \mid \lambda < 1\}$ *and* $\mathcal{L}_2 = C\operatorname{Lin}\{(I - E_\mu)\mathcal{H} \mid \mu > 1\}$ *are neutral mutually orthogonal subspaces;*

b) $\mathcal{H} = \operatorname{Ker}(U - I) [\oplus] (\mathcal{L}_1 \oplus \mathcal{L}_2)$;

c) $J(\mathcal{L}_1 \oplus \mathcal{L}_2) = \mathcal{L}_1 \oplus \mathcal{L}_2,\ J\operatorname{Ker}(U - I) = \operatorname{Ker}(U - I)$;

d) *every definite invariant subspace of the operator U lies in* $\operatorname{Ker}(U - I)$.

□ a) Since $E_\lambda\mathcal{H} \subset E_{\tilde{\lambda}}\mathcal{H}$ when $\tilde{\lambda} > \mathcal{L}$, and $\tilde{I} - E_{\tilde{\mu}})\mathcal{H} \subset (I - E_\mu)\mathcal{H}$ when $\tilde{\mu} > \mu$, to prove the neutrality of $\mathscr{L}_1$ and $\mathscr{L}_2$ it suffices to verify the neutrality of the subspace $E_\lambda\mathscr{H}$ when $\mathscr{L} < 1$ and the subspace $(I - E_\mu)\mathscr{H}$ when $\mu > 1$ respectively. From the spectral theory of self-adjoint operators it follows that $\sigma(U \mid E_\lambda\mathscr{H}) \subset [\lambda_{\min}, \lambda]$, where $\lambda_{\min} = \inf\{(Ux, x) \mid \|x\| = 1\}$; moreover $0 \in \rho(U)$ because U is a J-unitary positive operator, and therefore $\lambda_{\min} > 0$. The subspace $E_\lambda\mathscr{H}$ is also invariant relative to the operator U^c $(= U^{-1})$, and $\sigma(U^c \mid E_\lambda\mathscr{H}) \subset [1/\lambda, 1/\lambda_{\min}]$. Since $\lambda < 1$ by hypothesis, so $[\lambda_{\min}, \lambda] \cap [1/\lambda, 1/\lambda_{\min}] = \emptyset$. The neutrality of $E_\lambda\mathscr{H}$ now follows from 2.Theorem 1.13. The neutrality of $\tilde{I} = E_\mu)\mathscr{H}$ when $\mu > 1$ is verified similarly.

The orthogonality of $\mathscr{L}_1$ and $\mathscr{L}_2$ follows from the orthogonality property of the spectral function: $E_\lambda(I - E_\mu) = 0$ when $\lambda <$ and $\mu > 1$.

b) The orthogonality of $\operatorname{Ker}(U - I)$ to $\mathscr{L}_1$ and $\mathscr{L}_2$ follows from the orthogonality of $\operatorname{Ker}(U - I)$ to $E_\lambda\mathscr{H}$ when $\lambda < 1$ and to $(I - E_\mu)\mathscr{H}$ when $\mu > 1$.

We now verify that $\operatorname{Ker}(U - I)\,[\perp]\,\mathscr{L}_1$ and $\operatorname{Ker}(U - I)\,[\perp]\,\mathscr{L}_2$. Because of the continuity of the J-metric it suffices to verify that $\operatorname{Ker}(U - I)\,[\perp]\,E_\lambda\mathscr{H}$ when $\lambda < 1$ and that $\operatorname{Ker}(U - I)\,[\perp]\,(I - E_\mu)\mathscr{H}$ when $\mu > 1$ respectively, but this in turn follows directly from 2.Theorem 1.13 when we take into account that $\sigma(U^c \mid \operatorname{Ker}(U - I)) = \{1\}$.

We consider the subspace $\mathscr{H}' = \operatorname{Ker}(U - I)\,[\oplus]\,(\mathscr{L}_1 \oplus \mathscr{L}_2)$ which is invariant relative to U; we shall prove that $\mathscr{H}' = \mathscr{H}$. For otherwise we would have $\mathscr{H} = \mathscr{H}' \oplus \mathscr{H}'^{\perp}$ where $\mathscr{H}'^{\perp}$ $(\neq \theta)$ is a subspace invariant relative to U, and $U' = U \mid \mathscr{H}'^{\perp}$ is a positive operator. Let E_λ' be its spectral function. It is easy to see that $E_\nu'\mathscr{H}'^{\perp} \subset E_\nu\mathscr{H}$ for all $\nu \in \mathbb{R}$. Therefore, since $E_\lambda'\mathscr{H}'^{\perp}$ is orthogonal to $E_\lambda\mathscr{H}$ $(\subset \mathscr{L}_1 \subset \mathscr{H}')$ when $\lambda < 1$ and $(I - E_\mu')\mathscr{H}'^{\perp}$ is orthogonal to $(I - E_\mu)\mathscr{H}$ $(\subset \mathscr{L}_2 \subset \mathscr{H}')$ when $\mu > 1$, it follows that $E_\lambda'\mathscr{H}'^{\perp} = \{\theta\}$ for $\lambda < 1$ and $(I - E_\mu')\mathscr{H}^{\perp} = \{\theta\}$ for $\mu > 1$, and so $\sigma(V') = \{1\}$. Hence we conclude that $V' = I \mid \mathscr{H}'^{\perp}$, which contradicts the orthogonality of $\mathscr{H}'^{\perp}$ to $\operatorname{Ker}(U - I)$. Consequently $\mathscr{H}' = \mathscr{H}$.

c) Taking into account 1.Formula (7.1), this follows from b).

d) Let $\mathscr{L}$ be an invariant subspace of the operator U. It is well-known (see, e.g., [XXII] and *cf.* 4.Remark 1.8) that then $E_\lambda\mathscr{L} \subset \mathscr{L}$. Since $E_\lambda\mathscr{L}$ is a neutral subspace when $\lambda < 1$, the definiteness of $\mathscr{L}$ implies $E_\lambda\mathscr{L} = \{\theta\}$. Similarly $(I - E_\mu)\mathscr{L} = \{\theta\}$ when $\mu > 1$. Therefore $\mathscr{L} \perp (\mathscr{L}_1 \oplus \mathscr{L}_2)$, i.e., $\mathscr{L} \subset \operatorname{Ker}(U - I)$. ■

Corollary 4.3: *Let the operator U satisfy the conditions of Lemma 4.2, and let $\mathscr{B}_U$ be the algebra of all continuous operators $A\colon \mathscr{H} \to \mathscr{H}$ defined on $\mathscr{H}$ which commute with U.* Then *$\mathscr{B}_U$ has a common non-trivial neutral invariant subspace if and only if $U \neq I$. Each such subspace is J-orthogonal to* $\operatorname{Ker}(U - I)$.

□ If $U = I$, then $\mathscr{B}_U$ coincides with the algebra of all continuous operators $A\colon \mathscr{H} \to \mathscr{H}$, which, as is easy to see, has non-trivial (and including also neutral) invariant subspaces.

But if $U \neq I$, then, for example the $\mathscr{L}_1$ and $\mathscr{L}_2$ appearing in Lemma 4.2 are neutral and invariant relative to $\mathscr{B}_U$, since each of the operators of this algebra commutes with E_λ for all $\lambda \in \mathbb{R}$ (see, e.g., [XXII] and *cf* 4Theorem1.5).

Let $\mathscr{L}$ be any non-trivial neutral subspace invariant relative to $\mathscr{B}_U$, and let P be the J-orthogonal projector from $\mathscr{H}$ on to $\operatorname{Ker}(U-I)$. Then, as is easily verified. $P\mathscr{L}$ is an invariant subspace of the algebra $\mathscr{B}_U \mid \operatorname{Ker}(U-I)$, which coincides with the algebra of all continuous operators acting in $\operatorname{Ker}(U-I)$. Therefore either $P\mathscr{L} = \{\theta\}$ or $P\mathscr{L} = \operatorname{Ker}(U-I)$. Since $(P^c =)P \in \mathscr{B}_U$, so $P\mathscr{L} \subset \mathscr{L}$ and therefore $P\mathscr{L}$ is a neutral subspace. It follows from assertion b) in Lemma 4.2 that $\operatorname{Ker}(U-I)$ is a projectionally complete subspace, and therefore $PL = \{\theta\}$, i.e., $\mathscr{L} [\perp] \operatorname{Ker}(U-\mathrm{I})$. ■

The following lemma is of a general character and seems to be well-known.

Lemma 4.4: *If $\mathscr{V} = \{V\}$ is a group consisting of normal operators and containing the conjugate V^* whenever it contains V, then $VU^*U = U^*UV$ for any $U, V \in \mathscr{V}$.*

□ By hypothesis the operator $W = V^*VU^*U \in \mathscr{V}$ and therefore it is normal. Since W is a (U^*U)-selfadjoint operator and the scalar product $(U^*U\cdot, \cdot)$ is equivalent (in the sense of equivalence of the corresponding norms) to the original one, so $\sigma(W) \subset \mathbb{R}$, which, taking the normality of W into account, is equivalent to its selfadjointness: $W = W^*$, i.e., $V^*VU^*U = U^*UV^*V$. As is well-known (see. e.g., [XXII]) it follows from this that $(V^*V)^{1/2}U^*U = U^*U(V^*V)^{1/2}$. Let $V = S(V^*V)^{1/2}$ be the polar representation of the normal operator V; here S is a unitary operator commuting with $(V^*V)^{1/2}$. Since

$$\begin{aligned} S(U^*U)^2S^{-1}V^*V &= (V^*V)^{1/2}S(U^*U)^2S^{-1}(V^*V)^{1/2} = (VU^*U)(U^*UV) \\ &= (U^*UV^*)(VU^*U) = (U^*U)(V^*V)(U^*U) \\ &= (U^*U)^2V^*V, \end{aligned}$$

so $S(U^*U)^2 = (U^*U)^2S$. We conclude, as above, that $SU^*U = U^*US$. Consequently $VU^*U = U^*UV$. ■

We turn now to the formulation and proof of the main result of this paragraph.

Theorem 4.5: *Let $\mathscr{H}$ be a J-space and $\mathscr{U} = \{U\}$ be a group consisting of normal J-unitary operators and containing the operator U^* whenever it contains U. Then $\mathscr{U}$ has the property $\tilde{\Phi}^{[\perp]}$.*

□ Let $(\mathscr{L}_+, \mathscr{L}_-)$ be an invariant dual pair of the group $\mathscr{U}$, and let $(\tilde{\mathscr{L}}_+, \tilde{\mathscr{L}}_-)$ be its extension into a maximal invariant dual pair. We use Exercise 6 on §1 and we shall suppose that $(\tilde{\mathscr{L}}_+, \tilde{\mathscr{L}}_-)$ is a definite dual pair. By assertion d) of Lemma 4.2 we have $\tilde{\mathscr{L}}_\pm \subset \cap\{\operatorname{Ker}(U^*U-I) \mid U \in \mathscr{U}\}$. Moreover, the maximality of $\tilde{\mathscr{L}}_\pm$ implies the equality $U^*U = I$ for all $U \in \mathscr{U}$. For, if we had $U_0^*U_0 \neq I$ for some $U_0 \in \mathscr{U}$, then it would follow from Lemma 4.4 and

Corollary 4.3 that the group $\mathscr{U}$ has a common non-trivial neutral subspace J-orthogonal to $\mathrm{Ker}(U_0^*U_0 - I)$ and all the more to $\tilde{\mathscr{L}}_\pm$—so we have a contradiction. Thus $U^*U = I$ for all $U \in \mathscr{U}$, i.e., $\mathscr{U}$ is a group consisting of operators simultaneously J-unitary and unitary. By Theorem 4.1 it has the property $\tilde{\Phi}^{[\perp]}$, and therefore $\tilde{\mathscr{L}}_\pm \in \mathscr{M}^\pm$. ■

Corollary 4.6: *If $\mathscr{U} = \{U\}$ is a commutative group consisting of normal J-unitary operators, then the minimal group containing $\mathscr{U}$ and $\mathscr{U}^*$ has the property $\tilde{\Phi}^{[\perp]}$.*

□ The set $\mathscr{U}^* = \{U^* \mid U \in \mathscr{U}\}$ is also a commutative group consisting of normal J-unitary operators. Moreover, if $U, V \in \mathscr{U}$, then $UV = VU$, and therefore by a well-known theorem of Fuglede $U^*V = VU^*$, i.e., the elements of the groups $\mathscr{U}$ and $\mathscr{U}^*$ commute with one another. Consequently, the group $\tilde{\mathscr{U}}$ generated by the union of $\mathscr{U}$ and $\mathscr{U}^*$ is commutative and consists of normal J-unitary operators. Moreover, if $\tilde{U} \in \tilde{\mathscr{U}}$, then $\tilde{U}^* \in \tilde{\mathscr{U}}$. It only remains to use Theorem 4.5. ■

3 □ The operator ball $\mathscr{K}^+$, as was shown in 1.Proposition 8.20, is bicompact in the weak operator topology, and therefore (see 1.Proposition 8.21) the centralized system of its closed subsets has a non-empty intersection—on this is based the proof of the following key proposition.

Theorem 4.7: *Let $\mathscr{V} = \{V\}$ be a family of J-bi-non-contractive operators, let $F_{\mathscr{V}}^+ = \{F_V^+ \mid V \in \mathscr{V}\}$ be the corresponding linear-fractional transformations of the ball $\mathscr{K}$ (see (3.1)), and let $\mathscr{K}_V^+$ be the closed set of fixed points of the transformation F_V^+. If for each finite set $\{V_i\} \subset \mathscr{V}$ we have $\bigcap_i \mathscr{K}_{V_i}^+ \neq \emptyset$, then $\bigcap\{\mathscr{K}_V^+ \mid V \in \mathscr{V} \in \mathscr{V}\} \neq \emptyset$.* ■

Corollary 4.8: *Let $\mathscr{V} = \{V\}$ be a family of J-bi-non-contractive operators V with $V_{12} \in \mathscr{S}_\infty$ (respectively, $V_{21} \in \mathscr{S}_\infty$). If each finite subset of $\mathscr{V}$ has the property Φ_+ or $\tilde{\Phi}_+$ (respectively, Φ_- or $\tilde{\Phi}_+$), then the whole family $\mathscr{V}$ has the same property.*

If $\mathscr{V}$ is a family of J-unitary operators with $V_{12} \in \mathscr{S}_\infty$ (or, what is equivalent, $V_{21} \in \mathscr{S}_\infty$), then similar assertions hold also regarding the properties $\Phi^{[\perp]}$ and $\tilde{\Phi}^{[\perp]}$.

□ Let V be a J-bi-non-contractive operator, and let $\mathscr{K}_V^+$ be the set of fixed points of the function F_V^+. By Theorem 3.8, when $V_{12} \in \mathscr{S}_\infty$, this set $\mathscr{K}_V^+$ is closed in the weak operator topology. By Proposition 3.4 each finite subset $\mathscr{V}'$ from $\mathscr{V}$ has the property Φ_+ if and only if $\bigcap\{\mathscr{K}_V^+ \mid V \in \mathscr{V}^-\} \neq \emptyset$, and $\mathscr{V}$ having the property Φ_+ is equivalent to $\bigcap\{\mathscr{K}_V^+ \mid V \in \mathscr{V}\} \neq \emptyset$. it only remains to use Theorem 4.7.

The remaining assertions are proved similarly. ■

Corollary 4.9: *If $\mathscr{U}=\{U\}$ is a commutative family of stable J-unitary operators and if $U_{12}\in\mathscr{S}_\infty$ for all $U\in\mathscr{U}$, then $\mathscr{U}$ has the property $\tilde{\Phi}^{[\perp]}$.*

☐ Let $U_1, U_2, \ldots, U_n$ be an arbitrary finite set of operators from $\mathscr{U}$, and let $\mathscr{U}_n$ be the group generated by these operators. From the commutativity of $\mathscr{U}$ and the stability of the operators comprising $\mathscr{U}$, we conclude (see II, items 5.17–5.20) that $\mathscr{U}_n$ is a bounded amenable group. Therefore (see §4.1) $\mathscr{U}_n$ has the property $\tilde{\Phi}^{[\perp]}$. It remains to apply Corollary 4.8. ■

Under the conditions of Corollary 4.9 the whole family $\mathscr{U}$ is not necessarily bounded. The following example confirms this.

☐ *Example 4.10:* Let $\Pi_1=\Pi_+\oplus\Pi_-$ be a Pontryagin space, $\Pi_+=\text{Lin}\{e^+\}$, and $\Pi_-=C\,\text{Lin}\{e^-\}_1^\infty$, $(e_i^+, e_j^-)=\delta_{ij}$ $(i, j=1, 2, \ldots)$, and let the sequence $\{\xi_i\}_1^\infty$ of complex numbers be such that $\sum_{i=1}^\infty |\xi_i|^2=1$, $\xi_i\neq 0$ $(i=1, 2, \ldots)$. We consider the subspaces

$$\Pi_1^{(n)}=C\,\text{Lin}\left\{e^+ + \sum_{i=1}^n \xi_i e_i^-,\ e_{n+1}^-, e_{n+2}^-, \ldots\right\}.$$

These too are Pontryagin spaces with $\varkappa=1$, $\Pi_1^{(n)}\supset\Pi_1^{(n+1)}$, $\Pi_1^{(n)[\perp]}$ are negative subspaces, and $\Pi_1=\Pi_1^{(n)}[+]\Pi_1^{(n)[\perp]}$ $(n=1, 2, \ldots)$. We define operators U_n:

$$U_n x=x \quad \text{when } x\in\Pi_1^{(n)}, \qquad U_n y=-y \quad \text{when } y\in\Pi_1^{[\perp]}.$$

It is easy to see that these are π-unitary stable operators (because $U_n^2=I$) and $U_nU_m=U_mU_n$ $(n, m=1, 2, \ldots)$. It can be seen from the construction of the operators U_n that all their non-negative eigenvectors lie in $\Pi_1^{(n)}$, and therefore the non-negative eigenvectors of the family $\mathscr{U}=\{U_n\}_1^\infty$ lie in $\cap\{\Pi_1^{(n)}\}_1^\infty=\text{Lin}\{e^+ + \sum_{i=1}^\infty \xi_i e_i^-\}$. Thefore $\mathscr{U}$ has a single non-negative invariant subspace $\text{Lin}\{e^+\sum_1^\infty \xi_i e_i^-\}$ and it is neutral.

But if the family $\mathscr{U}$ were bounded, then the group generated by it would also be bounded. By 2.Theorem 5.18 this group, and therefore also the family $\mathscr{U}$, would have to have a positive invariant subspace—a contradiction. ■

4 We pass on now to the investigation of the question of invariant subspaces for π-non-contractive operators.

Theorem 4.11: *Let $\mathscr{V}=\{V\}$ be a commutative family of π-non-contractive operators in $\Pi_\varkappa$. Then it has the property $\tilde{\Phi}$.*

☐ In accordance with Corollary 4.8 it suffices to verify the assertion for a family $\mathscr{V}$ consisting of a finite set of operators. We verify that such a $\mathscr{V}$ has the property $\tilde{\Phi}_+$. Let $\mathscr{L}_+$ $(\subset\mathscr{P}^+)$ be an invariant subspace of this family, and let $\tilde{\mathscr{L}}_+$ be its maximal invariant non-negative subspace containing $\mathscr{L}_+$. We shall use Exercise 4 on §1 and we shall suppose that $\tilde{\mathscr{L}}_+$ is a positive subspace.

Then $\Pi_\varkappa = \tilde{\mathscr{L}}_+ [+] \tilde{\mathscr{L}}_+^{[\perp]}$, where $\tilde{\mathscr{L}}_+^{[\perp]}$ is again a Pontryagin space $\Pi_{\varkappa_1}$ with $\varkappa_1 = \varkappa - \dim \tilde{\mathscr{L}}_+$. We shall prove that $\varkappa_1 = 0$, i.e., that $\tilde{\mathscr{L}}_+ \in \mathscr{M}^+$.

Suppose $\varkappa_1 > 0$ and let P be the π-orthogonal projector on to $\tilde{\mathscr{L}}_+^{[\perp]}$. Then, as is easily verified, the operators $V_1 = PV | \mathscr{L}_+^{[\perp]}$ form a commutative family $\mathscr{V}_1 = \{V_1\}$, and moreover the V_1 are π-non-contractive operators in $\mathscr{L}_+^{[\perp]}$ (see the more general Lemma 5.9 below. Since $\tilde{\mathscr{L}}_+$ is the maximal invariant non-negative subspace of the family $\mathscr{V}$, it follows, on the one hand, that $\mathscr{V}_1$ has no non-negative invariant subspace $\mathscr{L}_1 \neq \{\theta\}$: for otherwise $\text{Lin}\{\tilde{\mathscr{L}}_+, \mathscr{L}_1\}$ would be an extension of $\tilde{\mathscr{L}}_+$ into a non-negative invariant subspace of the family $\mathscr{V}$; and on the other hand, by Corollary 2.9, each of the operators V_1 has in $\mathscr{L}_+^{[\perp]}$ a $\varkappa_1$-dimensional non-negative invariant subspace. Moreover, if $V_1 x_0 = \lambda x_0$, with $(\theta \neq) x_0 \in \mathscr{P}^+$, then $|\lambda| = 1$. For, if $|\lambda| > 1$, then, by Exercise 5 on 2.§6, $\mathscr{L}_\lambda(V_1)$ is a non-negative subspace and it is invariant relative to $\mathscr{V}_1$—which is impossible. But if we had $|\lambda| < 1$, then (see Exercise 5 on 2.§6) $x_0 \in \mathscr{P}^-$, i.e., $x_0 \in \mathscr{P}^0$, and the isotropic part of $\mathscr{L}_\lambda(V_1)$ would again be invariant relative to $\mathscr{V}$, which once more is impossible. Therefore, $|\lambda| = 1$ and $\text{Ker}(V_1 - \lambda I)$ is non-degenerate. Consequently (see Exercise 23 on 2.§4) the operator V_1 has no associated vectors corresponding to this λ. Let $\lambda_1, \lambda_2, \ldots, \lambda_p$ be all such points from $\sigma_p(V_1)$ to which correspond non-negative eigenvectors. It is clear that $\mathscr{H}_1 = \text{Lin}\{\text{Ker}(V_1 - \lambda_i I)\}^p$ is a certain Pontryagin space $\Pi'_{\tilde{\varkappa}}$. From Corollary 2.9 we conclude that $\tilde{\varkappa} = \varkappa_1$. Moreover, $\mathscr{H}_1$ is invariant relative to $\mathscr{V}_1$. We consider the family $\mathscr{V}_1^{(1)} = \mathscr{V}_1 | \mathscr{H}_1 = \{V | \mathscr{H}_1\}$. It again is commutative and consists of π-non-contractive operators one of which is, as is easily seen, a stable π-unitary operator. By carrying out the procedure indicated as many times as there are operators in $\mathscr{V}$, we obtain a family $\mathscr{V}' = \{V'\}$ consisting of π-unitary operators which are the restrictions of the original operators on to an invariant subspace $\Pi''_{\varkappa'}$ with $\varkappa' = \varkappa_1$. According to Corollary 4.9 the family $\mathscr{V}'$ has a $\varkappa_1$-dimensional non-negative subspace—we have obtained a contradiction.

We now verify that the family $\mathscr{V}$ has the property $\tilde{\Phi}_-$. Let $\tilde{\mathscr{L}}_-$ be the maximal invariant non-positive subspace of the family, containing the original one. Again by virtue of the result of Exercise 4 on §1 we can suppose that $\tilde{\mathscr{L}}_-$ is a negative subspace. Then $\tilde{\mathscr{L}}_-^{[\perp]}$ is a Pontryagin space $\Pi'_{\varkappa'}$ with $\varkappa' = \varkappa$ and it is invariant relative to the family $\mathscr{V}^c = \{V^c \mid V \in \mathscr{V}\}$ consisting of π-non-contractive operators. In accordance with what we have proved above, $\mathscr{V}^c$ has a $\varkappa$-dimensional non-negative invariant subspace $\mathscr{L}_+$ in $\tilde{\mathscr{L}}_-^{[\perp]}$; but then $\mathscr{L}_+^{[\perp]}$ $(\in \mathscr{M}^-)$ is invariant relative to $\mathscr{V}$ and $\mathscr{L}_+^{[\perp]} \supset \tilde{\mathscr{L}}_-$, and this is possible only when $\mathscr{L}_+^{[\perp]}$ coincides with $\tilde{\mathscr{L}}_-$. ■

Corollary 4.12: *If $\mathscr{U} = \{U\}$ is a family of pairwise commutating π-unitary operators, then it has the property $\tilde{\Phi}^{[\perp]}$.*

□ Let $(\tilde{\mathscr{L}}_+, \tilde{\mathscr{L}}_-)$ be the maximal invariant dual pair of the family $\mathscr{U}$, containing the original one. In accordance with Exercise 6 on §1 we can suppose that $(\tilde{\mathscr{L}}_+, \tilde{\mathscr{L}}_-)$ is a definite pair, and therefore $[\tilde{\mathscr{L}}_+ [+] \tilde{\mathscr{L}}_+]^{[\perp]}$ is a Pontryagin space $\Pi'_{\varkappa'}$ invariant relative to $\mathscr{U}$ with $\varkappa' = \varkappa - \dim \tilde{\mathscr{L}}_+$. By Theorem 4.11 when $\varkappa' > 0$ the family $\mathscr{U}$ has in $\Pi'_{\varkappa'}$ a $\varkappa'$-dimensional non-negative invariant subspace $\mathscr{L}_1$ and a maximal non-positive invariant subspace $\mathscr{L}_2 = \mathscr{L}_1^{[\perp]} \cap \Pi'_{\varkappa'}$. Consequently $(\text{Lin}\{\tilde{\mathscr{L}}_+, \mathscr{L}_1\}, \text{Lin}\{\tilde{\mathscr{L}}_-, \mathscr{L}_2\})$ is an extension of the dual pair $(\tilde{\mathscr{L}}_+, \tilde{\mathscr{L}}_-)$ preserving invariance relative to $\mathscr{U}$, and this is possible only when $\mathscr{L}_1 = \mathscr{L}_2 = \{\theta\}$, i.e., $\tilde{\mathscr{L}}_\pm \in \mathscr{M}^\pm$. ■

Corollary 4.13: *Let $\mathscr{H}$ be a $G^{(\varkappa)}$-space, let $\mathscr{U} = \{U\}$ be a commutative family of $G^{(\varkappa)}$-unitary operators, and let $\mathscr{L}_-$ be its maximal invariant non-positive subspace. Then $\mathscr{L}_- \in \mathscr{M}^-$.*

□ We use the results of §3.5 and canonically embed the $G^{(\varkappa)}$-space $\mathscr{H}$ in $\Pi_\varkappa$, and we also extend the family $\mathscr{U}$ by continuity into a commutative family $\tilde{\mathscr{U}} = \{\tilde{U}\}$ of π-unitary operators. In accordance with Corollary 4.12 $\tilde{\mathscr{U}}$ has the property $\tilde{\Phi}_-$, and therefore there is an $\tilde{\mathscr{L}} \in \mathscr{M}^-(\Pi_\varkappa)$ which contains $\mathscr{L}_-$ and is invariant relative to $\tilde{\mathscr{U}}$. Then (*cf.* 1.Proposition 8.18) $\tilde{\mathscr{L}} \cap \mathscr{H} \in \mathscr{M}^-(\mathscr{H})$, $\tilde{\mathscr{L}} \cap \mathscr{H} \supset \mathscr{L}_-$, and $\tilde{\mathscr{L}} \cap \mathscr{H}$ is invariant relative to $\mathscr{U}$. Since $\mathscr{L}_-$ is maximal it follows that $\mathscr{L}_- = \tilde{\mathscr{L}} \cap \mathscr{H}$. ■

The result obtained enables us to prove a series of propositions about the existence of invariant subspaces; one of them is

Theorem 4.14: *Let $\mathscr{U} = \{U\}$ be a commutative family of J-unitary operators, let $(\mathscr{L}_+, \mathscr{L}_-)$ be an invariant dual pair of $\mathscr{U}$ with* def $\mathscr{L}_+ < \infty$ *or* def $\mathscr{L}_- < \infty$, *and let $(\tilde{\mathscr{L}}_+, \tilde{\mathscr{L}}_-)$ be its extension into a maximal invariant dual pair. Then $\tilde{\mathscr{L}}_\pm \in \mathscr{M}^\pm$.*

□ Without loss of generality we shall suppose that def $\mathscr{L}_+ < \infty$, and so def $\tilde{\mathscr{L}}_+ < \infty$ also. In accordance with Exercise 6 on §1 we can suppose that $(\tilde{\mathscr{L}}_+, \tilde{\mathscr{L}}_-)$ is a definite dual pair. Consequently $\tilde{\mathscr{L}}_+^{[\perp]}$ is a $G^{(\varkappa)}$-space with $\varkappa = \text{def}\, \tilde{\mathscr{L}}_+$, the $\mathscr{U} \mid \tilde{\mathscr{L}}_+^{[\perp]}$ are $G^{(\varkappa)}$-unitary operators, and $\tilde{\mathscr{L}}$ is their maximal invariant non-positive subspace. We conclude from Corollary 4.13 that $\tilde{\mathscr{L}}_- \in \mathscr{M}^-(\tilde{\mathscr{L}}_+^{[\perp]})$, and so (*cf.* 1.Theorem 10.2) $\mathscr{L}_- \in \mathscr{M}^-(\mathscr{H})$. Therefore $(\tilde{\mathscr{L}}_-^{[\perp]}, \tilde{\mathscr{L}}_-)$ is an invariant dual pair containing $(\tilde{\mathscr{L}}_+, \tilde{\mathscr{L}}_-)$, and since the latter is maximal we obtain $\tilde{\mathscr{L}}_+ = \tilde{\mathscr{L}}_-^{[\perp]} \in \mathscr{M}^+$. ■

Exercises and problems

1 Investigate whether in Theorem 4.5 the condition $\{U \in \mathscr{U}\} \Rightarrow \{U^* \in \mathscr{U}\}$ can be omitted.

2 Let $\mathscr{V} = \{V\}$ be a commutative family consisting of π-non-contractive and π-bi-non-expansive operators. Then it has the property Φ (Azizov [6]).

3 Generalize Corollary 4.13 to the case where $\mathscr{U} = \{U\}$ is a commutative family of $G^{(\varkappa)}$-non-contractive operators and $\mathscr{D}_U = \mathscr{D}_{U^c} = \mathscr{H}$ (Azizov).

4 Prove that if $\mathscr{U} = \{U\}$ is a commutative family of J-bi-non-contractive operators, if $\mathscr{L}_+$ (respectively, $\mathscr{L}_-$) is a maximal completely invariant (respectively, maximal invariant) non-negative (respectively, non-positive) subspace of the family $\mathscr{U}$ and def $\mathscr{L}_+ < \infty$ (respectively, def $\mathscr{L}_- < \infty$), then $\mathscr{L}^+ \in \mathscr{M}^+$ (respectively, $\mathscr{L}^- \in \mathscr{M}^-$) (Azizov).

5 Let $\mathscr{H} = \text{Lin}\{e, f\}$, $\|e\| = \|f\| = 1$, $(e, f) = 0$. We introduce into $\mathscr{H}$ a J-metric by means of the operator J: $J(\alpha e + \beta f) = \beta e + \alpha f$. Prove that the group generated by the operators J and U: $U(\alpha e + \beta f) = \lambda\alpha e + \lambda^{-1}\beta f$ ($|\lambda| \neq 1$) is soluble, consists of π-unitary operators, and has no common non-trivial invariant subspaces (*cf.* 2.Theorem 5.18) (Azizov).

6 Let $\mathscr{A}$ be a commutative algebra of operators acting in a J-space and closed relative to the operations of conjugation and J-conjugation, i.e., $A \in \mathscr{A} \Rightarrow A^* \in \mathscr{A}$, $A^c \in \mathscr{A}$. Prove that if $(\mathscr{L}^+, \mathscr{L}^-)$ is any maximal dual pair invariant relative to $\mathscr{A}$, then $\mathscr{L}^\pm \in \mathscr{M}^\pm$ (Phillips [3]).
Hint: Use Theorems 1.13 and 4.5.

§5 Operators of the classes H and K(H)

1 In 1.§5.4 we introduced the concepts of the classes $h^\pm$ with which we shall operate in this section.

Definition 5.1: We shall say that a bounded operator T *belongs to the class* **H** ($T \in \mathbf{H}$) if it has at least one pair of invariant subspaces $\mathscr{L}^+ \in \mathscr{M}^+$ and $\mathscr{L}^- \in \mathscr{M}^-$ and every maximal semi-definite subspace $\mathscr{L}^\pm$ invariant relative to T belongs to $\mathscr{H}^\pm$ respectively.

From this definition and 2.Proposition 1.11 the implication

$$T \in \mathbf{H} \Leftrightarrow T^c \in \mathbf{H} \tag{5.1}$$

follows immediately.

We now investigate a number of other properties of operators of the class **H**.

Theorem 5.2: *If an operator T has an invariant subspace of the class $\mathscr{M}^+ \cap h^+$ (respectively, $\mathscr{M}^- \cap h^-$), then $T \in \Lambda_+$ (respectively, $T \in \Lambda_-$). In particular, $T \in \mathbf{H} \Rightarrow T \in \Lambda_+ \cap \Lambda_-$.*

□ Let K^+ be the angular operator of the invariant subspace $\mathscr{L}^+ \in \mathscr{M}^+$ of the operator T. If $\mathscr{L}^+ \in h^+$, then K^+ can be expressed in the form of a sum $K^+ = K_1^+ + K_2^+$, where $\|K_1^+\| < 1$, and K_2^+ is a finite-dimensional partially isometric operator. By Lemma 2.2 $G_T^+(K^+) = 0$, and therefore

$$\begin{aligned} G_T^+(K_1^+) &= (K^+ - K_2^+)T_{11} + (K^+ - K_2^+)T_{12}(K^+ - K_2^+)T_{21} - T_{22}(K^+ - K_2^+) \\ &= -K_2{}^+T_{11} - K_2^+T_{12}K^+ - K^+T_{12}K_2^+ - K_2{}^+T_{12}K_2^+ + T_{22}K_2^+ \in \mathscr{S}_\infty \end{aligned}$$

and by Definition 2.3 $T \in \Lambda_+$. Similarly one proves that if the operator T has an invariant subspace $\mathscr{L}^- \in \mathscr{M}^- \cap h^-$, then $T \in \Lambda_-$.

From Definition 5.1 and what has been proved, it follows that $T \in \mathbf{H} \Rightarrow T \in \Lambda_+ \cap \Lambda_-$. ■

It follows from this theorem that the propositions proved earlier (see §§2–3) for operators of the class $\Lambda_{\pm}$ hold also for operators of the class **H**.

□ In particular, from Theorem 3.9 and Exercise 8 on §3 we obtain

Corollary 5.3: *If T ($\in$**H**) is a J-bi-non-contractive operator, (respectively, a J-semi-unitary J-bi-non-contractive operator or, in particular, a J-unitary operator), then T has the property $\tilde{\Phi}$ (respectively, $\tilde{\Phi}^{[\perp]}$).* ■

Corollary 5.4: Let T ($\in$**H**) *be a J-bi-non-contractive operator. Then each of its completely invariant non-negative (respectively, invariant non-positive) subspaces belonging to the class h^+ (respectively, h^-).*

□ It is sufficient to use Corollary 5.3, the Definition 5.1, and the simple fact that a subspace of a subspace of the class $h^{\pm}$ belongs to $h^{\pm}$. ■

Corollary 5.5: *Let T be a J-bi-non-contractive operator of the class* **H**. *Then there is a constant $\varkappa_T < \infty$ such that the dimension of each of the neutral invariant subspaces of the operator T does not exceed $\varkappa_T$.*

□ Let $\{\mathscr{L}\}$ be the set of all neutral invariant subspaces of the operator T. It follows from Corollary 5.4 that $\dim \mathscr{L} < \infty$ for all $\mathscr{L} \in \{\mathscr{L}\}$. If we assume that there is no constant $\varkappa_T < \infty$ bounding the dimension of the subspaces $\mathscr{L}$, then among them, there could be found a sequence $\{\mathscr{L}_n'\}_1^\infty$ with $\dim \mathscr{L}_n' \to \infty$, $\dim \mathscr{L}_n' \leqslant \dim \mathscr{L}_{n+1}'$. We put

$$\mathscr{L}_1 = \mathscr{L}_1', \qquad \mathscr{L}_n = \operatorname{Lin}\{\mathscr{L}_{n-1}, \mathscr{L}_n' \cap \mathscr{L}_{n-1}^{[\perp]}\} \qquad (n = 2, 3, \ldots).$$

By construction $\{\mathscr{L}_n\}_1^\infty$ is a monotonely non-decreasing sequence of neutral invariant subspaces of the operator T, and $\dim \mathscr{L}_n \to \infty$ as $n \to \infty$. But then $C \operatorname{Lin}\{\mathscr{L}_n\}_1^\infty$ is also a neutral invariant subspace of the operator T, and $\dim C \operatorname{Lin}\{\mathscr{L}_n\}_1^\infty = \infty$—we have obtained a contradiction. ■

Corollary 5.6: *Let T ($\in$**H**) be a J-bi-non-contractive operator, and let $\operatorname{Ker}(T - \lambda I) = \mathscr{N}_\lambda^0 [+] \mathscr{N}_\lambda^+ [+] \mathscr{N}_\lambda^-$ be the canonical decomposition of the kernel of the operator $T - \lambda I$ into an isotropic component $\mathscr{N}_\lambda^0$, a positive component $\mathscr{N}_\lambda^+$, and a negative component $\mathscr{N}_\lambda^-$. Then $\mathscr{N}_\lambda^{\pm}$ are uniformly difinite subspaces and*

$$\min\{\dim(\mathscr{N}_\lambda^0 [+] \mathscr{N}_\lambda^+), \dim(\mathscr{N}_\lambda^0 [+] \mathscr{N}_\lambda^-)\} < \infty.$$

□ From Corollary 5.4 we obtain that $\dim \mathscr{N}_\lambda^0 < \infty$, that $\mathscr{N}_\lambda^{\pm}$ are uniformly definite subspaces, and that any definite subspaces in $\mathscr{N}_\lambda^+ [+] \mathscr{N}_\lambda^-$ are uniformly definite. Therefore by virtue of 1.Theorem 9.11 $\min\{\dim \mathscr{N}_\lambda^+, \dim \mathscr{N}_\lambda^-\} < \infty$, which implies the inequality $\min\{\dim(\mathscr{N}_\lambda^0 [+] \mathscr{N}_\lambda^+), \dim(\mathscr{N}_\lambda^0 [+] \mathscr{N}_\lambda^-)\} < \infty$. ■

Lemma 5.7: *Let T (λ**H**) be a J-bi-non-contractive operator. If T has no neutral eigenvectors, then it has a single pair $\mathscr{L}^+ \in \mathscr{M}^+$ and $\mathscr{L}^- \in \mathscr{M}^-$ of* *invariant subspaces and they are uniformly definite,* $\mathcal{H} = \mathcal{L}^+ \dotplus \mathcal{L}^-$. *Furthermore, if T is a J-semi-unitary or J-unitary operator, then* $\mathcal{L}^- = \mathcal{L}^{+[\perp]}$ *and* $\mathcal{H} = \mathcal{L}^+ [+] \mathcal{L}^-$.

□ Let $\mathcal{L}^{\pm}$ $(\in \mathcal{M}^{\pm})$ be invariant subspaces of the operator T. By definition 5.1 $\mathcal{L}^{\pm} \in h^{\pm}$, and therefore its isotropic part $\mathcal{L}_0^{\pm}$ is finite-dimensional and invariant relative to T. Therefore if T has no neutral eigenvectors, then $\mathcal{L}_0^{\pm} = \{\theta\}$, i.e., the subspaces $\mathcal{L}^{\pm}$ are uniformly definite. By Theorem 3.14 and 1.Corollary 8.16 it follows that the $\mathcal{L}^{\pm}$ are unique. Therefore if, in addition, T is a J-semi-unitary operator, then $T\mathcal{L}^+ = \mathcal{L}^+$ implies $T^c\mathcal{L}^+ = \mathcal{L}^+$, and so $T\mathcal{L}^{+[\perp]} \subset \mathcal{L}^{+[\perp]}$. Since the pair $\mathcal{L}^+$ and $\mathcal{L}^-$ is unique we obtain $\mathcal{L}^- = \mathcal{L}^{+[\perp]}$. It follows from 1.Corollary 8.6 that $\mathcal{H} = \mathcal{L}^+ \dotplus \mathcal{L}^-$ or, if $\mathcal{L}^- = \mathcal{L}^{+[\perp]}$, then $\mathcal{H} = \mathcal{L}^+ [\dotplus] \mathcal{L}^-$. ■

Lemma 5.8: *For J-bi-non-contractive operators membership of the class* **H** *is a* Φ*-invariant property.*

□ Let V be a J-bi-non-contractive operator, let $V \in \mathbf{H}$, and let $\mathcal{L}_0$ be a neutral invariant subspace of V. By Corollary 5.4 $\dim \mathcal{L}_0 < \infty$, and therefore by virtue of 1.Corollary4.14 $V\mathcal{L}_0 = \mathcal{L}_0$. It follows from the result of Exercise 20 on 2.§4 that the operator $\hat{V}$ induced by the operator V in the factor-space $\hat{\mathcal{H}} = \mathcal{L}_0^{[\perp]}/\mathcal{L}_0$ is J-bi-non-contractive. It is necessary to prove that $\hat{V} \in \mathbf{H}$. By Corollary 5.3 there are invariant subspaces $\mathcal{L}_{\pm} \in \mathcal{M}^{\pm} \cap h^{\pm}$ of the operator V which contain $\mathcal{L}_0$. But then $\hat{\mathcal{L}}_{\pm} = \mathcal{L}_{\pm}/\mathcal{L}_0 \in \mathcal{M}^{\pm}(\hat{\mathcal{H}}) \cap h^{\pm}$ (*cf.* Exercises 21, 22 on 1.§8), and $\hat{\mathcal{L}}_{\pm}$ are invariant relative to $\hat{V}$. Let $\hat{\mathcal{N}}_{\pm} \in \mathcal{M}^{\pm}(\mathcal{H})$ be arbitrary invariant subspaces of the operator $\hat{V}$. Then $\mathcal{N}_{\pm} = \{x_{\pm} \in \hat{x}_{\pm} \mid \hat{x}_{\pm} \in \mathcal{N}_{\pm}\}$ are maximal semi-definite invariant subspaces of the operator V, and therefore $\mathcal{N}_{\pm} \in h^{\pm}$, which implies that $\hat{\mathcal{N}}_{\pm}$ $(= \mathcal{N}_{\pm}/\mathcal{L}_0)$ are members of the classes $h^{\pm}$ respectively. By Definition 5.1 then, $\hat{V} \in \mathbf{H}$. ■

□ Let $\mathcal{L}$ be a uniformly definite invariant subspace of a J-bi-non-contractive operator V, and let $V\mathcal{L} = \mathcal{L}$ if $\mathcal{L} \subset \mathcal{P}^+$. Without loss of generality we suppose that $\mathcal{L} \subset \mathcal{H}^+$ or $\mathcal{L} \subset \mathcal{H}^-$ depending on the sign of $\mathcal{L}$. Then $\mathcal{H} = \mathcal{L} [\oplus] \mathcal{L}^{[\perp]}$ and relative to this decomposition the operators V and J can be expressed in matrix form:

$$V = \| V_{ij} \|_{i,j=1}^{2}, \qquad V_{21} = 0, \qquad 0 \in \rho(V_{11}),$$

and

$$J = \| J_{ij} \|_{i,j=1}^{2}, \qquad J_{21} = 0, \qquad J_{12} = 0, \qquad J_{22} = J^* = J_{22}^{-1}$$

and

$J_{11} = I \mid \mathcal{L}$ if $\mathcal{L} \subset \mathcal{P}^+$

or

$$J_{11} = -I \mid \mathcal{L} \text{ if } \mathcal{L} \subset \mathcal{P}^-;$$

and

$$J_{22} = J^* = J_{22}^{-1}.$$

If $\mathscr{L} \in \mathscr{P}^-$, then it follows from Exercise 34 on 2.§4 that V_{22} is a J-bi-non-contractive operator.

Suppose

$$\mathscr{L} \subset \mathscr{P}^+, x_2 \in \mathscr{L}^{[\perp]}, \quad \text{and} \quad x_1 = -V^{-1}V_{12}x_2.$$

Then

$$\begin{aligned}[V_{22}x_2, V_{22}x_2] &= [V(x_1 + x_2), V(x_1 + x_2)] \geqslant [x_1 + x_2, x_1 + x_2] \\ &= [x_1, x_1] + [x_2, x_2] \geqslant [x_2, x_2],\end{aligned}$$

i.e., V_{22} is a J_{22}-non-contractive operator. Since $\mathscr{L}^{[\perp]}$ is invariant relative to the J-non-contractive operator V^c it follows that $(V_{22})^c$ $(=(V^c)_{22})$ is a J_{22}-non-contractive operator, i.e., V_{22} is a J_{22}-bi-non-contractive operator. Thus, in both cases—whether $\mathscr{L} \subset \mathscr{P}^+$ or $\mathscr{L} \subset \mathscr{P}^-$—the operator V_{22} is J-bi-non-contractive.

Now suppose, moreover, that $V \in \mathbf{H}$. We verify that then $V_{22} \in \mathbf{H}$ also. Suppose, for example, that $\mathscr{L} \subset \mathscr{P}^+$, and that $\mathscr{L}^+$ $(\in \mathscr{M}^+)$, existing by virtue of Corollary 5.3, is an invariant subspace of the operator V containing $\mathscr{L}$, and that $\mathscr{L}^+ = \mathscr{L} [\dot{+}] \mathscr{L}_2^+$. $\mathscr{L}$ is completely invariant relative to V, so $V_{22}\mathscr{L}_2 = \mathscr{L}_2^+$. It follows from $\mathscr{L}^+ \in \mathscr{M}^+ \cap h^+$ that $\mathscr{L}_2^+ \in \mathscr{M}^+(\mathscr{L}^{[\perp]}) \cap h^+$, i.e., the operator V_{22} has at least one invariant subspace from $\mathscr{M}^+(\mathscr{L}^{[\perp]}) \cap h^+$. We now assume that $\mathscr{N}_2^+ \in \mathscr{M}^+(\mathscr{L}^{[\perp]})$ and $V_{22}\mathscr{N}_2^+ = \mathscr{N}_2^+$, and $\mathscr{N} = \text{Lin}\{\mathscr{L}_1 \mathscr{N}_2^+\}$. Then $\mathscr{N} \in \mathscr{M}^+(\mathscr{H})$ and $V\mathscr{N} = \mathscr{N}$. By Definition 5.1 $\mathscr{N} \in h^+$ and so $\mathscr{N}_2^+ \in h^+$ also, i.e., all maximal non-negative invariant subspaces of the operator V_{22} belong to the class h^+.

By Theorem 5.2 $V_{22} \in \Lambda_+$ and therefore V_{22} has at least one maximal non-positive invariant subspace (in $\mathscr{L}^{[\perp]}$ and therefore also in $\mathscr{H}$). Let $\mathscr{L}^-$ be such a subspace. Then $\mathscr{L}_+ = \mathscr{L}^{-[\perp]} \cap \mathscr{L}^{[\perp]}$ $(\in \mathscr{M}^+\mathscr{L}^{[\perp]})$ is an invariant subspace of the operator V^c. Hence $\mathscr{L}_+ \in h^+$, which implies (see Exercise 12 on 1.§8) that $\mathscr{L}^- \in h^-$, i.e., $V_{22} \in \mathbf{H}$.

A similar assertion in the case $\mathscr{L} \subset \mathscr{P}^-$ is proved in the same way.

Thus we have proved

Lemma 5.9: *Let $\mathscr{L}$ be a uniformly definite invariant subspace of a J-bi-non-contractive operator V, with $V\mathscr{L} = \mathscr{L}$ if $\mathscr{L} \in \mathscr{P}^+$, let $P_{\mathscr{L}}$ be the J-orthoprojector from $\mathscr{H}$ on to $\mathscr{L}$, and let $V_{22} = (I - P_{\mathscr{L}})V(I - P_{\mathscr{L}}) \mid (I - P_{\mathscr{L}})\mathscr{H}$. Then V_{22} is an $(I - P_{\mathscr{L}})^*J(I - P_{\mathscr{L}})$-bi-non-contractive operator, and if $V \in \mathbf{H}$, then $V_{22} \in \mathbf{H}$ also.* ■

2 *Definition 5.10:* We shall say that a family of operators $\mathscr{X} = \{X\}$ which commute with a J-bi-non-contractive operator V_0 of the class $\mathbf{H}$ (i.e.,

$XV_0 \supset V_0X$), with $\rho(X) \cap \mathbb{C}^+ \neq \emptyset$ for every $X \in \mathscr{X}$ *belongs to the class* **K(H)** and we shall write $\mathscr{X} \in \mathbf{K}(\mathbf{H})$.

We remark that $\mathscr{X}$ may consist of a single operator X and we shall then say that the operator X belongs to the class **K(H)** ($X \in \mathbf{K}(\mathbf{K})$).

As well as the notation $\mathscr{X} \in \mathbf{K}(\mathbf{H})$ we shall also use the symbols $\mathscr{X} \in \mathbf{K}(\mathbf{H}, V_0)$ to show precisely with which operator of the class **H** it is that $\mathscr{X}$ commutes.

Lemma 5.11: *Let* $\mathscr{V} = \{V\}$ *be a commutative family of J-bi-non-contractive operators with* $\mathscr{V} \in \mathbf{K}(\mathbf{H}, V_0)$. *Then when* $J \neq I$ (respectively, $J \neq -I$) *the family* $\mathscr{V}_0 = \mathscr{V} \cup V_0$ *has a non-trivial completely invariant non-negative subspace* $\mathscr{L}_+$ (*respectively, invariant non-positive subspace* $\mathscr{L}_-$).

□ We consider first the case when the operator V_0 has no neutral eigenvectors. Then, in accordance with Lemma 5.7, there is a single pair $\mathscr{L}^+ \in \mathscr{M}^+$ and $\mathscr{L}^- \in \mathscr{M}^-$ of subspaces invariant relative to V_0, and $V_0\mathscr{L}^+ = \mathscr{L}^+$. It follows from the uniqueness of $\mathscr{L}^+$ and $\mathscr{L}^-$ that $V\mathscr{L}^+ = \mathscr{L}^+$ and $V\mathscr{L}^- \subset \mathscr{L}^-$ for any $V \in \mathscr{V}$, i.e., $\mathscr{V}_0\mathscr{L}^+ = \mathscr{L}^+$, $\mathscr{V}_0\mathscr{L}^- \subset \mathscr{L}^-$.

We put $\mathscr{L}_\pm = \mathscr{L}^\pm$.

We now assume that the operator V_0 has at least one neutral eigenvector x_0: $V_0x_0 = \lambda_0x_0$. Then $\operatorname{Ker}(V_0 - \lambda_0I)$ is an invariant subspace of the family $\mathscr{V}$. If $\operatorname{Ker}(V_0 - \lambda_0I)$ is degenerate, then, in accordance with Corollary 5.6, its isotropic part $\mathscr{N}^0_{\lambda_0}$ is finite-dimensional. We consider the three possible cases: a) $|\lambda_0| < 1$; b) $|\lambda_0| > 1$; c) $|\lambda_0| = 1$.

a) $|\lambda_0| < 1$. In accordance with Exercise 6 on 2.§6, $\operatorname{Ker}(V_0 - \lambda_0I)$ is a non-positive subspace, and therefore (see Exercise 17 on 2.§4) $\mathscr{N}^0_{\lambda_0}$ is invariant relative to each of the operators $V \in \mathscr{V}$. Moreover, by 2.Proposition 4.14, $V\mathscr{N}^0_{\lambda_0} = \mathscr{N}^0_{\lambda_0}$, since $\dim \mathscr{N}^0_{\lambda_0} < \infty$. We put $\mathscr{L}_+ = \mathscr{L}_- = \mathscr{N}^0_{\lambda_0}$.

b) $|\lambda_0| > 1$. In accordance with Exercise 5 on 2.§6, $\operatorname{Ker}(V_0 - \lambda_0I)$ is a non-negative subspace. It follows from Exercise 17 on 2.§4 that $\mathscr{N}^0_{\lambda_0}$ is the isotropic part of $\operatorname{Ker}[V_0^c - (1/\bar{\lambda}_0)I]$ and therefore $V^c\mathscr{N}^0_{\lambda_0} = \mathscr{N}^0_{\lambda_0}$ for all $V \in \mathscr{V}$. We again use Exercise 17 on 2.§4 and we obtain that $V\mathscr{N}^0_{\lambda_0} = \mathscr{N}^0_{\lambda_0}$ for all $V \in \mathscr{V}$. We again put $\mathscr{L}_+ = \mathscr{L}_- = \mathscr{N}^0_{\lambda_0}$.

c) $|\lambda_0| = 1$. In accordance with Exercise 19 on 2.§4, $\operatorname{Ker}(V_0 - \lambda_0I) = \operatorname{Ker}(V_0^c - \bar{\lambda}_0I)$, and therefore $\operatorname{Ker}(V_0 - \lambda_0I)$ is invariant relative to both V and V^c for all $V \in \mathscr{V}$. Consequently, if $y_0 \in \mathscr{N}^0_{\lambda_0}$, then $[Vy_0, x] = [y_0, V_0^cx] = 0$ for all $x \in \operatorname{Ker}(V_0 - \lambda_0I)$, i.e., $V\mathscr{N}^0_{\lambda_0}$ is a subspace in $\mathscr{N}^0_{\lambda_0}$. From 2.Proposition 4.14 and the fact that $\mathscr{N}^0_{\lambda_0}$ is finite-dimensional, we obtain that $V\mathscr{N}^0_{\lambda_0} = \mathscr{N}^0_{\lambda_0}$ for all $V \in \mathscr{V}$. Again we put $\mathscr{L}_+ = \mathscr{L}_- = \mathscr{N}^0_{\lambda_0}$.

We now assume that $\operatorname{Ker}(V_0 - \lambda_0I)$ is non-degenerate. In accordance with Corollary 5.6 $\operatorname{Ker}(V_0 - \lambda_0I)$ is a Pontryagin space. From the consideration of the cases a)–c) we see that $|\lambda_0| = 1$, and the family $\mathscr{V}$ induces in Ker $(V_0 - \lambda_0I)$

bi-non-contractive operators relative to the form $[\cdot,\cdot]\,|\,\mathrm{Ker}(V_0-\lambda_0 I)$. By virtue of Exercise 2 on §4 we obtain that these operators have common non-trivial semi-definite invariant subspaces $\mathscr{L}_+$ and $\mathscr{L}_-$, and that $(V\,|\,\mathrm{Ker}(V_0-\lambda_0 I))\mathscr{L}_+ = \mathscr{L}_+$. ■

Theorem 5.12: *Let a commutative family $\mathscr{V}=\{V\}$ of J-bi-non-contractive operators belong to the class* $\mathbf{K}(\mathbf{H}, V_0)$. *Then the family $\mathscr{V}_0=\mathscr{V}\cup V_0$ has the property $\tilde{\Phi}$.*

□ Since the property {*a family of J-bi-non-contractive operators is commutative*} is Φ-invariant, it follows from Lemma 5.8 that the property {$\mathscr{V}_0$ belongs to the class $\mathbf{K}(\mathbf{H})$} is also Φ-invariant. It therefore suffices to prove that every maximal completely invariant subspace $\mathscr{L}_+\subset\mathscr{P}^{++}\cup\{\theta\}$ and every maximal invariant subspace $\mathscr{L}_-\subset\mathscr{P}^{--}\cup\{\theta\}$ of the family $\mathscr{V}_0$ belong to $\mathscr{M}^+$ and $\mathscr{M}^-$ respectively. Let $\mathscr{L}$ be one of such subspaces. By Corollary 5.4 $\mathscr{L}$ is uniformly definite, and therefore $\mathscr{H}=\mathscr{L}\,[\dot{+}]\,\mathscr{L}^{[\perp]}$. let $P_{\mathscr{L}}$ be the orthoprojector from $\mathscr{H}$ on to $\mathscr{L}$. We consider the family $\mathscr{V}_{02}=\{V_{22}\}\cup(V_0)_{22}$ of $(I-P_{\mathscr{L}})^*J(I-P_{\mathscr{L}})$-bi-non-contractive operators of the class $\mathbf{K}(\mathbf{H},(V_0)_{22})$ (see Lemma 5.9), where

$$V_{22}=(I-P_{\mathscr{L}})V(I-P_{\mathscr{L}})\,|\,\mathscr{L}^{[\perp]},\qquad (V_0)_{22}=(I-P_{\mathscr{L}})V_0(I-P_{\mathscr{L}})\,|\,\mathscr{L}^{[\perp]}.$$

From the maximality of $\mathscr{L}$ and Lemma 5.1 applied to the family $\mathscr{V}_{02}$ we conclude that $\mathscr{L}^{[\perp]}$ has the sign opposite to that of $\mathscr{L}$. In accordance with 1.Proposition 1.25 $\mathscr{L}$ is a maximal semi-definite subspace. ■

□ **Corollary 5.13:** *A J-bi-non-contractive operator V of the class* $\mathbf{K}(\mathbf{H})$ *belongs to $\Lambda_+\cap\Lambda_-$ and has the property $\tilde{\Phi}$.*

This follows immediately from Theorems 5.12 and 5.2. ■

3 Let a J-bi-non-contractive operator V belong to the class $\mathbf{K}(\mathbf{H}, V_0)$, $\mathscr{L}_0$ be the maximal invariant neutral subspace of the operators V_0 and V, and let $\mathscr{L}^{\pm}(\in\mathscr{M}^{\pm})$ be invariant subspaces of these operators containing $\mathscr{L}_0$. Let $\mathscr{L}^+=\mathscr{L}_0\,[\oplus]\,\mathscr{L}_+$, $\mathscr{L}^-=\mathscr{L}_0\,[\oplus]\,\mathscr{L}_-$ be the decompositions of the subspaces $\mathscr{L}^{\pm}$ into the sum of a neutral subspace $\mathscr{L}_0$ and uniformly definite subspaces $\mathscr{L}_{\pm}$. Since $\mathscr{L}^{\pm}\in\mathscr{M}^{\pm}\cap h^{\pm}$ we have $\mathscr{L}_0^{[\perp]}=\mathrm{Lin}\{\mathscr{L}^+,\mathscr{L}^-\}$, and therefore, by virtue of the formula 1.(7.4) and 1.Corollary 8.16

$$\mathscr{H}=\mathscr{L}_0\oplus(\mathscr{L}_+\dot{+}\mathscr{L}_-)\oplus J\mathscr{L}_0. \tag{5.2}$$

Relative to this decomposition the operator V is represented by the matrix

$$V=\begin{Vmatrix} V_{11} & V_{12} & V_{13} & V_{14}\\ 0 & V_{22} & 0 & V_{24}\\ 0 & 0 & V_{33} & V_{34}\\ 0 & 0 & 0 & V_{44}\end{Vmatrix} \tag{5.3}$$

We mention that in a number of cases later (for example, in the proof of Proposition 5.14) we shall make use of the condition $V \in \mathbf{K}(\mathbf{H})$ in order to have the expansion (5.3), and therefore it would be possible to demand the existence of the expansion (5.3) instead of the condition $V \in \mathbf{K}(\mathbf{H})$. However, in order to avoid awkwardness of presentation we shall not do this.

5.14 Let J-bi-non-contractive operator V belong to $\mathbf{K}(\mathbf{H})$, *and let* $\varepsilon \in \sigma_p(V) \cap \mathbb{T}$. *Then the root lineal* $\mathscr{L}_\varepsilon(V)$ *is a subspace, and it splits up into a J-orthogonal sum* $\mathscr{L}_\varepsilon(V) = \mathscr{N}_\varepsilon [+] \mathscr{M}_\varepsilon$, *where* $\dim \mathscr{N}_\varepsilon < \infty, V\mathscr{N}_\varepsilon = \mathscr{N}_\varepsilon$, $\mathscr{M}_\varepsilon \subset \mathrm{Ker}(V - \varepsilon I)$ and $\mathscr{M}_\varepsilon$ is a projectionally complete subspace or, in particular, $\mathscr{M}_\varepsilon = \{\theta\}$.

□ We consider the decomposition (5.3) of the operator V. The subspace $\mathscr{M}_\varepsilon \equiv \overline{\mathscr{L}_\varepsilon(V) \cap \mathrm{Ker}\ V_{12}} \dotplus \overline{\mathscr{L}_\varepsilon(V) \cap \mathrm{Ker}\ V_{13}}$ is regular and moreover $\mathscr{L}_\varepsilon(V) \cap \mathrm{Ker}\, V_{1j} \subset \mathscr{L}_\varepsilon(V_{ii})$, $i = 1, 2, 3$. The operator V_{22} is non-contractive relative to the scalar product $[x, y]$ $(x, y \in \mathscr{L}_+)$, and V_{33} is non-expansive relative to the scalar product $-[x, y]$ $(x, y \in \mathscr{L}_-)$. Therefore (*cf.* Exercise 23 on 2.§4) $\mathscr{L}_\varepsilon(V_{ii}) = \mathrm{Ker}(V_{ii} - \varepsilon I_i)$, $i = 2, 3$, which is equivalent in the present case to the inclusion $\mathscr{M}_\varepsilon \subset \mathrm{Ker}(V - \varepsilon I)$.

Since by virtue of Exercise 19 on 2.§4 $\mathscr{M}_\varepsilon \subset \mathrm{Ker}(V^c - \varepsilon I)$, and $\mathscr{L}_\rho(V) = \mathscr{N}_\varepsilon [+] \mathscr{M}_\varepsilon$, where $\mathscr{N}_\varepsilon = \mathscr{M}_2^{[\perp]} \cap \mathscr{L}_\varepsilon(V)$, it follows that $\mathscr{N}_\varepsilon$ is invariant relative to the operator V. By construction $\mathscr{N}_\varepsilon [\perp]\ \mathscr{L}_\varepsilon(V) \cap \mathrm{Ker}\ V_{1j}$ $(i = 2, 3)$. It remains to verify that $\mathscr{N}_\varepsilon$ is finite-dimensional, from which it will also follow that it is completely invariant. We assume the contrary: $\mathscr{N}_\varepsilon$ is an infinite-dimensional subspace, and $\mathscr{L}_0$, $\mathscr{L}_+$ and $\mathscr{L}_-$ are the same as in the decomposition (5.2). Since $\dim(\mathscr{L}_0 \oplus J\mathscr{L}_0) < \infty$, so $\mathscr{N}_\varepsilon \cap (\mathscr{L}_+ \dotplus \mathscr{L}_-)$ is an infinite-dimensional subspace, and therefore at least one of the subspaces $\mathscr{N}_\varepsilon \cap \mathscr{L}_+$ or $\mathscr{N}_\varepsilon \cap \mathscr{L}_-$ will be infinite-dimensional. Suppose, for example, that $\dim \mathscr{N}_\varepsilon \cap \mathscr{L}_+ = \infty$. Since $\dim \mathscr{L}_+/\mathrm{Ker}\ V_{12} < \infty$, it follows that $\mathscr{N}_\varepsilon \cap \mathrm{Ker}\ V_{12} \neq \{\theta\}$, and we obtain a contradiction with the fact that $\mathscr{N}_\varepsilon [\perp]\ \mathscr{L}_\varepsilon(V) \cap \mathrm{Ker}\ V_{12}$. ■

□ From Corollary 5.6 and Proposition 5.14 we obtain

Corollary 5.15: *It under the conditions of Proposition* 5.14 *the operator V belongs to* **H**, *then, when* $\dim \mathscr{L}_\varepsilon(V) = \infty$, $\mathscr{M}_\varepsilon$ *is a Pontryagin space with a finite number of either negative or positive squares.* ■

4 We now consider a family $\mathscr{A} = \{A\}$ of maximal *J*-dissipative operators of the class $\mathbf{K}(\mathbf{H})$ and we investigate for it the questions of the existence of maximal semi-definite invariant subspaces.

Theorem 5.16: *Let* $\mathscr{A} = \{A\}$ *be a set of maximal J-dissipative operators, let* $\mathscr{A} \in \mathbf{K}(\mathbf{H}, V_0)$, *and let* $\mathscr{L}_\pm$ $(\subset \mathscr{P}^\pm)$ *be invariant subspaces of the family* $\mathscr{A} \cup V_0$ *with* $\mathscr{L}_+$ *completely invariant relative to* V_0; *and suppose there is a* *point* $\lambda_A \in \rho(A) \cap \mathbb{C}^+$ *such that* $\lambda_A, \bar{\lambda}_A \in \rho(A \mid \mathscr{L}_+)$ *and* $\lambda_A \in \rho(A \mid \mathscr{L}_-)$. *Then, if the resolvents of the operators* $A \in \mathscr{A}$ *commute pairwise, there are extensions* $\tilde{\mathscr{L}}_\pm \in \mathscr{M}^\pm$ *of the subspaces* $\mathscr{L}_\pm$ *which have the same properties.*

□ We consider the family of operators $\mathscr{V} = \{V = \mathbf{K}_\lambda(A) \mid A \in \mathscr{A}\}$. In accordance with 2.Theorem 6.13 the V are J-bi-non-contractive operators. It follows from the inclusion $\mathscr{A} \in \mathbf{K}(\mathbf{H}, V_0)$ that $\mathscr{V} \in \mathbf{K}(\mathbf{H}, V_0)$. Since $\lambda_A, \bar{\lambda}_A \in \rho(A \mid \mathscr{L}_+)$, the subspace $\mathscr{L}_+$ is completely invariant relative to $\mathscr{V}$, and the condition $\lambda_A \in \rho(A \mid \mathscr{L}_-)$ implies the invariance of $\mathscr{L}_-$ relative to $\mathscr{V}$. Consequently, by virtue of Theorem 5.12, there are $\tilde{\mathscr{L}}_\pm \in \mathscr{M}^\pm \cap h^\pm$ containing $\mathscr{L}_\pm$ and such that $V\tilde{\mathscr{L}}_+ = \tilde{\mathscr{L}}_+$, $V\tilde{\mathscr{L}}_- \subset \tilde{\mathscr{L}}_-$ for all $V \in \mathscr{V}_0 \equiv \mathscr{V} \cup V_0$. We now prove that $\tilde{\mathscr{L}}_\pm$ are invariant relative to $\mathscr{A}$, and that $\lambda_A, \bar{\lambda}_A \in \rho(A \mid \tilde{\mathscr{L}}_+)$ and $\lambda_A \in \rho(A \mid \tilde{\mathscr{L}}_-)$. To do this we note that $\mathscr{L}_0 = \tilde{\mathscr{L}}_+ \cap \tilde{\mathscr{L}}_-$ is an invariant finite-dimensional neutral subspace of the family $\mathscr{V}$, and therefore, by virtue of Exercise 13 on 2.§6, $\mathscr{L}_0 \subset \{\cap \mathscr{D}_A \mid A \in \mathscr{A}\}$ and $\mathscr{A}\mathscr{L}_0 \subset \mathscr{L}_0$. We bring into consideration the $\hat{J}$-space $\hat{\mathscr{H}} = \mathscr{L}_0^{[\perp]}/\mathscr{L}_0$ and the families of operators acting in it $\hat{\mathscr{A}} = \{\hat{A}\}$ and $\hat{\mathscr{V}}_0 = \{\hat{V}\} \cup \hat{V}_0$ generated by the families $\mathscr{A}$ and $\mathscr{V}_0$. Moreover, $\mathscr{D}_{\hat{A}} = (\mathscr{D}_A \cap \mathscr{L}_0^{[\perp]})/\mathscr{L}_0$, $\lambda_A \in \rho(\hat{A})$ and $\widehat{\mathbf{K}_{\lambda_A}(A)} = \mathbf{K}_{\lambda_A}(\hat{A})$. We note that because $\tilde{\mathscr{L}}_\pm$ are maximal semi-definite the equality $(\tilde{\mathscr{L}}_+ + \tilde{\mathscr{L}}_-)^{[\perp]} = \mathscr{L}_0$ follows, and since $\tilde{\mathscr{L}}_\pm \in h^\pm$, so (see Exercise 19 on 1.§8) $\mathscr{L}_0^{[\perp]} = \tilde{\mathscr{L}}_+ + \tilde{\mathscr{L}}_-$, and therefore $\hat{\mathscr{H}} = \tilde{\mathscr{L}}_+/\mathscr{L}_0 \dotplus \tilde{\mathscr{L}}_-/\mathscr{L}_0$. Relative to this decomposition of $\hat{\mathscr{H}}$ the operators $\hat{V} \in \hat{\mathscr{V}}_0$ have the diagonal form

$$\hat{Y} = \left\| \begin{matrix} \hat{V}_{11} & 0 \\ 0 & \hat{V}_{22} \end{matrix} \right\|,$$

and therefore

$$\hat{A} = \mathbf{K}_{\bar{\lambda}_A}^{-1}(\hat{V}) = \left\| \begin{matrix} \mathbf{K}_{\bar{\lambda}_A}^{-1}(\hat{V}_{11}) & 0 \\ 0 & \mathbf{K}_{\bar{\lambda}_A}^{-1}(\hat{V}_{22}) \end{matrix} \right\|,$$

which implies that $\tilde{\mathscr{L}}_\pm/\mathscr{L}_0$ are invariant relative to $\hat{\mathscr{A}}$ and that λ_A and λ_A' belong to $\rho(A \mid \tilde{\mathscr{L}}_+/\mathscr{L}_0)$ and λ_A belongs to $\rho(\hat{A} \mid \tilde{\mathscr{L}}_-/\mathscr{L}_0)$. It only remains to use the fact that $\lambda_A, \lambda_A \in \rho(A \mid \mathscr{L}_0)$. ■

This theorem enables us to prove the following assertion which will be useful later.

Theorem 5.17: *Let* $A \in \mathbf{K}(\mathbf{H}, V_0)$ *be a maximal J-dissipative operator, let* $\lambda = \bar{\lambda} \in \rho_p(A)$, *let* $\mathscr{H} = \mathrm{Ker}(A - \lambda)/\mathrm{Ker}_0(A - \lambda I)$ *be a* $\hat{J}$*-space, where* $\mathrm{Ker}_0(A - \lambda I)$ *is the isotropic part of the subspace* $\mathrm{Ker}(A - \lambda I)$, *and let* $\hat{V}_0$ *be the* $\hat{J}$*-bi-non-contractive operator induced in* $\hat{\mathscr{H}}$ *by the operator* V_0. *Then* $\hat{V}_0 \in \mathbf{H}$.

□ In accordance with 2.Corollary 2.17 $\mathrm{Ker}(A - \lambda I) = \mathrm{Ker}(A^c - \lambda I)$, and therefore $\mathrm{Ker}(A - \lambda I)$ is a common invariant subspace of the operators V_0 and V_0^c. Hence, it follows, in particular, that $\mathrm{Ker}_0(A - \alpha I)$ also is their common completely invariant subspace. We conclude from 2.Theorem 1.17 that $\hat{V}_0$ is a $\hat{J}$-bi-non-contractive operator. We now verify that $\hat{V}_0 \in \mathbf{H}$. Since

$V_0 \in \mathbf{H}$ it follows that it is sufficient to find at least one maximal non-negative invariant subspace of the operator $\hat{V}_0$. Let $\mathscr{L}^+$ be the common maximal non-negative invariant subspace of the operators A and V_0 (we know that there is such a subspace from Theorem 5.16). We consider the subspace $\mathscr{L}_\lambda = \mathscr{L}^+ \cap \operatorname{Ker}(A - \lambda I)$. We first assume, and later prove, that the deficiency (see 2.Theorem 4.15) def $\hat{\mathscr{L}}_\lambda$ of the subspace
$\hat{\mathscr{L}}_\lambda = \operatorname{Lin}\{\mathscr{L}_\lambda, \operatorname{Ker}_0(A - \lambda I)\}/\operatorname{Ker}_0(A - \lambda I)$ is finite. By virtue of Exercise 4 on §4 the operator $\hat{V}_0$ has at least one maximal non-negative invariant subspace, and therefore, as proved above, $\hat{V}_0 \in \mathbf{H}$.

It remains to verify that def $\hat{\mathscr{L}}_\lambda < \infty$. We consider the decomposition of the root subspace $\mathscr{L}_\lambda(A) = \mathscr{N}_\lambda [\dot{+}] \mathscr{M}_\lambda$, where $\mathscr{N}_\lambda$ and $\mathscr{M}_\lambda$ are invariant relative to A, $\dim \mathscr{N}_\lambda < \infty$, and $\mathscr{M}_\lambda \subset \operatorname{Ker}(A - \lambda I)$. The existence of such a decomposition follows from Proposition 5.14 and 2.Theorem 6.13. We assume that def $\hat{\mathscr{L}}_\lambda = \infty$. Since $\dim \mathscr{N}_\lambda < \infty$, there is an $x_0 \in \operatorname{Ker}(A - \lambda I)$ which is J-orthogonal to $\mathscr{L}_\lambda(A) \cap \mathscr{L}^+$. In accordance with 2.Corollary 2.18 x_0 is also J-orthogonal to all the subspaces $\mathscr{L}_\mu(A) \cap \mathscr{L}^+$. Let $\mathscr{L}_0^+$ be the isotropic part of the subspace $\mathscr{L}^+$. Since $\dim \mathscr{L}_0^+ < \infty$, we have $\mathscr{L}_0^+ \subset \operatorname{Lin}\{\lambda_\mu(A) \cap \mathscr{L}^+ \mid \mu \in \sigma(A)\}$, and therefore $x_0 [\perp] \mathscr{L}_0^+$. We consider the $\hat{J}_1$-subspace $\hat{\mathscr{H}}_1 = \operatorname{Lin}\{\mathscr{L}^+, x_0\}/\mathscr{L}_0^+$. Since $\mathscr{L}^+ \in h^+$ and $x_0 \notin \mathscr{L}^+$, it follows that $\hat{\mathscr{H}}_1$ is Pontryagin space with one negative square. Since $\mathscr{L}_0^+$ and $\operatorname{Lin}\{\mathscr{L}^+, x_0\}$ are invariant relative to A it follows that A induces in $\hat{\mathscr{H}}_1$ a J_1-dissipative operator $\hat{A}$. Since $\mathscr{L}_0^+$ is the linear envelope of the isotropic parts of the subspaces $\mathscr{L}_\lambda(A) \cap \mathscr{L}^+ (\mu \in \sigma_p(A))$, the root subspaces of the operator $\hat{A}$ are positive—which contradicts Corollary 2.9 and Theorem 1.13, according to which the operator $\hat{A}$ must have at least one non-positive non-zero eigenvector. ■

Corollary 5.18: *Under the conditions of Theorem* 5.17, *if* $\operatorname{Ker}(A - \lambda I)$ *is not semi-definite, then the operator* $A \mid \operatorname{Ker}(A - \lambda I)$ *has semi-definite invariant subspaces* $\mathscr{L}_\lambda^\pm$ *of both signs, which are maximal in* $\operatorname{Ker}(A - \lambda I)$. *Moreover, for all* $\lambda = \bar{\lambda}$ *except possibly a finite number of points we have* $\mathscr{L}_\lambda^+ + \mathscr{L}_\lambda^- = \operatorname{Ker}(A - \lambda I)$, *and for the remaining* $\lambda \in \mathbb{R}$ *we have* $\dim\text{-}\operatorname{Ker}(A - \lambda I)/\mathscr{L}_\lambda^+ + \mathscr{L}_\lambda^- < \infty$.

□ Let $\operatorname{Ker}(A - \lambda I)$ be not semi-definite, let $\hat{\mathscr{L}}_\lambda^\pm$ be the semi-definite invariant subspaces, maximal in $\hat{\mathscr{H}}$, of an operator $\hat{V}_0$ belonging to the class **H**, and let $\mathscr{L}_\lambda^\pm$ be their generating subspaces in $\operatorname{Ker}(A - \lambda I)$. The maximality of $\hat{\mathscr{L}}_\lambda^\pm$ in $\hat{\mathscr{H}}$ implies the maximality of $\mathscr{L}_\lambda^\pm$ in $\operatorname{Ker}(A - \lambda I)$, and the invariance of $\hat{\mathscr{L}}_\lambda^\pm$ relative to $\hat{V}_0$ implies the invariance of $\mathscr{L}_\lambda^\pm$ relative to V_0. Since $\mathscr{L}_\lambda^\pm \in h^\pm$, it follows (see Exercise 19 on I.§8) that $\dim \hat{\mathscr{H}}/\hat{\mathscr{L}}_\lambda^+ + \hat{\mathscr{L}}_\lambda^- = \dim \hat{\mathscr{L}}_\lambda^{\mp} \cap \hat{\mathscr{L}}_\lambda^- < \infty$, which in turn implies that
$(\dim \hat{\mathscr{H}}/\hat{\mathscr{L}}_\lambda^+ + \hat{\mathscr{L}}_\lambda^- =) \dim \operatorname{Ker}(A - \lambda I)/\mathscr{L}_\lambda^+ + \mathscr{L}_\lambda^- < \infty$.

From the J-orthogonality of $\operatorname{Ker}(A - \lambda I)$ and $\operatorname{Ker}(A - \mu I)$ when $\lambda \neq \mu$, $\lambda, \mu \in \mathbb{R}$, and from their invariance relative to V_0, V_0^c ($\in \mathbf{H}$), it follows that only a finite number of the subspaces $\operatorname{Ker}(A-\lambda I)$ and $\mathscr{L}_\lambda^\pm$ can be degenerate. Moreover, again from the fact that $\mathscr{L}_\lambda^\pm \in h^\pm$ we obtain that

$$\dim \operatorname{Ker}(A-\lambda I)/\mathscr{L}_\lambda^+ + \mathscr{L}_\lambda^- = \dim \mathscr{L}_\lambda^+ \cap \mathscr{L}_\lambda^- - \dim K_0(A-\lambda I),$$

and the equalities

$$\dim \operatorname{Ker}(A-\lambda I)/\mathscr{L}_\lambda^+ + \mathscr{L}_\lambda^- = 0 \quad \text{and} \quad \mathscr{L}_\lambda^+ + \mathscr{L}_\lambda^- = \operatorname{Ker}(A-\lambda I)$$

are equivalent. ■

5 Let $\mathscr{U} = \{U\}$ be a commutative group of J-unitary operators, and $\mathscr{U} \in \mathbf{K}(\mathbf{H}, V_0)$. If V_0 is also a J-unitary operator, then, by carrying out an argument similar to that used in the proof of Theorem 5.12, we can satisfy ourselves that the family $\mathscr{U}_0 = \mathscr{U} \cup V_0$ has the property $\tilde{\Phi}^{[\perp]}$. But if V_0 is not a J-unitary operator, we can at present assert only that the family $\mathscr{U}_0$ has the property $\tilde{\Phi}$. However, in any case the group $\mathscr{U}$ does have the property $\tilde{\Phi}^{[\perp]}$. For, let

$$\mathscr{L}^+ = \mathscr{L}_0 [+] \mathscr{L}_+ \in \mathscr{M}^+ \cap h^+ \qquad (\mathscr{L}_0 \subset \mathscr{P}_0, \mathscr{L}_+ \subset \mathscr{P}^{++} \cup \{\theta\})$$

be an invariant subspace of the group $\mathscr{U}$. Then

$$\mathscr{L}^- = \mathscr{L}^{+[\perp]} = \mathscr{L}_0 [+] \mathscr{L}_- \in \mathscr{M}^- \cap \mathscr{H}^-$$

is also an invariant subspace of this group. Without loss of generality we can suppose that $\mathscr{L}_\pm \subset \mathscr{H}^\pm$ and $\mathscr{L}_0 \perp (\mathscr{L}_+ [\oplus] \mathscr{L}_-)$. Therefore in the decomposition (5.2) of the space $\mathscr{H}$ all the components are pairwise orthogonal to one another and $(\mathscr{L}_0 \oplus J\mathscr{L}_0) [\perp] (\mathscr{L}_+ [\oplus] \mathscr{L}_-)$. Let $\mathscr{N}_\pm$ $(\subset \mathscr{P}^\pm)$ be arbitrary maximal invariant subspaces of the group $\mathscr{U}$, with $\mathscr{N}_+ [\perp] \mathscr{N}_-$, i.e., $(\mathscr{N}_+, \mathscr{N}_-)$ is a dual pair. We verify that $\mathscr{N}_\pm \in \mathscr{M}^\pm$. To do this we consider the subspaces $\mathscr{N}'_\pm = \mathscr{N}_\pm \cap \mathscr{L}_0^{[\perp]}$. This too is a dual pair, invariant relative to $\mathscr{U}$, and $\dim \mathscr{N}^\pm \mid \mathscr{N}'_\pm \leqslant \dim \mathscr{L}_0 < \infty$. Together with $(\mathscr{N}'_+, \mathscr{N}'_-)$, $(\mathscr{N}''_+, \mathscr{N}'_-)$ will also be an invariant dual pair if $\mathscr{N}''_+ = \operatorname{Lin}\{\mathscr{L}_0, \mathscr{N}'_+\}$. We consider the $\hat{J}$-space $\hat{\mathscr{H}} = \mathscr{L}_0^{[\perp]}/\mathscr{L}_0$ and the group $\hat{\mathscr{U}}$ induced in it by $\mathscr{U}$. Since $\mathscr{L}_\pm \subset \mathscr{H}^\pm$, we have $\hat{\mathscr{H}} = \hat{\mathscr{L}}^+ [\oplus] \hat{\mathscr{L}}^-$, $\hat{\mathscr{U}}\hat{\mathscr{L}}^\pm = \hat{\mathscr{L}}^\pm$, and therefore $\hat{\mathscr{U}}$ is a group of operators which are simultaneously $\hat{J}$-unitary and unitary. By Theorem 4.1 $\hat{\mathscr{U}}$ has the property $\tilde{\Phi}^{[\perp]}$, and therefore (see Exercise 7 on 1.§5) there is an $\bar{\mathscr{N}}''_+ \in \mathscr{M}^+$ such that $(\mathscr{N}''_+, \mathscr{N}'_-)$ is a dual pair invariant relative to $\mathscr{U}$, and moreover $\mathscr{N}''_+ \subset \bar{\mathscr{N}}_+$. The subspace $\mathscr{N}_1 = \operatorname{Lin}\{\bar{\mathscr{N}}''_+, \mathscr{N}_+\}$ is invariant relative to $\mathscr{U}$, is J-orthogonal to $\mathscr{N}'$ and contains at least one maximal non-negative subspace $(\bar{\mathscr{N}}_+)$. By construction it is a $W^{(\varkappa)}$-space with $\varkappa$ $(\leqslant \dim \mathscr{L}_0)$ negative squares. Consequently (*cf.* 1.Theorem 10.2 and I.Proposition 8.18) every non-negative subspace which is maximal in $\mathscr{N}_1$ is also maximal non-negative in $\mathscr{H}$.

We verify that there is in $\mathscr{N}_1$ a maximal non-negative subspace $\bar{\mathscr{N}}_+$ containing $\mathscr{N}_+$ and invariant relative to $\mathscr{U}$. To do this it suffices to note that in the $\hat{G}^{(\varkappa)}$-space $\hat{\mathscr{N}}_1 = \mathscr{N}_1/\mathscr{N}_0$ (where $\mathscr{N}_0$ is the isotropic part of $\mathscr{N}_1$) the group $\hat{\mathscr{U}}_1$, induced in $\hat{\mathscr{N}}_1$ by the group $\mathscr{U} \mid \mathscr{N}_1$, has (by Corollary 4.13) a maximal non-negative invariant subspace containing $\mathrm{Lin}(\hat{\mathcal{N}}_+, \mathcal{N}_-)$, and then to make use of the second part of the proof of 1.Theorem 10.2.

Since $\bar{\mathcal{N}}_+ \subset \mathcal{N}'^{[\perp]}$ and $\dim \mathcal{N}_-/\mathcal{N}'_- < \infty$ it follows that $\dim \bar{\mathcal{N}}_+/\bar{\mathcal{N}}_+ \cap \mathcal{N}_-^{[\perp]} < \infty$. Since $(\bar{\mathcal{N}}_+ \cap \mathcal{N}_-^{[\perp]}, \mathcal{N}_-)$ is an invariant dual pair of the group $\mathcal{U}$ and contains $(\mathcal{N}_+, \mathcal{N}_-)$, we have $\bar{\mathcal{N}}_+ \cap \mathcal{N}_-^{[\perp]} = \mathcal{N}_+$ and $\operatorname{def} \mathcal{N}_+ < \infty$. By Theorem 4.14 $\mathcal{N}_\pm \in \mathcal{M}^\pm$.

Thus we have proved

Theorem 5.19: *A commutative group of J-unitary operators belonging to the class* **K(H)** *has the property* $\tilde{\Phi}^{[\perp]}$. ■

6 In this paragraph we describe the structure of the spectra of J-unitary and J-self-adjoint operators of the class **K(H)**.

Theorem 5.20: *Let U be a J-unitary operator and $U \in$* **K(H**, V_0), *and let* $\varkappa_{V_0} = \max\{\dim \mathcal{L}_0 \mid \mathcal{L}_0 \subset \mathcal{P}^0, V_0\mathcal{L}_0 = \mathcal{L}_0\}$. *Then the non-unitary spectrum of the operator U consists of a finite number $\varkappa_1$ ($\geqslant 0$) of pairs of normal eigenvalues disposed as mirror-images relative to the unit circle $\mathbb{T}$, and $\sigma_p(U) \cap \mathbb{T}$ contains a finite number $\varkappa_2$ ($\geqslant 0$) of points to which correspond the non-prime elementary divisors, and $\varkappa_1 + \varkappa_2 \leqslant \varkappa_{V_0}$*

□ By virtue of Corollary 5.13 $U \in \Lambda_+ \cap \Lambda_-$, and from Remark 2.4 and 2.Corollary 5.13 we conclude that the non-unitary spectrum of the operator U consists of normal eigenvalues. We use Exercise 6 on 2.§6 and obtain that $\mathcal{L}_1 \equiv \mathrm{Lin}\{\mathcal{L}_\lambda(U) \mid |\lambda| > \}$ is a neutral lineal. Since $V_0\mathcal{L}_1 \subset \mathcal{L}_1$, we have $\dim \mathcal{L}_1 \equiv \varkappa_1' \leqslant \varkappa_{V_0}$. Since the spectrum of a J-unitary operator is symmetric relative to the unit circle, the first part of the theorem has been proved.

Let $\mathcal{E} = \{\varepsilon \in \mathbb{T} \mid U \mid \mathcal{L}_\varepsilon(U)$ has non-prime elementary divisors$\}$. Then, by virtue of Exercise 14 on 2.§6, $\mathrm{Ker}(U - \varepsilon I)$ is a degenerate subspace; let $\mathcal{L}_\varepsilon^0$ denote its isotropic part. Since $\mathrm{Ker}(U - \varepsilon I)$ is invariant relative to V_0 and V_0^c, we have $V_0\mathcal{L}_\varepsilon^0 = \mathcal{L}_\varepsilon^0$, and therefore $\mathcal{L}_2 \equiv \mathrm{Lin}\{\mathcal{L}_\varepsilon^0 \mid \varepsilon \in \mathcal{E}\}$ is a neutral lineal invariant relative to V_0. Consequently $\dim \mathcal{L}_2 \equiv \varkappa_1' \leqslant \varkappa_{V_0}$, and so $\mathcal{E}$ consists of a finite number $\varkappa_2$ of points and $\varkappa_2 \leqslant \varkappa_2{}'$. Since $\mathrm{Lin}(\mathcal{L}_1, \mathcal{L}_2)$ is also a neutral subspace invariant relative to V_0, we have $\varkappa_1 + \varkappa_2 \leqslant \varkappa_1' + \varkappa_2' \leqslant \varkappa_{V_0}$. ■

Corollary 5.21: *Let A be a J-self-adjoint operator and $A \in$* **K(H**, V_0), *and let $\varkappa_{V_0}$ be the same as in Theorem 5.20. Then the non-real spectrum of the operator A consists of a finite number $\varkappa_1$ ($\geqslant 0$) pairs of normal eigenvalues symmetrically situated relative to $\mathbb{R}$, and $\sigma_p(A) \cap \mathbb{R}$ contains a finite number $\varkappa_2$ ($\geqslant 0$) of points to which correspond to non-prime elementary divisors, and $\varkappa_1 + \varkappa_2 \leqslant \varkappa_{V_0}$.*

□ It follows from the definition of the class **K(H)** that the Cayley–Neyman transform $\mathbf{K}_\lambda(A)$ of an operator A belongs, when $\lambda \in \rho(A)$, to the class

$\mathbf{K}(\mathbf{H}, V_0)$ whenever A does. It remains only to apply Theorem 5.20 and II.Corollary 6.18.

Thus, if $\lambda_1, \bar{\lambda}_1; \lambda_2, \bar{\lambda}_2; \ldots \lambda_{\varkappa_1}, \bar{\lambda}_{\varkappa_1}$ are the different non-real eigenvalues of a J-self-adjoint operator $A \in \mathbf{K}(\mathbf{H})$, then

$$\mathscr{H} = \mathop{[\dot{+}]}_{i=1}^{\varkappa_1} (\mathscr{L}_{\lambda_i}(A) \dotplus \mathscr{L}_{\bar{\lambda}_i}(A)) [\dot{+}] \mathscr{H}', \tag{5.4}$$

where $\mathscr{H}'$ is the J-orthogonal complement of $\Sigma_{i=1}^{\varkappa} [\dot{+}] (\mathscr{L}_{\lambda_i}(A) \dotplus \mathscr{L}_{\bar{\lambda}_i}(A))$, and the operator A is expressed relative to (5.4) by a diagonal matrix which we write symbolically as

$$A = \mathop{[\dot{+}]}_{i=1}^{\varkappa_1} (A_{\lambda_i} \dotplus A_{\bar{\lambda}_i}) [\dot{+}] A', \tag{5.5}$$

where $A_\mu = A \mid \mathscr{L}_\mu(A)$, and $A' = A \mid \mathscr{H}'$, and moreover the spectrum of the operator A' is real, and $\sigma(A_\mu) = \{\mu\}$ $(\mu = \lambda_1, \bar{\lambda}_1; \lambda_2, \bar{\lambda}_2; \ldots; \lambda_{\varkappa_1}, \bar{\lambda}_{\varkappa_1})$.

Similar expansions

$$\mathscr{H} = \mathop{[\dot{+}]}_{i=1}^{\varkappa_1} (\mathscr{L}_{\varepsilon_i}(U) \dotplus \mathscr{L}_{\bar{\varepsilon}_i^{-1}}(U)) [\dot{+}] \mathscr{H}', \tag{5.6}$$

$$U = \mathop{[\dot{+}]}_{i=1}^{\varkappa_1} (U_{\varepsilon_i} + U_{\bar{\varepsilon}_i^{-1}}) [\dot{+}] U' \tag{5.7}$$

also hold for J-unitary operators U of the class $\mathbf{K}(\mathbf{H})$, where $\{\varepsilon_i, \bar{\varepsilon}_i^{-1}\}$ are all the different non-unitary eigenvalues of the operator U, $\sigma(U_\zeta) = \{\zeta\}$, $\zeta \in \{\varepsilon_i, \bar{\varepsilon}_i^{-1}\}$, and the spectrum $\sigma(U')$ is unitary.

In particular, the expansion (5.6) and (5.7) are valid also for J-unitary operators of the class $\mathbf{H}$.

7 In conclusion we give some examples of operators of the classes $\mathbf{H}$ and $\mathbf{K}(\mathbf{H})$.

□ *Example 5.22: Every J-non-contractive or J-bi-non-expansive operator in* $\Pi_\varkappa$ *belongs to the class* $\mathbf{H}$ (*cf.* Definition 5.1 with Corollary 2.9). ■

Example 5.23: Let V *be a J-bi-non-contractive operator and let* $V = V' + V''$, *where* $V'' = \| V_{ij}' \|_{i,j=1}^2$ *is a diagonal operator relative to the decomposition* $\mathscr{H} = \mathscr{H}^+ [\oplus] \mathscr{H}^-$; *let* $\rho(V_{11}') \cap \rho(V_{22}') = \emptyset$, *and* $V'' \in \mathscr{S}_\infty$. *Then* $V \in \mathbf{H}$.

□ We conclude from the representation $V = V' + V''$ that $V \in \Lambda_+ \cap \Lambda_-$. Consequently, by virtue of Theorem 2.8, it has at least one pair $\mathscr{L}^+$ and $\mathscr{L}^-$ of maximal semi-definite invariant subspaces. Let $K^\pm$ be the angular operators of these subspaces. In accordance with Lemma 2.2

$$K^+ V_{11} - V_{22} K^+ = -G_{V''}^+(K^+) \quad \text{and} \quad V_{11}' K^- - K^- V_{22}' = G_{V''}^-(K^-)$$

Since $\rho(V'_{11}) \cap \rho(V'_{22}) = \emptyset$, we obtain from [VI] that these equations can be rewritten in the equivalent form

$$K^{+} = -\frac{1}{4\pi^{2}} \oint_{\Gamma_{V'_{11}}} \oint_{\Gamma_{V'_{22}}} \frac{(V'_{22} - \lambda I^{-})^{-1} G^{+}_{V''}(K^{+})(V_{11} - \mu I^{+})^{+1}}{\lambda^{-} - \mu} \, d\lambda \, d\mu \qquad (5.8)$$

$$K^{-} = -\frac{1}{4\pi^{2}} \oint_{\Gamma_{V'_{11}}} \oint_{\Gamma_{V'_{22}}} \frac{(V'_{11} - \mu I^{+})^{-1} G^{-}_{V''}(K^{-})(V'_{22} - \lambda I^{-})^{-1}}{\mu - \lambda} \, d\lambda \, d\mu \qquad (5.9)$$

where $\Gamma_{V'_{11}}$ and $\Gamma_{V'_{22}}$ are non-intersecting Jordon contours surrounding $\sigma(V'_{11})$ and $\sigma(V'_{22})$ respectively. Since $V'' \in \mathscr{S}_{\infty}$, so $G^{\pm}_{V''}(K^{\pm}) \in \mathscr{S}_{\infty}$, and therefore we obtain from the equations (5.8) and (5.9) that $K^{\pm} \in \mathscr{S}_{\infty}$. It only remains to notice (see Exercise 18 on I.§8) that this is sufficient for the inclusions $\mathscr{L}^{\pm} \in h^{\pm}$.

Example 5.24: An amenable uniformly bounded group $\mathscr{U}$ *of J-unitary operators belongs to the class* **K(H)**.

☐ As the operator V_0 with which all the operators $U \in \mathscr{U}$ will commute we may take (see Exercise 1 below) $V_0 = p_{+} - p_{-}$, where $P_{\pm}$ are the J-orthoprojectors on to the components of the maximal uniformly definite dual pair $(\mathscr{L}_{+}, \mathscr{L}_{-})$ invariant relative to $\mathscr{U}$ (the existence of such a pair is guaranteed by II.Theorem 5.18).

Exercises and problems

1 Verify that the operator $J \in$ **H**.

2 Prove that every uniformly J-bi-expansive operator belongs to the class **H**. *Hint:* Use Theorem 2.1.

3 Prove that a J-unitary operator U belongs to the class **H** if and only if there is a canonical decomposition $\mathscr{H} = \mathscr{H}_1^{+} [+] \mathscr{H}_1^{-}$ such that the angular operators (with respect to this decomposition) of all invariant subspaces $\mathscr{L}^{\pm} \in \mathscr{M}^{\pm}$ of the operator U are completely continuous (Azizov [9]).

4 Prove that, under the conditions of Proposition 5.14, if $V \in$ **K(H)** and if $d_1(\varepsilon)$, $d_2(\varepsilon), \ldots, d_{r_\varepsilon}(\varepsilon)$ are the orders of the elementary divisors of the operator $V \mid \mathscr{R}_{\varepsilon}$, then

$$\sum_{\varepsilon \in \mathbb{T}} \sum_{i=1}^{r_\varepsilon} \left[\frac{d_i(\varepsilon)}{2}\right] \leqslant \varkappa_{V_0},$$

where $\varkappa_{V_0}$ is the same as in Corollary 5.5. In particular, if V is a J-unitary operator, then

$$\sum_{\varepsilon \in \mathbb{T}} \sum_{i=1}^{r_\varepsilon} \left[\frac{d_i(\varepsilon)}{2}\right] + \sum_{|\lambda| > 1} \dim \mathscr{L}_{\lambda}(V) \leqslant \varkappa_{V_0}$$

(Azizov: *cf.* Usuyatsova [2]).

Hint: Similarly to that was done in proving 2.Theorem 2.26 show that there is in $\mathscr{L}_\varepsilon(V)$ a $\Sigma_{i=1}^{r_\varepsilon}\{[d_i(\varepsilon)]/2\rangle$-dimensional neutral subspace invariant relative to V_0. In the case of a J-unitary operator use Theorem 5.20.

5 Formulate and prove an analogue of Proposition 5.14 and Exercise 4 for maximal J-dissipative operators and, in particular, for J-self-adjoint operators of the class **K**(**H**) (for the latter taking Corollary 5.21 into account).

6 Prove that is a J-bi-non-contractive operator V_0 belongs to **H**, and if $\mathscr{L}_0$ is its finite-dimensional non-degenerate invariant subspace, then the operator $V_0^c \mid \mathscr{L}_0^{[\perp]} \in \mathbf{H}$ (Azizov).
Hint: Use Lemmas 5.8 and 5.9.

7 Prove that in the expansion (5.7) $U' \in \mathbf{K}(\mathbf{H})$ (Azizov).
Hint: Verify that if $U \in \mathbf{K}(\mathbf{H}, V_0)$, then $V_0\mathscr{H}' \subset \mathscr{H}'$, and use the result of Exercise 6.

8 Prove that if $U_0 \in \mathbf{H}$ is a J-unitary operator and if all its root subspaces $\mathscr{L}_\varepsilon(U)$ with $|\varepsilon| = 1$ are non-degenerate, then the space $\mathscr{H}'$ in the expansion (5.6) can be represented as the J-orthogonal sum of subspaces invariant relative to U_0 and any operator community with it:

$$\mathscr{H}' = \underset{i=1}{\overset{k}{[\dot{+}]}} \mathscr{L}_{z_i}(U_0)\,[\dot{+}]\,\mathscr{H}'' \tag{5.10}$$

where the ζ_i are the eigenvalues of the operator U_0 to which corresponds at least one non-trivial neutral vector $(i = 1, 2, \ldots, k)$, and $\mathscr{H}''$ is the J-orthogonal complement to the first k terms of the expansion (5.10) (Azizov).
Hint: Use Theorem 5.20 and Corollary 5.15.

9 Let $\mathscr{A} = \{A\} \in \mathbf{K}(\mathbf{H})$ be a family of maximal J-dissipative operators, let $\mathscr{A} \in (L)$ and the resolvents of the operators of this family commute pairwise, and let $\mathscr{L}_+ \subset \mathscr{P}^+ \cap \{\cap \mathscr{D}_A \mid A \in \mathscr{A}\}$ be an invariant subspace of the family $\mathscr{A}$. Prove that then there is a subspace

$$\tilde{\mathscr{L}}_+ \supset \mathscr{L}_+ : \tilde{\mathscr{L}}_+ \in \mathscr{M}^+,\ \tilde{\mathscr{L}}_+ \subset \{\cap \mathscr{D}_A \mid A \in \mathscr{A}\},\ \mathscr{A}\tilde{\mathscr{L}}_+ \subset \tilde{\mathscr{L}}_+$$

(Azizov [9]).
Hint: Use Theorems 1.13 and 5.12.

10 Let $\mathscr{B} = \{B\}$ be a commutative family of J-bi-non-contractive operators containing an operator $B_0 \in \mathbf{H}$, and let $\mathscr{A} = \{A\}$ be a family of J-bi-non-contractive operators which has the property that $AB = BA$ for any $A \in \mathscr{A}$, $B \in \mathscr{B}$ and for arbitrary A_1, $A_2 \in \mathscr{A}$ there is a $B_{12} \in \mathscr{B}$ such that $A_1A_2 = B_{12}A_2A_1$. Prove that then the family $\mathscr{A} \cup \mathscr{B}$ has the property Φ (Azizov and Dragileva [1]).

11 Prove that a metabelian group of J-unitary operators of the class **K**(**H**) has the property $\tilde{\Phi}^{[\perp]}$ (Azizov; see [IV]).

12 Let $\mathscr{V} = \{V\}$ be a scalarly commutative family (i.e., for every V_1, $V_2 \in \mathscr{V}$ there is a constant λ_{12} such that $V_1V_2 = \lambda_{12}V_2V_1$) of J-bi-non-contractive operators containing an operator $V_0 \in \mathbf{H}$. Then $\mathscr{V}$ has the property $\tilde{\Phi}$ (Azizov).
Hint: Use the scheme of the proof of Theorem 5.12.

13 Prove that if, under the conditions of Exercise 12, $\mathscr{V}$ is a group of J-unitary operators, then it has the property $\tilde{\Phi}^{[\perp]}$ (Azizov [9], Khatskevich [4], *cf.* Helton [2]).

14 Prove that a bounded J-selfadjoint operator A belongs to the class **K**(**H**) if and only if it has at least one invariant subspace $\mathscr{L}^+ \in \mathscr{M}^+ \cap h^+$ (Azizov [13]).

15 Let $A \in \mathbf{K}(\mathbf{H})$ be a bounded J-selfadjoint operator, and let $\mathscr{L}$ be a finite-dimensional non-degenerate subspace invariant relative to A. Prove that $A \mid \mathscr{L}^{[\perp]} \in \mathbf{K}(\mathbf{H})$ (Azizov). *Hint:* Let $\mathscr{L}^+ \in \mathscr{M}^+ \cap h^+$, $A\mathscr{L}^+ \subset \mathscr{L}^+$. Verify that $\det(\mathscr{L}^+ \cap \mathscr{L}^{[\perp]}) < \infty$, use Theorem 4.14 and Exercise 14.

16 Prove that if V is a J-bi-non-contractive operator of the class $\mathbf{H}$, if $\mathscr{L}$ is its projectionally complete invariant subspace, and if the operator $V \mid \mathscr{L}$ has a maximal (in $\mathscr{L}$) completely invariant non-negative subspace $\mathscr{L}_+$, then $V \mid \mathscr{L} \in \mathbf{H}$ (Azizov). *Hint:* Use Corollary 5.4 and Theorems 5.2 and 2.8.

17 Let the conditions of Problem 16 hold, and let P be the J-orthoprojector on to $\mathscr{L}^{[\perp]}$. Prove that the operator $PV \mid \mathscr{L}^{[\perp]} \in \mathbf{H}$ and that it is P^*JP-bi-non-contractive (Azizov). *Hint:* Consider the subspace $\mathscr{L}^+ \in \mathscr{M}^+ : \mathscr{L}_+ \subset \mathscr{L}^+$, $V\mathscr{L}^+ = \mathscr{L}^+$ and prove that $P\mathscr{L}^+ \in h^+$, $(PV \mid \mathscr{L}^{[\perp]})P\mathscr{L}^+ = P\mathscr{L}^+$ and that $V^c \mid \mathscr{L}^{[\perp]}$ is a P^*JP-bi-non-contractive operator.

Remarks and bibliographical indication on chapter 3

§1.1–1.3. In the formulation of Problems 1.2, 1.3, 1.6, 1.7, 1.10, and 1.11 their sources were indicated. To what was said there it should be added that S. L. Sobolev proved that a π-self-adjoint operator in Π_1 has a non-negative eigenvector. The examples 1.4 and 1.8 were constructed by Azizov. J-self-adjoint operators A with $\mathscr{H}^+ \subset \mathscr{D}_A$ were first considered by Langer [2], [3] (see Definition 1.5). The Remark 1.12 is also due to him (see Langer [13]. The reduction of Problem 1.10 to Problem 1.11 was borrowed from Phillips [3].

§1.4. Theorem 1.13 was proved by Azizov [9]

§1.5–1.6. The definitions and propositions are due to Azizov. We mention that the idea of formulating Lemma 1.20 in terms of 'a complex $\mathscr{K}$' (see also 4.Lemma2.5 below) was borroed from [XI].

§2.1. Theorem 2.1 is due to Azizov (see Azizov and Khoroshavin [1]). The existence of the property Φ for a uniformly J-bi-expansive operator was proved earlier by a different method in [*VI*]. Similar results for doubly strict focusing plus-operators (*cf.* 2.§4) in the cases of real and complex J-spaces by the methods of compressed mappings in the metric mentioned in the Exericses on 1.§8 were obtained by Khatskevitch [6], [11], [15], A. V. Sobolev and Khatskevich [1], [2], and in the case of real $\Pi_\varkappa$ earlier by Krasnosel'skiy and A. V. Sobolev [1].

§2.2. Lemma 2.2 was established by M. Krein [4]. Regarding Definition 2.3, see Azizov and Khoroshavin [1] and also Azizov and Kondras [1]. In the last paper the Remark 2.4 was also made, a remark stimulated by Noel's article [1].

§2.3. Proposition 2.6 was borrowed from Langer (see [XVI]). Theorem 2.8 is due to Azizov and Khoroshavich [1]. Its first part was essentially proved earlier by I. Iokhvidov [10]. The remaining assertions are encountered in particular cases and in parts earlier in the papers of M. Krein [4], [5], [XIV], Brodskiy [1], Langer [1]–[3], Azizov [4]–[8], Azizov and E. Iokhvidov [1],

Azizov and Usvyatsova [1], I. Iokhvidov, M. Krein, and Langer [XVI]. The same also applies to Corollary 2.9. We remark that our proof of these propositions is based on the application of a device of M. Krein [4] of approximating to a J-non-contractive operator V by uniformly J-expansive operators VI_n. Corollary 2.10 was formulated by Azizov.

§3.1. Krein–Shmul'yan linear-fractional transformations were first brought into consideration by M. Krein [4]. They and their generalisation were systematically studied in the articles of M. Krein and Shmul'yan [4], Shmul'yan [2], [6], [8], Larionov [4], Helton [1]–[3], Khatskevich [5], [13], [14], and others. Propositions 3.1 and 3.4 are due to M. Krein [4], [5].

§3.2. Theorem 3.6 plays in our exposition the same rôle as was played earlier in these questions by Tikhonov's principle and its generalization given by I. Iokhvidov [10]. Theorem 3.8 is a natural generalization of the corresponding propositions of M. Krein [5] and I. Iokhvidov [10]. Its simple proof given here, based on Glicksberg's thgeorem, is by Azizov.

§3.3. Theorem 3.9—a direct consequence of Theorem 3.8—was formulated and proved here by Azizov. In the case when $V_{12} \in \mathscr{S}_\infty$ the first part was proved by Wittstock [1] and, essentially, earlier by I. Iokhvidov [10]; the second and third parts were obtained by M. Krein [5] (*cf.* Langer [3]). Corollary 3.10—see M. Krein [5], I. Iokhvidov [10], Whittstock [1], and Brodskiy [1]. Corollary 3.11 was noted by Azizov. Theorem 3.12 is due to Langer [3], M. Krein [5]; the short proof of it in the text is due to Azizov.

§3.4. Lemma 3.13 and Theorem 3.14 were proved by Azizov. Earlier results close to Theorem 3.14 were published by Larionov [4] and Khatskevich [5].

§3.5. Theorem 3.20 and Corollary 3.21 follow from Wittstock's results [1]; Azizov gave the proofs given in the text.

§4.1. Theorem 4.1 is due to Phillips [3]; the proof of it given in the text is due to Azizov.

§4.2. The results in this section are due to Azizov. Lemma 4.4 seems to be well-known. Theorem 4.5 in the case where $\mathscr{U}$ is commutative constitutes the main results of Phillip's work [3]. As Shul'man noted, Corollary 4.6 is also true without the requirement for $\mathscr{U}$ to be commutative.

§4.3. The idea of applying Theorem 4.7 in the questions examined here is due to Naymark [2]. Esentially Corollary 4.8 is also due to him. Corollary 4.9 was obtained by Larionov [2], [3]. Example 4.10 was constructed by Azizov.

§4.4, §4.5. Theorem 4.11 was proved by Azizov [6]. Corollary 4.12 was proved earlier by Naymark [2]. Corollary 4.13 is due to Azizov. Theorem 4.14 is also due to him.

§5.1–5.4. Helton in his article [2] considered a commutative group of J-unitary operators containing an operator U_0 which has at least one maximal non-negative invariant subspace and the angular operator of each such subspace is completely continuous. This provided the stimulus for the introduction by Azizov [9] of operators of the class $\mathbf{H}$ and $\mathbf{K(H)}$. All the results of these sections, except Proposition 5.14, are due to Azizov. Proposition 5.14 was mentioned in Usvyatsova's paper [2]; the proof in the text was given by Azizov.

§5.5. Theorem 5.19 (see also Exercise 13 on this section), which is due to Azizov [9] and Khatskevich [4], generalizes a corresponding result of Helton [2]; in this paper the property $\Phi^{[\perp]}$ is proved by a different method.

§5.6–5.7. The results in these sections are due to Azizov.

4 SPECTRAL TOPICS AND SOME APPLICATIONS

In §1 the concept of a J-spectral function is introduced and some classes of operators for which it exists are distinguished. Some of its applications to the solution of problems of invariant subspaces for families of 'definitizable' operators and to the construction of the (J_1, J_2)-polar decomposition of (J_1, J_2)-bi-non-expansive operators are also indicated.

§2 deals with questions of completeness and basicity of systems of eigenvectors and principal vectors of J-dissipative and, in particular, J-selfadjoint operators.

In §3 we give some examples and applications.

§1 The spectral function

1 Let E be a homomorphism mapping a ring $\mathscr{R}$, generated by certain intervals of the real axis and containing the interval $(-\infty, \infty)$, into a set of J-selfadjoint projectors $E(\Delta)$ acting in a J-space $\mathscr{H}$, and let $E(\emptyset) = 0$; $E(-\infty, \infty) = I$; $E(\Delta \cap \Delta') = E(\Delta)E/\Delta')$; $E(\Delta \cup \Delta') = E(\Delta) + E(\Delta')$ if $\Delta \cap \Delta' = \emptyset$ $(\Delta, \Delta' \in \mathscr{R})$. We denote the carrier of this homomorphism by $\sigma(E)$.

A point $\lambda \in \sigma(E)$ is called *a point of definite* (*positive* or *negative*) *type* if there is an interval $\Delta \in \mathscr{R}$ containing λ and such that $E(\Delta)\mathscr{H} \subset \mathscr{P}^+$ or $E(\Delta)\mathscr{H} \subset \mathscr{P}^-$ respectively. The sets of points of positive and negative type are denoted by $\sigma_+(E)$ and $\sigma_-(E)$ respectively.

Let $s = \{\alpha_i\}_1^n$ be a finite set of real points, and let $\mathscr{R} = \mathscr{R}(s)$ be the ring generated by the intervals for which points from s are not boundary points.

Definition 1.1: The homomorphism E described above, defined on $\mathscr{R}(s)$ is called *a J-spectral function with a set s = s(E) of critical points* if, when

$\lambda_0, \mu_0 \notin s(E), \mu_0 < \lambda_{01}$ and $\mu \downarrow \mu_{01}$ the strong limit of the operators $E((\mu, \lambda_0])$. exists and it coincides with $E((\mu_0, \lambda_0])$.

Definition 1.2: A point $\alpha \in s(E)$ is called *a regular critical point of a J-spectral function E* if the strong limits $s\text{-lim}_{\lambda \uparrow \alpha} E((-\infty, \lambda])$ and $s\text{-lim}_{\lambda \downarrow \alpha} E((\lambda, \infty))$ exist. Otherwise α is called *a singular critical point*. If $s(E)$ contains no singular critical points, then E is called *a regular J-spectral function*.

Definition 1.3: A J-spectral function E with a set $s(E)$ of critical points is called an *eigen spectral function of a J-selfadjoint operator A* if:

1) $AE(\Delta) = E(\Delta)A$; 2) $\sigma(A \mid E(\Delta)\mathscr{H}) \subset \bar{\Delta}$;
3) when $\Delta \in \mathscr{R}(s)$ and $\Delta \cap s(E) = \emptyset$ we have $AE(\Delta) = \int_\Delta t \, dE_t$,

where the integral converges in norm if Δ is finite, and strongly if Δ is infinite. The set $s(E)$ is called the *set of critical points* of the operator A and is denoted by $s(A)$, i.e., $s(E) = s(A)$.

2 We now indicate some classes of J-selfadjoint operators which have a J-spectral function. One of them is the class of operators **K(H)** and, in particular, **H** (see Exercise 1 below).

'Definitizable operators' form another class.

Definition 1.4: A J-selfadjoint operator A with $\rho(A) \neq \emptyset$ is said to be *definitizable* if there is a polynomial $p(\lambda)$ such that $[p(A)x, x] \geqslant 0$ when $x \in \mathscr{D}_{A^k}$, where k is the degree of the polynomial.

In particular we observe that *a J-non-negative operator A with $\rho(A) \neq \emptyset$ is definitized by the polynomial $p(\lambda) = \lambda$.*

We shall not stop now to prove that definitizable operators have a J-spectral function; this result is formulated later as an Exercise. Instead we shall prove the key theorem in the theory of definitizable operators.

Theorem 1.5 *Let A be a bounded J-non-negative operator. Then it is possible to set up in correspondence with every real number $\lambda \neq 0$ one and only one J-selfadjoint projector E_λ such that the function $\lambda \to E_\lambda$ satisfies the following conditions:*

1) *if $\lambda \leqslant \mu$, then $E_\lambda E_\mu = E_\mu E_\lambda = E_\lambda$;*
2) *if $\lambda < \mu < 0$, then $[E_\lambda x, x] \geqslant [E_\mu x, x]$, and if $\mu > \lambda > 0$, then $[E_\lambda x, x] \leqslant [E_\mu x, x]$ for all $x \in \mathscr{H}$;*
3) *if $\lambda < -\| A \|$, then $E_\lambda = 0$, and if $\lambda \geqslant \| A \|$, then $E_\lambda = T$;*
4) *if $\lambda \neq 0$, then the strong limit $s\text{-lim}_{\mu \downarrow \lambda} E_\mu = E_{\lambda+0}$ exists and coincides with E_λ;*
5) *If T is a bounded operator which commutes with A, then $TE_\lambda = E_\lambda T$;*

6) *the spectrum* $\sigma(A \mid E_\lambda \mathscr{H})$ *lies in* $(-\infty, \lambda]$ *and* $\sigma(A \mid (I - E_\lambda)\mathscr{H}) \subset [\lambda, \infty)$.

Moreover, $\int_{-\|A\|-0}^{\|A\|} \nu \, dE_2$, *converges in the strong operator topology as an improper integral with a singular point* $\lambda = 0$, *and the operator* $S \equiv A - \int_{\|A\|-0}^{\|A\|} \nu \, dE_\nu$ *is a bounded J-non-negative operator, and* $S^2 = 0$, $SE_\lambda = E_\lambda S = 0$ *when* $\lambda < 0$, *and* $S(T - E_\lambda) = (T - E_\lambda)S = 0$ *when* $\lambda > 0$.

□ We put $B = JA$. This is a non-negative operator, and therefore the operator $C = B^{1/2} J B^{1/2} = C^*$ satisfying the condition $CB^{1/2} = B^{1/2}A$ is properly defined. Since $\| C \| \leqslant \| B^{1/2} \|^2 = \| B \| = \| A \|$, the spectral expansion of the selfadjoint operator C can be written in the form $C = \int_{-\|A\|-0}^{\|A\|} \lambda \, dF_\lambda$, where $\{F_\lambda\}_{-\infty}^{\infty}$ is the spectral function, continuous on the right, of the operator C. We put $C_\lambda = C \mid F_\lambda \mathscr{H}$ when $\lambda < 0$, and $C_\lambda = C \mid (I - F_\lambda)\mathscr{H}$ when $\lambda > 0$, and $E_\lambda = JB^{1/2} C_\lambda^{-1} F_\lambda B^{1/2}$ when $\lambda < 0$ and $I - E_\lambda = JB^{1/2} C_\lambda^{-1} (I - F_\lambda) B^{1/2}$ when $\lambda > 0$.

1) The operators E_λ are bounded for every $\lambda \neq 0$. Moreover, if $\lambda < 0$ (respectively, $\lambda > 0$), then $C_\lambda^{-1} F_\lambda$ (respectively, $C_\lambda^{-1} (I - F_\lambda)$) are selfadjoint operators, and therefore the E_λ are J-selfadjoint with $\lambda \neq 0$. Since $CC_\lambda^{-1} F_\lambda = F_\lambda$ when $\lambda < 0$, and $CC_\lambda^{-1} (I - F_\lambda) = I - F_\lambda$ when $\lambda > 0$, it follows that $E_\lambda E_\mu = E_\mu E_\lambda = E_\lambda$ when $\lambda \leqslant \mu$.

Hence, in particular, we conclude that the E_λ are projectors.

2) If $\lambda < \mu < 0$, then

$$[E_\mu x, x] - [E_\lambda x, x] = (C_\mu^{-1} F_\mu B^{1/2} x, B^{1/2} x) - (C_\lambda^{-1} F_\lambda B^{1/2} x, B^{1/2} x)$$

$$= \int_\lambda^\mu \frac{1}{\nu} \, dF_\lambda (B^{1/2} x, B^{1/2} x) \leqslant 0 \quad \text{for every } x \in \mathscr{H}.$$

Similarly it can be verified that $[E_\lambda x, x] \leqslant [E_\mu x, x]$ when $0 < \lambda < \mu$.

Assertion 3) follows directly from the definition of the function $\lambda \to E_\lambda$.

4) Let $\lambda < \mu < 0$. Then

$$\| E_\mu x - E_\lambda x \|^2 = \| JB^{1/2} (C_\mu^{-1} F_\mu - C_\lambda^{-1} F_\lambda) B^{1/2} x \|^2$$

$$\leqslant \| B \| \int_\lambda^\mu \frac{1}{\nu^2} \, d(F_\mu B^{1/2} x, B^{1/2} x) \leqslant \| B \| \frac{1}{\mu^2} \| (F_\mu - F_\lambda) B^{1/2} x \|^2,$$

and since by hypothesis $\{F_\lambda\}_{-\infty}^{\infty}$ is a function continuous on the right, $s = \lim_{\lambda \downarrow \mu} E_\mu = E_\lambda$. The assertion is verified similarly when $0 < \lambda < \mu$.

5) Let T be a continuous operator and let $TA = AT$. We verify that $TE_\lambda = E_\lambda T$ when $\lambda < 0$. By definition, $E_\lambda = JB^{1/2} \int_{-\|A\|-0}^{\lambda} 1/\nu \, dF_\mu B^{1/2}$ when $\lambda < 0$. Let ${}_2\{p_n\}_1^\infty$ be a sequence of polynomials with

$$\sup_n \{p_n(\lambda) \mid \lambda \in [-\|A\|, \|A\|\} \leqslant M < \infty,$$

and

$$\lim_{n \to \infty} p_n(\nu) = \frac{1}{\nu} \text{ if } -\|A\| < \nu \leqslant \lambda, \text{ and } \lim_{n \to \infty} p_n(\nu) = 0 \text{ if } \lambda < \nu \leqslant \|A\|.$$

Then

$$s\text{-}\lim_{n\to\infty} JB^{1/2}p_n(C)BV^{1/2} = E_\lambda.$$

Since $JB^{1/2}C^mB^{1/2} = A^{m+1}$ $(m = 0, 1, 2, \ldots)$, we have $TJB^{1/2}p_n(C)B^{1/2} = JB^{1/2}p_n(C)B^{1/2}T$, and therefore $TE_\lambda = E_\lambda T$. Similarly one verifies that $TE_\lambda = E_\lambda T$ when $\lambda > 0$. In particular, $AE_\lambda = E_\lambda A$.

6) We observe that, for any bounded operators T_1 and T_2 the sets $\sigma(T_1T_2) \cup \{0\}$ and $\sigma(T_2T_1) \cup \{0\}$ coincide. This follows immediately from the formula

$$(T_2T_1 - \lambda I)^{-1} = \frac{1}{\lambda}(T_2(T_1T_2 - \lambda I)^{-1}T_1 - I)$$

which holds when $0 \neq \lambda \in \rho(T_1T_2)$. Consequently, since

$$AE_\lambda = JB^{1/2}F_\lambda B^{1/2} \quad \text{and} \quad A(I - E_\lambda) = JB^{1/2}(I - F_\lambda)B^{1/2} \qquad (\lambda \neq 0)$$

we have

$$\sigma(AE_\lambda) \cup \{0\} = \sigma(B^{1/2}JB^{1/2}F_\lambda) \cup \{0\} = \sigma(CF_\lambda) \cup \{0\},$$

and

$$\sigma(A(I - E_\lambda)) \cup \{0\} = \sigma(B^{1/2}JB^{1/2}(I - F_\lambda)) \cup \{0\} = \sigma(C(I - F_\lambda)) \cup \{0\}.$$

Hence

$$\sigma(A \mid E_\lambda\mathscr{H}) \subset \{0\} \cup (-\infty, \lambda], \ \sigma(A \mid (I - E_\lambda)\mathscr{H}) \subset \{0\} \cup [\lambda, \infty).$$

It remains to verify that if $\lambda < 0$, then $0 \notin \sigma(A \mid E_\lambda\mathscr{H})$, and if $\lambda > 0$, then $0 \notin \sigma(A \mid (I - E_\lambda)\mathscr{H})$. We shall verify the first of these statements; the second is verified similarly. Suppose that $\lambda < 0$ and $0 \in \sigma(A \mid E_\lambda\mathscr{H})$. Since $A \mid E_\lambda\mathscr{H}$ is a non-negative operator relative to the form $[\,\cdot\,,\,\cdot\,] \mid E_\lambda\mathscr{H}$ and $E_\lambda\mathscr{H}$ is a Krein space, it follows by virtue of 2.Corollary 3.26 that $\sigma_r(A \mid E_\lambda\mathscr{H}) = \emptyset$, and therefore $0 \in \sigma_c(A \mid E_\lambda\mathscr{H}) \cup \sigma_p(A \mid E_\lambda\mathscr{H})$. In both cases there is a sequence $\{x_n\} \in \mathscr{H}$ such that $\|E_\lambda x_n\| = 1$ and $AE_\lambda x_n \to \theta$. But then $B^{1/2}AE_\lambda x_n = (B^{1/2}E_\lambda x_n = CB^{1/2}JB^{1/2}C_\lambda^{-1}F_\lambda B^{1/2}x_n = C_\lambda F_\lambda B^{1/2}x_n \to \theta$ also. Consequently $JB^{1/2}C_\lambda^{-2}C_\lambda F_\lambda B^{1/2}x_n = E_\lambda x_n \to \theta$ also—we have a contradiction.

We now verify that the function $\lambda \to E_\lambda$ satisfying the conditions 1)–6) is uniquely defined. Let $\lambda \to E'_\lambda$ be another such function with $\lambda < \mu < 0$ or $0 < \lambda < \mu$. It is clear that $E'_\lambda = E'_\lambda E_\mu + E'_\lambda(I - E_\mu)$. By condition 5) the operators E'_λ and E_μ commute. Consequently, $E'_x(I - E_\mu)$ is a J-orthogonal projector on to the subspace $E'_\lambda\mathscr{H} \cap (I - E_\mu)\mathscr{H}$ which is invariant relative to A. By condition 6) we obtain $\sigma(A \mid E'_\lambda\mathscr{H} \cap (I - E_\mu)\mathscr{H}) \subset (-\infty, \lambda] \cap [\mu, \infty) = \emptyset$, i.e., $E'_\lambda\mathscr{H} \cap (I - E_\mu)\mathscr{H} = \{\theta\}$, and therefore $E'_\lambda = E'_\lambda E_\mu$. Similarly $E_\lambda = E_\lambda E'_\mu$. When $\mu \downarrow \lambda$ we conclude that $E'_\lambda = E'_\lambda E_\lambda = E_\lambda E'_\lambda = E_\lambda$, i.e., we have proved 1)–6).

Now let $\varepsilon > 0$. Since

$$\int_{-\|A\|-0}^{-\varepsilon} \nu \, dE_\nu = JB^{1/2}C_{-\varepsilon}^{-1} \int_{-\|A\|-0}^{-\varepsilon} \nu \, dF_\nu B^{1/2}$$
$$= JB^{1/2}C_{-\varepsilon}^{-1}F_{-\varepsilon}B^{1/2} = JB^{1/2}F_{-\varepsilon}B^{1/2},$$

and

$$\int_{\varepsilon}^{\|A\|} \nu \, dE_\nu = -\int_{\varepsilon}^{\|A\|} \nu \, d(I - E_\nu) = -JB^{1/2}C_\varepsilon^{-1}\int_{\varepsilon}^{\|A\|} \nu \, d(I - F_\nu)B^{1/2}$$
$$= JB^{1/2}C_\varepsilon^{-1}\int_{\varepsilon}^{\|A\|} \nu F_\nu B^{1/2} = JB^{1/2}C_\varepsilon^{-1}C_\varepsilon(I - F_\varepsilon)B^{1/2}$$
$$= JB^{1/2}(I - F_\varepsilon)B^{1/2} = A - JB^{1/2}F_\varepsilon B^{1/2},$$

it follows that

$$\int_{-\|A\|-0}^{\|A\|} \nu \, dE_\nu = A - JB^{1/2}(F_0 - F_{-0})B^{1/2}.$$

We put $S = JB^{1/2}(F_0 - F_{-0})B^{1/2}$. Since $F_0 - F_{-0}$ is a projector on to Ker C, we have

$$S^2 = JB^{1/2}(F_0 - F_{-0})B^{1/2}JB^{1/2}(F_0 - F_{-0})B^{1/2}$$
$$= JB^{1/2}(F_0 - F_{-0})C(F_0 - F_{-0})B^{1/2} = 0.$$

Similarly,

$$SE_\lambda = E_\lambda S = 0 \quad \text{when} \quad \lambda < 0 \quad \text{and} \quad S(I - F_\lambda) = (I - E_\lambda)S = 0 \quad \text{when} \quad \lambda > 0. \quad \blacksquare$$

Corollary 1.6: *A bounded J-non-negative operator A has a J-spectral function E with a single critical point $\lambda = 0$. Moreover,*

$$\sigma(E)\backslash\{0\} = \sigma_+(E) \cup \sigma_-(E) \quad \text{and} \quad \sigma_+(E) = (0, \infty) \cap \sigma(E),$$
$$\sigma_-(E) = (-\infty, 0) \cap \sigma(E).$$

□ As $E(\Delta)$, when $\Delta = (\alpha, \beta], \alpha \neq 0, \beta \neq 0$, we put $E(\Delta) = E_\beta - E\alpha$. The fact that this determines a J-spectral function of the operator A with a single critical point $\lambda = 0$ can be seen from the properties of the function E_λ investigated in Theorem 1.5. In particular, it follows from properties 2) and 3) that $\sigma(E)\backslash\{0\} = \sigma_+(E) \cup \sigma_-(E)$, and $\sigma_+(E) = (0, \infty) \cap \sigma(E)$, $\sigma_-(E) = (-\infty, 0) \cap \sigma(E)$. ■

Remark 1.7: It follows from condition 5) of Theorem 1.5 that if T is a bounded operator which commutes with A, then it also commutes with S. ■

Remark 1.8: Since (see the proof of property 5) in Theorem 1.5) the operators E_λ ($\lambda \neq 0$) are the strong limits of certain polynomials of A, every subspace $\mathscr{L}$ which is invariant relative to A is invariant relative to E_λ and S. Moreover, $\mathscr{L} = E_\lambda \mathscr{L} [+] (I - E_\lambda)\mathscr{L}$, $E_\lambda \mathscr{L}$ and $(I - E_\lambda)\mathscr{L}$ are invariant relative to A, and $\sigma(A \mid E_\lambda \mathscr{L}) \subset (-\infty, \lambda]$, and $\sigma(A \mid (I - E_\lambda)\mathscr{L}] \subset [\lambda, \infty)$. If, moreover, $\mathscr{L}$ is invariant relative to a bounded operator T which commutes with A, then $E_\lambda \mathscr{L}$, $(I - E_\lambda)\mathscr{L}$ and $S\mathscr{L}$ are also invariant relative to this operator. ■

We note also that by virtue of condition 3) in Theorem 1.5 the integral $\int_{-\|A\|-0}^{\|A\|} \nu \, dE_\nu$ can be written in the form $\int_{-\infty}^{\infty} \nu \, dE_\nu$. In future, where it is necessary to do so, the operators E_ν and S corresponding to the operator A will be denoted by the symbols $E_\mu^{(A)}$ and S_A.

Corollary 1.9: *If A and D are commuting J-non-negative operators, then $E_\lambda^{(A)} E_\lambda^{(D)} = E_\lambda^{(D)} E_\lambda^{(A)}$; $S_A E_\lambda^{(D)} = E_\lambda^{(D)} S_A = 0$ when $\lambda < 0$ and $S_A(I - E_\lambda^{(D)}) = (I - E_\lambda^{(D)}) S_A = 0$ when $\lambda > 0$; $S_A S_D = S_D S_A = 0$; $DS_A = S_A D = 0$.*

□ The permutability of the operators $E_\lambda^{(A)}$, $E_\lambda^{(D)}$, S_A and S_D follows from condition 5) in Theorem 1.5 and Remark 1.7. In accordance with Exercise 2 on 2.§2, $\mathscr{R}_{S_A}$ is a neutral lineal. On the other hand, we conclude from the commutativity of the operators S_A and $E_\lambda^{(D)}$ when $\lambda \neq 0$ that it is invariant relative to $E_\lambda^{(D)}$. Consequently $E_\lambda^{(D)} \mathscr{R}_{S_A}$ is a negative lineal when $\lambda < 0$, and $(I - E_\lambda^{(D)})\mathscr{R}_{S_A}$ is a positive lineal when $\lambda > 0$ (see Theorem 1.5 condition 2)). Hence it follows that $E_\lambda^{(D)} \mathscr{R}_{S_A} = \{\theta\}$ when $\lambda < 0$ and $(I - E_\lambda^{(D)})\mathscr{R}_{S_A} = \{\theta\}$ when $\lambda > 0$, i.e., $E_\lambda^{(D)} S_A = 0$ when $\lambda < 0$ and $(I - E_\lambda^{(D)}) S_A = 0$ when $\lambda > 0$.

By virtue of condition S in Theorem 1.5 the operators D and S_A commute, and therefore the lineal $\mathscr{R}_{S_A}$ is invariant relative to D. Since $\mathscr{R}_{S_A} \subset \mathscr{P}^0$, it follows from the inequality $|[DS_A x, y]|^2 \leqslant [DS_A x, S_A x][Dy, y] = 0$ that $DS_A = 0$.

It remains to verify the equality $S_A S_D = 0$. Since

$$\int_{-\infty}^{\infty} \nu E_\nu^{(D)} = s\text{-}\lim_{\varepsilon \downarrow 0} (DE_{-\varepsilon}^{(D)} + D(I - E_\varepsilon^{(D)})), \text{ we have } \int_{-\infty}^{\infty} \nu \, dE_\nu^{(D)} S_A = 0.$$

We now use the fact $DS_A = 0$ and $S_D = D - \int_{-\infty}^{\infty} \nu \, dE_\nu^{(D)}$. ■

3 Let $\mathscr{A} = \{A\}$ be a commutative family of bounded J-selfadjoint operators. We shall say that it has the property $\tilde{\Phi}^{[\perp]}$ if every invariant dual pair of this family admits extension into a maximal dual pair invariant relative to $\mathscr{A}$.

Later in Exercise 8 it will be proved that a family of definitizable J-selfadjoint operators has the property $\tilde{\Phi}^{[\perp]}$ if this family is either finite, or if for all the operator entering into it the 'corners' $P^+ AP^- \mid \mathscr{H}^-$ are completely continuous. But here now we shall prove the validity of the following result.

Theorem 1.10: *If $\mathscr{A} = \{A\}$ is a commutative family of bounded J-non-negative operators, then $\mathscr{A}$ has the property $\tilde{\Phi}^{[\perp]}$.*

□ Let $(\mathscr{L}_+, \mathscr{L}_-)$ be an invariant dual pair of the family $\mathscr{A}$. We have to prove that there are $\mathscr{L}^\pm \in \mathscr{M}^\pm$ such that $\mathscr{L}^\pm \supset \mathscr{L}_\pm$, $\mathscr{L}^+ [\perp] \mathscr{L}^-$, and $\mathscr{A}\mathscr{L}^\pm \subset \mathscr{L}^\pm$.

We first verify this assertion for the case when the $\mathscr{L}_\pm$ are definite. Let $\mathscr{E}_+(A) = C\ \operatorname{Lin}\{(I - E_\lambda^{(A)})\mathscr{H} \mid \lambda > 0\}$, $\mathscr{E}_-(A) = C\ \operatorname{Lin}\{E_\lambda^{(A)}\mathscr{H} \mid \lambda < 0\}$, $\mathscr{E}_\pm(\mathscr{A}) = C\ \operatorname{Lin}\{\mathscr{E}_\pm(A) \mid A \in \mathscr{A}\}$, and $\mathscr{N} = C\ \operatorname{Lin}\{\mathscr{R}_{S_A} \mid A \in \mathscr{A}\}$. It follows from Corollary 1.9 that $\mathscr{E}_+(\mathscr{A}) [\perp] \mathscr{E}_-(\mathscr{A})$ and $\mathscr{N} [\perp] \mathscr{E}_\pm(\mathscr{A})$, and from Remark 1.8 that $\mathscr{L}_\pm \subset (\mathscr{E}_\mp(\mathscr{A}))^{[\perp]}$. By virtue of this same Remark 1.8, $S_A\mathscr{L}_\pm \subset \mathscr{L}_\pm$ $(A \in \mathscr{A})$. But since the $\mathscr{L}_\pm$ are definite and $\mathscr{R}_{A_A} \subset \mathscr{P}^0$, we have $S_A\mathscr{L}_\pm = \{\theta\}$ $(A \in \mathscr{A})$, and therefore $\mathscr{L}_\pm [\perp] \mathscr{N}$.

We consider the subspaces $\mathscr{L}'_\pm = C\ \operatorname{Lin}\{\mathscr{L}_\pm, \mathscr{E}_\pm(\mathscr{A}), \mathscr{N}\}$. From what has been proved above it follows that $\mathscr{L}'_+ [\perp] \mathscr{L}'_-$. We verify the inclusions $\mathscr{L}'_\pm \subset \mathscr{P}^\pm$. In accordance with Exercise 11 on II.§3 the subspaces $C\ \operatorname{Lin}\{\mathscr{L}_\pm, \mathscr{E}_\pm\mathscr{A})\} \subset \mathscr{P}^\pm$. It remains to use the fact that $\mathscr{N} [\perp] C\ \operatorname{Lin}\{\mathscr{L}_\pm, \mathscr{E}_\pm(\mathscr{A})\}$. Thus $(\mathscr{L}'_+, \mathscr{L}'_-)$ is an invariant dual pair of the family $\mathscr{A}$ and it contains $(\mathscr{L}_+, \mathscr{L}_-)$.

Let $(\mathscr{L}^+, \mathscr{L}^-)$ be an arbitrary maximal dual pair containing $(\mathscr{L}'_+, \mathscr{L}'_-)$. We shall prove that $A\mathscr{L}^\pm \subset \mathscr{L}^\pm$ for all $A \in \mathscr{A}$. By construction $\mathscr{L}^+ \subset \mathscr{L}'^{\,1[\perp]}_- \subset \mathscr{E}_-(A)^{[\perp]}$ and therefore $A\mathscr{L}^+ \subset A\mathscr{E}_-(A)^{[\perp]}$. Since $A = S_A + \int_{-\infty}^{\infty} \nu\, dE_\nu^{(A)}$, we have
$A\mathscr{E}_-(A)^{[\perp]} \subset C\ \operatorname{Lin}\{\mathscr{R}_{S_A}, \mathscr{E}_+(A)\} \subset \mathscr{L}'_+ \subset \mathscr{L}^+$.
Consequently, $A\mathscr{L}^+ \subset \mathscr{L}^+$. Similarly $A\mathscr{L}^- \subset \mathscr{L}^-$ $(A \in \mathscr{A})$. Thus the theorem is proved when $\mathscr{L}_\pm$ are definite.

Let $(\mathscr{L}_+, \mathscr{L}_-)$ be an arbitrary dual pair invariant relative to $\mathscr{A}$, and let $\mathscr{L}_\pm^{(1)} = C\ \operatorname{Lin}\{\mathscr{L}_\pm, \mathscr{L}_\mp \cap \mathscr{P}^0\}$. Then $(\mathscr{L}_+^{(1)}, \mathscr{L}_-^{(1)})$ is an extension of the pair $(\mathscr{L}_+, \mathscr{L}_-)$, and $\mathscr{L}_+^{(1)} [\perp] \mathscr{L}_-^{(1)}$, $\mathscr{L}_+^{(1)} \cap \mathscr{L}_-^{(1)} = \mathscr{L}_+^{(1)} \cap \mathscr{L}_+^{(1)[\perp]} = \mathscr{L}_-^{(1)} \cap \mathscr{L}_-^{(1)[\perp]} \equiv \mathscr{L}_0$, and $\mathscr{A}\mathscr{L}_\pm^{(1)} \subset \mathscr{L}_\pm^{(1)}$.

We pass to the consideration of the factor-space $\hat{\mathscr{H}} = \mathscr{L}_0^{[\perp]}/\mathscr{L}_0$, and of the family of $\hat{J}$-non-negative operators $\hat{\mathscr{A}} = \{\hat{A}\}$ induced in it by the family $\mathscr{A}$, and its invariant dual definite pair $(\hat{\mathscr{L}}_+^{(1)}, \hat{\mathscr{L}}_-^{(1)})$, where $\hat{\mathscr{L}}_\pm^{(1)} = \mathscr{L}_\pm^{(1)}/\mathscr{L}_0$. In accordance with what has been proved above, the family $\hat{\mathscr{A}}$ has an invariant maximal dual pair $(\hat{\mathscr{L}}_+, \hat{\mathscr{L}}_-)$ containing $(\hat{\mathscr{L}}_+^{(1)}, \hat{\mathscr{L}}_-^{(1)})$. In accordance with Exercise 7 on 1.§5 the subspaces $\mathscr{L}^\pm = \{x_\pm \in \hat{x}_\pm \mid x_\pm \in \hat{\mathscr{L}}^\pm\}$ form a maximal dual pair in $\mathscr{H}$, and $\mathscr{A}\mathscr{L}^\pm \subset \mathscr{L}^\pm \supset \mathscr{L}_\pm$. ■

4 In conclusion we show how the results in §1.2 can be applied to obtain a (J_1, J_2)-polar decomposition of a (J_1, J_2)-bi-non-expansive operator. But first we give a definition.

Definition 1.11: A bounded J_1-selfadjoint operator R is called a *J_1-module* of a bounded operator T acting from a J_1-space $\mathscr{H}_1$ into a J_2-space $\mathscr{H}_2$ if $\sigma(R) \subset [0, \infty)$, $R^2 = T^cT$ and $\operatorname{Ker} R = \operatorname{Ker} T^cT$.

Theorem 1.12: *Every (J_1, I_2)-bi-non-expansive operator V has at least one J_1-module R and admits a '(J_1, J_2)-polar decomposition' $V = WR$, where W is a (J_1, J_2)-bi-non-expansive operator which is simultaneously a (J_1, J_2)-semi-unitary operator or is conjugate to a (J_1, J_2)-semi-unitary operator.*

☐ The operator $A = I_1 - V^c V$ is a bounded J_1-non-negative operator. Since from 2.Theorem 4.30 and 2.Remark 4.29 we have $\sigma(V^c V) \subset [0, \infty)$, it follows that $\sigma(A) \subset (-\infty, 1)$. Let $\Delta_0 = (\frac{1}{2}, 1]$, and let $E(\Delta_0)$ be the corresponding value of the J_1-spectral function of the operator A. From the definition of the J_1-spectral function it follows that $E(\Delta_0)\mathscr{H}_1$ and $(I_1 - E(\Delta_0))\mathscr{H}_1$ are invariant relative to the operator $V^c V$, and moreover $\sigma(V^c V | E(\Delta_0)\mathscr{H}_1) \subset [0, \frac{1}{2}]$ and $\sigma(V^c V | (I_1 - E(\Delta_0))\mathscr{H}_1) \subset [\frac{1}{2}, \| V^c V \|]$. In accordance with Corollary 1.6 the subspace $E(\Delta_0)\mathscr{H}_1$ is uniformly positive. Since $V^c V | E(\Delta_0)\mathscr{H}_1$ is an operator which is selfadjoint relative to the form $[\cdot, \cdot]_1 | E(\Delta_0)\mathscr{H}_1$ and has a non-negative spectrum, it is non-negative and therefore it has a non-negative square root R_1. Let Γ be a Jordan contour, enclosing the interval $[\frac{1}{2}, \| V^c V \|]$, symmetric relative to $\mathbb{R}$ and not containing $\lambda = 0$ within itself, and let $\sqrt{\lambda}$ be an analytic function taking positive values for positive λ. Then (see [XXII]) the operator $R_2 = -(1/2\pi i)\oint_\Gamma \sqrt{\lambda}(V^c V|(I_1 - E(\Delta_0))\mathscr{H}_1 - \lambda I_1)^{-1}\, d\lambda$ is bounded, $R_2^2 = V^c V | (I_1 - E(\Delta_0))\mathscr{H}_1$, $\sigma(R_2) \subset [1/\sqrt{2}, \sqrt{\| V^c V \|}]$, and, because Γ is symmetric relative to $\mathbb{R}$, the operator R_2 will be J_1-Hermitian. Since $E(\Delta_0)\mathscr{H}_1 [+] (I_1 - E(\Delta_0))\mathscr{H}_1 = \mathscr{H}_1$, the operator R such that $Rx = R_1 x$ if $x \in E(\Delta_0)\mathscr{H}_1$ and $Rx = R_2 x$ if $x \in (I_1 - E(\Delta_0))\mathscr{H}_1$ is J_1-selfajoint and bounded, $R^2 = V^c V$, Ker R = Ker $V^c V$, $\sigma(R) \subset [0, \infty)$, i.e., R is the J_1-module of the operator V.

We now consider the operators $V_1 = V | E(\Delta_0)\mathscr{H}_1$ and $V_2 = V | (I_1 - E(\Delta_0))\mathscr{H}_1$. Since $E(\Delta_0)\mathscr{H}_1$ is uniformly positive, $(I_1 - E/\Delta_0))\mathscr{H}_1$ contains the uniformly negative subspace which is maximal in $\mathscr{H}_1$, and therefore $V(I_1 - E(\Delta_0))\mathscr{H}_1$ contains the uniformly negative subspace maximal in $\mathscr{H}_2$. From 1. Corollary 8.14 we conclude that $\mathscr{R}_{V_2}$ $(= \bar{\mathscr{R}}_{V_2})$ is projectionally complete, and $\mathscr{R}_{V_2}^{[\perp]}$ is uniformly positive. Since $E(\Delta_0)\mathscr{H}_1$ and $(I_1 - E(\Delta_0))\mathscr{H}_1$ are invariant relative to $V^c V$ it follows that $V^c \mathscr{R}_{V_2}^{[\perp]} \subset E(\Delta_0)\mathscr{H}_1$, $V^c \mathscr{R}_{V_2} \subset (I_1 - E(\Delta_0))\mathscr{H}_1$, and $\mathscr{R}_{V_1} \subset \mathscr{R}_{V_2}^{[\perp]}$. The operator V_1 as an operator acting from the Hilbert space $E(\Delta_0)\mathscr{H}_1$ with the scalar product $[\cdot, \cdot]_1$ into the Hilbert space $\mathscr{R}_{V_2}^{[\perp]}$ with the scalar product $[\cdot, \cdot]_2$ has the decomposition $V_1 = W_1 R_1$, where $W_1 : E(\Delta_0)\mathscr{H}_1 \to \mathscr{R}_{V_2}^{[\perp]}$ is semi-unitary or is adjoint to a semi-unitary operator. It is now sufficient to put $Wx = W_1 x$ if $x \in E(\Delta_0)\mathscr{H}_1$, and $Wx = V_2 R_2^{-1} x$ if $x \in (I_1 - E(\Delta_0))\mathscr{H}_1$, and we obtain that $V = WR$ and W satisfies the requirements of the theorem. ■

Exercises and problems

1 Prove that J-selfadjoint operators A of the class $\mathbf{K}(\mathbf{H})$ with a real spectrum have a unique J-spectral function with a finite number of critical points $\{\lambda_i\}$, where λ_i are the eigenvalues of the operator A with a degenerate Ker$(A - \lambda_i I)$ (Azizov).
Hint: Use the result of Exercise 15 on 3.§5 and the decompositon analogous to 3.(5.3) for a bounded J-selfadjoint operator of the class $\mathbf{K}(\mathbf{H})$.

2 Similarly to Definition 1.3 introduce the concept of an eigen J-spectral function of a J-unitary operator and prove that J-unitary operators $U \in \mathbf{K}(\mathbf{H})$ have a J-spectral function, which is, moreover, unique (Azizov).

3 Let $\Pi_\varkappa$ be a space with $\varkappa$ negative (respectively, positive) squares, let $\mathscr{L}$ be a $\varkappa$-dimensional non-positive (respectively, non-negative) invariant subspace of a π-selfadjoint operator A, and let $\lambda_1, \lambda_2, \ldots, \lambda_\varkappa$ be the eigenvalues (taking multiplicity into account) of the operator $A \mid \mathscr{L}$. Prove that the polynomial $p(\lambda) = \prod_{k=1}^{\varkappa}(\lambda - \lambda_\varkappa)$ (respectively, $-p(\lambda)$) definitizes the operator A (I. Iokhvidov and M. Krein [XV]).

4 Prove that if A is a definitizable J-selfadjoint operator, and $\{\lambda_i, \bar{\lambda}_i\}_1^k$ are all the non-real roots of the definitizing polynomial p, and $\{\lambda_i, \bar{\lambda}_i\} \subset \sigma(A)$, then $\sigma(A) \backslash \mathbb{R} = \{\lambda_i, \bar{\lambda}_i\}_1^k$, $\{\lambda_i, \bar{\lambda}_i\} \subset \sigma_p(A)$, $\mathscr{L}_{\lambda_i}(A)$ and $\mathscr{L}_{\bar{\lambda}_i}(A)$ are subspaces $(i = 1, 2, \ldots, k)$, and

$$\mathscr{H} = \mathop{[\dot{+}]}_{i=1}^{k} (\mathscr{L}_{\lambda_i}(A) \dotplus \mathscr{L}_{\bar{\lambda}_i}(A)) \,[\dot{+}]\, \mathscr{H}',$$

and $A : \mathscr{H}' \to \mathscr{H}'$ is an operator definitizable in $\mathscr{H}'$ with a real spectrum (Langer [4], [5]).

5 Prove that a definitizable J-selfadjoint operator with a real spectrum has a J-spectral function E which is, moreover, unique. Further, $\sigma(E) = \sigma_+(E) \cup \sigma_-(E) \cup s(E)$, and when $\Delta \cap s(E) = \emptyset$ and the Δ are finite the operators $E(\Delta)$ are the strong limits of polynomials of A (Langer [4], [5]).

6 Prove that operators A of the class $\mathbf{K(H)}$ or definitizable operators are bounded if and only if $\sigma(A)$ is a bounded set (Azizov; Langer [4], [5]).
Hint: Use the decomposition 3.(5.2) and the results of Exercises 1, 4, and 5 respectively.

7 Prove that if $A \in \mathbf{K(H)}$ or if A is a definitizable operator, and if $\sigma(A) = \{\lambda_i\}_1^k$, then $\mathscr{H} = \mathscr{L}_{\lambda_1}(A) \dotplus \mathscr{L}_{\lambda_2}(A) \dotplus \ldots \dotplus \mathscr{L}_{\lambda_k}(A)$, and the lengths of the Jordan chains of the operators $A \mid \mathscr{L}_{\lambda_i}(A)$ are bounded (Azizov; Langer [4], [5]).
Hint: Use the expansion 3.(5.4) and the results of Exercises 1, 4, 5.

8 Prove that if $\mathscr{A}$ is a commutative family of bounded definitizable J-selfadjoint operators and either $\mathscr{A}$ is finite or the 'corners' $P^+AP^- \mid \mathscr{H} \in \mathscr{S}_\infty$ for all $A \in \mathscr{A}$, then $\mathscr{A}$ has the property $\tilde{\Phi}^{[\perp]}$ (Langer [9], [13]).

9 Prove that if A is a definitizable J-selfadjoint operator, and if D is a finite-dimensional J-selfadjoint operator, then $A + D$ is a definitizable operator (Jonas and Langer [1]).

10 Prove that a J-selfadjoint operator of the class $\mathbf{K(H)}$ is definitizable if and only if its J-spectral function E (see Exercise 1) is such that $\sigma(E) = \sigma_+(E) \cup \sigma_-(E) \cup s(E)$ (Azizov).
Hint: Use Exercise 15 on 3.§ and the results of Exercises 1 and 5 taking into account a decomposition similar to 3.(5.2) for a bounded J-selfadjoint operator of the class $\mathbf{K(H)}$.

11 Prove that the J-spectral function of a J-selfadjoint operator $A \in \mathbf{K(H)}$ with $\sigma(A) \subset \mathbb{R}$ is regular if and only if all the $\mathscr{L}_\lambda(A)$ are non-degenerate (Azizov).
Hint: Use the expansion 3.(5.4).

12 Prove that if a J-selfadjoint operator A satisfies the condition $[Ax, x] \neq 0$ when $[x, x] = 0$, then there is a constant α such that either $A - \alpha I$ or $\alpha I - A$ is a J-non-negative operator with a definite kernel and $A = \alpha I + \int_{-\infty}^{\infty}(\mu - \alpha)\, dE_\mu$. (Personen [1], M. Krein and Shmul'yan [3]).

13 Let T be a strict plus-operator acting from a J_1-space $\mathscr{H}_1$ into a J_2-space $\mathscr{H}_2$, and

let $\sigma(T^c T) \geqslant 0$. Then the operator T has a J_1-module and only one (*cf.* M. Krein and Shmul'yan [3]).
Hint: See the proof of Theorem 1.14.

14 Prove that a J-selfadjoint operator A is definitizable if and only if there is a function f, holomorphic in the neighbourhood of its spectrum and of the point $\lambda = \infty$, such that $f(A)$ is a non-negative operator (Langer [4], [5]).

15 Prove that if A is a J-selfadjoint operator of the class $\mathbf{K(H)}$, then there is a decomposition $\mathscr{H} = \mathscr{H}_1 [\dot{+}] \mathscr{H}_2$, where $\mathscr{H}_1 \subset \mathscr{D}_A$, $A\mathscr{H}_i \subset \mathscr{H}_i$ $(i = 1, 2)$, $\sigma(A \mid \mathscr{H}_1) \subset [m, M]$, $-\infty < m < M < \infty$, and $A \mid \mathscr{H}_2$ is similar to a selfadjoint operator and $\sigma(A \mid \mathscr{H}_2) \subset (-\infty, m] \cup [M, \infty)$ (Azizov, Gordienko and I. Iokhvidov [1]).
Hint: Use the results of Exercise 1.

16 Let $\{U_t\}_{-\infty}^{\infty}$ be a one-parameter group of the class (C_0), (i.e., $U_{t+s} = U_t U_s$, $U_0 = I$, and $U_t x \to U_{t_0} x$ when $t \to t_0$, and for all $x \in \mathscr{H}$) consisting of J-unitary operators, and let iA be an arbitrary operator of this group. Prove that $(A^c =)$ $A \in \mathbf{K(H)}$ if and only if $\{U_t\}_{-\infty}^{\infty} \in \mathbf{K(H)}$ and then $U_t = \exp(itA)$ (Azizov, Gordienko and I. Iokhvidov [1]).
Hint: Use the results of Exercises 1, 2, 15 and also Naymark's article [3].

§2. Completeness and basicity of a system of root vectors of J-dissipative operators

1 In this section, if nothing to the contrary is stated, all the operators are continuous and are defined on the whole space. The spaces are assumed, generally speaking, to be non-separable. Our references to the monograph [XI] in which the spaces are assumed to be separable are valid in this respect, that either separability is not essential for the validity of the actual results or our operators here are acting in a separable space.

We recall that a system of vectors $\{e_\alpha\}$ is said to be *complete in a space* $\mathscr{H}$ if $\mathscr{H} = C\ \mathrm{Lin}\{e_\alpha\}$. Everywhere it is to be understood that $\alpha \in A$, where A is a certain set of indices.

A basis $\{f_\alpha\}$ of a space $\mathscr{H}$ is said to be *orthonormalized* if $(f_\alpha, f_\beta) = \delta(\alpha, \beta)$ $(\alpha, \beta \in A)$. If T is a bounded and boundedly invertible operator, then the system $\{Tf_\alpha\}$ will also be a basis in $\mathscr{H}$ and it is called a *Riesz basis.*

Here we shall investigate the questions of the completeness and basicity of the roof vectors (i.e., the eigenvectors and principal vectors) of J-dissipative, and in particular, of J-selfadjoint operators of the class $\mathbf{K(H)}$ and of J-selfadjoint definitizable operators. We introduce the notation

$$\mathscr{E}(A) = C\ \mathrm{Lin}\{\mathscr{L}_\lambda(A) \mid \lambda \in \sigma_p(A)\}, \tag{2.1}$$

$$\mathscr{E}_0(A) = C\ \mathrm{Lin}\{\mathrm{Ker}(A - \lambda I) \mid \lambda \in \sigma_p(A)\}. \tag{2.2}$$

By definition $\mathscr{E}_0(A) \subset \mathscr{E}(A)$ and therefore $\dim \mathscr{H}/\mathscr{E}(A) \leqslant \dim \mathscr{H}/\mathscr{E}_0(A)$.

We shall first investigate the given questions in the case of J-dissipative operators of the $\mathbf{K(H)}$ class. We note that *if* $A \in \mathbf{K(H)}$ *and if* $\lambda = \bar{\lambda} \in \sigma_p(A)$, *then the root lineal* $\mathscr{L}_\lambda(A)$ *is closed and its non-degeneracy is equivalent to its projectional completeness* (Exercise 5 on 3.§5). Since $A \in \mathbf{K}(\mathbf{H})$, there is not more than a finite number of real points λ to which correpond degenerate kernels $\operatorname{Ker}(A - \lambda T)$ of this operator. We shall call the set of such points, as also in the case of J-selfadjoint operators of the class $\mathbf{K}(\mathbf{H})$ (see Exercise 1 on §1), the *set of critical points* and denote it by the symbol $s(A)$.

Lemma 2.1: *Let A be a bounded J-dissipative operator of the class $\mathbf{K}(\mathbf{H})$ whose non-real spectrum consists of normal points. Then non-degeneracy of $\mathscr{E}(A)$ and non-degeneracy of* $\operatorname{Lin}\{\mathscr{L}_\lambda(A) \mid \lambda \in s(A)\}$ *are equivalent. Moreover, if $\mathscr{E}(A)$ is non-degenerate, then it is projectionally complete.*

□ Since (see Exercise 5 on 3.§5) $\mathscr{L}_\lambda(A)$ when $\lambda = \bar{\lambda}$ can be expressed in the form of a sum of subspaces invariant relative to $A : \mathscr{L}_\lambda(A) = \mathscr{N}_\lambda [\dot{+}] \mathscr{M}_\lambda$, where $\dim \mathscr{N}_\lambda < \infty$, and $\mathscr{M}_\lambda$ is a projectionally complete subspace and $\mathscr{M}_\lambda \subset \operatorname{Ker}(A - \lambda I)$, we conclude from 2.Corollaries 2.16 and 2.18 that $\operatorname{Lin}\{\mathscr{L}_\lambda(A) \mid \lambda \in s(A)\}$ is a subspace and its non-degeneracy is equivalent to its projectional completeness.

Now let $\operatorname{Lin}\{\mathscr{L}_\lambda(A) \mid \lambda \in s(A)\}$ be degenerate, and let $\mathscr{L}$ be its isotropic part. Then $\mathscr{L} \subset \operatorname{Lin}\{\mathscr{N}_\lambda \mid \lambda \in s(A)\}$, and therefore $\mathscr{L}$ is finite-dimensional. By virtue of 2.Lemma 2.19 we have $A\mathscr{L} \subset \mathscr{L}$. Consequently there is in $\mathscr{L}$ at least one eigenvector x_0 of the operator A. By its choice $x_0 [\perp] \operatorname{Lin}\{\mathscr{L}_\lambda(A) \mid \lambda \in s(A)\}$, and by virtue of 2.Corollaries 2.16 and 2.18 we have $x_0 [\perp] \mathscr{L}_\mu(A)$ when $\mu \notin s(A)$, i.e., $x_0 [\perp] \mathscr{E}(A)$. Since $x_0 \in \mathscr{E}(A)$, we see that degeneracy of $\operatorname{Lin}\{\mathscr{L}_\lambda(A) \mid \lambda \in s(A)\}$ implies the degeneracy of $\mathscr{E}(A)$.

Conversely, let $\mathscr{E}(A)$ be degenerate, and let $\mathscr{L}_1$ be its isotropic part. Then $A\mathscr{L}_1 \subset \mathscr{L}_1$, by virtue of 2.Lemma 2.19. We first assume, and later prove, that $\dim \mathscr{L}_1 < \infty$. If it is, then there is in $\mathscr{L}_1$ an eigenvector y_0 of the operator $A : Ay_0 = \lambda_0 y_0$. From 2.Corollary 2.25 and the definition of the set of critical points we conclude that $\lambda_0 \in s(A)$, and therefore $\operatorname{Lin}\{\mathscr{L}_\lambda(A) \mid \lambda \in s(A)\}$ is a degenerate subspace.

We now verify that $\mathscr{L}_1$ is finite-dimensional, and at the same time that if $\mathscr{E}(A)$ is non-degenerate, then it is projectionally complete. To do this it suffices to show that there is in $\mathscr{E}(A)$ a projectionally complete subspace $\mathscr{N}$ such that $\dim \mathscr{E}(A) \mid \mathscr{N} < \infty$. Since $A \in \mathbf{K}(\mathbf{H})$, by virtue of 3.Corollary 5.18 there are in every subspace $\operatorname{Ker}(A - \lambda I)$ when $\lambda = \bar{\lambda}$ semi-definite invariance subspaces $\mathscr{L}_\lambda^{\pm}$, maximal (in $\operatorname{Ker}(A - \lambda I)$), of the operator V_0. Then $\mathscr{L}_+ = C\operatorname{Lin}\{\mathscr{L}_\mu(A), \mathscr{L}_\lambda^+ \mid \operatorname{Im} \mu > 0, \lambda = \bar{\lambda}\}$ is a subspace completely invariant relative to V_0 and invariant relative to A, and $\mathscr{L}_- = C\operatorname{Lin}\{\mathscr{L}_\mu(A), \mathscr{L}^- \mid \operatorname{Im} \mu < 0, \lambda = \bar{\lambda}\}$ is a subspace invariant relative to V_0 and A. By virture of 2.Corollaries 2.18 and 2.22 it follows that the first of these subspaces is non-negative and the second is non-positive. Taking 3.Corollary 5.4 into account this implies the inclusions $\mathscr{L}_\pm \in h^{\pm}$. Let $\mathscr{L}_0 = \mathscr{L}_+ \cap \mathscr{L}_-$. Then $\mathscr{L}_+ + \mathscr{L}_- = \mathscr{L}_0 [\dot{+}] \mathscr{N}$, where $\mathscr{N}$ is a projectionally complete subspace. It remains to notice that $\dim \mathscr{E}(A) \mid \mathscr{L}_+ + \mathscr{L}_- < \infty$, and therefore if $\mathscr{E}(A)$ is degenerate, then its isotropic part is finite-dimensional, and if it is non-degenerate, then it is projectionally complete. ■

□ *Remark 2.2:* The condition in Lemma 2.1 about the normality of the non-real points of the spectrum of the operator A is used only in the proof of the fact that degeneracy of $\mathscr{E}(A)$ implies the degeneracy of $\text{Lin}\{\mathscr{L}_\lambda(A) \mid \lambda \in s(A)\}$. Therefore the remaining assertions are valid even without this condition, and in particular, if $\mathscr{E}(A)$ is non-degenerate and therefore projectionally complete, then the operator $V_0 \mid \mathscr{E}(A)$ belongs to the class $\mathbf{H}$. In fact, the subspace $\mathscr{L}_+$ constructed in the proof of Lemma 2.1 has in $\mathscr{E}(A)$ the property def $\mathscr{L}_+ < \infty$. From Exercise 9 on 3.§3 we conclude that $V_0 \mid \mathscr{E}(A)$ is a bi-non-contractive operator relative to the form $[\cdot, \cdot] \mid \mathscr{E}(A)$ and it has a completely invariant subspace $\widetilde{\mathscr{L}}_+ \in \mathscr{M}^+(\mathscr{E}(A))$. In combination with 3.Theorem5.2 and the fact that $V_0 \in \mathbf{H}$, this is sufficient to prove the validity of our assertion. Moreover, from Exercise 17 on 3.§5 it follows that if P is the J-orthoprojector on to $\mathscr{E}(A)^{[\perp]}$, then the operator $PV_0 \mid \mathscr{E}(A)^{[\perp]} \in \mathbf{H}$ and it is P^*JP-bi-non-contractive. ■

Corollary 2.3: *Let $A \in \mathbf{K}(\mathbf{H})$ be a completely continuous J-dissipative operator. Then non-degenerary of $\mathscr{E}(A)$ and $\mathscr{L}_0(A)$ are equivalent.*

□ The operator A satisfies the conditions of Lemma 2.1. Moreover, since all $\lambda \neq 0$ are normal points of the operator A, it follows by virtue of 2.Corollary 2.25 that all the $\mathscr{L}_\lambda(A)$ when $\lambda = \bar{\lambda} \neq 0$ are non-degenerate. Consequently, degeneracy of $\mathscr{L}_0(A)$ and degeneracy of $\text{Lin}\{\mathscr{L}_\lambda(A) \mid \lambda \in s(A)\}$ are equivalent. It remains to use Lemma 2.1. ■

2 The following two lemmas are key results in solving questions about the completeness of systems of root vectors of J-dissipative operators.

Lemma 2.4: *Let the operator satisfy the conditions of Lemma 2.1, let $\mathscr{E}(A)$ be non-degenerate, and let P be the J-orthoprojector on to $\mathscr{E}(A)^{[\perp]}$. Then $A_1 \equiv PA \mid \mathscr{E}(A)^{[\perp]} \in \mathbf{K}(\mathbf{H})$ and $\sigma_p(A_1) = \emptyset$.*

□ That the operator A_1 belongs to the class $\mathbf{K}(\mathbf{H})$ follows from the fact that if $A \in \mathbf{K}(\mathbf{H}, V_0)$, then by virtue of Exercise 17 on 3.§5 $PV_0 \mid \mathscr{E}(A)^{[\perp]} \in \mathbf{H}$ and therefore $A_1 \in \mathbf{K}(\mathbf{H}, PV_0 \mid \mathscr{E}(A)^{[\perp]})$.

Now let $\bar{\lambda}_0 = \lambda \in \sigma_p(A)$. Since $\mathscr{E}(A)^{[\perp]}$ is a P^*JP-space, and A_1 is a P^*JP-dissipative operator, it follows from $\lambda_0 = \bar{\lambda}_0$ that $\lambda_0 \in \sigma_p(C^c)$. Let $A_1^c x_0 = \lambda_0 x_0$. By virtue of 2.Corollary 2.17 $A_1 x_0 = \lambda_0 x_0$. On the other hand, $A_1^c = A^c \mid \mathscr{E}(A)^{[\perp]}$, and again by virtue of 2.Corollary 2.17 $Ax_0 = \lambda_0 x_0$; but this contradicts the fact that $x_0 \, [\perp] \, \mathscr{E}(A)$ and $\mathscr{E}(A)$ is non-degenerate, and therefore $\sigma_p(A_1) \cap \mathbb{R} = \emptyset$.

Now suppose that $\bar{\lambda}_0 \neq \lambda_0 \in \sigma_p(A_1)$. By virtue of 2.Theorem 1.16 $\bar{\lambda}_0 \in \sigma_p(A_1^c) \cup \sigma_r(A_1^c)$. We again use the fact that $A_1^c = A^c \mid \mathscr{E}(A)^{[\perp]}$. Since by hypothesis all the non-real points of the spectrum of the operator A (and therefore, by virtue of 2.Theorem 1.16, also of A^c) are normal eigenvalues, it follows that $\bar{\lambda}_0 \in \sigma_p(A^c)$, which implies in the present case the inclusion

$\lambda_0 \in \lambda_p(A)$, i.e., λ_0 is a normal eigenvalue of the operator A and therefore also of the operator $A \mid \mathscr{E}(A)$. Consequently, there is a decomposition $\mathscr{E}(A) = \mathscr{L}_{\lambda_0}(A) + \mathscr{E}'$ such that $A\mathscr{E}' \subset \mathscr{E}'$ and $\lambda_0 \in \rho(A \mid \mathscr{E}')$. Then $\mathscr{H} = \mathscr{L}_{\lambda_0}(A) \dot{+} (\mathscr{E}' [\dot{+}] \mathscr{E}(A)^{[\perp]})$. Since $\lambda_0 \in \sigma_p(A_1)$ it follows from $\lambda_0 \in \rho(A \mid \mathscr{E}')$ that $\lambda_0 \in \sigma_p/A_2)$, where $A_2 = QA \mid (\mathscr{E}' [\dot{+}] \mathscr{E}(A)^{[\perp]})$, and Q is the projector on to $\mathscr{E}' [\dot{+}] \mathscr{E}(A)^{[\perp]}$ parallel to $\mathscr{L}_{\lambda_0}(A)$. Let $A_2 x_0 = \lambda_0 x_0$. Then $(A - \lambda_0 I)x_0 \in \mathscr{L}_{\lambda_0}(A)$, i.e., x_0 is a root-vector of the operator A corresponding to λ_0 and not lying in $\mathscr{L}_{\lambda_0}(A)$—we have obtained a contradiction ■

Lemma 2.5: *Let $\mathscr{K}$ be a certain complex of properties invariant relative to a bounded projection and an equivalent renormalization of the space, and let each bounded dissipative operator having this complex of properties have a complete system of root vectors. Then every J-dissipative operator $A \in \mathbf{K}(\mathbf{H})$ with a non-degenerate $\mathscr{E}(A)$ also has a complete system of root vectors in $\mathscr{H}$.*

□ We suppose the opposite. Since the properties in the complex $\mathscr{K}$ are invariant relative to a bounded projection, they are in particular invariant also relative to a J-orthogonal projection. It follows from Lemma 2.4 that either Lemma 2.5 is true, or there is a bounded J-dissipative operator $A \in \mathbf{K}(\mathbf{H})$ having this complex of properties and such that $\sigma_p(A) = \emptyset$. We suppose the latter holds. From 3.Theorem 5.1b we have that the operator A has at least one pair of invariant subspaces $\mathscr{L}^{\pm} \in \mathscr{M}^{\pm} \cap h^{\pm}$. Since $\sigma_p(A) = \emptyset$, the $\mathscr{L}^{\pm}$ are uniformly definite subspaces, and therefore $\mathscr{H} = \mathscr{L}^+ + \mathscr{L}^-$. Since the operators $\pm A \mid \mathscr{L}^{\pm}$ have the properties of the complex $\mathscr{K}$ and are dissipative in relation to the scalar product $\pm[\cdot,\cdot] \mid \mathscr{L}^{\pm}$ which is equivalent to the original one, it follows by the hypothesis of the theorem that $\mathscr{E}(A \mid \mathscr{L}^{\pm}) = \mathscr{L}^{\pm}$, and therefore $\sigma_p(A) \neq \emptyset$—we have obtained a contradiction. ■

Lemma 2.5 enables us to transfer to the case of bounded J-dissipative operators of the class $\mathbf{K}(\mathbf{H})$ a whole series of assertions about the completeness of the system of root vectors for ordinary dissipative operators since, as a rule, the conditions in these assertions are invariant relative to the operations of bounded projection and equivalent renormalization of the space (see, e.g., [XI]). We shall not cite all these assertions in the main text, but we shall introduce some of them later in the form of exercises. We show by the example of Theorem 2.6 how to carry out these exercises. We also prove Theorem 2.8; this is interesting because it cannot be generalized, not even in the case of operators of the class $\mathbf{H}$ (see Example 3.3 below). But first we recall some familiar terms and introduce some new ones.

Let $A : \mathscr{H}_1 \to \mathscr{H}_2$ be a completely continuous operator, and let $\{s_\alpha(A)\}$ be the set of eigenvalues of the operator $(A^*A)^{1/2}$ (taking multiplicity into account), or in other words, *$\{s_\alpha(A)\}$ is the s-number of the operator A* ([XI]). We say that $A \in \mathscr{S}_p$ if $\Sigma s_\alpha^p(A) < \infty$. In particular, if $p = 1$, then A is called a *nuclear operator*, and if $p = 2$, A *Hilbert-Schmidt operator*. We remark that even if $\mathscr{H}_1$ is non-separable, all the s-numbers except perhaps for a countable set are equal to zero. It will therefore be convenient to us to suppose below that $s_k(A) \neq 0$ when $k = 1, 2, \ldots, \nu$; $\nu \leqslant \infty$. Usually the s-numbers $s_k(A)$ are numbered in the order 'biggest first', and then the formula $s_{n+1} = \min\{\| AK \| \mid \dim K \leqslant n\}$ holds ([XI]). Hence, it follows in particular that if $\{\tilde{s}_n(A)\}$ are the s-numbers of the operator A in another norm equivalent to the first one, then there is an $m > 0$ such that $(1/m)s_n(A) \leqslant \tilde{s}_n(A) \leqslant ms_n(A)$ $(n = 1, 2, \ldots)$.

Let $\{e_\alpha\}$ be an orthonormalized basis, and $\{f_\alpha\}$ be a J-orthonormalized Riesz basis of a J-space $\mathscr{H}$. The number sp $A = \Sigma_\alpha (Ae_\alpha, e_\alpha)$, where $A \in \mathscr{S}_1$, is called the *trace of the operator A*. As is well-known (see, e.g., [XI]) sp $A = \rho_\alpha \lambda_\alpha$, where $\{\lambda_\alpha\}$ is the set of eigenvalues of the operator A taking multiplicity into account, and therefore sp A does not depend on the choice of the orthonormalized basis and the equivalent renormalization of the space. This enables us to write the formula for the trace in the equivalent form: sp $A = \Sigma_\alpha [Af_\alpha, f_\alpha]$ sign $[f_\alpha, f_\alpha]$. All three of these formulae will be used later.

Theorem 2.6: *Let $A \in \mathbf{K}(\mathbf{H})$ be a completely continuous J-dissipative operator with a nuclear J-imaginary component A_I, and let $\lim_\rho \to \infty n(\rho; A_R)/\rho = 0$, where A_R is the J-real part of the operator A, $n(\rho; A_R)$ is the number of numbers of the form $1/s_n(A_R)$ $(n = 1, ,2, \ldots)$* situated in the segment $[0, \rho]$. *Then $\mathscr{E}(A) = \mathscr{H}$ if and only if $\mathscr{L}_0(A)$ is non-degenerate.*

□ The *necessity* for non-degeneracy follows from Corollary 2.2.

Sufficiency: From Corollary 2.2 we obtain that if $\mathscr{L}_0(A)$ is non-degenerate, then $\mathscr{E}(A)$ is also non-degenerate. Let P be the J-orthoprojector on to $\mathscr{E}(A)^{[\perp]}$, and let $A_1 = PA \mid \mathscr{E}(A)^{[\perp]}$. Since $s_n((A_1)_R) \leqslant \| P \|^2 s_n(A_R)$ (see, e.g., [XI], 2.2.1), we have $\lim_{\rho\to\infty} n(\rho; (A_1)_R)/\rho = 0$. Therefore if $\mathscr{E}(A) \neq \mathscr{H}$, we can by virtue of Lemma 2.4 suppose without loss of generality that the original operator A is a Volterra (i.e., $A \in \mathscr{S}_\infty$ and $\sigma(A) = \{0\}$) J-dissipative operator of the class $\mathbf{K}(\mathbf{H})$ with $0 \notin \sigma_p(A)$ and we shall prove that the equality $\lim_{\rho\to\infty} n(\rho; A_R)\rho = 0$ is impossible. Indeed, it follows from 3.Theorem 5.16 that the operator A has invariant subspaces $\mathscr{L}^\pm \in \mathscr{M}^\pm \cap h^\pm$. They are non-degenerate; for otherwise it would follow from 2.Lemma 2.19 that $\sigma_p(A) \neq \emptyset$. Consequently the $\mathscr{L}^\pm$ are uniformly definite and therefore $\mathscr{H} = \mathscr{L}^+ + \mathscr{L}^-$. Suppose, for example, that $\mathscr{L}^+ \neq \{\theta\}$. We consider the operator $A^+ = A \mid \mathscr{L}^+$. It is a Volterra dissipative operator relative to the (definite) scalar product $[\cdot, \cdot] \mid \mathscr{L}^+$, and $\lim_{\rho\to\infty} n(\rho; A_R^+)/\rho = 0$. Since the scalar product $[\cdot, \cdot] \mid \mathscr{L}^+$ is equivalent on $\mathscr{L}^+$ to the original one, it follows that $\lim_{\rho\to\infty} \tilde{n}(\rho; A_R^+)/\rho = 0$, where $\tilde{n}(\rho_1 A_R^+)$ is the number of numbers of the form $1/\tilde{s}_n(A_R^+)$ in the interval $[0, \rho]$, and the $\{\tilde{s}_n(A_R^+)\}$ are the s-numbers of the operator A_R^+ relative to the scalar product $[\cdot, \cdot] \mid \mathscr{L}^+$. Let $n_+(\rho; A_R^+)$ be the number of numbers of the form $1/\lambda_n^+(A_R^+)$ in the interval $[0, \rho]$, where the $\{\lambda_n^+(A_R^+)\}$ are the positive eigenvalues of the selfadjoint operator A_R^+. Since

$0 \leqslant n_+(\rho; A_R^+) \leqslant \tilde{n}(\rho; A_R^+)$, we have $\lim_{\rho\to\infty} n_+(\rho; A_R^+)/\rho = 0$. It follows from [XI] 4.7.2 in this case that $\lim_{\rho\to\infty} n_+(\rho; A_R^+)/\rho = (1/\pi)$ sp A_I^+. Consequently sp $A_I^+ = 0$. Since A_I^+ is a non-negative operator, sp $A_I^+ = 0$ implies $A_I^+ = 0$, i.e., $A^+ = A_R^+$ is a completely continuous self-adjoint operator, and therefore $\sigma_P(A^+) \neq \emptyset$—we have obtained a contradiction. We arrive similary at a contradiction if $\mathscr{L}^- \neq \{\theta\}$. ■

Later we shall more than once make use of the following simple proposition.

Lemma 2.7: *If A is a selfadjoint operator with a spectrum having no more than a countable set of points of condensation, then $\mathscr{E}_0(A) = \mathscr{H}$, and in $\mathscr{H}$ there is an orthonormalized basis composed of the eigenvectors of the operator A.*

□ Since $\mathscr{L}_\lambda(A) = \mathrm{Ker}(A - \lambda I)$ and $\mathrm{Ker}(A - \lambda I) \perp \mathrm{Kerr}(A - \mu I)$ when $\lambda \neq \mu$, there is in $\mathscr{E}_0(A)$ an orthonormalized basis composed of eigenvectors of the operator A. It remains to verify that $\mathscr{E}_0(A)^\perp = \{\theta\}$. Suppose this is not so. Since $\sigma(A \mid \mathscr{E}_0(A)^\perp)$ contains no eigenvalues, and the set of points of condensation of $\sigma(A \mid \mathscr{E}_0(A)^\perp)$ can be no more than countable, there is at least one isolated point of the spectrum of the operator $A(\mathscr{E}_0(A)^\perp$. This point must, as is well-known (see, e.g., [I]), be an eigenvalue—we obtain a contradiction. ■

A J-dissipative operator A will be called *simple* if there is no subspace invariant relative to A and A^c on which these operators coincide.

Theorem 2.8: *If an operator A with a nuclear J-imaginary component A_I is a simple π-dissipative operator or if $\sigma(A)$ has no more than a countable set of points of condensation, then $\mathscr{E}(A) = \Pi_\varkappa$ if and only if the lineal* $\mathrm{Lin}\{\mathscr{L}_\lambda(A) \mid \lambda \in s(A)\}$ *is non-degenerate and* $\Sigma_\alpha \operatorname{Im} \lambda_\alpha = \mathrm{sp}\, A_I$, *where* λ_α *traverses the set of all eigenvalues of the operator A taking multiplicity into account.*

□ Since $A = A_R + iA_I$ and the non-real spectrum of the π-selfadjoint operator A_R consists of normal eigenvalues, it follows from $A_I \in \mathscr{S}_1$ that the non-real spectrum of the operator A consists of normal eigenvalues. Therefore all the points of condensation of $\sigma(A)$ lie in $\mathbb{R}$. Moreover, by virtue of 2.Theorem 2.26, $\mathscr{L}_\lambda(A) = \mathscr{N}_\lambda [\dot{+}] \mathscr{M}_\lambda$, where $A\mathscr{N}_\lambda \subset \mathscr{N}_\lambda$, $\dim \mathscr{N}_\lambda < \infty$, and $\mathscr{M}_\lambda \subset \mathrm{Ker}(A - \lambda I)$ and $\mathscr{M}_\lambda$ is non-degenerate, and therefore $\mathrm{Lin}\{\mathscr{L}_\lambda(A) \mid \lambda \in s(A)\}$ is a subspace.

Suppose that $\mathscr{E}(A) = \Pi_\varkappa$. We then obtain from Lemma 2.1 that $\mathrm{Lin}\{\mathscr{L}_\lambda(A) \mid \lambda \in s(A)\}$ is non-degenerate. We consider the subspace $\mathscr{L} = C\, \mathrm{Lin}\{\mathscr{M}_\lambda \mid \lambda \in \sigma(A) \cap \mathbb{R}\}$. This is a non-degenerate subspace: for otherwise it would follow from 2.Lemma 2.19 that the operator $A \mid \mathscr{L}$ has an eigenvector $x_0 \neq \theta$ isotropic in $\mathscr{L}$, and by virtue of the fact that $\mathscr{L}$ is π-orthogonal to $C\, \mathrm{Lin}\{\mathscr{N}_\lambda, \mathscr{L}_\mu(A) \mid \lambda = \bar{\lambda}, \mu \neq \bar{\mu}\}$ it would follow that $x_0 [\perp] \mathscr{E}(A)$—we obtain a contradiction. We consider the decomposition

$\Pi_\varkappa = \mathscr{L}[\dot{+}]\mathscr{L}^{[\perp]}$. Since $A\mathscr{L} \subset \mathscr{L}$ and $A^c\mathscr{L} \subset \mathscr{L}$, it follows that $A\mathscr{L}^{[\perp]} \subset \mathscr{L}^{[\perp]}$ and the operator $A \mid \mathscr{L}^{[\perp]}$ satisfies the same conditions as A but with this difference that all the root subspaces of the operator $A \mid \mathscr{L}^{[\perp]}$ are finite-dimensional. Since $\mathscr{L} \subset \mathscr{E}(A)$ we shall suppose without loss of generality that $\mathscr{L} = \{\theta\}$. Since (by construction) $\mathscr{N}_\lambda \neq \{\theta\}$ only when $\lambda \in s(A)$, so $\sigma_p(A) \cap \mathbb{R}$ has not more than a finite number of points, and therefore the operator has not more than a countable set of eigenvalues. Consequently, just as in [XI], 1.4.1, an orthonormalized basis $\{e_k\}$ can be constructed in $\Pi_\varkappa$ such that $(Ae_k, e_k) = \lambda_k$ and λ_k runs through the set of eigenvalues of the operator A taking multiplicity into account. Hence, $[(1/2i)(A - A^*)e_k, e_k] = \operatorname{Im} \lambda_k$. Since $A_I \in \mathscr{S}_1$, and $A_I - (1/2i)(A - A^*) = (1/2i)(-A^c + A^*) = (1/2i)(-JA^*J + A^*)$ is a finite-dimensional operator, and moreover

$$\operatorname{sp}\left(\frac{1}{2i}(JA^*J - A^*)\right) = \operatorname{sp}\left(\frac{1}{2i}J(A^* - JA^*J)J\right) = \operatorname{sp}\left(\frac{1}{2i}(A^* - JA^*J)\right)$$
$$= -\operatorname{sp}\left(\frac{1}{2i}(JA^*J - A^*)\right),$$

it follows that

$$\operatorname{sp}\left(\frac{1}{2i}(JA^*J - A^*)\right) = 0, \frac{1}{2i}(A - A^*) \in \mathscr{S}_1 \text{ and } \operatorname{sp} A_I = \operatorname{sp}\left(\frac{1}{2i}(A - A^*)\right).$$

Conversely, let $\operatorname{Lin}\{\mathscr{L}_\lambda(A) \mid \lambda \in s(A)\}$ be non-degenerate and let $\operatorname{sp} A_I = \Sigma_\alpha \operatorname{Im} \lambda_\alpha$. Again by virtue of Lemma 2.1. $\mathscr{E}(A)$ is non-degenerate. Again without loss of generality we suppose that $\mathscr{L} \equiv C\operatorname{Lin}\{\mathscr{M}_\lambda \mid \lambda = \bar{\lambda}\} = \{\theta\}$. We consider the operators $A' = A \mid \mathscr{E}(A)$ and $A_1 = PA \mid \mathscr{E}(A)^{[\perp]}$, where P is the π-orthoprojector on to $\mathscr{E}(A)^{[\perp]}$. Since the operator A' satisfies the same conditions as the operator A, we have $\operatorname{sp} A_I' = \Sigma_\alpha \operatorname{Im} \lambda_\alpha$. Let $\{f_k^{(1)}\}$ and $\{f_\alpha^{(2)}\}$ be π-orthonormalized based in $\mathscr{E}(A)$ and $\mathscr{E}(A)^{[\perp]}$ respectively. Then

$$\operatorname{sp} A_I = \sum_k [A_I f_k^{(1)}, f_k^{(1)}] \operatorname{sign}[f_k^{(1)}, f_k^{(1)}] + \sum_\alpha [A_I f_\alpha^{(2)}, f_\alpha^{(2)}] \operatorname{sign}[f_\alpha^{(2)}, f_\alpha^{(2)}]$$

$$= \sum_k [A_I' f_k^{(1)}, f_k^{(1)}] \operatorname{sign}[f_k^{(1)}, f_k^{(1)}] + \sum_\alpha [(A_1)_I f_\alpha^{(2)}, f_\alpha^{(2)}] \operatorname{sign}[f_\alpha^{(2)}, f_\alpha^{(2)}]$$

$$= \sum_k \operatorname{Im} \lambda_k + \sum_\alpha [(A_1)_I f_\alpha^{(2)}, f_\alpha^{(2)}] \operatorname{sign}[f_\alpha^{(2)}, f_\alpha^{(2)}],$$

which, by virtue of the equality $\operatorname{sp} A_I = \Sigma_k \operatorname{Im} \lambda_k$ implies the equality

$$\sum_\alpha [(A_1)_I f_\alpha^{(2)}, f_\alpha^{(2)}] \operatorname{sign}[f_\alpha^{(2)}, f_\alpha^{(2)}] = 0.$$

Suppose for definiteness that $\Pi_\varkappa$ is a Pontryagin space with $\varkappa$ negative squares. Then $\mathscr{E}(A)$ is also a Pontryagin space with $\varkappa$ negative squares, and therefore

$\mathscr{E}(A)^{[\perp]}$ is a positive subspace. Hence it follows from

$$\sum_{\alpha} [(A_1)_I f_\alpha^{(2)}, f_\alpha^{(2)}]\operatorname{sign}[f_\alpha^{(2)}, f_\alpha^{(2)}] = 0$$

that $(A_1)_I = 0$, i.e., $A_1 = A^c$. Since $A_1^c = A^c \mid \mathscr{E}(A)^{[\perp]}$, we conclude from 2. Theorem 2.15 that $A\mathscr{E}(A)^{[\perp]} \subset \mathscr{E}(A)^{[\perp]}$ and $A \mid \mathscr{E}(A)^{[\perp]} = A^c \mid \mathscr{E}(A)^{[\perp]} = A_1$. If A is a simple π-dissipative operator, it follows from this that $\mathscr{E}(A)^{[\perp]} = \{\theta\}$. If, however A is not simple but $\sigma(A)$ has no more than a countable set of points of condensation, then the operator A_1, which is self-adjoint relative to the definite scalar product $[\cdot, \cdot] \mid \mathscr{E}(A)^{[\perp]}$, also has this property. Therefore, by virtue of Lemma 2.7, $\sigma_p(A_1) \neq \emptyset$ if $\mathscr{E}(A)^{[\perp]} \neq \{\theta\}$; but, by Lemma 2.1, $\sigma_p(A) = \emptyset$ which implies that $\mathscr{E}(A)^{[\perp]} = \{\theta\}$. ■

Corollary 2.9: *If $A \in \mathbf{K}(\mathbf{H})$ is a nuclear J-dissipative operator, then non-degeneracy of $\mathscr{L}_0(A)$* is equivalent to the equality $\mathscr{E}(A) = \mathscr{H}$.

□ It follows from Corollary 2.2 that if $\mathscr{E}(A) = \mathscr{H}$, then $\mathscr{L}_0(A)$ is non-degenerate.

Conversely, let $\mathscr{L}_0(A)$ be non-degenerate. Then again from Corollary 2.2 we obtain that $\mathscr{E}(A)$ is also non-degenerate. We make use of Lemma 2.1, by virtue of which it suffices to prove that when $\mathscr{H} \neq \{\theta\}$ there are no Volterra nuclear J-dissipative operators A of the class $\mathbf{K}(\mathbf{H})$ with $0 \notin \sigma_p(A)$. Let us suppose the contrary. As in the proof of Theorem 2.6 we conclude from $\sigma_p(A) = \emptyset$ that the operator A has a maximal uniformly definite invariant subspace $\mathscr{L}$. Suppose for definiteness that $\mathscr{L} \in \mathscr{M}^+$ (the case $\mathscr{L} \in \mathscr{M}^-$ is verified similarly). We consider the operator $A \mid \mathscr{L}$. This is a nuclear dissipative operator relative to the form $[\cdot, \cdot] \mid \mathscr{L}$. Since $\operatorname{sp}(A \mid \mathscr{L}) = \Sigma_\alpha \lambda_\alpha(A \mid \mathscr{L})$, we have $\operatorname{sp}(A \mid \mathscr{L})_I = \Sigma_\alpha \operatorname{Im} \lambda_\alpha(A \mid \mathscr{L})$. By Theorem 2.8 $\sigma_p(A \mid \mathscr{L}) \neq \emptyset$—and we have a contradiction. ■

3 Before we present results about the basicity of systems of root vectors of J-selfadjoint operators, we introduce.

Definition 2.10: A basis $\{f_\alpha\}$ of a J-space $\mathscr{H}$ is said to be *almost J-orthonormalized* if it can be presented as the union of a finite subset of vectors and a J-orthonormalized subset, these subsets being J-orthogonal to one another.

Definition 2.11: A basis $\{f_\alpha\}$ is called a *p-basis* if there is an orthonormalized basis $\{e_\alpha\}$ and an operator $T \in \mathscr{S}_p$ such that $f_\alpha = (I + T)e_\alpha$ $(\alpha \in A)$.

Theorem 2.12: *Let A be a continuous J-selfadjoint operator of the class $\mathbf{K}(\mathbf{H})$, and let $\sigma(A)$ have no more than a countable set of points of* *condensation. Then:*

a) $\dim \mathscr{H}/\mathscr{E}(A) \leqslant \dim \mathscr{H}/\mathscr{E}_0(A) < \infty$;
b) $\mathscr{E}_0(A) = \mathscr{H}$ *if and only if* $s(A) = \emptyset$ *and* $\mathscr{L}_\lambda(A) = \operatorname{Ker}(A - \lambda I)$ *when* $\lambda \neq \bar{\lambda}$;
c) $\mathscr{E}(A) = \mathscr{H}$ *if and only if* $Lin\{\mathscr{L}_\lambda(A) \mid \lambda \in s(A)\}$ *is a non-degenerate subspace;*
d) *if* $\mathscr{E}_0(A) = \mathscr{H}$ *(respectively,* $\mathscr{E}(A) = \mathscr{H}$*), then there is in* $\mathscr{H}$ *an almost J-orthonormalized Riesz basis composed of eigenvectors (respectively, root vectors) of the operator* A;
e) *if* $\mathscr{E}_0(A) = \mathscr{H}$, *then there is in* $\mathscr{H}$ *a J-orthonormalized basis composed of eigenvectors of the operator A if and only if* $\sigma(A) \subset \mathbb{R}$;
f) *the bases mentioned above can be chosen as p-bases if and only if the operator A has an invariant subspace* $\mathscr{L}_+ \in \mathscr{M}^+$ *with an angular operator* $K_{\mathscr{L}_+} \in \mathscr{L}_p$.

□ a) Since $\mathscr{E}_0(A) \subset \mathscr{E}(A)$ it follows that $\dim \mathscr{H}/\mathscr{E}(A) \leqslant \dim \mathscr{H}/\mathscr{E}_0(A)$. We use 3.Corollary 5.21 and an analogue of 3.Proposition 5.14 for J-selfadjoint operators of the class $\mathbf{K(H)}$ and we obtain that $\dim \mathscr{E}(A)/\mathscr{E}_0(A) < \infty$. Consequently to prove the inequality $\dim \mathscr{H}/\mathscr{E}_0(A) < \infty$ it suffices to show that $\dim \mathscr{H}/\mathscr{E}(A) < \infty$.

Let $\mathscr{L}_+ \in \mathscr{M}^+ \cap h^+$ be an invariant subspace, existing by virtue of 3.Theorem 5.16, of the operator A, let $\mathscr{L}_- = \mathscr{L}_-^{[\perp]}$, and $\mathscr{L}_0 = \mathscr{L}_+ \cap \mathscr{L}_-$. Since $\mathscr{L}_0^{[\perp]} = \operatorname{Lin}\{\mathscr{L}_+, \mathscr{L}_-\}$ and $\dim \mathscr{H}/\mathscr{L}_0^{[\perp]} = \dim \mathscr{L}_0 < \infty$, to prove the inequality under discussion it is sufficient to establish that $\mathscr{L}_0^{[\perp]} = \mathscr{E}(A \mid \mathscr{L}_0^{[\perp]})$, or, equivalently, that the $\hat{J}$-space $\hat{\mathscr{H}} = \mathscr{L}_0^{[\perp]}/\mathscr{L}_0$ coincides with $\mathscr{E}(\hat{A})$, where $\hat{A}$ is the $\hat{J}$-selfadjoint operator generated by the operator A. But the latter follows from the fact that $\hat{\mathscr{H}} = \hat{\mathscr{L}}_+ [+] \hat{\mathscr{L}}_\pm = \mathscr{L}_\pm/\mathscr{L}_0$ are uniformly definite subspaces invariant relative to $\hat{A}$, and the operators $\hat{A}/\hat{\mathscr{L}}_\pm$ satisfy the conditions of Lemma 2.7.

b) Let $\mathscr{E}_0(A) = \mathscr{H}$. It follows immediately from 3.Formula (5.4), from the definition of the set $s(A)$ and the fact that $\mathscr{L}_\lambda(A)$ is J-orthogonal to $\mathscr{L}_\mu(A)$ when $\lambda \neq \bar{\mu}$, that $s(A) = \emptyset$ and $\mathscr{L}_\lambda(A) = \operatorname{Ker}(A - \lambda I)$ when $\lambda \neq \bar{\lambda}$.

Conversely, let $s(A) = \emptyset$ and $\mathscr{L}_\lambda(A) = \operatorname{Ker}(A - \lambda I)$ when $\lambda \neq \bar{\lambda}$. We again use 3.Formula (5.4) and without loss of generality we shall suppose that $\sigma(A) \subset \mathbb{R}$. Since $s(A) = \emptyset$, all the kernels $\operatorname{Ker}(A - \lambda I)$ are non-degenerate, and therefore $\mathscr{L}_\lambda(A) = \operatorname{Ker}(A - \lambda I)$. Hence, by virtue of Lemma 2.1, $\mathscr{E}(A)$ $(= \mathscr{E}_0(A))$ is also non-degenerate. But since $\dim \mathscr{E}_0(A)^{[\perp]} = \dim \mathscr{H}/\mathscr{E}_0(A) < \infty$ and $A\mathscr{E}_0(A)^{[\perp]} \subset \mathscr{E}_0(A)^{[\perp]}$, we have $\mathscr{E}_0(A)^{[\perp]} = \{\theta\}$, i.e. $\mathscr{H} = \mathscr{E}_0(A)$.

c) If $\mathscr{E}(A) = \mathscr{H}$, then $\mathscr{E}(A)$ is non-degenerate, and by Lemma 2.1 $\operatorname{Lin}\{\mathscr{L}_\lambda(A) \mid \lambda \in s(A)\}$ is a non-degenerate subspace.

Conversely, let $\operatorname{Lin}\{\mathscr{L}_\lambda(A) \mid \lambda \in s(A)\}$ be a non-degenerate subspace. Again by virtue of Lemma 2.1 $\mathscr{E}(A)$ is non-degenerate and $\mathscr{H} = \mathscr{E}(A) [\dot{+}] \mathscr{E}(A)^{[\perp]}$, moreover $A\mathscr{E}(A)^{[\perp]} \subset \mathscr{E}(A)^{[\perp]}$, $A \mid \mathscr{E}(A)^{[\perp]} \in \mathbf{K(H)}$, and $\sigma_p(A \mid \mathscr{E}(A)^{[\perp]}) = \emptyset$.

Hence it follows that $\mathscr{E}(A)^{[\perp]} = L'_+ [\dot{+}] \mathscr{L}'_-$, where $\mathscr{L}'_\pm$ are maximal (in $\mathscr{E}(A)^{[\perp]}$) uniformly definite subspaces invariant relative to A. Since $\sigma(A \mid \mathscr{L}'_\pm) \subset \sigma(A)$, the sets $\sigma(A \mid \mathscr{L}'_\pm)$ have no more than a countable set of points of condensation. By Lemma 2.7 $\sigma_p(A \mid \mathscr{L}'_\pm) = \emptyset$—we obtain a contraction with the fact that

$$((\sigma_p(A) \mid \mathscr{L}'_+) \cup \sigma_p(A \mid \mathscr{L}'_-) = (\sigma_p(A)\mathscr{E}(A)^{[\perp]}) = \emptyset.$$

d), e). First of all we note that if an operator $A \in \mathbf{K}(\mathbf{H})$ has non-real eigenvalues, then, by virtue of the neutrality of the eigenvectors corresponding to them and 3.Formula (5.4), there is in $\mathscr{H}$ no J-orthonormalized basis composed of eigenvectors of the operator A.

Now let $\mathscr{E}(a) = \mathscr{H}$, and $\mathscr{L}_\lambda(A) = \mathscr{N}_\lambda [\dot{+}] \mathscr{M}_\lambda$ when $\lambda \in s(A)$, where $A\mathscr{N}_\lambda \subset \mathscr{N}_\lambda$, $\dim \mathscr{N}_\lambda < \infty$, $\mathscr{M}_\lambda \subset \operatorname{Ker}(A - \lambda I)$ and $\mathscr{M}_\lambda$ is projectionally complete (see Exercise 5 on 3.§5). Since $s(A)$ consists of a finite number of points, and by virtue of Lemma 2.1 all the $\mathscr{L}_\lambda(A)$ are non-degenerate when $\lambda \in s(A)$, so, taking 3.(5.4) into account, $\mathscr{H} = \mathscr{H}_1 [\dot{+}] \mathscr{H}_2$, where $\mathscr{H}_1$ and $\mathscr{H}_2$ are invariant relative to A, and $\mathscr{H}_1 = \operatorname{Lin}\{\mathscr{L}_\mu(A), \mathscr{N}_\lambda \mid \mu \neq \bar{\mu}, \lambda \in s(A)\}$. Consequently $\dim \mathscr{H}_1 < \infty$, and $\sigma(A \mid \mathscr{H}_2) \subset \mathbb{R}_1$ and moreover $\mathscr{E}_0(A \mid \mathscr{H}_2) = \mathscr{E}(A \mid \mathscr{H}_2) = \mathscr{H}_2$. By virtue of Exercise 15 on 3.§5 the operator $A \mid \mathscr{H}_2 \in \mathbf{K}(\mathbf{H})$. Therefore assertions d) and e) of the theorem will be proved if we show that the conditions $\sigma(A) \subset \mathscr{R}$ and $\mathscr{E}_0(A) = \mathscr{H}$ imply the existence in $\mathscr{H}$ of a J-orthonormalized Riesz basis composed of eigenvectors of the operator A. We verify this.

Let $(\mathscr{L}_+, \mathscr{L}_-)$ be a maximal dual pair invariant relative to A, and let $\mathscr{L}_\pm \in h^\pm$, and $\mathscr{L}_\lambda^\pm = \operatorname{Ker}(A - \lambda I) \cap \mathscr{L}_\pm$. It can be verified, just as in the proof of assertion a), that $\mathscr{E}_0(A \mid \mathscr{L}_\pm) = \mathscr{L}_\pm$. Since $\operatorname{Ker}(A - \lambda I) [\perp] \operatorname{Ker}(A - \mu I)$ when $\lambda \neq \mu$, and $\mathscr{L}_\pm \in h^\pm$, it follows that among the subspaces $\mathscr{L}_\lambda^\pm$ for different λ there are only a finite number of degenerate ones. Let these be $\mathscr{L}_{\lambda_1}^\pm, \mathscr{L}_{\lambda_2}^\pm, \ldots, \mathscr{L}_{\lambda_n}^\pm$ and let $\mathscr{L}_{\lambda_i}^\pm = \mathscr{L}_{0,\lambda_i} + \mathscr{L}_{1,\lambda_i}^\pm$, where $\mathscr{L}_{0,\lambda_i}$ is the isotropic part of $\mathscr{L}_{\lambda_i}^\pm$, i.e., $\mathscr{L}_{0,\lambda_i} = \mathscr{L}_{\lambda_i}^+ \cap \mathscr{L}_{\lambda_i}^-$, and the subspace $\mathscr{L}_{1,\lambda_i}^\pm$ is definite and completes $\mathscr{L}_{0,\lambda_i}$ into $\mathscr{L}_{\lambda_i}^\pm$. Since $\operatorname{Lin}\{\mathscr{L}_{0,\lambda_i}\}_1^n$ is a finite-dimensional subspace, it follows that $\dim \mathscr{H}/\mathscr{E}_1(A) < \infty$, where $\mathscr{E}_1(A) = C \operatorname{Lin}\{\mathscr{L}_\mu^+, \mathscr{L}_\mu^-, \mathscr{L}_{1,\lambda_i}^+, \mathscr{L}_{1,\lambda_i}^- \mid \mu \neq \lambda_i, i = 1, 2, \ldots, n\}$ is a non-degenerate subspace relative to A, and moreover the maximal (in $\mathscr{E}_1(A)$) uniformly definite subspaces $\mathscr{E}_1^\pm(A) = C \operatorname{Lin}\{\mathscr{L}_\mu^\pm, \mathscr{L}_{1,\lambda_i}^\pm \mid \mu \neq \lambda_i, i = 1, 2, \ldots, n\}$ are invariant relative to A. We now note that the operators $A \mid \mathscr{E}_1^\pm(A)$ satisfy the conditions of Lemma 2.7 relative to the scalar products $\pm [\cdot, \cdot] \mid \mathscr{E}_1^\pm(A)$ respectively, and $\dim \mathscr{E}_1(A)^{[\perp]} < \infty$, and moreover $\mathscr{E}_1(A)^{[\perp]} = \mathscr{E}_0(A \mid \mathscr{E}_1(A)^{[\perp]})$ and the kernels $\operatorname{Ker}(A \mid \mathscr{E}_1(A)^{[\perp]} - \lambda I)$ of the operators $(A \mid \mathscr{E}_1(A)^{[\perp]} - \lambda I)$ are non-degenerate. Consequently in each of the subspaces $\mathscr{E}_1^+(A)$, $\mathscr{E}_1^-(A)$, and $\mathscr{L}_\lambda(A \mid \mathscr{E}_1(A)^{[\perp]})$ there are J-orthonormalized Riesz bases composed of eigenvectors of the operator A. The union of these bases will then be the required basis.

f) Before proving this assertion we point out that, if two bases differ on a finite number of elements and if one of the bases is a p-basis, then the second one will also be a p-basis. In proving assertions d) and e) it was essentially established that if $\mathscr{E}(A)$ (respectively $\mathscr{E}_0(A)$) coincides with $\mathscr{H}$, then there is in $\mathscr{H}$ an almost J-orthonormalized basis composed of root vectors (respectively, eigenvectors) of the operator A, and all these vectors, with the exception, perhaps, of a finite number, are characteristic vectors for A and lie in pre-assigned invariant subspaces $\mathscr{L}_\pm \in \mathscr{M}^\pm \cap h^\pm$ of the operator A. Let this operator have the invariant subspace $\mathscr{L}_+ \in \mathscr{M}^+ \cap h^+$ with the angular operator $K_{\mathscr{L}_+} \in \mathscr{S}_p$, let $\{f_\alpha^\pm\}$ be the J-orthonormalized part of a Riesz basis composed of root vectors (or eigenvectors) of the operator A, and let $\{f_\alpha^\pm\} \subset \mathscr{L}_\pm$, where $\mathscr{L}_- = \mathscr{L}_+^{[\perp]}$. Since the system $\{f_\alpha^+\} \cup \{f_\alpha^-\}$ differs from a basis in $\mathscr{H}$ on a finite number of elements, it can be constructed into a J-orthonormalized basis in $\mathscr{H}$ which will differ from the original one on a finite number of elements. Therefore without loss of generality we shall suppose that $\{f_\alpha^+\} \cup \{f_\alpha^+\}$ is a J-orthonormalized basis in $\mathscr{H}$ and we shall prove that it is a p-basis. Since $\{f_\alpha^+\} \cup \{f_\alpha^-\}$ is a basis, we have $\mathscr{L}_\pm = C\,\text{Lin}\{f_\alpha^\pm\}$, and therefore $\| K_{\mathscr{L}_+} \| < 1$. From 2.Formula (5.3) we construct the operator $U(K_{\mathscr{L}_+})$, which is positive, J-unitary, and such that $U(K_{\mathscr{L}_+}) - I \in \mathscr{S}_p$. Since $\{f_\alpha^+\} \cup \{f_\alpha^-\}$ is a J-orthonormalized basis, $\{e_\alpha^+\} \cup \{e_\alpha^-\}$, where $e_\alpha^\pm = U^{-1}(K_{\mathscr{L}_+})f_\alpha^\pm$, also is a J-orthonormalized basis. But since $e_\alpha^\pm$ by construction lie in $\mathscr{H}^\pm$, it follows that $\{e_\alpha^+\} \cup \{e_\alpha^-\}$ is an orthonormalized basis. It remains to notice that the inclusion $U(K_{\mathscr{L}_+}) - I \in \mathscr{S}_p$ implies the inclusion $U^{-1}(K_{\mathscr{L}_+}) - I \in \mathscr{S}_p$, and therefore $\{f_\alpha^+\} \cup \{f_\alpha^-\}$ is a p-basis.

Conversely, suppose there is in $\mathscr{H}$ an almost J-orthonormalized basis composed of root vectors (or eigenvectors) of A and that it is a p-basis. We shall prove that there is an invariant subspace $\mathscr{L}_+ \in \mathscr{M}^+ \cap h^+$ of the operator A with an angular operator $K_{\mathscr{L}_+} \in \mathscr{S}_p$. To do this we separate from the basis the J-orthonormalized system $\{f_\alpha^+\} \cup \{f_\alpha^-\}$ composed of the definite eigenvectors of the operator A, with $\{f_\alpha^\pm\} \subset \mathscr{P}^{\pm\pm}$. We form the subspaces $\mathscr{L}_\pm^{(1)} = C\,\text{Lin}\{f_\alpha^\pm\}$. They are uniformly definite, invariant relative to A, and such that def $\mathscr{L}_\pm^{(1)} < \infty$. By 3.Theorem 4.14 the dual pair $(\mathscr{L}_+^{(1)}, \mathscr{L}_-^{(1)})$ can be extended into a maximal dual pair $(\mathscr{L}_+, \mathscr{L}_-)$ invariant relative to A. It is clear from the construction that $\mathscr{L}_+ \in \mathscr{M}^+ \cap h^+$. We verify that $K_{\mathscr{L}_+} \in \mathscr{S}_p$. Since $\dim \mathscr{L}_\pm / \mathscr{L}_\pm^{(1)} < \infty$, we can suppose without loss of generality that $\mathscr{L}_\pm^{(1)} = \mathscr{L}_\pm$, i.e., $\{f_\alpha^+\} \cup \{f_\alpha^-\}$ is a J-othonormalized p-basis. Let $f_\alpha^\pm = (I + T)g_\alpha^\pm$, where $\{g_\alpha^\pm\}$ is an orthonormalized basis in $\mathscr{H}$, $T \in \mathscr{S}_p$, and $e_\alpha^\pm = U^{-1}(K_{\mathscr{L}_+})f_\alpha^\pm$ is an orthonormalized basis in $\mathscr{H}$ composed of vectors $e_\alpha^\pm \in \mathscr{H}^\pm$. From 2.Formula (5.3) it can be seen that $K_{\mathscr{L}_+} \in \mathscr{S}_p$ if and only if $U(K_{\mathscr{L}_+}) - T \in \mathscr{S}_p$. We establish this lost result. Since $e_\alpha^\pm = U^{-1}(K_{\mathscr{L}_+})(I + T)g_\alpha^\pm$, it follows that $U^{-1}(K_{\mathscr{L}_+})(I + T) \equiv V$ is a unitary operator, and therefore $I + T = V(V^*U(K_{\mathscr{L}_-})V)$ is the polar decomposition of the operator $I + T$. Hence, $(I + T^*)(I + T) = V^*U^2(K_{\mathscr{L}_+})V$, and therefore $U^2(K_{\mathscr{L}_-}) - \in \mathscr{S}_p$, which implies the inclusion $(U(K_{\mathscr{L}_+}) + I)^{-1}(U^2(K_{\mathscr{L}_-}) - I) = U(K_{\mathscr{L}_-}) - I \in \mathscr{S}_p$. ∎

Remark 2.13: Since completely continuous J-selfadjoint operators of the class $\mathbf{K(H)}$ and all π-selfadjoint operators with a spectrum having no more than a countable set of points of condensation satisfy the conditions of Theorem 2.12, the assertions in it also hold for them. Moreover, for completely continuous operators it is possible, by Corollary 2.3, to write $\mathscr{L}_0(A)$ instead of $\mathrm{Lin}\{\mathscr{L}_\lambda(A) \mid \lambda \in s(A)]$ in the formulation of Theorem 2.12.

4 In conclusion we investigate the question of the completeness of the system of root vectors of definitizable J-selfadjoint operators.

Lemma 2.14: *Let A be a bounded J-non-negative operator having a spectrum with no more than a countable set of points of condensation. Then the equality $\mathscr{E}(A) = \mathscr{H}$ is equivalent to non-degeneracy of $\mathscr{L}_0(A)$.*

□ Since $\mathscr{L}_0(A)\,[\perp]\,\mathscr{L}_\mu(A)$ when $\mu \neq 0$, it is clear that non-degeneracy of $\mathscr{E}(A)$ $(=\mathscr{H})$ implies non-degeneracy of $\mathscr{L}_0(A)$.

Conversely, let $\mathscr{L}_0(A)$ be non-degenerate. Then $\mathscr{E}(A)$ is also non-degenerate. For, if $\mathscr{E}(A)$ is degenerate and if $x_0 \in \mathscr{E}(A) \cap \mathscr{E}(A)^{[\perp]}$, then $[Ax_0, x_0] = 0$. Consequently $x_0 \in \mathrm{Ker}\,A$, which implies the degeneracy of $\mathscr{L}_0(A)$.

Let us assume that $\mathscr{E}(A) \neq \mathscr{H}$, i.e., $\mathscr{E}(A)^{[\perp]} \neq \{\theta\}$. We consider the operator $A' \equiv A \mid \mathscr{E}(A)^{[\perp]}$. This operator has the following properties: it is positive relative to the G-metric $[\,\cdot\,,\,\cdot\,]\mid \mathscr{E}(A)^{[\perp]}$; $\sigma_p(A' \mid = \emptyset$; and $\sigma(A')$ consists of not more than a countable set of points. The first two of these assertions are trivial, and so we verify the third. To do this we note first that if $\lambda \in \rho(A)$, then $\lambda \in \rho(A')$. From 2.Corollary 3.25 and Exercise 7 on §1 it follows that if $\lambda_0 (\neq 0)$ is an isolated point of the spectrum of the operator A, then the range of values of the operator $A - \lambda_0 I$ is closed, $\mathrm{Ker}(A - \lambda_0 I)$ is projectionally complete, and $\mathscr{H} = \mathrm{Ker}(A - \lambda_0 I)\,[\dotplus]\,\mathrm{Ker}(A - \lambda_0 I)^{[\perp]}$, and moreover $\lambda_0 \in \rho(A \mid \mathrm{Ker}(A - \lambda_0 I)^{[\perp]})$. Hence we conclude that if $\lambda_0 (\neq 0)$ is an isolated point in $\sigma(A)$, then $\lambda_0 \in \rho(A')$. By hypothesis $\sigma(A)$ has no more than a countable set of points of condensation and therefore $\sigma(A')$ consists of not more than a countable set of points.

Let μ_0 be an isolated point of the spectrum of the operator A', and let γ_{μ_0} be a circle of sufficiently small radius with centre at the point μ_0. Then $P_{\mu_0} = (1/2\pi i)\oint_{\gamma_{\mu_0}}(A' - \lambda I)^{-1}\,d\lambda$ is a G-orthogonal projector on to the 'G_{μ_0}-space' $\mathscr{H}_{\mu_0} = P_{\mu_0}\mathscr{E}(A)^{[\perp]}$ invariant relative to A', and $\sigma(A' \mid \mathscr{H}_{\mu_0}) = \{\mu_0\}$. The operator $A_{\mu_0} \equiv A' \mid \mathscr{H}_{\mu_0}$ is G_{μ_0}-positive, and $\mu_0 \notin \sigma_p(A_{\mu_0})$. By 3.Lemma 3.16 we extend A_{μ_0} into a $\tilde{J}$-non-negative operator $\tilde{A}_{\mu_0}$ which has μ_0 as the only point of its spectrum. In accordance with Exercise 7 on §1 $\tilde{A}_{\mu_0} = \mu_0 \tilde{I}_{\mu_0}$ when $\mu_0 \neq 0$ and $\tilde{A}^2_{\mu_0} = 0$ when $\mu_0 = 0$, i.e., $\mu_0 \in \sigma_p(A')$—but $\mu_p(A') = \emptyset$, so we have obtained a contradiction. ■

Theorem 2.15: *Let A be a bounded J-selfadjoint operator, definitizable by the polynomial $p(\lambda)$ with the roots $\{\lambda_i\}_1^n$ and let $\sigma(A)$ have no more than a* *countable set of points of condensation. Then the equality* $\mathscr{E}(A)=\mathscr{H}$ *is equivalent to non-degeneracy of* $\operatorname{Lin}\{\mathscr{L}_{\lambda_i}(A)\}_1^n$.

□ Since $\mathscr{E}(A)=\mathscr{E}(p(A))$ and $\mathscr{L}_0(p(A))=\operatorname{Lin}\{\mathscr{L}_{\lambda_i}(A)\}_1^n$, it is sufficient to use Lemma 2.14. ■

Remark 2.16: Since for a J-non-negative operator A the equality $\mathscr{L}_\lambda(A)=\operatorname{Ker}(A-\lambda I)$ holds for all $\lambda\neq 0$, so, on replacing in the formulations of Lemma 2.14 and Theorem 2.15 the root subspaces by the corresponding kernels $\operatorname{Ker}(A-\lambda I)$ of the operator A, we obtain a criterion for the coincidence of $\mathscr{E}_0(A)$ with $\mathscr{H}$. ■

In contrast to the case of operators of the class $\mathbf{k}(\mathbf{H})$ completeness of the system of root vectors or even of the eigenvectors of a J-non-negative operator does not guarantee the existence of a basis composed of such vectors. However, the following theorem holds.

Theorem 2.17: *Let a bounded J-non-negative operator A satisfy the conditions of Lemma* 2.14. *Then* $A^nx=\Sigma_\alpha\lambda_\alpha^n[x,f_\alpha]f_\alpha\operatorname{sign}\lambda_\alpha$, *where* $n\geqslant 2$, $\{\lambda_\alpha\}\subset\sigma_p(A)$, $\lambda_\alpha\neq 0$, *and* $\{f_\alpha\}$ *is a J-orthonormalized system composed of the eigenvectors of operator A*: $Af_\alpha=\lambda_\alpha f_\alpha$; *if moreover A is a J-positive operator, then* $Ax=\Sigma_\alpha\lambda_\alpha[x,f_\alpha]f_\alpha\operatorname{sign}\lambda_\alpha$ (*here the series converge with respect to the norm of the space* $\mathscr{H}$).

□ In accordance with Theorem 1.5 $Ax=Sx+\int_{-\infty}^{\infty}\nu\,dE_\nu x$. From Lemma 2.7 we conclude that

$$I-E_\lambda=\sum_{\alpha\geqslant\lambda}[\cdot,f_\alpha]f_\alpha,\qquad E_{-\lambda}=-\sum_{\alpha<-\lambda}[\cdot,f_\alpha]f_\alpha\quad\text{when }\lambda>0,$$

where $\{f_\alpha\}$ is the J-orthonormalized system of eigenvectors of the operator A corresponding to the eigenvalues $\{\lambda_\alpha\}$: $Af_\alpha=\lambda_\alpha f_\alpha$, $\lambda_\alpha\neq 0$. Consequently $\int_{-\infty}^{\infty}\nu\,dE_\nu x=\Sigma_\alpha\lambda_\alpha[x_1f_\alpha]f_\alpha\operatorname{sign}\lambda_\alpha$, where the series converges with respect to the norm of $\mathscr{H}$ (we notice at once that if A is a J-positive operator, then $S=0$ and therefore $Ax=\Sigma_\alpha\lambda_\alpha[x,f_\alpha]f_\alpha\operatorname{sign}\lambda_\alpha$). We use Theorem 1.5:

$$A^nx=A^{n-1}\left(S+\int_{-\infty}^{\infty}\nu\,dE_\nu\right)x=A^{n-1}\int_{-\infty}^{\infty}\nu\,dE_\nu x$$

$$=\sum_\alpha\lambda_\alpha^n[x,f_\alpha]f_\alpha\operatorname{sign}\lambda_\alpha\quad\text{when }n=2,3,\ldots.$$ ■

Exercises and problems

1 Let A be a completely continuous J-dissipative operator of the class $\mathbf{K}(\mathbf{H})$, let $A_J\in\mathscr{S}_1$ and $\underline{\lim}_{n\to\infty}ns_n(A)=0$. Prove that $\mathscr{E}(A)=\mathscr{H}$ if and only if $\mathscr{L}_0(A)$ is non-degenerate (Azizov [4], [8], Azizov and Usvyatsova [2]).

Hint: Use Lemma 2.4 and the definite analogue of this assertion ([XI], Theorem 4.4.2).

2 Let $A \in \mathscr{S}_\infty \cap \mathbf{K}(\mathbf{H})$, $\theta_A = \pi/p$, $p \geqslant 1$, and $s_n(A) = o(n^{-1/p})$ when $n \to \infty$. Prove that $\mathscr{E}(A) = \mathscr{H}$ if and only if $\mathscr{L}_0(A)$ is non-degenerate [Azizov [4], [8], Azizov and Usvyatsova [2]).
Hint: Use Exercise 6 on 2.§2, Lemma 2.4 and the definite analogue of this assertion ([XI], Theorem 5.6.1).

3 Let $A \in \mathscr{S}_\infty \cap \mathbf{K}(\mathbf{H})$, $\theta_A = \pi/p$, $p > 1$, and for some α for the operator $B = [e^{i\alpha}A]_I$ let $s_n(B) = o(n^{-1/p})$ when $n \to \infty$. Prove that the equality $\mathscr{E}(A) = \mathscr{H}$ is equivalent to the non-degeneracy of $\mathscr{L}_0(A)$ (Azizov [4], [8], Azizov and Usvyatsova [2]).
Hint: The same as for Exercise 2 except that [XI] 5.6.1 is replaced by [XI] 5.6.2.

4 Let $\mathscr{H}$ be a Pontryagin $G^{(\varkappa)}$-space (i.e., $0 \in \rho(G^{(\varkappa)})$). Prove that there is in $\mathscr{H}$ at least one 'almost $G^{(\varkappa)}$-orthonormalized p-basis' if and only if $|G^{(\varkappa)}|^{1/2} - I \in \mathscr{S}_p$ (Azizov and Kuznetsova [1]).

5 Prove that if in a Pontryagin $G^{(\varkappa)}$-space $\mathscr{H}$ there is at least one almost $G^{(\varkappa)}$-orthonormalized p-basis, then any other almost $G^{(\varkappa)}$-orthonormalized basis is also a p-basis. In particular, if $G^{(\varkappa)2} = I$, then any $G^{(\varkappa)}$-orthonormalized basis is a p-basis for any $p > 0$ (Azizov and Kuznetsova [1]).

6 Give an example of a J-non-negative bounded operator $B \notin \mathscr{S}_\infty$ such that $B^2 \in \mathscr{S}_p$ $(0 < p \leqslant \infty)$. (I. Iokhvidov [8], Azizov and Shlyakman [1]).
Hint: In 2 Example 3.36 put $B_1 \in \mathscr{S}_p$ and take as B_2 the linear homeomorphism mapping $\mathscr{H}_1$ on the $\mathscr{H}_2$, $B_1 \geqslant 0$, $B_2 \geqslant 0$.

7 Prove that if A is a bounded J-selfadjoint operator definitizable by the polynomial $p(\lambda)$ with the roots $\{\lambda_i\}_1^n$, then the following assertions are equivalent:
a) $\mathbb{C}\backslash\{\lambda_i\}_1^n \subset \tilde{\rho}(A)$;
b) there is an integer $q > 0$ such that $A^q p(A) \in \mathscr{S}_\infty$;
c) $Ap(A) \in \mathscr{S}_\infty$. (V. Shtraus [3]; Azizov and Shlyakman [1]).

8 Let $A \in \mathscr{S}_\infty$ be a $G^{(\varkappa)}$-selfadjoint operator in $\mathscr{H}$ $(0 \leqslant \varkappa < \infty)$, and let $0 \in \sigma_c(A)$. Prove that then $\mathscr{E}(A^*) = \mathscr{H}^{(\varkappa)}$ and that there is in $\mathscr{H}^{(\varkappa)}$ a Riesz basis relative to the norm $\|x\|_1 = \||G|^{1/2}x\|$ composed of root vectors of the operator A ([3]).

9 Let $A \in \mathscr{S}_\infty$ be a G-selfadjoint operator in $\mathscr{H}$, let $0 \in \sigma_c(A)$, and let at least one of the sets $(-\infty, 0) \cap \sigma(GA)$ or $\sigma(GA) \cap (0, \infty)$ consist of a finite number of normal eigenvalues. Prove that then $\mathscr{E}(A^*) = \mathscr{H}$ and that there is in $\mathscr{H}$ a Riesz basis relative to the norm $\|x\|_1 = \||GA|^{1/2}x\|$ composed of root vectors of the operator A ([III]).
Hint: Relative to the indefinite form $[x, y]_1 = [Ax, y]$ the space $\mathscr{H}$ is a $G^{(\varkappa)}$-space with $G^{(\varkappa)} = GA$. Use the result of Exercise 8.

10 Let $\mathscr{H}$ be a $W^{(\varkappa)}$-space, with $0 \in \tilde{\rho}(W^{(\varkappa)})$, let $\mathscr{H} \mid \operatorname{Ker} W^{(\varkappa)}$ be a Pontryagin space, and let A be a completely continuous $W^{(\varkappa)}$-selfadjoint operator. Prove that there is in $\mathscr{H}$ a Riesz basis composed of root vectors of the operator A if and only if $\mathscr{L}_0(A) \cap C\operatorname{Lin}\{\mathscr{L}_\lambda(A) | \lambda \neq 0\} = \{\theta\}$ (Azizov [7]).

11 Prove that if $A \in \mathscr{S}_\infty$ is a π-selfadjoint operator and $\mathscr{E}(A) \neq \Pi_\varkappa$, then it is impossible to choose in the subspace $\mathscr{E}(A)$ a Riesz basis composed of root vectors of the operator A (Azizov [7]).
Hint: Prove that the inequality $\mathscr{E}(A) \neq \Pi_\varkappa$ implies the inequality $\mathscr{L}_0(A) \cap C\operatorname{Lin}\{\mathscr{L}_\lambda(A) \mid \lambda \neq 0\} \neq \{\theta\}$, and use the result of Exercise 10.

§3 Examples and applications

1 In this section we give some examples showing the impossibility of weakening the conditions in some of the theorems given in §2, and also examples showing some applications of the results in the preceding section. For our first purpose we need the following.

Theorem 3.1: *Let $A = (A_R + iA_I)$ be a continuous operator acting in a Pontryagin space $\Pi_\varkappa$, and let A_R and A_I be π-non-negative operators. Then non-degeneracy of $\mathscr{L}_0(A)$ is equivalent to the inclusion* $\operatorname{Ker} A \cap \bar{\mathscr{R}}_A \subset \mathscr{R}_A$.

□ Let $\mathscr{L}_0(A)$ be non-degenerate. In accordance with 2.Theorem 2.26 $\mathscr{L}_0(A) = \mathscr{N}_0 [\dot{+}] \mathscr{M}_0$, where $\mathscr{N}_0$ is finite-dimensional and invariant relative to A, and $\mathscr{M}_0 \subset \operatorname{Ker} A$ is non-degenerate. Consequently $\mathscr{N}_0$ is a non-degenerate subspace. By construction (see the proof of 2.Theorem 2.26) the kernel of the operator $A_1 = A \mid \mathscr{N}_0$ is neutral and it is the isotropic part of the kernel of the operator A. Since A and A_1 are π-dissipative operators, it follows from 2.Corollary 2.17 that $\operatorname{Ker} A = \operatorname{Ker} A^c$, and $\operatorname{Ker} A_1 = \operatorname{Ker} A_1^c$. This implies the equality

$$\begin{aligned}\operatorname{Ker} A \cap \bar{\mathscr{R}}_A &= \operatorname{Ker} A \cap (\operatorname{Ker} A^c)^{[\perp]} = \operatorname{Ker} A \cap (\operatorname{Ker} A)^{[\perp]} \\ &= \operatorname{Ker} A_1 = \operatorname{Ker} A_1 \cap \mathscr{R}_{A_1}.\end{aligned}$$

Since $\mathscr{R}_{A_1} \subset \mathscr{R}_A$, it follows that $\operatorname{Ker} A \cap \bar{\mathscr{R}}_A \subset \mathscr{R}_A$.

Conversely, let $\operatorname{Ker} A \cap \mathscr{R}_A \subset \mathscr{R}_A$. If $x_0 \in \mathscr{L}_0(A) \cap \mathscr{L}_0(A)^{[\perp]}$, then $[Ax_0, x_0] = 0$. Using the fact that A_R and A_T are π-non-negative we obtain that $x_0 \in \operatorname{Ker} A$. Consequently, since $x_0 [\perp] \mathscr{L}_0(A)$, we have $x_0 [\perp] \operatorname{Ker} A (= \operatorname{Ker} A^c)$, and therefore $x_0 \in \bar{\mathscr{R}}_A$, i.e., $x_0 \in \operatorname{Ker} A \cap \bar{\mathscr{R}}_A$. By hypothesis $\operatorname{Ker} A \cap \bar{\mathscr{R}}_A \subset \mathscr{R}_A$, and therefore there is a vector y_0 such that $x_0 = Ay_0$. The vector $y_0 \in \mathscr{L}_0(A)$ and therefore $0 = [x_0, y_0] = [Ay_0, y_0]$. Again using the fact that A_R and A_I are π-non-negative, we obtain that $(A_{y0} =)x_0 = \theta$, i.e., $\mathscr{L}_0(A)$ is non-degenerate. ■

The theorem just proved enables us easily to construct examples demonstrating that in the theorems about completeness (see §2) the condition of non-degeneracy of $\operatorname{Lin}\{\mathscr{L}_\lambda(A) \mid \lambda \in s(A)\}$ does not follow from any of the other conditions. We give one such example, and leave the reader to construct others.

Example 3.2: (*cf.* Corollary 2.9 and Exercise 3 on §2 when $p \geqslant 2$). Let B be a nuclear operator in an infinite-dimensional space $\mathscr{H}$, with $\operatorname{Ker} B \neq \{\theta\}$, and let the operators $\frac{1}{2}(B + B^*)$ and $(1/2i)(B - B^*)$ be non-negative. In $\operatorname{Ker} B$ and $\bar{\mathscr{R}}_B \mid \mathscr{R}_B$ we fix on vectors x_0 ($\| x_0 \| = 1$) and y_0 ($\| y_0 \| = 1$) respectively. Since B is a dissipative operator, $x_0 \perp y_0$. This in turn implies that the operator

$$J\colon J(\alpha x_0 + \beta y_0) = \beta x_0 + \alpha y_0, \qquad J \mid (\operatorname{Lin}\{x_0, y_0\})^\perp = -I$$

is selfadjoint and unitary. By means of this operator we introduce the form $[x, y] = (Jx, y)$, turning the space $\mathscr{H}$ into a Pontryagin space with one positive square. The operator $A = JB$ is nuclear, and $A_R = J[(B + B^*)/2]$ and $A_J = J[(B - B^*)/2i]$, and so A_R and A_I are π-non-negative operators. Since $\operatorname{Ker} A = \operatorname{Ker} B$ and $y_0 \perp \operatorname{Ker} B$, so $x_0 [\perp] \operatorname{Ker} A$, i.e., x_0 is an isotropic vector in $\operatorname{Ker} A$, and therefore $x_0 \in \operatorname{Ker} A \cap \bar{\mathscr{R}}_A$. But $x_0 \notin \mathscr{R}_A$, for otherwise y_0 would be in $\mathscr{R}_B$. Consequently, by Theorem 3.1, $\mathscr{L}_0(A)$ is degenerate. ■

2 The following example shows that in Theorem 2.8 the condition that the operator A be π-dissipative cannot be replaced by the more general condition that A is a J-dissipative operator of the class **H**.

□ *Example 3.3: A Volterra J-dissipative operator A of the class* **H** *with a nuclear J-imaginary component and sp* $A_I = 0$.

Let $\mathscr{H} = \mathscr{H}^+ \oplus \mathscr{H}^-$ be a J-space, where $\mathscr{H}^\pm \bar{L}^2(0, 1)$. We put

$$A = \| A_{ij} \|_{i,j=1}^{2}, \qquad A_{ij} = 0 \text{ when } i \neq j, \qquad A_{11} = -A_{22} = 2i \int_0^t \cdot \, ds$$

As is well-known (see, e.g., [XI], 4.7.4) A_{11} is a Volterra disipative operator and $(1/2i)(A_{11} - A_{11}^*)$ is a one-dimensional operator. So A is a Volterra J-dissipative operator with a two-dimensional J-imaginary component A_I and $\operatorname{sp} A_I = \operatorname{sp}(A_{11})_I + \operatorname{sp}(A_{22})_J = 0$. It remains to verify that $A \in \mathbf{H}$. To do this it suffices to establish that $\mathscr{H}^\pm$ are its only maximal semi-definite invariant subspaces. In the present case the latter is equivalent to the fact that for any bounded operator B it follows from $BA_{11} = A_{22}B$ that $B = 0$ (see 3.Lemma 2.2).

Let $2iB\int_0^t f(s)\, ds = -2i\int_0^t (Bf)(s)\, ds$. Differentiating both sides of this equality with respect to t we obtain $Bf = -Bf$, i.e., $Bf = 0$ and therefore $B = 0$. ■

3 We recall that a function $K(s, t)$ of two variables defined on a square $[a, b] \times [a, b]$ is called a *Hermitian non-negative kernel* if for any finite set of points $\{t_i\}_1^n$ of $[a, b]$ and for any complex numbers $\{\xi_i\}_1^n$ the sum $\sum_{i,j=1}^n K(t_i, t_j)\xi_i\bar{\xi}_j$ is non-negative, and it is called a *Hermitian positive kernel* if $\sum_{i,j=1}^n K(t_i, t_j)\xi_i\bar{\xi}_j = 0$ if and only if $\xi_i = 0$, $i = 1, 2, \ldots, n$ (for a more general definition see 4.§3.11 below). The function $K(s, t)$ is called a *dissipative kernel* if $(1/2i)(K(s, t) - \overline{K(t, s)})$ is a Hermitian non-negative kernel, and a *strictly dissipative kernel* if $(1/2i)(K(s, t) - \overline{K(t, s)})$ is a Hermitian positive kernel.

Let $\mathscr{H} = L_\omega^2(a, b)$ (*cf.* Exercise 5 on 1.§2), i.e., the space of all ω-measurable functions φ such that $|\int_a^b |\varphi(t)|^2\, d\omega(t)| < \infty$, where ω is a function either non-decreasing or non-increasing on $[a, b]$, and the scalar product (φ, ψ) is given, up to sign, by the relation $(\varphi, \psi) = \int_a^b \varphi(t)\bar{\psi}(t)\, d\omega(t)$. Moreover, we regard $\varphi = \psi$ if

$$\int_a^b |\varphi(t) - \psi(t)|^2 \, d\omega(t) = 0 \qquad (\varphi, \psi \in L^2_\omega(a, b)).$$

Let $\sigma(t)$ be a function of bounded variation on $[a, b]$. In Exercises 5 and 6 on 1.§2 it was proved that the space $L^2_\omega(a, b)$, where $\omega(t) = \omega_1(t) + \omega_2(t)$, and $\sigma(t) = \omega_1(t) - \omega_2(t)$ is the canonical representation of $\sigma(t)$ in the form of the difference of two non-decreasing functions, and the space is provided with the form $[\varphi, \psi] = \int_a^b \varphi(t)\overline{\psi(t)} \, d\sigma(t)$ is a J-space. In particular, if $\omega_1(t)$ is a piecewise-constant function with $\varkappa$ points of growth, then $L^2_\omega(a, b)$ is a Pontryagin space with $\varkappa$ positive squares (see Exercise 7 on 3.§9).

In this and the following paragraphs we apply the results obtained in §2 to the investigation of integral operators $A = \int_a^b K(s, t) \cdot d\sigma(t)$.

Theorem 3.4: *Let $K(s, t)$ be a dissipative kernel continuous on $[a, b] \times [a, b]$, let A and A_1 be integral operators defined by the relations*

$$A = \int_a^b K(s, t) \, d\sigma(t), \qquad A_1 = \int_a^b K(s, t) \, dt,$$

let $\sigma(t) = \omega_1(t) - \omega_2(t)$, and let $\omega_1(t)$ be a piecewise-constant function with $\varkappa$ points of growth. Then if the system of root vectors of the operator A_1 corresponding to its non-zero eigenvalues is complete in $C[a, b]$, then the system of root vectors of the operator A corresponding to its non-zero eigenvalues is complete in $L^2_\omega(a, b)$. If in addition $\sigma(t)$ has no intervals of constancy, then the system mentioned of root vectors of the operator A is complete in $C[a, b]$.

□ It follows from the definition of the operator A that it is a nuclear π-dissipative operator. Therefore by virtue of Corollary 2.9 the root vectors of the operator A corresponding to its non-zero eigenvalues will be complete in $L^2_\omega(a, b)$ if and only if $0 \notin \sigma_p(A)$. We prove that $0 \notin \sigma_p(A)$. Let $\varphi_0 \in \operatorname{Ker} A$. Then $\varphi_0 \in \operatorname{Ker} A^c$, i.e., $\int_a^b \overline{K(s, t)} \varphi_0(s) \, d\sigma(s) = 0$. Let $\zeta_1(t), \zeta_2(t), \ldots$ be the root vectors of the operator A_1 corresponding to the non-zero eigenvalues and forming a system complete in $C[a, b]$. It follows from the definition of root vectors that $\zeta_i(t) \in \mathscr{R}_{A_1}$, i.e., there are functions $\xi_i(t) \in C[a, b]$ such that $\zeta_i(t) = (A_1 \xi_i)(t)$ $(i = 1, 2, \ldots)$. Then

$$\begin{aligned} [\varphi_0, \zeta_1] &= \int_a^b \varphi_0(s)\overline{\zeta_i(s)} \, d\sigma(s) = \int_a^b \varphi_0(s)\overline{(A_1\xi_i)(s)} \, d\sigma(s) \\ &= \int_a^b \varphi_0(s) \int_a^b \overline{K(s, t)\xi_i(t)} \, dt \, d\sigma(s) \\ &= \int_a^b \overline{\xi_i(t)} \int_a^b \overline{K(s, t)} \varphi_0(s) \, d\sigma(s) \, dt = 0. \end{aligned}$$

Since $C[a, b]$ is denose in $L^2_\omega(a, b)$, we have $\varphi_0 = \theta$, i.e., $0 \notin \sigma_p(A)$.

Now let $\sigma(t)$ have no intervals of constancy, and let $\eta_1, \eta_2, \ldots$ be a system, complete in $\mathscr{L}^2_\omega(a, b)$, of root vectors of the operator A. Since $\eta_i \in \mathscr{R}_A$, the η_i are continuous functions, and in $L^2_\omega(a, b)$ there are functions ψ_i such that $\eta_i = A\psi_i$ $(i = 1, 2, \ldots)$. Let Φ be a continuous linear functional on $C[a, b]$ such that $\Phi(\eta_i) = 0$ $(i = 1, 2, \ldots)$. By virtue of Riez's theorem on the integral representation of a linear functional on $C[a, b]$ there is a complex function of bounded variation $\tilde{\omega}$ such that

$$\Phi(\eta_i) = \Phi(A\psi_i) = \int_a^b (A\psi_i)(s)\, d\tilde{\omega}(s)$$

$$= \int_a^b \int_a^b K(s, t)\psi_i(t)\, d\sigma(t)\, d\tilde{\omega}(s)$$

$$= \int_a^b \psi_i(t) \int_a^b K(s, t)\, d\tilde{\omega}(s)\, d\sigma(t) \qquad (i = 1, 2, \ldots)$$

Since $\mathscr{E}(A) = L^2_\omega(a, b)$, and the function $\omega = \omega_1 + \omega_2$ like σ has no intervals of constancy, $\int_a^b K(s, t)\ d\tilde{\omega}(s) = 0$ for all $t \in [a, b]$. But then $\Phi(\zeta_i) = \int_a^b \xi_i(t) \int_a^b K(s, t)\, d\tilde{\omega}(s)\, dt = 0$ $(i = 1, 2, \ldots)$. Since $\{\zeta_i\}$ is complete in $C[a, b]$ we conclude that $\Phi = \theta$. This is equivalent to the completeness of $\{\eta_i\}$ in $C[a, b]$. ■

4 □ Now let $\mathbf{K}(s, t)$ be a Hermitian positive kernel, and let $\sigma(t)$ be an arbitrary function of bounded variation on $[a, b]$. We bring into consideration the iterated kernels

$$K^{(n)}(s, t) = K(s, t), K^{(n)}(s, t) = \int_a^b K^{(n-1)}(s, l)K(l, t)\, d\sigma(l) \ (n = 2, 3, \ldots).$$

Moreover, we assume that the kernel $K(s, t)$ generates a bounded operator $A = \int_a^b K(s, t) \cdot d\sigma(t)$ having no more than a countable set of points of condensation of the spectrum. For example, if $|K(s, t)| \leqslant c < \infty$ and if the function $K(s, t)$ is continuous in each variable when the other is fixed, then (see, e.g., I. Iokhvidov and Ektov [1], [2]) A is a completely continuous operator and therefore has a single point of condensation of the spectrum. Let η_α denote the eigenvectors of the operator A corresponding to $\lambda_\alpha \neq 0$. In accordance with Theorem 2.17 for every function $x \in L^2_\omega(a, b)$ we have

$$A^n x = \sum_\alpha \lambda^n_\alpha\ [x, \eta_\alpha]\eta_\alpha(t)\text{sign } \lambda_\alpha$$

$$= \sum_\alpha \lambda^n_\alpha \eta_\alpha(t) \int_a^b x(\tau)\overline{\eta_\alpha(\tau)}\, d\sigma(\tau)\text{sign } \lambda_\alpha.$$

The function $K^{(n)}(s, t)$ belongs to $L^2_\omega(a, b)$ with respect to each of the variables. Consequently

$$AK(s, t) = \sum_\alpha \lambda_\alpha \eta_\alpha(t) \int_a^b K(s, \tau)\overline{\eta_\alpha(\tau)}\, d\sigma(\tau)\text{sign } \lambda_\alpha = \sum_\alpha \lambda^2_\alpha \eta_\alpha(t)\overline{\eta_\alpha(s)}\text{sign } \lambda_\alpha.$$

Similarly

$$AK^{(n)}(s,t) = \sum_{\alpha} \lambda_\alpha^{n+1} \eta_\alpha(t)\overline{\eta_\alpha(s)} \operatorname{sign} \lambda_\alpha \qquad (n = 2, 3, \ldots).$$

Here the series converge in the norm of $L^2_\omega(a, b)$. We note also that, by definition, $K^{(n)} = AK^{(n-1)}$. So we have proved

Theorem 3.5: *If a Hermitian positive kernel $K(s,t)$ and its iterations $K^{(n)}(s,t)$ belong to $L^2_\omega(a,b)$, and if the operator $A = \int_a^b K(s,t) \cdot d\sigma(t)$ is continuous with not more than a countable set of points of condensation of the spectrum, then when $n \geqslant 2$*

$$K^{(n)}(s,t) = \sum_{\alpha} \lambda_\alpha^{n} \eta_\alpha(t)\overline{\eta_\alpha(s)} \operatorname{sign} \lambda_\alpha, \tag{3.1}$$

where $\{\eta_\alpha\}$ is a J-orthonormalized system of eigenvectors of the operator A corresponding to $\lambda_\alpha \neq 0$, and the series (3.1) *converges in the norm of the space $L^2_\omega(a,b)$.* ■

Theorem 3.5 has been obtained as a simple consequence of Theorem 2.17. By applying additional and different methods, going beyond the scope of our book, for an integral operator and iterations of the kernels generated by it more precise results can be obtained (see, for example, Exercises 3–5 below).

5 In S. Krein's article [1] it is shown that the problem of the oscillations of a heavy viscous fluid in an open fixed container reduces to the study of an operator-valued function

$$L(\lambda) = \lambda G + \lambda^{-1} H - I, \tag{3.2}$$

where G and H are completely continuous selfadjoint operators of finite order, i.e. $G \in \mathscr{S}_p$, $H \in \mathscr{S}_q$, $p, q < \infty$, and $G > 0$, and $H \geqslant 0$. A whole series of general non-selfadjoint boundary problems with a parameter λ in the equation and in the boundary conditions reduce to the spectral analysis of a similar operator-valued function. Here we show one of the ways of analyzing an equation (3.2) based on applying Theorem 2.13. For other results in this direction see Exercises 6 and 7 below.

Definition 3.6: In equation (3.2) let $G = G^*$ and $H = H^*$ be bounded operates acting in a Hilbert space $\mathscr{H}'$. A point $\lambda_0 \in \mathbb{C}$ is said to be a *regular point for the function* $L(\lambda)$ if $0 \in \rho(L(\lambda_0))$; otherwise it is a *point of the spectrum of the function* $L(\lambda)$. A vector x_0 is called on *eigenvector of the function* $L(\lambda)$ if there is a $\lambda_0 \in \mathbb{C}$ (λ_0 is an eigenvalue) such that $L(\lambda_0)x_0 = \theta$. The vectors $x_1, x_2, \ldots, x_m$ are said to be *associated with the eigenvector* x_0 and the set $\{x_i\}_0^m$ is called a *Jordan chain* if

$$\sum_{j=0}^{k} \frac{\alpha_{jL}(\lambda_0)}{j!\alpha\lambda j} x_{k-j} = \theta \qquad (k = 0, 1, 2, \ldots, m).$$

□ Following M. Krein and Langer [2] we introduce a scheme of argumentation which reduces the spectral analysis of the function $L(\lambda)$ to the spectral analysis of a certain J-selfadjoint operator. We shall suppose that $G > 0$. After replacement of the variables $\lambda = -\mu^{-1} - a$ the function $L(\lambda)$ becomes the function

$$L_1(\mu) = -\frac{1}{\mu(1+\mu a)}\,[\mu^2(a^2G + H + aI) + \mu(2aG + I) + G].$$

We put

$$a > \inf\{b > 0 \mid F_b \equiv b^2G + H + bI \gg 0\}.$$

Then

$$L_1(\mu) = -\frac{1}{\mu(1+\mu a)}\,F_a^{1/2}(\mu^2 I + \mu B_a + C_a)F_a^{1/2},$$

where

$$B_a = F_a^{-1/2}(2aG + I)F_a^{-1/2} \gg 0, \qquad C_a = F_a^{-1/2}GF_a^{-1/2}.$$

It can be verified immediately that λ_0 is an eigenvalue of the function $L(\lambda)$ if and only if $\mu_0 = -(\lambda_0 + a)^{-1}$ is an eigenvalue of the function $L_2(\mu) = \mu^2 I + \mu B_a + C_a$, called a *quadratic bundle*. Moreover, $(x_0, x_1, \ldots, x_m)$ is a Jordan chain of the function $L(\lambda)$ if and only if $(F_\alpha^{1/2}x_0, F_\alpha^{1/2}x_1, \ldots, F_\alpha^{1/2}x_m)$ is a Jordan chain of the bundle $L_2(\mu)$.

We bring into consideration the J-space $\mathscr{H} = \mathscr{H}^+ \oplus \mathscr{H}^-, \mathscr{H}^\pm = \mathscr{H}'$ and the J-selfadjoint operator

$$A_a = \begin{Vmatrix} 0 & C_a^{1/2} \\ -C_a^{1/2} & -B_a \end{Vmatrix} \tag{3.3}$$

acting in it. It is easy to see that the regular points, the spectrum, and in particular the eigenvalues of the bundle $L_2(\mu)$ and of the operator A_a coincide. Moreover, if $(y_0, y_1, \ldots, y_m)$ is the Jordan chain of the bundle $L_2(\mu)$ corresponding to the eigenvalue μ_0, then the vectors

$$\begin{Vmatrix} C_a^{1/2}y_0 \\ \mu_0 y_0 \end{Vmatrix}, \begin{Vmatrix} C_a^{1/2}y_1 \\ \mu_0 y_1 + y_0 \end{Vmatrix}, \ldots, \begin{Vmatrix} C_a^{1/2} \\ \mu_0 y_m + y_{m-1} \end{Vmatrix}$$

form a Jordan chain of the operator A_a, and conversely if

$$\begin{Vmatrix} z_0^{(1)} \\ z_0^{(2)} \end{Vmatrix}, \begin{Vmatrix} z_1^{(1)} \\ z_1^{(2)} \end{Vmatrix}, \ldots, \begin{Vmatrix} z_m^{(1)} \\ z_m^{(2)} \end{Vmatrix}$$

is a Jordan chain of the operator A_α, then

$$y_0 = \frac{1}{\mu_0}\,z_0^{(2)}, \qquad y_i = \frac{1}{\mu_0}\,(z_1^{(2)} - y_0), \ldots, y_m = \frac{1}{\mu_0}\,(z_m^{(2)} - y_{m-1})$$

is a Jordan chain of the bundle $L_2(\mu)$.

Thus, a one-to-one correspondence has been established between the Jordan chains of the function $L(\lambda)$ and the operator A. ■

Here we denote by symbols $\mathcal{N}_\lambda$ and $\mathcal{M}_\lambda$

$$\operatorname{Ker} L(\bar{\lambda}) \cap ((2\lambda G - I)\operatorname{Ker} L(\lambda))^{\perp}$$
$$\text{and} \quad \operatorname{Ker} L(\bar{\lambda}) \cap (\operatorname{Lin}\{(2\lambda G - I)x_k + Gx_{k-1}\}_0^{m})^{\perp}$$

respectively, where $(x_0, x_1, \ldots, x_m)$ are all possible Jordan chains of the bundle $L(\lambda)$ corresponding to the eigenvalue λ, and $x_{-1} = \theta$.

Theorem 3.7: *Let the function* (3.2) *be given, where* $G > 0$, $G \in \mathcal{S}_\infty$, $H = H^*$ *is a bounded operator, and the set* $\sigma(H)$ *is no more than countable. Then*:

1) *each of the operators* (3.3) *generated by the function* $L(\lambda)$ *belongs to the class* **H**;
2) $\dim \mathcal{H}/\mathcal{E}(A_0) \leqslant \dim \mathcal{H}/\mathcal{E}_0(A_a) < \infty$;
3) $\mathcal{E}_0(A_a) = \mathcal{H}$ *if and only if* $\mathcal{N}_\lambda = \{\theta\}$ *for all* λ *such that* $(2\|G\|)^{-1} \leqslant |\lambda| \leqslant 2\|H\|$;
4) $\mathcal{E}(A_a) = \mathcal{H}$ *if and only if* $\mathcal{N}_\lambda = \{\theta\}$ *for all* λ *such that* $(2\|G\|)^{-1} \leqslant |\lambda| \leqslant 2\|H\|$;
5) *If* $\mathcal{E}_0(A_a) = \mathcal{H}$ *(respectively,* $\mathcal{E}(A_a) = \mathcal{H}$*), then there is in* $\mathcal{H}$ *an almost J-orthonormalized Riesz basis composed of eigenvectors (respectively, root vectors) of the operator* A_a. If $\mathcal{E}_0(A_a) = \mathcal{H}$, *then there is in* $\mathcal{H}$ *a J-orthonormalized Riesz basis composed of eigenvectors of the operator* A_a *if and only if the function* $L(\lambda)$ *has no non-real eigenvalues. Moreover, the bases mentioned above can be chosen as p-bases if and only if* $G \in \mathcal{S}_{p/2}$.

□ 1) Since it follows from $G \in \mathcal{S}_\infty$ that $C_a^{1/2} \in \mathcal{S}_\infty$, so (*cf.* 3.Theorem 1.13, 2.Remark 6.14, and 3.Theorem 2.8) the operator A has at least one maximal non-negative invariant subspace $\mathcal{L}^+$ with an angular operator $K_{\mathcal{L}^+}$. In accordance with 3.Lemma 2.2 $K_{\mathcal{L}^+} C_a^{1/2} K_{\mathcal{L}^+} + C_a^{1/2} + B_a K_{\mathcal{L}^+} = 0$, and since $B_a \gg 0$, so $K_{\mathcal{L}^+} = -B_a^{-1}(C_a^{1/2} + K_{\mathcal{L}^+} C_a^{1/2} K_{\mathcal{L}^+}) \in \mathcal{S}_\infty$. Consequently (*cf.* Exercise 3 on 3.§5), $A \in \mathbf{H}$, and moreover, $K_{\mathcal{L}^+} \in \mathcal{S}_p$ if and only if $C_a^{1/2} \in \mathcal{S}_p$, which in turn is equivalent to G belonging to the class $\mathcal{S}_{p/2}$ (see [XI], 3.7.3).

Assertions 2) to 5) will now follow from Theorem 2.13 if we show that the set $\sigma(A_a)$ is no more than countable, that the non-real eigenvalues of the operator A_a and the set $s(A_a)$ lie in the set $\{\mu = -(\lambda + a)^{-1} \mid (2\|G\|^{-1} \leqslant |\lambda| \leqslant 2\|H\|\}$, and that $\mathcal{N}_\lambda = \{\theta\}$ and $\mathcal{M}_\lambda = \{\theta\}$ are equivalent to non-degeneracy of $\operatorname{Ker}(A_a - \mu I) + \operatorname{Ker}(A_a - \bar{\mu} I)$ and $\mathcal{L}_\mu(A_a) + \mathcal{L}_{\bar{\mu}}(A_a)$ respectively when $\mu = (\lambda + a)^{-1}$.

That the set $\sigma(A_a)$ is no more than countable follows from the fact that the operator A_a is the sum of the operator

$$\begin{Vmatrix} 0 & 0 \\ 0 & -F_a^{-1} \end{Vmatrix}$$

and the completely continuous operator

$$\left\| \begin{matrix} 0 & C_a^{1/2} \\ -C_a^{1/2} & -2aF_a^{-1/2}GF_a^{-1/2} \end{matrix} \right\|,$$

and the spectrum of the first of these consists of not more than a countable set of points, since the operator H, and therefore also the operator $F_a = a^2G + aI + H$, has this property (see 2.Theorem 2.11).

We now verify that all the non-real points of the spectrum of the function $\mathscr{L}(\lambda)$ are situated in the ring $(2\|G\|)^{-1} \leqslant |\lambda| \leqslant 2\|H\|$. For, if $L(\lambda_0)x_0 = \theta$ and $\lambda_0 \neq \bar{\lambda}_0$, then, solving the quadratic equation

$$\lambda_0^2(Gx_0, x_0) - \lambda_0(x_0, x_0) + (Hx_0, x_0) = 0,$$

we obtain

$$\lambda_0 = \frac{(x_0, x_0) \pm \sqrt{(x_0, x_0)^2 - 4(Gx_0, x_0)H(x_0, x_0)}}{2(Gx_0, x_0)}$$

Hence,

$$|\lambda_0|^2 = \frac{(Hx_0, x_0)}{(Gx_0, x_0)} = \frac{2(Hx_0, x_a)}{(x_0, x_0)} \cdot \frac{(x_0, x_0)}{2(Gx_0, x_0)}.$$

But since the discriminant of the equation is less than zero, we have

$$\frac{(x_0, x_0)}{2(Gx_a x_0}) \leqslant \frac{2(Hx_0, x_0)}{(x_0, x_0)},$$

and therefore

$$\left(\frac{(x_0, x_0)}{2(Gx_0, x_0)}\right)^2 \leqslant |\lambda_0|^2 \leqslant \left(\frac{2(Hx_0, x_0)}{(x_0, x_0)}\right)^2.$$

Hence, we conclude that λ_0 is situated in the ring $(2\|G\|)^{-1} \leqslant |\lambda| \leqslant 2\|H\|$. Consequently the non-real spectrum of the operator A coincides with points μ_0 of the form $\mu_0 = -(\lambda_0 + \alpha)^{-1}$.

We now show that the set $\sigma(A_a)$ is situated in the same ring. Let $\mu_1 = \bar{\mu}_1 \in \sigma_p(A_a)$ and let $\operatorname{Ker}(A_a - \mu I)$ be degenerate. As proved earlier this is equivalent to the fact that there exists in $\operatorname{Ker} L(\lambda_1)$ $(\lambda_1 = -\mu_1^{-1} - a)$ a vector x_0 such that the vector

$$\left\| \begin{matrix} C_a^{1/2}F_a^{1/2}x_0 \\ -\dfrac{1}{\lambda_1 + a} F_a^{1/2}x_0 \end{matrix} \right\|$$

is J-orthogonal to all vectors of the form

$$\left\| \begin{matrix} C_a^{1/2}F_a^{1/2}y \\ -\dfrac{1}{\lambda_1 + a} F_a^{1/2}y \end{matrix} \right\|,$$

where $y \in \text{Ker } L(\lambda_1)$, i.e.,

$$(C_a^{1/2} F_a^{1/2} x_0, C_a^{1/2} F_a^{1/2} y) - \frac{1}{(\lambda_1 + a)^2} (F_a^{1/2} x_0, F_a^{1/2} y) = 0.$$

Starting from the definitions of the operators C_a and F_a we obtain as an equivalent form of this equation

$$((\lambda_1^2 G + 2a\lambda_1 G - I - H)x_0, y) = 0. \tag{3.4}$$

We substitute in this equation the vector $\lambda_1 x_0 - \lambda_1^2 G x_0$ instead of the vector Hx_0 and, after cancellation by $\lambda_1 + a$ we obtain $((2\lambda_1 G - I)x_0, y = 0$. In particular, when $y = x_0$, we have $\lambda = (x_0, x_0)/2(Gx_0, x_0)$. But if in (3.4) we substitute instead of $\lambda_1^2 G x_0$ the vector $\lambda_1 x_0 - Hx_0$ and instead of $\lambda_1 G x_0$ the vector $x_0 - (1/\lambda)Hx_0$, and cancel out $\lambda_1 + a$ (when $y = x_0$) we obtain that $\lambda_1 = 2(Hx_0, x_0)/(x_0, x_0)$. From this we conclude as before that λ_1 lies in the ring $(2\|G\|)^{-1} \leqslant |\lambda| \leqslant 2\|H\|$.

The validity of the interpretation given above of the equalities $\mathcal{N}_\lambda = \{\theta\}$ and $\mathcal{M}_\lambda = \{\theta\}$ is proved by entirely similar considerations. ∎

□ *Remark 3.8:* In the proof of the equality $\mathcal{E}(A_a) = \mathcal{H}$ it is necessary to verify the triviality of $\mathcal{M}_\lambda$ only for those points of $s(A_a)$ which are points of condensation of the spectrum of the operator A_a. But the latter coincide with the set of points of the form $\mu = -(\lambda + a)^{-1}$ where λ traverses the set of points of condensation of the spectrum of the function (3.2). If $G, H \in \mathcal{S}_\infty$, then it follows from the form of the operator A_a that it has two points of condensation of its spectrum: $\mu_1 = 0$ and $\mu_2 = -a^{-1}$. Consequently the function (3.2) also has only two points of condensation of its spectrum: $\lambda_1 = \infty$ and $\lambda_2 = 0$ respectively. Both these points do not lie in the ring $(2\|G\|)^{-1} \leqslant |\lambda| \leqslant 2\|H\|$, and therefore when $G, H \in \mathcal{S}_\infty$ we always have $\mathcal{E}(A) = \mathcal{H}$. ∎

Definition 3.9: The function $L(\lambda) = \lambda G + (1/\lambda)H - I$ is called a *strongly damped bundle* if $(x, x)^2 > 4(Gx, x)(Hx, x)$ for all $x \neq \theta$.

Corollary 3.10: *If under the conditions of Theorem* 3.7 *the function* (3.2) *is a strongly damped bundle, then there is in $\mathcal{H}$ a J-orthonormalized Riesz basis composed of eigenvectors of the operator A, and it will be a p-basis if $G \in \mathcal{S}_{p/2}$.*

□ By Theorem 3.7 it suffices to verify that the function (3.2) has no non-real eigenvectors and that $\mathcal{N}_\lambda = \{\theta\}$ for all λ. The first follows immediately from the fact that if λ is an eigenvalue of the function (3.2) and if x_0 is the corresponding eigenvector, then

$$\lambda_0 = \frac{(x_0, x_0) \pm \sqrt{(x_0, x_0)^2 - 4(Gx_0, x_0)(Hx_0, x_0)}}{2(Gx_0, x_a)} = \bar{\lambda}_0.$$

Let us suppose that for some λ_0 the subspace $\mathcal{N}_\lambda \neq \{\theta\}$. Then, in particular, $((2\lambda_0 G - I)x_0, x_0) = 0$ for some $x_0 \in \text{Ker } L(\lambda_0)$, and therefore $\lambda_0 = (x_0, x_0)/2(Gx_0, x_0)$, which in turn implies the equality $(x_0, x_0)^2 = 4(Gx_0, x_0)(Hx_0, x_0)$, and we have a contradiction with the fact that the function (3.2) is strongly damped. ■

Exercises and problems

1 Investigate the possibility of generalizing Theorem 3.1 to the case of a Krein space.

2 let $K(s, t)$ be a Hermitian positive kernel continuous on the square $0 \leqslant s \leqslant t \leqslant b$, let $R(s, t)$ be a function bounded on this same square and continuous with respect to each variable when the other is fixed, let $R(s, t) = \overline{R(s, t)}$, and let $\sigma(t)$ generate a Pontryagin space according to the form $[\varphi, \psi] = \int_a^b \varphi(t)\overline{\psi(t)}\, d\phi(t)$, and let $A \cdot = \int_a^b (R(s, t) + iK(s, t)) \cdot d\sigma(t)$. Prove that the equalities $\mathscr{E}(A) = L^2_\omega(a, b)$ and $\int_a^b K(t, t)\, d\sigma(t) = \sum_j \text{Im } \lambda_j$, where $\{\lambda_j\}$ is the set of eigenvalues of the operator A, are equivalent. (Azizov [8]).

3 Consider on $[a, b] \times [a, b]$ a Hermitian bounded kernel $K(s, t)$ continuous with respect to each variable when the other is fixed, and a function of bounded variation $\sigma(t) \not\equiv$ const. We shall say that the kernel $K(s, t)$ is *σ-non-negative* if

$$\int_a^b \int_a^b K(s, t) g(s)\overline{g(t)}\, d\sigma(s)\, d\sigma(t) \geqslant 0 \text{ for all } g \in \mathscr{V},$$

where $\mathscr{V}$ is the lineal of all those functions from $L^2_\omega(a, b)$ ($\omega = \text{Var } \sigma$) which are generated by continuous functions from $C[a, b]$. Prove that σ-non-negativity is equivalent to either of the conditions:

a) $\int_a^b\int_a^b K(s, t)\, d\tau(s)\, d\tau(t) \geqslant 0$ for any (complex) function of bounded variation τ on $[a, b]$ with $E_\tau \subset E_\sigma$ (E_τ and E_σ are the sets of points of variation of the functions τ and σ respectively);

b) $\sum_{j,k=1}^n K(t_j, t_u)\xi_k\bar{\xi}_k \geqslant 0$ for all $n \in \mathbb{N}$, $t_1, t_2, \ldots, {}_n \in E_\sigma$, and $\xi_1, \xi_2, \ldots, \xi_n \in \mathbb{C}$ (I. Iokhvidov and Ektov [1], [2]).

4 Let $K(s, t)$ be a σ-non-negative kernel satisfying the conditions of Exercise 3. Prove that the assertions '$K^{(2)}(s, t) \equiv 0$ on $E_\sigma > E_\sigma$' and '$K^{(2)}(s, t) \equiv 0$ on $[a, b] \times [a, b]$' are equivalent (I. Iokhvidov and Ektov [1], [2]).

5 Prove that the integral equation $\varphi(s) = \lambda \int_a^b K(s, t)\varphi(t)\, d\sigma(t)$ with σ a function of bounded variation, and a σ-positive kernel, has $\varkappa < \infty$ positive characteristic points if and only if the positive variation of the function σ has exactly $\varkappa$ points of growth (M. Krein [1], I. Iokvidov and Ektov [1]).

6 Let $S = S^* \in \mathscr{S}_\infty$, $T = T^* \in \mathscr{S}_\infty$, $-1 \in \rho(s)$, $0 \notin \sigma_p(T)$ and $A = (I + S)T$. Prove that $\mathscr{E}(A) = \mathscr{H}$ and that in $\mathscr{H}$ there is a Riesz basis composed of root vectors of the operator A ([XI]); if $S \in \mathscr{S}_p$, then there is in $\mathscr{H}$ a p-basis composed of root vectors of the operator A (Kopachevskiy [1]–[3]).
Hint: The operator A is π-selfadjoint relative to the form $[\cdot, \cdot] = ((I + S)^{-1} \cdot, \cdot)$. It only remains to use Remark 2.13 and Exercise 4 on §2.

7 Prove that if $L(\lambda)$ is the function (3.2), $G = G^* \in \mathscr{S}_\infty$, $H = H^* \in \mathscr{S}_\infty$, and Ker G = Ker $H = \{\theta\}$, then:

1) the system

$$\left\{ \left\| \begin{array}{c} x_k(\lambda) \\ \sum_{j=0}^{k} \frac{(-1)^j}{\lambda^{j+1}} x_{k-j}(l) \end{array} \right\| \right\}$$

of 'special vectors', where $\{x_0(\lambda), \ldots, x_k(\lambda)\}$ is the Jordan chain of the function $L(\lambda)$ corresponding to the eigenvalues λ, is complete in $\mathscr{H}$ (Askerov, S. Krein and Laptev [1]);

2) in $\mathscr{H}$ there is a Riesz basis composed of vectors of the special form (Larionov [6], [7]);

3) If $G \in \mathscr{S}_p$, $H \in \mathscr{S}_q$, and $r = \max(p, q)$, then there is in $\mathscr{H}$ an r-basis composed vectors of the special form (Kopachevskiy [1]–[3]).

Remarks and bibliographical indications on chapter IV

§1.1. The J-spectral function was introduced by M. Krein and Langer [1] for the case of $\Pi_\varkappa$, and later Langer [4], [5] transferred the investigation to the case of Krein spaces. The definitions in the text, given by Azizov, modify Krein and Langer's definitions in order to accommodate them to operators of the **K(H)** class (see Exercises 1 and 2).

§1.2–1.3. Definitizable operators in $\Pi_\varkappa$ were introduced by I. Iokhvidov and M. Krein [XV], and in a J-space by Langer [4], [5], [9]. All the results of these sections are due to him. We borrowed from Bognar [5] the elegant proof of Theorem 1.5. In Bognar [5] these are also given bibliographical references to other proofs and, in particular, references to M. Krein and Shmul'yan's proof [3] based on considerations from the problem of moments. In connection with criteria for the regularity of the J-spectral function see Langer [4], [5], Jonas [1]–[7], Jonas and Langer [1], [2], Akopyan [1]–[3], and Spitovskiy [1].

§1.4. Definition 1.11 and Theorem 1.12 for the case of operators acting from one space into another are modifications of corresponding results of Potapov [1], Ginzburg [2], M. Krein and Shmul'yan [3]. On square roots of J-selfadjoint operators see Bognar [1] and, more fully, Bayasgalan [1].

§2.1. The results in this paragraph are due to Azizov. Corollary 2.3 was published in the paper by Azizov and Usvyatsova [2].

§2.2. Lemmas 2.4 and 2.5, Theorem 2.6 and Corollary 2.9 in the case $A \in \mathscr{S}_\infty$ were published in the paper by Azizov and Usvyatsova [2]. The presentation in the text is due to Azizov. To him also is due Theorem 2.8, which is a transfer to $\Pi_\varkappa$ of a corresponding result from [XI]. Lemma 2.7 is, apparently, well-known, though we have not found its formulation in print.

§2.3. Theorem 2.12 is due to Azizov [13]. For the formulation of the problems about p-basicity he is obliged to Kopachevskiy, who first began the study of this question for the case of indefinite spaces in connection with problems of hydrodynamics (see, e.g., Kopachevskry [1]–[3]). The concept of p-basicity itself was introduced by Prigorskiy [1]. We mention also that the questions of completeness of a system of root vectors of π-selfadjoint operators $A \in \mathscr{S}_\infty$ with $0 \notin \sigma_p(A)$ were investigated for the first time by I. Iokhvidov [2]; the existence of a Riesz basis of root vectors of such operators was proved essentially in [XI]; the criterion for completeness and basicity of these vectors without the condition $0 \notin \sigma_p(A)$ was given by Azizov and I. Iokhvidov [1]. On other conditions for completeness and basicity and for a historical survey see [IV] and also the Exercises.

§2.4. The results of this paragraph in so general a formulation are due to Azizov. In the case when the set $\sigma(A)$ has no more than a finite number of points of condensation, Theorem 2.15 and Remark 2.16 (even in the case of Banach spaces with an indefinite metric) were obtained earlier by Azizov and Shtraus [1], and Theorem 2.17 when $A^2 \in \mathscr{S}_\infty$ is due to Kühne [1] (see I. Iokhvidov [8], Ektov [1]). As above, for a historical survey and for other results we refer the reader to the survey [IV] and to the Exercises on this section.

§3.1–3.3. All the results in these paragraphs are due to Azizov. Some of them were published in Azizov's paper [8]. Theorem 3.4 for the case of a Hermitian kernel was obtained earlier by I. Iokhvidov [2].

§3.4. Theorem 3.5 in the text was proved by Azizov; for the case $A \in \mathscr{S}_\infty$ see M. Krein [1], I. Iokhvidov and Ektov [1].

§3.5. Theorem 3.7 and Corollary 3.10 were proved by Azizov [13] (*cf.* Askerov, S. Krein and Laptev [1], M. Krein and Langer [2], Larionov [6], [7], Kopachevskry [1]–[3], Azizov and Usvyatsova [2]). As regards other investigations of operator bundles and application, of an indefinite metric see Langer [4], [6], [10], [12], Kostyuchenko and Orazov [1], and others.

5 THEORY OF EXTENSIONS OF ISOMETRIC AND SYMMETRIC OPERATORS IN SPACES WITH AN INDEFINITE METRIC

In §1 the apparatus of Potapov–Ginzburg transformations is developed, and its application to the theory of extensions is demonstrated.

§2 is devoted to another approach to the theory of extensions of isometric operators in Krein spaces.

In §3 generalized resolvents of J-symmetric operators are described.

§1 Potapov–Ginzburg linear-fractional transformations and extensions of operators

1 One of the methods allowing extensions of (J_1, J_2)-isometric operators to be constructed is the application of *Potapov-Ginzburg transformations* or, briefly, *PG-transformations.*

Definition 1.1: Let $\mathscr{H}_1 = \mathscr{H}_1^+ \oplus \mathscr{H}_1^-$ and $\mathscr{H}_2 = \mathscr{H}_2^+ \oplus \mathscr{H}_2^-$ be the canonical decompositions of a J_1-space $\mathscr{H}_1$ and a J_2-space $\mathscr{H}_2$ respectively, and let $\mathscr{H}_3 = \mathscr{H}_1^+ \oplus \mathscr{H}_2^-$ and $\mathscr{H}_4 = \mathscr{H}_2^+ \oplus \mathscr{H}_1^-$ be a J_3-space and a J_4-space constructed from them with $J_3 = I_1^+ \oplus -I_2^-$ and $J_4 = I_2^+ \oplus -I_1^-$. A transformation $\omega^+ : \mathscr{H}_1 \oplus \mathscr{H}_2 \to H_3 \oplus \mathscr{H}_4$ carrying a vector $\langle x_1, x_2 \rangle$ into a vector

$\langle x_3, x_4 \rangle$, where

$$x_1 = x_1^+ + x_1^-, \qquad x_2 = x_2^+ + x_2^-, \qquad x_3 = x_1^+ + x_2^-, \qquad x_4 = x_2^+ + x_1^-,$$

$x_i^\pm \in \mathscr{H}_i^\pm$ $(i = 1, 2)$, is called a *PG-transformation.*

As well as the transformation ω^+ having the form

$$\omega^+ : \mathscr{H}_1 \oplus \mathscr{H}_2 \to \mathscr{H}_3 \oplus \mathscr{H}_4, \qquad \omega^+\langle x_1, x_2 \rangle = \langle x_1^+ + x_x^-, x_2^+ + x_1^- \rangle,$$

other *PG*-transformations can be considered:

$$\omega^- : \mathscr{H}_1 \oplus \mathscr{H}_2 \to \mathscr{H}_3 \oplus \mathscr{H}_4, \qquad \omega^-\langle x_1, x_2 \rangle = \langle -x_1^+ + x_2^-, x_2^+ + x_1^- \rangle,$$
$$\omega_\pm : \mathscr{H}_1 \oplus \mathscr{H}_2 \to \mathscr{H}_4 \oplus \mathscr{H}_3, \qquad \omega_\pm\langle x_1, x_2 \rangle = \langle x_2^+ \pm x_1^-, \pm x_1^+ + x_2^- \rangle.$$

□ Below we shall study the properties of the transformation ω^+. The properties of ω^- and $\omega_\pm$ are similar, and we leave it to the reader to formulate them and prove them by the same scheme as for ω^+ or relying on the following obvious proposition:

1.2.

$$\omega^- = \omega^+ j, \qquad \omega_- = \omega_+ j, \qquad \omega^\pm\langle x_1, x_2 \rangle = (\omega_\pm \langle x_1, x_2 \rangle)^{-1},$$

where

$$j\langle x, y \rangle = \langle -x, y \rangle \quad \textit{and} \quad (\langle x, y \rangle)^{-1} = \langle y, x \rangle. \qquad ■$$

We note also that the spaces $\mathscr{H}_3$, $\mathscr{H}_4$ and $\mathscr{H}_1$, $\mathscr{H}_2$ may interchange rôles, and the transformation ω^+ can be applied to a vector $\langle x_3, x_4 \rangle \in \mathscr{H}_3 \oplus \mathscr{H}_4$. In what follows we shall cite definitions and propositions for one of these variants, remembering that they are symmetric.

1.3. *The transformation* ω^+ *maps* $\mathscr{H}_1 \oplus \mathscr{H}_2$ *one-to-one on to* $\mathscr{H}_3 \oplus \mathscr{H}_4$, *and it is involutory*: $\omega^+(\omega^+\langle x_1, x_2 \rangle) = \langle x_1, x_2 \rangle$, *and if*

$\langle x_1, x_2 \rangle, \langle y_1, y_2 \rangle \in \mathscr{H}_1 \oplus H_2$, *and*

$$\langle x_3, x_4 \rangle = \omega^+\langle x_1, x_2 \rangle, \langle y_3, y_4 \rangle = \omega^+\langle y_1, y_2 \rangle,$$

then

$$[x_1, y_1]_1 - [x_2, y_2]_2 = (x_3, y_3)_3 - (x_4, y_4)_4 \tag{1.1}$$

here $(\cdot, \cdot)_3$ *and* $(\cdot, \cdot)_4$ *denote the scalar products in* $\mathscr{H}_3$ *and* $\mathscr{H}_4$ *respectively.*

□ The vector $\langle x_3, x_4 \rangle = \omega^+\langle x_1, x_2 \rangle$ is defined uniquely by the pair $\langle x_1, x_2 \rangle$ because projection of x_i on to $\mathscr{H}_i^\pm$ $(i = 1, 2)$ is single-valued. The existence of the inverse transformation, its definition on the whole space $\mathscr{H}_3 \oplus \mathscr{H}_4$, and the equality $\omega^+(\omega^+\langle x_1, x_2 \rangle$ are obvious. It remains to verify the identity (1.1):

$$\begin{aligned}[x_1, y_1]_1 - [x_2, y_2]_2 &= (x_1^+, y_1^+)_1 - (x_1^-, y_2^-)_1 - (x_2^+, y_2^+)_2 + (x_2^-, y_2^-)_2 \\ &= (x_1^+ + x_2^-, y_1^+ + y_2^-)_3 - (x_2^+ + x_1{}^-, y_2^+ + y_1^-)_4 \\ &= (x_3, y_3)_3 - (x_4, y_4)_4. \qquad ■\end{aligned}$$

2 We define the PG-transformation ω^+ on sets $\mathcal{N} \subset \mathcal{H}_1 \oplus \mathcal{H}_2$:

$$\omega^+(\mathcal{N}) = \{\omega^+\langle x_1, x_2\rangle \mid \langle x_1, x_2\rangle \in \mathcal{N}\}. \tag{1.2}$$

□ From Definition 1.1 and formula (1.2) it follows immediately that 1.4 *If $\mathcal{L}$ is a lineal in $\mathcal{H}_1 \oplus \mathcal{H}_2$, then $\omega^+(\mathcal{L})$ is a lineal in $\mathcal{H}_3 + \mathcal{H}_4$.* ■

Definition 1.5: Let $T: \mathcal{H}_1 \to \mathcal{H}_2$ be a linear operator. We shall say that $T \in \mathcal{T}$ if $\operatorname{Ker}(P_2^- T P_1^- \mid \mathcal{H}_1^-) = \{\theta\}$, and we introduce the *PG-transformation $\omega^+(T): \mathcal{H}_3 \to \mathcal{H}_4$ for such operators* by the formula

$$\omega^+(T) = (P_1^- + P_2^- T)(P_1^+ + P_2^- T)^{-1} \tag{1.3}$$

(the existence of the operator $(P_1^+ + P_2^- T)^{-1}$ follows from Exercise 6 on 2.§1).

Theorem 1.6: *If the operator $T \in \mathcal{T}$ and if Γ_T is its graph, then*

$$\omega^+(\Gamma_T) = \Gamma_{\omega^+(T)} \quad and \quad \omega^+(T) \in \mathcal{T}.$$

□ Since $\mathcal{D}_{\omega^+(T)} = \{(P_1^+ + P_2^- T)x_1 \mid x_1 \in \mathcal{D}_T\}$ and $\Gamma_T = \{\langle x_1, Tx_1\rangle \mid x_1 \in \mathcal{D}_T)$, we have, using (1.2) and the Definition 1.1,

$$\begin{aligned}\omega^+(\Gamma_T) &= \{\omega^+\langle x_1, Tx_1\rangle \mid x_1 \in \mathcal{D}_T\} \\ &= \{\langle (P_1^+ + P_2^- T)x_1, (P_1^- + P_2^- T)x_1\rangle \mid x_1 \in \mathcal{D}_T\} \\ &= \{\langle (y_3, (P_1^- + P_2^+ T)(P_1^+ + P_2^- T)^{-1} y_3\rangle \mid y_3 = (P_1^+ + P_2^-)x_1, x_1 \in \mathcal{D}_T\} \\ &= \Gamma_{\omega^+(T)}.\end{aligned}$$

Since $\omega^+(T)$ acts from $\mathcal{H}_3$ into $\mathcal{H}_4$, so, in accordance with Definition 1.5 and Exercise 6 on 2.§1, $\omega^+(T) \in \mathcal{T}$ if $\operatorname{Ker}(P_3^+ + P_4^- \omega^+(T)) = \{\theta\}$.

Let

$$y_3 \in \operatorname{Ker}(P_3^+ + P_4^- \omega^+(T)); \qquad y_3 = (P_1^+ + P_2^- T)x_1 \in \mathcal{D}_{\omega^+(T)}, x_1 \in \mathcal{D}_T.$$

Then

$$\begin{aligned}\theta = (P_3^+ + P_4^- \omega^+(T))y_3 &= P_3^+(P_1^+ + P_2^- T)x_1 + P_4^-(P_1^- + P_2^+ T)x_1 \\ &= x_1^+ + \ldots x_1^- = x_1.\end{aligned}$$

Consequently, $y_3 = \theta$, i.e., $\operatorname{Ker}(P_3^+ + P_4^- \omega^+(T)) = \{\theta\}$, and therefore $\omega^+(T) \in \mathcal{T}$. ■

Corollary 1.7: *The transformation* (1.3) *establishes a one-to-one correspondence between (J_1, J_2)-non-expansive operators $V: \mathcal{H}_1 \to \mathcal{H}_2$ and contractions $W: \mathcal{H}_3 \to \mathcal{H}_4$, $W \in \mathcal{T}$. Moreover, V is a (J_1, J_2)-bi-non-expansive operator (respectively, a (J, J_2)-isometric operator; a (J_1, J_2)-semi-unitary operator; a (J_1, J_2)-unitary operator) if and only if $W \equiv \omega^+(V)$ is a contraction, $\mathcal{D}_W = \mathcal{H}_3$ and $R_{P_1^- W P_2^-} = \mathcal{H}_1^-$ (respectively, W is an isometry; W is an isometry and $\mathcal{R}_{P_1^- W P_2^-} = \mathcal{H}_1^-$; W is a unitary operator and $\mathcal{R}_{P_1^- W P_2^-} = \mathcal{H}_1^-$).*

□ It follows from Exercise 33 on 2.§4 in conjunction with 2. Remark 4.29 that strict minus-operators (including (J_1, J_2)-non-expansive operators) $V \in \mathcal{T}$. It only remains to use Theorem 1.6, the Formulae (1.1) and (1.3), the definitions of the corresponding classes of operators, and the results of Exercise 33 on 2.§4 and Exercise 7 on 2.§1. For, let us carry out, for example, the appropriate argument for the case when V is a (J_1, J_2)-unitary operator. It follows from Theorem 1.6 that $W \in \mathcal{T}$. From (1.1) we conclude that W is an isometry, and from (1.3) in combination with Exercise 33 on 2.§4 that $\mathcal{D}_W = \mathcal{H}_3$. Similar considerations show that the domain of definition of the operator

$$\omega_+(V) \equiv (P_1^+ + P_2^- V)(P_1^- + P_2^+ V)^{-1} : \mathcal{H}_4 \to \mathcal{H}_3$$

is $\mathcal{H}_4$. It remains to observe that (*cf.* Proposition 1.2) $\omega_+(V) = (\omega^+(V))^{-1} = W^{-1}$, and therefore W is a unitary operator. Since $P_1^- WP_2^- \mid \mathcal{H}_2^- = (P_2^- VP_1^- \mid \mathcal{H}_1^-)^{-1}$ (see Exercise 1 below), $\mathcal{R}_{P_1^- WP_2^-} = \mathcal{H}_1^-$.

Conversely, let W be a unitary operator and let $\mathcal{R}_{P_1^- WP_2} = \mathcal{H}_1^-$. One proves, as we did above for W, that the operator V is defined on $\mathcal{H}_1$, and V is a (J_1, J_2)-isometric operator, i.e., V is a (J_1, J_2)-semi-unitary operator. From the facts that the operator $W \in \mathcal{T}$ is unitary and that $\mathcal{R}_{P_1^- WP_2^-} = \mathcal{H}_1^-$ it is obtained immediately that

$$\operatorname{Ker}(P_2^+ WP_1^+ \mid \mathcal{H}_1^+)^2 = \{\theta\}, \qquad \mathcal{R}_{P_2^+ WP_1^+} = \mathcal{H}_2^+,$$

and, using the same considerations as above for W, we obtain that the operator V^{-1} is (J_1, J_2)-semi-unitary, and therefore V is a (J_1, J_2)-unitary operator. ■

3 Again let Γ_T be the graph of the operator $T: \mathcal{H}_1 \to \mathcal{H}_2$. We recall that an operator $\tilde{T}$ is an extension of the operator T if and only if $\Gamma_T \subset \Gamma_{\tilde{T}}$. It therefore follows from Theorem 1.6 and Corollary 1.7 that:

1.8 An operator $T(\in \mathcal{T})$ admits an extenson $\tilde{T} \in \mathcal{T}$ if and only if the operator $\omega^+(T)$ has an extension $\widetilde{\omega^+(T)} \in \mathcal{T}$, and then $\widetilde{\omega^+(T)} = \omega^+(\tilde{T})$.

In particular, a (J_1, J_2)-non-expansive operator (a (J_1, J_2)-isometric operator) V admits a (J_1, J_2)-non-expansive extension (a (J_1, J_2)-isometric extension) $\tilde{V}$ if and only if the contraction (isometry) $\omega^+(V)$ admits a contractive (isometric) extension $\widetilde{\omega^+(V)} \in \mathcal{T}$, and then $\widetilde{\omega^+(V)} = \omega^+(\tilde{V})$. ■

We now pass on to the description of special extensions of (J_1, J_2)-isometric operators.

Definition 1.9: An extension T of a (J_1, J_2)-isometric operator $V: \mathcal{H}_1 \to \mathcal{H}_2$ is called its (J_1, J_2)-*bi-extension* if Γ_V is isotropic in Γ_T relative to the form $|\langle x_1, x_2\rangle, \langle y_1, y_2\rangle|_{\hat{\Gamma}} = [x_2, y_2]_2 - [x_1, y_1]_1$ (2.Formula (4.11)).

□ From Definition 1.9 and the fact that graphs of (J_1, J_2)-non-expansive operators and (J_1, J_2)-non-contractive operators are semi-definite lineals relative to the form 2.(4.11), and the graph of a (J_1, J_2)-isometric operator is a neutral lineal, we obtain, taking 1.Proposition 1.17 into accounts.

1.10 *(J_1, J_2)-non-expansive, (J_1, J_2)-non-contractive, and, in particular, (J_1, J_2)-isometric extensions of a (J_1, J_2)-isometric operator are (J_1, J_2)-bi-extensions of it.* ■

□ The following proposition also follows directly from the Definition 1.9:

1.11 Let T be an extension of a (J_1, J_2)-isometric operator V. If T is a (J_1, J_2)-bi-extension of it, then $T(\mathscr{D}_T \cap \mathscr{D}_V^{[\perp]}) \subset \mathscr{R}_V^{[\perp]}$. Conversely, if $\mathscr{D}_V$ is a projectionally complete subspace and $T(\mathscr{D}_T \cap \mathscr{D}_V^{[\perp]}) \subset \mathscr{R}_V^{[\perp]}$, then T is a (J_1, J_2)-bi-extension of the operator V. ■

□ From the Definition 1.9, from 2.Formula (4.14), and from the fact that the graphs $\Gamma_V = \{\langle x, Vx\rangle \mid x \in \mathscr{D}_V\}$ and $\Gamma_{V^{-1}} = \{\langle V^{-1}y, y\rangle \mid y \in \mathscr{D}_{V^{-1}}\}$ of the operators V and V^{-1} respectively, regarded as lineals in $\mathscr{H}_1 \oplus \mathscr{H}_2$, coincide, it follows that the following proposition holds:

1.12. *Let V be a (J_1, J_2)-isometric operator with* Ker $V = \{\theta\}$, *and let T be an extension densely defined in $\mathscr{H}_1$ of the operator V. Then the operator T is a (J_1, J_2)-bi-extension of the operator V if and only if $V^{-1} \subset T^c$ and therefore the operator T^c is a (J_2, J_1)-bi-extension of the operator V^{-1}.* ■

The following theorem establishes a connection between (J_1, J_2)-bi-extensions of a (J_1, J_2)-isometric operator and bi-extensions of its *PG*-transformation.

Theorem 1.13: *An extension T of a (J_1, J_2)-isometric operator V is a (J_1, J_2)-bi-extension of V if and only if $\omega^+(\Gamma_V)$ $(\subset \mathscr{H}_3 \oplus \mathscr{H}_4)$ is isotropic in $\omega^+(\Gamma_T)$ $(\subset \mathscr{H}_3 \oplus \mathscr{H}_4)$ relative to the form $(x_3, y_3)_3 - (x_4, y_4)_4$. In particular, if $T \in \mathscr{T}$, then T is a (J_1, J_2)-bi-extension of a (J_1, J_2)-isometric operator V if and only if the operator $\omega^+(T)$ is a bi-extension of the isometric operator $\omega^+(V): \mathscr{H}_3 \to \mathscr{H}_4$.*

□ We use the Definition 1.9 and we also note that the left-hand side of the formula (1.1) coincides with the $(-J_{\hat{\Gamma}})$-metric. Therefore $\langle x_1, x_2\rangle$ is an isotropic vector in Γ_T if and only if $\omega_+\langle x_1, x_2\rangle = \langle x_3, x_4\rangle$ is an isotropic vector in $\omega^+(\Gamma_T)$ relative to the form $(x_3, y_3)_3 - (x_4, y_4)_4$. From this the first assertion of Theorem 1.13 is obtained. The second assertion is a direct consequence of the first, taking Theorem 1.6 into account. ■

Corollary 1.7, Proposition 1.8, and Theorem 1.13 show us the way to describe special extensions of (J_1, J_2)-isometric operators (see Exercises 5–11).

4 It is well known that in extensions of symmetric operators in a Hilbert space the method of Cayley–Neyman transforms is used, reducing the given problem to the problem of extending isometric operators. In J-spaces such a method is not always applicable, since (*cf.* 2.Example 3.36) it is not excluded that $\sigma_p(A) = \mathbb{C}$. However, in 2.Proposition 2.3 we mentioned another method of extending J-dissipative, and therefore also J-symmetric, operators: $A \to JA \to \widetilde{JA} \to J\widetilde{JA}$. Finally, in the preceding paragraphs a method of extension by means of PG-transforms has been indicated.

It turns out that the three transformations—the Cayley–Newman transform, multiplication by J, and the PG-transformation—are closely connected.

Theorem 1.14: *Let $\mathcal{H}$ be a J-space and $\mathcal{H}_\Gamma = \mathcal{H} \oplus \mathcal{H}$ be the space of graphs. Then $\mathsf{K}_i \circ [J, I] \circ \mathsf{K}_i^{-1} \langle x, y \rangle = \omega^+ \langle x, y \rangle$ for all $\langle x, y \rangle \in \mathcal{H}_\Gamma$, here* $[J, I] \circ \langle u, v \rangle = \langle Ju, v \rangle$ ($\langle u, v \rangle \in \mathcal{H}_\Gamma$).

□ $$\mathsf{K}_i \circ [J, I] \circ \mathsf{K}_i^{-1} \langle x, y \rangle = \mathsf{K}_i \circ [J, I] \circ \left\langle \frac{-x+y}{2i}, \frac{x+y}{2i} \right\rangle$$
$$= \mathsf{K}_i \left\langle \frac{-Jx+Jy}{2i}, \frac{x+y}{2i} \right\rangle = \langle x^+ + y^-, x^- + y^+ \rangle = \omega^+ \langle x, y \rangle. \quad ■$$

□ From this follows directly.

Corollary 1.15: *Let $T: \mathcal{H} \to \mathcal{H}$ be a linear operator: Then:*
a) $T \in \mathcal{T}$, $1 \notin \sigma_p(T) \Rightarrow i \notin \sigma_p(\mathsf{K}_i^{-1}(T)J)$;
b) $T \in \mathcal{T}$, $1 \notin \sigma_p(\omega^+ T) \Rightarrow 1 \notin \sigma_p(T)$:
c) $1 \notin \sigma_p(T)$, $i \notin \sigma_p(\mathsf{K}_i^{-1}(T)J) \Rightarrow T \in \mathcal{T}$.

If the premise of at least one of the conditions a), b), *or* b) *holds, then* $\mathsf{K}_i(\mathsf{K}_i^{-1}(T)J) = \omega^+(T)$. ■

§ Exercises and problems

1 Let $T: \mathcal{H}_1 \; (= \mathcal{H}_1^+ \oplus \mathcal{H}_1^-) \to \mathcal{H}_2 \; (= \mathcal{H}_2^+ \oplus \mathcal{H}_2^-)$, $\mathcal{D}_T = \mathcal{H}_1$, $T \in \mathcal{T}$, $\mathcal{R}_{P_2^- T P_1^-} = \mathcal{H}_2^-$. Prove that then $\mathcal{D}_{\omega^+(T)} = \mathcal{H}_3 \; (= \mathcal{H}_1^+ \oplus \mathcal{H}_2^-)$, and if $T = \| T_{ij} \|_{i,j=2}^2$ is the matrix representation of the operator T, then $\omega^+(T): \mathcal{H}_3 \to \mathcal{H}_4 \; (= \mathcal{H}_2^- \oplus \mathcal{H}_1^-)$ has the matrix representation $\omega^+(T) = \|(\omega^+(T))_{ij}\|_{i,j=1}^2$, where $(\omega^+(T))_{11} = T_{11} - T_{12}T_{22}^{-1}T_{21}$, $(\omega^+(T))_{12} = T_{12}T_{22}^{-1}$, $(\omega^+(T))_{21} = T_{22}^{-1}T_{21}$, $(\omega^+(T))_{22} = T_{22}^{-1}$ (Shmul'yan [5]).

2 Prove that the PG-transformation ω^+ establishes a one-to-one correspondence between all (J_1, J_2)-non-expansive linear relations $\mathcal{N} \subset \mathcal{H}_1 \oplus \mathcal{H}_2$ (i.e., $[x_2, x_2] \leqslant [x_1, x_1]$ if $\langle x_1, x_2 \rangle \in \mathcal{N}$) and all contractions $V: \mathcal{H}_3 \to \mathcal{H}_4$. Investigate what is the pre-image of isometries, of semi-unitary and of unitary operators (Shmul'yan [5], Ritsner [4]).
Hint: Use Formula (1.1).

3 Construct an example of a (J_1, J_2)-bi-extension of a (J_1, J_2)-isometric operator which is neither a (J_1, J_2)-non-expansive operator nor a (J_1, J_2)-non-contractive operator.
Hint: Use Proposition 1.11.

4 Derive formulae connecting the Cayley–Neyman transform K_ζ for an arbitrary $\zeta \neq \bar{\zeta}$ and the Potapov–Ginzburg transformations $\omega^{\pm}$, $\omega_{\pm}$ (Azizov, E. Iokhvidov, and I. Iokhvidov [1].
Hint. Use the scheme of the proof of Theorem 1.14.

5 Prove by means of *PG*-transformations that a continuous, continuously invertible, closed (J_1, J_2)-isometric operator with finite deficiency-numbers $p \equiv \dim \mathscr{D}_V^{[\perp]}$, $q \equiv \dim \mathscr{R}_V^{[\perp]}$ admits a (J_1, J_2)-unitary extension if and only if $\text{In}\, \mathscr{D}^{[\perp]} = \text{In}\, \mathscr{R}_V^{[\perp]}$ (*cf.* E. Iokhvidov [4], [5]).

6 Let V be a bounded (J_1, J_2)-isometric operator with $\mathscr{D}_V = \bar{\mathscr{D}}_V$, $p = \dim \mathscr{D}_V^{[\perp]}$, $q = \dim \mathscr{R}_V^{[\perp]}$, $r = \dim \text{Ker } V$, $m = \dim \mathscr{D}^{\perp}_{\omega^+(V)}$, $n = \dim \mathscr{R}^{\perp}_{\omega^+(V)}$. Then $p + q = m + n + r$ (Azizov, E. Iokhvidov, I. Iokhvidov [1]; *cf.* E. Iokhvidov).

7 Give examples of operators for which the numbers p, q, r, m, n in the equality in Example 6 take all the possible values admissible by this equality (Azizov; *cf.* E. Iokhvidov [4]).
Hint: Consider a J-space $\mathscr{H} = \mathscr{H}^+ \oplus \mathscr{H}^-$ (where $\mathscr{H}^+ = \mathscr{H}^-$ is an infinite-dimensional space), and a semi-unitary operator, acting in $\mathscr{H}$, $U = \| U_{ij} \|_{i,j=1}^2$, where

$$U_{11} = -K^* U_{21},\ K \in \mathscr{K}^+;\ U_{22} = KU_{12};\ U_{21} = S_{21}(S_{21}^* KK^* S_{21} + I^+)^{-1/2},$$

S_{21} is a semi-unitary operator carrying $\mathscr{H}^+$ into $\mathscr{H}^-$; $U_{12} = S_{12}(S_{12}^* K^* K S_{12} + I^-)^{-1/2}$, S_{12} is a semi-unitary operator carrying $\mathscr{H}^-$ into $\mathscr{H}^+$. To solve the problem it suffices to vary the operators K, S_{12} and S_{21}, and also the number of examples of the space $\mathscr{H}$, and to put $V = \omega^+(V)$.

8 Let (m, n) be an arbitrary pair of non-negative integers. Give an example of an unbounded closed (J_1, J_2)-isometric operator V with

$$\bar{\mathscr{D}}_V = \mathscr{H}, \bar{\mathscr{R}}_V = \mathscr{H} \quad \text{and} \quad m = \dim \mathscr{D}^{\perp}_{\omega^+(V)},\ n = \dim \mathscr{R}^{\perp}_{\omega^+(V)}$$

(essentiality of the condition for boundedness of V in Exercise 6).
Hint: Use the device given in the hint on Exercise 7.

9 Let V be a (J_1, J_2)-isometric relation, $\bar{\mathscr{D}}_V = \mathscr{D}_V$, $\bar{\mathscr{R}}_V = \mathscr{R}_V$, and let m, n, p, q, r be the same as in Exercise 6, and $s = \dim \text{Ker } V^{-1}$. Then $p + q = m + n + r + s$ (Azizov, E. Iokhvidov and I. Iokhvidov [1]).

10 Let $\mathscr{H}$ be a J-space, T a linear relation, and $n_\zeta(T) = \dim \mathscr{R}^{\perp}_{T - \zeta I}$. Prove that if T is a J-Hermitian linear relation, $\zeta \neq \bar{\zeta}$, and $\mathscr{R}_{T - \zeta I}$ and $\mathscr{R}_{T - \bar{\zeta} T}$ are subspaces in $\mathscr{H}$, then

$$n_\zeta(T) + n_{\bar{\zeta}}(T) = n_\zeta(JT) + n_{\bar{\zeta}}(JT) + \dim \text{Ker}(T - \zeta I) + \dim \text{Ker}(T - \bar{\zeta} I).$$

In particular, if ζ is a point of regular type for T, then the condition that $\mathscr{R}_{T - \bar{\zeta} I}$ be closed may be omitted (Azizov, E. Iokhvidov, I. Iokhvidov [1]).
Hint: Use the results of Exercises 6 and 9 and also Theorem 1.14.

11 Under the conditions of Exercise 10 let T be a J-Hermitian operator. Prove that $T = T_c$ if and only if

$$n_\zeta(T) + n_{\bar{\zeta}}(T) = \dim \text{Ker}(T - \zeta I) + \dim \text{Ker}(T - \bar{\zeta} I)$$

(Azizov, E. Iokhvodov, I. Iokhvidov [1]).
Hint: Use the results of Exercise 10.

12 Prove that if T is a J-non-expansive operator and $1 \notin \sigma_p(T)$, then $\mathbf{K}_i(\mathbf{K}_i^{-1}(T) J) = \omega^+(T)$.
Hint: Use Exercise 33 on 2.§4 and Corollary 1.15.

13 Let $\mathscr{D} = \mathscr{D}_0 [\dot{+}] \mathscr{D}_2$ and $\mathscr{D}^{\perp} = (\mathscr{D}^{\perp})_0 [\dot{+}] (\mathscr{D}^{\perp})_1 [\dot{+}] (\mathscr{D}^{\perp})_2$ be the $(\mathscr{H}^+, \mathscr{H}^-)$-decompositions of the subspaces $\mathscr{D}$ and $\mathscr{D}^+$, let $\mathscr{H} = \mathscr{D}_0 [\dot{+}] \mathscr{D}_1 [\dot{+}] \mathscr{D}_2 \oplus (\mathscr{D}^{\perp})_0 [\dot{+}] (\mathscr{D}^{\perp})_1 [\dot{+}] (\mathscr{D}^{\perp})_2$, and let $J = \| J_{ij} \|_{i,j=1}^6$ be the matrix representation of J (see Exercise 5 on 1.§10). Verify that a bounded

operator X with $\mathscr{D}_X = \mathscr{H}$ represented by the matrix $X = \| X_{ij} \|_{i,j=1}^{6}$ will be a J-bi-extension of the operator $I_{\mathscr{D}} = I \mid \mathscr{D}$ if and only if $\| X_{ij} \|_{i,j=1}^{3} = I_{\mathscr{D}}$, $X_{ij} = 0$ when $i = 4, 5, 6$, $j = 1, 2, 3$; $X_{2j} = 0$ when $j = 4, 5, 6$; $\mathscr{R}_{X_{6j}} \subset \mathscr{R}_{J_{66}}$ and $X_{3j} = -J_{36} F_{66}^{-1} X_{6j}$, $j = 4, 5$; $\mathscr{R}_{I_6 - X_{66}} \subset \mathscr{R}_{J_{66}}$ and $X_{36} = J_{36} J_{66}^{-1}(I - {}_{66})$; $X_{44} = I$, and all the remaining X_{ij} are arbitrary (Azizov [14]).
Hint: Write out the matrix X^c and use the inclusion $I_{\mathscr{D}} \subset X^c$ derived from the inclusion in Proposition 1.12.

14 Let $\mathscr{H}$ be a J-space, $\mathscr{D}$ a subspace of $\mathscr{H}$, and $J = \| J_{ij} \|_{i,j=1}^{2}$ be the matrix representation of an operator J relative to the decomposition $\mathscr{H} = \mathscr{D} \oplus \mathscr{D}^{\perp}$. Prove that an operator $Y(\in \mathscr{T})$ will be a J-bi-extension of the operator $I_{\mathscr{D}}$ if and only if there is an operator $Z_{22} : \mathscr{D}_2^{\perp} \to \mathscr{D}^{\perp}$ such that $0 \notin \sigma_p(I_2 + J_{22} + (I_2 - J_{22}) Z_{22})$ and

$$Y = \| Y_{ij} \|_{i,j=1}^{2}, \text{ where } Y_{11} = I_{\mathscr{D}}, Y_{21} = 0,$$
$$Y_{12} = -2 J_{12}(I_2 - Z_{22})(I_2 + J_{22} + (I_2 - J_{22}(Z_{22})^{-1},$$

and

$$Y_{22} = (I_2 - J_{22} + (I_2 + J_{22}) Z_{22})(I_2 + J_{22} + (I_2 - J_{22}) Z_{22})^{-1} \text{ (Azizov [14]).}$$

Hint: Verify that $\omega^{+}(I_{\mathscr{D}}) = I_{\mathscr{D}}$; use Proposition 1.11 and establish that bi-extensions of the isometric operator $I_{\mathscr{D}}$ coincide with the set of operators representable by matrices $Z = \| Z_{ij} \|_{i,j=1}^{2}$, where $Z_{11} = I_{\mathscr{D}}$, $Z_{12} = 0$, $Z_{21} = 0$, and $Z_{22} : \mathscr{D}^{\perp} \to \mathscr{D}^{\perp}$; verify that the condition $Z \in \mathscr{T}$ is equivalent to the condition $0 \notin \sigma_p(I_2 + J_{22} + (I_2 - J_{22}) Z_{22})$ and that $Y = \omega^{+}(Z)$.

15 Prove that under the conditions of Exercise 14 the operator Y will j-bi-non-expansive (respectively, J-semi-unitary J-bi-non-expansive, J-unitary) if and only if Z_{22} is a contractive (respectively, semi-unitary, unitary) operator with $\mathscr{D}_{Z_{22}} = \mathscr{D}^{\perp}$ and $0 \in \rho(I_2 + J_{22} + (I_2 - J_{22}) Z_{22})$ (Azizov [14]).
Hint: Use the hint on Exercise 14 and Corollary 1.7.

16 Let V be a (J_1, J_2)-isometric operator which has a continuous (J_1, J_2)-bi-extension V_0 with $\mathscr{D}_{V_0} = \mathscr{H}_1$, a projectionally complete range of values, and $\operatorname{Ker} V_0 = \{\theta\}$. Prove that the operator $\tilde{V}$ with $\mathscr{D}_{\tilde{V}} = \mathscr{H}_1$ is a (J_1, J_2)-bi-extension of the operator V if and only if $\tilde{V} = \tilde{I}_{\mathscr{R}} V_0 (V_0^c V_0)^{-1}$, where $\tilde{I}_{\mathscr{R}}$ is a certain J_2-bi-extension of the operator $I_{\mathscr{R}} \equiv I_2 \mid \mathscr{R}_{V_6}$; $\mathscr{D}_{\tilde{I}_{\mathscr{R}}} = \mathscr{H}_2$ (Azizov [14]).
Hint: Use the fact that $V_0^c V_0$ is a linear homeomorphism of the space $\mathscr{H}_1$ and is a J_1-bi-extension of the operator $I \mid \mathscr{D}_V$.

17 Prove that if, under the conditions of Exercise 16, the operator V_0 is (J_1, J_2)-semi-unitary, then the formula given there can be rewritten in the form $\tilde{V} = \tilde{I}_{\mathscr{R}} V_0$, and if V_0 is a (J_1, J_2)-unitary operator, then V is a (J_1, J_2)-unitary (respectively, (J_1, J_2)-semi-unitary, (J_1, J_2)-bi-non-expansive, (J_1, J_2)-bi-non-contractive) operator if and only if $\tilde{I}_{\mathscr{R}}$ belongs to the corresponding class (Azizov [14]).

§2 Extensions of standard isometric and symmetric operators

1 Let $\mathscr{H}_i = \mathscr{H}_i^{+} \oplus \mathscr{H}_i^{-}$ be J_i-spaces $(i = 1, 2)$. In this section we present another approach to the construction of a theory of extensions of (J_1, J_2)-isometric and J-symmetric operators and we describe all their (J_1, J_2)-bi-extensions and J-bi-extentions respectively.

Definition 2.1: We shall call an operator $V: \mathscr{H}_1 \to \mathscr{H}_2$ a *standard operator* and write $V \in \mathrm{St}(\mathscr{H}_1, \mathscr{H}_2)$ if V is a closed continuous and continuously invertible (J_1, J_2)-isometric operator. In particular, if $\mathscr{H}_1 = \mathscr{H}_2 = \mathscr{H}$ is a J-space, we shall write $V \in \mathrm{St}\, \mathscr{H}$.

We point out that the condition $V \in \mathrm{St}(\mathscr{H}_1, \mathscr{H}_2)$ implies that $\mathscr{D}_V$ and $\mathscr{R}_V$ are closed.

Using Zorn's lemma it is easy to prove that *every* (J_1, J_2)-*isometric operator* V *admits extension into a maximal* (J_1, J_2)-*isometric operator* $\tilde{V}$, i.e., $\tilde{V}$ now has no non-trivial $\tilde{J}_1, J_2)$-isometric extensions. Less obvious is the assertion that *if* $V \in \mathrm{St}(\mathscr{H}_1, \mathscr{H}_2)$, *then among its maximal extensions there is a* $\tilde{V} \in \mathrm{St}(\mathscr{H}_1, \mathscr{H}_2)$. Theorem 2.2 is devoted to the investigation of this question; in the proof of this theorem a construction for the corresponding extension is given which will be used more than once later.

Theorem 2.2: *Let* $V \in \mathrm{St}(\mathscr{H}_1, \mathscr{H}_2)$. *Then it has a maximal* (J_1, J_2)-*isometric extension* $\tilde{V} \in \mathrm{St}(\mathscr{H}_1, \mathscr{H}_2)$.

□ First of all we observe that if $\mathscr{D}_V^{[\perp]}$ and $\mathscr{R}_V^{[\perp]}$ are regular subspaces, then, as is easily verified, the operator V has an extension $\tilde{V} \in \mathrm{St}(\mathscr{H}_1, \mathscr{H}_2)$ such that its deficiency subspaces $\mathscr{D}_{\tilde{V}}^{[\perp]}$ and $\mathscr{R}_{\tilde{V}}^{[\perp]}$ are regular and either at least one of them is $\{\theta\}$ or they are uniformly definite and have different signs. The maximality of such (J_1, J_2)-isometric operators is obvious. Consequently the theorem will be proved if we verify that every operator $V \in \mathrm{St}(\mathscr{H}_1, \mathscr{H}_2)$ admits extension into an operator $V' \in \mathrm{St}(\mathscr{H}_1, \mathscr{H}_2)$ with regular deficiency subspaces $\mathscr{D}_{V'}^{[\perp]}$ and $\mathscr{R}_{V'}^{[\perp]}$.

We consider the $(\mathscr{H}_1^+, \mathscr{H}_1^-)$-decomposition of the subspace $\mathscr{D}_V$ (see 1.Definition 10.3):

$$\mathscr{D}_V = \mathscr{D}_0 \,[\dot{+}]\, \mathscr{D}_1 \,[\dot{+}]\, \mathscr{D}_2 \tag{2.1}$$

and the $(\mathscr{H}_2^+, \mathscr{H}_2^-)$-decomposition of the subspace $\mathscr{R}_V$:

$$\mathscr{R}_V = \mathscr{R}_0 \,[\dot{+}]\, \mathscr{R}_1 \,[\dot{+}]\, \mathscr{R}_2. \tag{2.2}$$

Without loss of generality we shall suppose that (2.1) and (2.2) are connected by the relations $V\mathscr{D}_i = \mathscr{R}_i$, $i = 0, 1, 2$. Indeed, $\mathscr{R}_0 = V\mathscr{D}_0$, since a (J_1, J_2)-isometric operator maps the isotropic part $\mathscr{D}_0$ of the subspace $\mathscr{D}_V$ on to the isotropic part $\mathscr{R}_0$ of the subspace $\mathscr{R}_V$. But if even one of the relations $\mathscr{R}_i = V\mathscr{D}_i$ $(i = 1, 2)$ is infringed, we proceed as follows. Taking 2.Corollary 4.13 into account we shall suppose, without loss of generality, that $V(\mathscr{D}_1 \cap \mathscr{H}_1^{\pm}) \subset \mathscr{H}_2^{\pm}$ and we consider the decomposition

$$\mathscr{D}_V = \mathscr{D}_0 \,[\dot{+}]\, V^{-1}\mathscr{R}_1 \,[\dot{+}]\, V^{-1}\mathscr{R}_2. \tag{2.3}$$

Hence

$$\mathscr{D}_1 \subset V^{-1}\mathscr{R}_1, \quad V^{-1}\mathscr{R}_1 = \mathscr{D}_1 \,[\dot{+}]\, \mathscr{D}_2', \quad \text{and} \quad \mathscr{D}_2 = \mathscr{D}_2' \,[\dot{+}]\, \mathscr{D}_2'',$$

where $\mathscr{D}_2'' = V^{-1}\mathscr{R}_2$. By virtue of 1.Theorem 10.4 we can find a maximal uniformly definite dual pair $(\mathscr{L}_1^+, \mathscr{L}_1^-)$ such that (2.3) will be the $(\mathscr{L}_1^+, \mathscr{L}_1^-)$-decomposition of the subspace $\mathscr{D}_V$. It only remains to put $\mathscr{L}_1^\pm = \mathscr{H}_1^\pm$.

Let

$$\mathscr{D}_V^\perp = (\mathscr{D}^\perp)_0 \,[\dot{+}]\, (\mathscr{D}^\perp)_1 \,[\dot{+}]\, (\mathscr{D}^\perp)_2$$

and

$$\mathscr{R}_V^\perp = (\mathscr{R}^\perp)_0 \,[\dot{+}]\, (\mathscr{R}^\perp)_1 \,[\dot{+}]\, (\mathscr{R}^\perp)_2$$

be the $(\eta_1^+, \mathscr{H}_1^-)$-decomposition and the $(\mathscr{H}_2^+, \mathscr{H}_2^-)$-decomposition of the subspaces $\mathscr{D}_V^\perp$ and $\mathscr{R}_V^\perp$ respectively. Then relative to the decomposition

$$\mathscr{H}_1 = \mathscr{D}_0 \,[\dot{+}]\, \mathscr{D}_1 \,[\dot{+}]\, \mathscr{D}_2 \oplus (\mathscr{D}^\perp)_0 \,[\dot{+}]\, (\mathscr{D}^\perp)_1 \,[\dot{+}]\, (\mathscr{D}^\perp)_2$$

(respectively, $\mathscr{H}_2 = \mathscr{R}_0 \,[\dot{+}]\, \mathscr{R}_1 \,[\dot{+}]\, \mathscr{R}_2 \oplus (\mathscr{R}^\perp)_0 \,[\dot{+}]\, (\mathscr{R}^\perp)_1 \,[\dot{+}]\, (\mathscr{R}^\perp)_2$) the operator J_1 (respectively, J_2) can be expressed in accordance with Exercise 5 on 1.§10 in the form of a matrix $J_1 = \| J_{ij}^{(1)} \|_{i,j=1}^6$ (respectively, $J_2 = \| J_{ij}^{(2)} \|_{i,j=1}^6$. The components of the expansions of the matrices J_1 and J_2 have the properties indicated in Exercise 5 on 1.§10. Thus, for example, $J_{36}^{(1)} = (I_3^{(1)} - J_{33}^{(1)2})^{1/2} V_{36}^{(1)}$, and $J_{36}^{(2)} = (I_3^{(2)} - J_{33}^{(2)2})^{1/2} V_{36}^{(2)}$, where $V_{36}^{(1)}$ is an isometry mapping $(\mathscr{D}^\perp)_2$ on to $\mathscr{D}_2$, and $V_{36}^{(2)}$ is an isometry mapping $(\mathscr{R}^\perp)_2$ on to $\mathscr{R}_2$. We introduce the notation $V_{i+1} = V \mid \mathscr{D}_i$ $(i = 0, 1, 2)$ and define an extension V' of the operator V in the following way

$$V' \mid \mathscr{D}_V = V;$$

$$V' \mid (\mathscr{D}^\perp)_0 \equiv V_4 : (\mathscr{D}^\perp)_0 \to (\mathscr{R}^\perp)_0, \qquad V_4 = J_{41}^{(2)} V_1^{*-1} J_{14}^{(1)};$$

$$V' \mid (\mathscr{D}^\perp)_2 \equiv V_6 : (\mathscr{D}^\perp)_2 \to \mathscr{R}_2 \oplus (\mathscr{R}^\perp)_2, \; V_6 x = V_{36}' x + V_6' x, \; x \in (\mathscr{D}^\perp)_2,$$

$$V_{36}' x \in \mathscr{R}_2, \; V_6' x \in (\mathscr{R}^\perp)_2,$$

$$\begin{aligned} V_{36}' = {} & [(I_3^{(2)} + (I_3^{(2)} - J_{33}^{(2)2})^{1/2})^{-1} J_{33}^{(2)} V^{*-1} \\ & - V_3 (I_3^{(1)} + (I_3^{(1)} - J_{33}^{(1)2})^{-1}) J_{33}^{(1)}] V_{36}^{(1)}, \end{aligned}$$

$$V_6' = V_{36}^{(2)*} V_3^{*-1} V_{36}^{(1)}.$$

The operator V' is continuous and continuously invertible;

$$\mathscr{D}_{V'} = \mathscr{D}_0 \,[\dot{+}]\, \mathscr{D}_1 \,[\dot{+}]\, \mathscr{D}_2 \oplus (\mathscr{D}^\perp)_0 \,[\dot{+}]\, (\mathscr{D}^\perp)_2,$$

and

$$\mathscr{R}_{V'} = \mathscr{R}_0 \,[\dot{+}]\, \mathscr{R}_1 \,[\dot{+}] + \mathscr{R}_2 \oplus (\mathscr{R}^\perp)_0 \,[\dot{+}]\, (\mathscr{R}^\perp)_2.$$

Moreover, $V'(\mathscr{D}_0 \oplus (\mathscr{D}^\perp)_0) = \mathscr{R}_0 \oplus (\mathscr{R}^\perp)_0$, $V'\mathscr{D}_1 = \mathscr{R}_1$, and $V'(\mathscr{D}_2 \oplus (\mathscr{D}^\perp)_2) = \mathscr{R}_2 \oplus (\mathscr{R}^\perp)_2$. In accordance with Exercise 4 on 1.§10 the subspaces $\mathscr{D}_0 \oplus (\mathscr{D}^\perp)_0$, $\mathscr{D}_1$, and $\mathscr{D}_2 \oplus (\mathscr{D}^\perp)_2$ (respectively, $\mathscr{R}_0 \oplus (\mathscr{R}^\perp)_0$, $\mathscr{R}_1$, and $\mathscr{R}_2 \oplus (\mathscr{R}^\perp)_2$) are pairwise J_1-orthogonal (respectively, J_2-orthogonal), and these subspaces are invariant relative to J_1 (respectively J_2). Therefore V' will be a (J_1, J_2)-isometric operator if and only if the operators $V' \mid \mathscr{D}_0 \oplus (\mathscr{D}^\perp)_0$, $V' \mid \mathscr{D}_1$, and $V' \mid \mathscr{D}_2 \oplus (\mathscr{D}^\perp)_2$ are (J_1, J_2)-isometric. Since $V' \mid \mathscr{D}_1 = V \mid \mathscr{D}_1$, the operator $V' \mid \mathscr{D}_1$ is (J_1, J_2)-isometric (by hypothesis). Let $x_1 \in \mathscr{D}_0$,

$x_4 \in (\mathscr{D}^{\perp})_0$. Then

$$V'(x_1 + x_4) = V_1 x_1 + J_{41}^{(2)} V_1^{*-1} J_{14}^{(1)} x_4,$$

and

$$J_2 V'(x_1 + x_4) = J_{41}^{(2)} V_1 x_1 + V_1^{*-1} J_{14}^{(1)} x_4.$$

Consequently

$$[V'(x_1 + x_4), V'(x_1 + x_4)]_2 = (J_{14}^{(1)*} x_1, x_4)_1 + (J_{14}^{(1)} x_4, x_1)_1 = [x_1 + x_4, x_1 + x_4]_1,$$

i.e., $V' \mid \mathscr{D}_0 \oplus (\mathscr{D}^{\perp})_0$ is a (J_1, J_2)-isometric operator.

We verify that the operator $V' \mid \mathscr{D}_2 \oplus (\mathscr{D}^{\perp})_2$ is also (J_1, J_2)-isometric. To do this we shall suppose, without loss of generality, that $\mathscr{H}_1 = \mathscr{D}_2 \oplus (\mathscr{D}^{\perp})_2$, $\mathscr{H}_2 = \mathscr{R} \oplus (\mathscr{R}^{\perp})_2$, and we shall prove that V' is a (J_1, J_2)-unitary operator, i.e., $V'^* J_2 V' = J_1$. It can be verified immediately that the operator $T = V'^* J_2 V'$ has the matrix representation

$$T = \begin{Vmatrix} T_{33} & T_{36} \\ T_{63} & T_{66} \end{Vmatrix}, \quad \text{where } T_{33} = V_3^* J_{33}^{(2)} V_3,$$

$$T_{63}^* = T_{36} = V_3^* J_{33}^{(2)} V_{36}' + V_3^* (I_3^{(2)} - J_{33}^{(2)2})^{1/2} V_6',$$

and

$$T_{66}' = V_{36}'^* J_{33}^{(2)} V_{36}' + V_6'^* V_{36}^{(2)*} (I_3^{(2)} - J_{33}^{(2)2})^{1/2} V_{36}'$$
$$+ V_{36}'^* (I_3^{(2)} - J_{33}^{(2)2})^{1/2} V_{36}^{(2)} V_6' - V_6'^* V_{36}^{(2)*} J_{33}^{(2)} V_{32}^2 V_6'.$$

Since the operator V_3 is (J_1, J_2) isometric, we obtain $T_{33} = J_{33}^{(1)}$. We substitute the values of the operators V_{36}' and V_6' in the expressions for T_{36} and T_{66}, and after some elementary transformations we find that $T_{36} = J_{36}^{(1)}$ and $T_{66} = J_{66}^{(1)}$, i.e., $T = J_1$. Thus $V' \mid \mathscr{D}_2 \oplus (\mathscr{D}')_2$ is a (J_1, J_2) isometric operator and $V' \in \mathrm{St}(\mathscr{H}_1, \mathscr{H}_2)$. Moreover, $(\mathscr{D}^{\perp})_1$ and $(\mathscr{R}^{\perp})_1$ are its deficiency (regular) subspaces and therefore V', and with it also V, has among its maximal (J_1, J_2)-isometric extensions also operators $\tilde{V} \in \mathrm{St}(\mathscr{H}_1, \mathscr{H}_2)$. ■

Corollary 2.3: *For an operator $V \in \mathrm{St}(\mathscr{H}_1, \mathscr{H}_2)$ to be a maximal (J_1, J_2)-isometric operator it is necessary and sufficient that at least one of the following conditions shall hold:*

a) $\mathscr{D}_V^{[\perp]} = \{\theta\}$, i.e., *$V$ is a (J_1, J_2)-semi-unitary operator;*
b) $\mathscr{R}_V^{[\perp]} = \{\theta\}$, i.e., *$V^{-1}$ is a (J_1, J_2)-semi-unitary operator,*
c) *$\mathscr{D}_V^{[\perp]}$ and $\mathscr{R}_V^{[\perp]}$ are uniformaly definite subspaces of different signs.*

□ The sufficiency of each of the conditions a)–c) for maximality of a (J_1, J_2)-isometric operator V is obvious.

Now let $V \in \mathrm{St}(\mathscr{H}_1, \mathscr{H}_2)$ be a maximal (J_1, J_2)-isometric operator. From the course of the proof of Theorem 2.2 it follows that $\mathscr{D}_V^{[\perp]}$ and $\mathscr{R}_V^{[\perp]}$ are regular subspaces. Let us suppose that none of the conditions a)–c) is satisfied. Then there are $\theta \neq x_0 \in \mathscr{D}_V^{[\perp]}$ and $\theta \neq y_0 \in \mathscr{R}_V^{[\perp]}$ such that $[x_0, x_0] = [y_0, y_0]$. We define an operator $V' : \mathrm{Lin}\{\mathscr{D}_V, x_0\} \to \mathrm{Lin}\{\mathscr{R}_V, y_0\}$, $V \subset V'$, $V' x_0 = y_0$. It is easy to see that $V' \in \mathrm{St}(\mathscr{H}_1, \mathscr{H}_2)$—and we have obtained a contradiction. ■

Corollary 2.4: *Every operator $V \in \mathrm{St}(\mathscr{H}_1, \mathscr{H}_2)$ admits a (J_1, J_2)-bi-extension V_0 such that $\mathscr{R}_{V_0}$ is a projectionally complete subspace and at least one of the operators V_0 or V_0^c has a trivial kernel.*

□ By virtue of Theorem 2.2 we can suppose that $V \in \mathrm{St}(\mathscr{H}_1, \mathscr{H}_2)$ is a maximal (J_1, J_2)-isometric operator. We now use Corollary 2.3 and Proposition 1.11. If V satisfies condition a), then we put $V_0 = V$; if V satisfies condition b), then we put $V_0 = (V^{-1})^c$; if V satisfies condition c) and $\dim \mathscr{D}_V^{[\perp]} \leqslant \dim \mathscr{R}_V^{[\perp]}$, then as V_0 we take any extension of the operator V on to $\mathscr{H}_1$ which maps $\mathscr{D}_V^{[\perp]}$, then as V_0 we take any extension of the operator V on to $\mathscr{H}_1$ which maps $\mathscr{D}_V^{[\perp]}$ injectively on to a subspace in $\mathscr{R}_V^{[\perp]}$, and if $\dim \mathscr{D}_V^{[\perp]} > \dim \mathscr{R}_V^{[\perp]}$, then we put $V_0 = (\tilde{V}^{-1})^c$, where $\tilde{V}^{-1}$ is an arbitrary extension of the operator V^{-1} on to $\mathscr{H}_2$ which maps $\mathscr{R}_V^{[\perp]}$ injectively on to a subspace in $\mathscr{D}_V^{[\perp]}$. ■

Thus, for the description of all (J_1, J_2)-bi-extensions of an operator $V \in \mathrm{St}(\mathscr{H}_1, \mathscr{H}_2)$ the results of Exercises 13–17 on §1 can be used.

2 We pass on now to study the problem of the possibility of (J_1, J_2) unitary extensions of operators $V \in \mathrm{St}(\mathscr{H}_1, \mathscr{H}_2)$. Example 2.8 given below of an operator $V \in \mathrm{St}(\mathscr{H}_1, \mathscr{H}_2)$ with $\mathscr{D}_V = \mathscr{R}_V$ (and therefore $\mathscr{D}_V^{[\perp]} = \mathscr{R}_V^{[\perp]}$) shows that, in contrast to the Hilbert space case (see, e.g., [I]), in the case of a Pontryagin space (see Exercise 3 below) information only about the properties of $\mathscr{D}_V^{[\perp]}$ and $\mathscr{R}_V^{[\perp]}$ is not always sufficient to make conclusions about the possibility of extending an operator V into a (J_1, J_2)-unitary operator. But first we introduce some preliminary materials.

Definition 2.5: Let $(\mathscr{L}^+, \mathscr{L}^-)$ be a maximal uniformly definite dual pair in a J-space $\mathscr{H}$, and suppose the subspaces $\mathscr{N}_+ \in \mathscr{M}^+$ and $\mathscr{N}_- = \mathscr{N}_+^{[\perp]}$ $(\in \mathscr{M}^-)$ contain no infinite-dimensional uniformly definite subspaces. Then the number $\nu(\mathscr{N}_+) = \dim(\mathscr{N}_+ \cap \mathscr{L}^+) - \dim(\mathscr{N}_- \cap \mathscr{L}^-)$ is called the *index of the subspace* $\mathscr{N}_+$.

Lemma 2.6: *In Definition 2.5 $\nu(\mathscr{N}_+)$ does not depend on the choice of the maximal uniformly definite dual pair $(\mathscr{L}^+, \mathscr{L}^-)$.*

□ Let

$$\nu_1(\mathscr{N}_+) = \dim(\mathscr{N}_+ \cap \mathscr{H}^+) - \dim(\mathscr{N}_- \cap \mathscr{H}^-)$$

and

$$\nu_2(\mathcal{N}_+) = \dim(\mathcal{N}_+ \cap \mathcal{L}^+) - \dim(\mathcal{N}_- \cap \mathcal{L}^-).$$

We verify that $\nu_1(\mathcal{N}_+) = \nu_2(\mathcal{N}_+)$.

Let $K_\pm$ be the angular operators of the subspaces $\mathcal{N}_\pm$ and $Q_\pm$ be the angular operators of the subspaces $\mathcal{L}^\pm$. From Exercise 17 on 1.§8 and Exercise 13 on 2.§2 it follows that K_+ is a Φ-operator and $|K_+| = I^+ + S$, where $S \in \mathcal{S}_\infty$. From Exercise 14 on 2.§2 it therefore follows that $K_+ - Q_+$ is also a Φ-operator and ind $K_+ = \operatorname{ind}(K_+ - Q_+)$. It remains to observe that $\mathcal{N}_\pm \cap \mathcal{H}^\pm = \operatorname{Ker} K_\pm$, $\mathcal{N}_\pm \cap \mathcal{L}^\pm = \operatorname{Ker}(K_\pm - Q_\pm)$, and $K_+ = K_-^*$, $K_+ - Q_+ = K_-^* - Q_-^*$, i.e., $\nu_1(\mathcal{N}_+) = -\operatorname{ind} K_+$ and $\nu_2(\mathcal{N}_+) = -\operatorname{ind}(K_+ - Q_+)$. ■

Lemma 2.7: *If $V \in \mathrm{St}(\mathcal{H}_1, \mathcal{H}_2)$, $\mathcal{D}_V \in M^+(\mathcal{H}_1)$, $\mathcal{R}_V \in \mathcal{M}^+(\mathcal{H}_2)$, and $\mathcal{D}_V$, $\mathcal{R}_V$, $\mathcal{D}_V^{[\perp]}$ and $\mathcal{R}_V^{[\perp]}$ contain no infinite-dimensional uniformly definite subspaces, then V admits a (J_1, J_2)-unitary extension if and only if $\nu(\mathcal{D}_V) = \nu(\mathcal{R}_V)$. Moreover every maximal extension $V' \in \mathrm{St}(\mathcal{H}_1, \mathcal{H}_2)$ of the operator V will be (J_1, J_2)-unitary.*

□ Let $\tilde{V}$ be a (J_1, J_2)-unitary extension of the operator V. We introduce in $\mathcal{H}_2$ a scalar product $(x, y)_2' = (\tilde{V}^{-1}x, \tilde{V}^{-1}y)_1$ $(x, y \in \mathcal{H}_2)$ equivalent to the original one. Since $[x, y]_2 = (J_2'x, y)_2$, where $J_2' = \tilde{V}J_1\tilde{V}^{-1} = J_2'^{-1}$, the components of the new canonical decomposition $\mathcal{H}_2 = \mathcal{H}_2^{+\prime} \oplus \mathcal{H}_2^{-\prime}$ will be the subspaces $\mathcal{H}_2^{\pm\prime} = \tilde{V}\mathcal{H}_1^\pm$. Consequently $\dim(\mathcal{D}_V \cap \mathcal{H}_1^+) = \dim(\tilde{V}(\mathcal{D}_V \cap \mathcal{H}_1^+)$ $= \dim(\mathcal{R}_V \cap \mathcal{H}_2^{+\prime})$ and $\dim(\mathcal{D}_V^{[\perp]} \cap \mathcal{H}_1^-) = \dim(\tilde{V}(\mathcal{D}_V^{[\perp]} \cap \mathcal{H}_1^{-1})) = \dim(\mathcal{R}_V^{[\perp]} \cap \mathcal{H}_2^{-\prime})$, and therefore $\nu(\mathcal{D}_V) = \nu(\mathcal{R}_V)$.

Now let $\nu(\mathcal{D}_V) = \nu(\mathcal{R}_V)$. Without loss of generality we shall suppose (see the proof of Theorem 2.2) that the decompositions (2.1) and (2.2) of the subspaces $\mathcal{D}_V$ and $\mathcal{R}_V$ are connected by the relations $V\mathcal{D}_i = \mathcal{R}_i$, $i = 0, 1, 2$. Since $\mathcal{D}_1 = \mathcal{D}_V \cap \mathcal{H}_1^+$ and $\mathcal{R}_1 = \mathcal{R}_V \cap \mathcal{H}_2^+$, we have $\dim(\mathcal{D}_V \cap \mathcal{H}_1^+) = \dim(\mathcal{R}_V \cap \mathcal{H}_2^+)$. Consequently $\dim(\mathcal{D}_V^\perp \cap \mathcal{H}_1^-) = \dim(\mathcal{D}_V^{[\perp]} \cap \mathcal{H}_1^-) = -\nu(\mathcal{D}_V) + \dim(\mathcal{D}_V \cap \mathcal{H}_1^+) = -\nu(\mathcal{R}_V) + \dim(\mathcal{R}_V \cap \mathcal{H}_2^+) = \dim(\mathcal{R}_V^{[\perp]} \cap \mathcal{H}_2^-)$ $= \dim(\mathcal{R}_V^\perp \cap \mathcal{H}_2^-)$.

Following the scheme of the proof of Theorem 2.2 we deduce that the operator V admits an extension $V' \in \mathrm{St}(\mathcal{H}_1, \mathcal{H}_2)$ with $\mathcal{D}_{V'}^{[\perp]} = \mathcal{D}_V^\perp \cap \mathcal{H}_1^-$ and $\mathcal{R}_{V'}^{[\perp]} = \mathcal{R}_V^\perp \cap \mathcal{H}_2^-$. Hence $\mathcal{D}_{V'}^{[\perp]}$ and $\mathcal{R}^{[\perp]}$ are negative subspaces of equal finite dimension and therefore it is now easy to construct a (J_1, J_2)-unitary extension of the operator V' and it will also be an extension of V.

Let $V' \in \mathrm{St}(\mathcal{H}_1, \mathcal{H}_2)$ be an arbitrary maximal (J_1, J_2)-isometric extension of the operator V, and suppose that V' is not a (J_1, J_2)-unitary operator. Since $\mathcal{D}_V^{[\perp]}$ and $\mathcal{R}_V^{[\perp]}$ have the same sign, we conclude from Corollary 2.3 that we can, without loss of generality, suppose the operator V' to be (J_1, J_2)-semi-unitary and $\dim R_{V'}^{[\perp]} \neq 0$.

We consider the J_2'-space $\mathcal{H}_2' = V'\mathcal{H}_1$. From what has been proved, $\nu(\mathcal{D}_V) = \nu'(\mathcal{R}_V)$, where $\nu'(\mathcal{R}_V)$ is the index of $\mathcal{R}_V$ in $\mathcal{H}_2'$. But by hypothesis

$\nu(\mathscr{D}_V)=\nu(\mathscr{R}_V)$, and it is easy to see that $\nu(\mathscr{R}_V)=\nu'(\mathscr{R}_V)-\dim \mathscr{R}_V^{[\perp]}$—so we have obtained a contradiction. ■

We pass on now to the construction of an operator $V \in \mathrm{St}\,\mathscr{H}$ with $\mathscr{D}_V=\mathscr{R}_V$, but which does not have J-unitary extensions.

□ *Example 2.8:* Let $\mathscr{D}_1, \mathscr{D}_2, (\mathscr{D}^\perp)_2$ be infinite-dimensional separable Hilbert spaces. We form the space $\mathscr{H}=\mathscr{D}_1 \oplus \mathscr{D}_2 \oplus (\mathscr{D}^\perp)_2$ and introduce in it a J-metric (see 1.Example 3.9) by means of the operator

$$J=\begin{Vmatrix} J_{22} & 0 & 0 \\ 0 & J_{33} & (J_3-J_{33}^2)^{1/2}V_{36} \\ 0 & V_{36}^*(I_3-J_{33}^2)^{1/2} & -V^*J_{03}V_{36} \end{Vmatrix},$$

constructed relative to this decomposition of the space $\mathscr{H}$, where $J_{22}=I_2$, $J_{33}\in\mathscr{S}_\infty$, $0<J_{33}<I_3$, and V_{36} is an isometry mapping $(\mathscr{D}^\perp)_2$ on to $\mathscr{D}_2$. Let $\lambda_1 \geqslant \lambda_2 \geqslant \cdots$ be the eigenvalues of the operator J_{33}, and let $\lim_{k\to\infty}\lambda_{k+1}/\lambda_k=\alpha>0$; let $\{f_k\}_1^\infty$ be an orthonormalized basis in $\mathscr{D}_2$ composed of the eigenvectors of the operator $J_{33}: J_{33}f_k=\lambda_k f_k$ $(k=1,2,\ldots)$; and let $\{e_k\}_1^\infty$ be an orthonormalized basis in $\mathscr{D}_1$. We define on $\mathscr{D}_V=\mathscr{D}_1\oplus\mathscr{D}_2$ an operator $V\in\mathrm{St}\,\mathscr{H}$ by putting

$$Ve_k=e_{k+1}(k=1,2,\ldots), \qquad Vf_1=\sqrt{\lambda_1}\,e_1, \qquad Vf_k=\sqrt{\lambda_{k+1}/\lambda_k}\,f_k \quad (k=1,2,\ldots).$$

By construction $\mathscr{R}_V=\mathscr{D}_V$. The operator V admits no J-unitary extensions $\tilde{V}$, for if it did the operator $V|\mathscr{D}_2$ $(\in \mathrm{St}(\mathscr{H}_1,\mathscr{H}_2))$ acting from the J_1-space $\mathscr{H}_1=\mathscr{D}_2 \oplus (\mathscr{D}^1)_2$ into the J_2-space $\mathscr{H}_2=\mathrm{Lin}\{e_1\}\oplus\mathscr{D}_2\oplus(\mathscr{D}^\perp)_2$ with $J_i=J|\mathscr{H}_i$, $i=1,2$, would have (J_1, J_2)-unitary extensions, and this is impossible by virtue of Lemma 2.7 since $\nu(\mathscr{D}_2)=0$ and $\nu(V\mathscr{D}_2)=1$. ■

The above example shows that in answering the question whether an operator $V\in\mathrm{St}(\mathscr{H}_1,\mathscr{H}_2)$ has (J_1, J_2)-unitary extensions one has to take into account not only the characteristics of the spaces $\mathscr{D}_V^{[\perp]}$ and $\mathscr{R}_V^{[\perp]}$ but also the 'action' of the operator V.

Let $(\mathscr{L}_1^+,\mathscr{L}_1^-)$ be a maximal uniformly definite dual pair in $\mathscr{H}_1$, let $\mathscr{D}_V=\mathscr{D}_0\,[\dot+]\,\mathscr{D}_1\,[\dot+]\,\mathscr{D}_2$ be the $(\mathscr{L}_1^+,\mathscr{L}_1^-)$-decomposition of the domain of definition of an operator $V\in\mathrm{St}(\mathscr{H}_1,\mathscr{H}_2)$, and $\mathscr{R}_V=\mathscr{R}_0\,[\dot+]\,V\mathscr{D}_1\,[\dot+]\,V\mathscr{D}_2$ be the decomposition of its range of values. We choose in $\mathscr{H}_2$ a maximal uniformly definite dual pair $(\mathscr{L}_2^+,\mathscr{L}_2^-)$ such that $V(\mathscr{D}_V\cap\mathscr{L}_1^\pm)\subset\mathscr{L}_2^\pm$, and we introduce the following constants $\nu_+(V)$ and $\nu_-(V)$:

$$\nu_\pm(V)=\begin{cases} 0 & \text{if there is in } \mathscr{D}_V^{[\perp]} \text{ or in } \mathscr{R}_V^{[\perp]} \text{ an infinite-dimensional} \\ & \text{uniformly negative (respectively, uniformly positive)} \\ & \text{subspace;} \\ \dim(\mathscr{D}_V^{[\perp]}\cap\mathscr{L}_1^\mp)+\dim(V\mathscr{D}_2\cap\mathscr{L}_2^\pm)-\dim(\mathscr{R}_V^{[\perp]}\cap\mathscr{L}_2^\mp) & \\ & \text{otherwise.} \end{cases} \tag{2.4}$$

Remark 2.9: If there are no infinite-dimensional uniformly definite subspaces in $\mathscr{D}_V$, $\mathscr{R}_V$, $\mathscr{D}_V^{[\perp]}$ and $\mathscr{R}_V^{[\perp]}$, and if $\mathscr{D}_V \in \mathscr{M}^+(\mathscr{H}_1)$, and $\mathscr{R}_V \in \mathscr{M}^+(\mathscr{H}_2)$, then $\nu_-(V) = 0$ and $\nu_+(V) = \nu(\mathscr{R}_V) - \nu(\mathscr{D}_V)$.

Our immediate purpose will be to prove that $\nu_\pm(V)$ are independent of the choice of the dual pairs $(\mathscr{L}_i^+, \mathscr{L}_i^-)$, $i = 1,2$. We preface this with the following proposition.

Lemma 2.10 *Let $\mathscr{H}$ be a J-space, $\mathscr{D}$ a subspace of it, and let U be a J-semi-unitary J-bi-non-contractive (respectively, J-bi-non-expansive) extension of the operator $I \mid \mathscr{D}$. Then the condition $\mathscr{R}_U^{[\perp]} \neq \{\theta\}$ implies that there is in $\mathscr{D}^{[\perp]}$ an infinite-dimensional uniformly negative (respectively, uniformly positive) subspace.*

□ Let $\mathscr{D} = \mathscr{D}_0 [\dotplus] \mathscr{D}_1 [\dotplus] \mathscr{D}_2$ be the $(\mathscr{H}^+, \mathscr{H}^-)$-decomposition of the subspace $\mathscr{D}$, and let U be a J-semi-unitary J-bi-non-contractive operator. Since $\mathscr{D}_0$ and $\mathscr{D}_0^{[\perp]}$ are invariant relative to the operators U and U^c, it follows in accordance with Exercise 20 on 2.§4 that the operator U induces in the $\hat{J}$-space $\hat{\mathscr{H}} = \mathscr{D}_0^{[\perp]}/\mathscr{D}_0$ a $\hat{J}$-bi-non-contractive $\hat{J}$-semi-unitary operator $\hat{U}$, and $\hat{U} \mid \hat{\mathscr{D}} = \hat{J} \mid \hat{\mathscr{D}}$, where $\hat{\mathscr{D}} = \mathscr{D}/\mathscr{D}_0$. Moreover, $\mathscr{R}_{\hat{U}}^{[1]} \neq \{\theta\}$ if and only if $\mathscr{R}_U^{[\perp]} \neq \{\theta\}$; and $\mathscr{D}^{[\perp]}$ contains an infinite-dimensional uniformly negative subspace if and only if $\hat{\mathscr{D}}^{[\perp]} = \mathscr{D}^{[\perp]}/\mathscr{D}_0$ has this property. Therefore we shall suppose without loss of generality that $\mathscr{D}_0 = \{\theta\}$, i.e., $\mathscr{D}$, and with it also $\mathscr{D}^{[\perp]}$ and $\mathscr{D}^\perp$, are non-degenerate subspaces.

In accordance with 2.Proposition 4.14 we have for the J-bi-non-contractive operator U that $\mathscr{R}_U^{[\perp]}$ (= Ker U^c) is a uniformly negative subspace, and so without loss of generality we shall suppose that $\mathscr{R}_U^{[\perp]} \subset \mathscr{H}^-$. We shall assume that $\mathscr{R}_U^{[\perp]} \neq \{\theta\}$ and that, contrary to the proposition, $\mathscr{D}^{[\perp]}$ does not contain infinite-dimensional uniformly negative subspaces, i.e., (see Exercise 7 on 1.§6) the Gram operator of every non-positive subspace from $\mathscr{D}^{[\perp]}$ is completely continuous. Let $U = \| U_{ij} \|_{i,j=1}^2$, $J = \| J_{ij} \|_{i,j=1}^2$ be the matrix representations of the operators U and J relative to the decomposition $\mathscr{H} = \mathscr{D} \oplus \mathscr{D}^\perp$. Since $\mathscr{R}_U^{[\perp]} \subset \mathscr{H}^-$ and $\mathscr{D} \subset \mathscr{R}_{\bar{U}}$, we have $\mathscr{R}_U^{[\perp]} \subset \mathscr{D}^\perp (= J\mathscr{D}^{[\perp]})$, and $\mathscr{R}_U^{[\perp]} = \mathscr{R}_{U_{22}}^\perp \cap \mathscr{D}^\perp = \mathscr{R}_{22}^{[\perp]} \cap \mathscr{D}^\perp$. By direct calculations taking Exercise 7 on 1.§3 into account we verify that the fact of the operator U being J-semi-unitary is equivalent to the relations

a) $\mathscr{R}_{I_2 - U_{22}} \subset \mathscr{R}_{J_{22}}$, $U_{12} = -J_{21} J_{22}^{-1} (I_2 - U_{22})$;
b) $U_{22}^* J_{22}^{-1} U_{22} J_{22} = I_2$.

Since Ker $U_{22}^* = \mathscr{R}_{U_{22}}^\perp \cap \mathscr{D}^\perp$, and condition b) is the condition for the operator $(U_{22}^*)^c = J_{22}^{-1} U_{22} J_{22}$ to be J_{22}-semi-unitary, so $\mathscr{R}_{(U_{22}^*)^c} = \mathscr{R}_{U_{22}}$ (= (Ker $U_{22}^*)^{[\perp]}$). Let $\mathscr{D}^\perp = \mathscr{D}'_+ [\dotplus] \mathscr{D}'_-$ be the canonical decomposition of $\mathscr{D}^\perp$, and let $P'_\pm$ be the projectors on to $\mathscr{D}'_\pm$, $P'_+ + P'_- = I_2$.

It follows from the relation a) and the complete continuity of $J'_{22} \mid \mathscr{D}'_-$ that the operator

$$T \equiv P'_- J_{22}^{-1}(I_2 - U_{22}) J_{22} \mid \mathscr{D}'_- = I'_- - P'_- J_{22}^{-1} U_{22} J_{22} \mid \mathscr{D}'_-$$

is completely continuous, and therefore $P'_- J_{22}^{-1} U_{22} J_{22} \mid \mathscr{D}'_- = I'_- - T$. Since $J_{22}^{-1} U_{22} J_{22}$ is a J_{22}-semi-unitary operator, we have $\operatorname{Ker}(P'_- J_{22}^{-1} U_{22} J_{22} \mid \mathscr{D}'_-) = \{\theta\}$ and therefore $\mathscr{R}_{I'_- - T} = \mathscr{D}'_-$. Hence we conclude that $J_{22}^{-1} U_{22} J_{22} \mathscr{D}'_-$ is the maximal non-negative subspace in $\mathscr{D}^{\perp}$. But the subspace $\operatorname{Ker} U_{22}^* \ (\neq \{\theta\})$, which is J-orthogonal to $\mathscr{R}_{(U_{22}^*)^c}$, is also negative—and we have obtained a contradiction.

The case when U is a bi-non-expansive operator is proved similarly. ■

Lemma 2.11: *In Definition* (2.4) $\nu_{\pm}(V)$ *do not depend on the choice of the dual pairs* $(\mathscr{L}_i^+, \mathscr{L}_i^-)$, $i = 1, 2$.

□ We shall verify that $\nu_+(V)$ does not depend on the choice of the dual pairs $(\mathscr{L}_i^+, \mathscr{L}_i^-)$; the argument for $\nu_-(V)$ is similar. If there is an infinite-dimensional uniformly negative subspace in $\mathscr{D}_V^{[\perp]}$, then $\nu_+(V) = 0$ by definition does not depend on the choice of the dual pairs. Suppose all the uniformly negative subspaces in $\mathscr{D}_V^{[\perp]}$ are finite-dimensional.

Without loss of generality we shall assume that the decompositions (2.1) and (2.2) are connected by the relations $V\mathscr{D}_i = \mathscr{R}_i$, $i = 0, 1, 2$, and therefore the constant $\nu_+(V) \equiv \nu_1$ calculated with respect to the dual pairs $(\mathscr{H}_i^+, \mathscr{H}_i^-)$, $i = 1, 2$, coincides with the difference $\dim(\mathscr{D}_V^{[\perp]} \cap \mathscr{H}_1^-) - \dim(\mathscr{R}_V^{[\perp]} \cap \mathscr{H}_2^-)$. We bring into consideration a $\tilde{J}_1$-space $\tilde{\mathscr{H}}_1$ and a $\tilde{J}_2$-space $\tilde{\mathscr{H}}_2$, putting

$$\tilde{\mathscr{H}}_1 = \begin{cases} \mathscr{H}_1 [\oplus] \mathscr{H}^+, & \tilde{J}_1 \mid \mathscr{H}^+ = I^+ \quad \text{if } \nu_1 \geqslant 0, \\ \mathscr{H}_1 [\oplus] \mathscr{H}^- [\oplus] \mathscr{H}^+, & \tilde{J}_1 \mid \mathscr{H}^{\pm} = \pm K^{\pm} \quad \text{if } \nu_1 < 0, \end{cases}$$

$$\tilde{\mathscr{H}}_2 = \begin{cases} \mathscr{H}_2 [\oplus] \mathscr{H}^- [\oplus] \mathscr{H}^+, & \tilde{J}_2 \mid \mathscr{H}^{\pm} = \pm I^{\pm} \quad \text{if } \nu_1 > 0, \\ \mathscr{H}_2 [\oplus] \mathscr{H}^+, & \tilde{J}_2 \mid \mathscr{H}^+ = I^+ \quad \text{if } \nu_1 \leqslant 0, \end{cases}$$

where $\dim \mathscr{H}^- = |\nu_1|$, and $\mathscr{H}^+$ is an infinite-dimensional space with $\dim \mathscr{H}^+ \geqslant \max\{\dim \mathscr{H}_1, \dim \mathscr{H}_2\}$. It is clear that the constant $\tilde{\nu}_+(V) \equiv \tilde{\nu}_1$ (see (2.4)) constructed relative to $(\tilde{\mathscr{H}}_i^+, \tilde{\mathscr{H}}_i^-)$, $i = 1, 2$, is equal to zero, and it is sufficient for us to verify that it does not change for any maximal uniformly definite dual pairs $(\tilde{\mathscr{L}}_i^+, \tilde{\mathscr{L}}_i^-)$, $i = 1, 2$ containing $(\mathscr{H}^+, \mathscr{H}_0^-)$, where $\mathscr{H}_0^- = \mathscr{H}^-$ if $\mathscr{H}^-$ enters into the considered space $\tilde{\mathscr{H}}_i$ $(i = 1, 2)$, and $\mathscr{H}_0 = \{\theta\}$ otherwise. Since $\tilde{\nu}_1 = 0$, we can show, by repeating the arguments used in the proof of Theorem 2.2, that the operator V has a $(\tilde{J}_1, \tilde{I}_2)$-unitary extension $\tilde{V}_1$. Let $\tilde{\nu}_2$ be the constant (2.4) calculated relative to some other of the dual pairs $(\tilde{\mathscr{L}}_i^+, \tilde{\mathscr{L}}_i^-)$, $i = 1, 2$, indicated above. Again reverting to the proof of Theorem 2.2 and using Lemma 2.7 it can be proved that the operator V has a maximal $(\tilde{J}_1, \tilde{J}_2)$-isometric extension $\tilde{V}_2 \in \operatorname{St}(\tilde{\mathscr{H}}_1, \tilde{\mathscr{H}}_2)$, and that one of the operators $\tilde{V}_2$ or $\tilde{V}_2^{-1}$ is defined on the whole space, and the orthogonal complement to its range of values is negative and has dimension $|\nu_2|$. Let $\tilde{V}_2$ be such an operator (for $\tilde{V}_2^{-1}$ the argument is similar). Then the operator $\tilde{V}_1^{-1}\tilde{V}_2$ is a $\tilde{J}_1$-semi-unitary $\tilde{J}_1$-bi-non-contractive extension of the operator $I \mid \mathscr{D}_V$ and $\mathscr{D}_V^{[\perp]}$ contains no infinite-dimensional uniformly negative subspaces. By virtue of Lemma 2.10

$\tilde{V}_1^{-1}\tilde{V}_2$ is a $\tilde{J}_1$-unitary operator, and therefore $\tilde{V}_1$ is a $(\tilde{J}_1, \tilde{J}_2)$-unitary operator, which implies the equality $\tilde{\nu}_2 = 0$.

■

We now pass on to the main result of this section.

Theorem 2.12: *An operator $V \in \mathrm{St}(\mathscr{H}_1, \mathscr{H}_2)$ admits a (J_1, J_2)-unitary extension $\tilde{V}$ if and only if $\nu_+(V) = \nu_-(V) = 0$ and there is an operator $V_1 \in \mathrm{St}(\mathscr{H}_1, \mathscr{H}_2)$ which maps $\mathscr{D}_V^{[\perp]}$ on to $\mathscr{R}_V^{[\perp]}$.*

□ Let $\tilde{V}$ be a $(\tilde{J}_1, J_2)$-unitary extension of the operator V. We then put $\tilde{V} | \mathscr{D}_V^{[\perp]} \equiv V_1 \in \mathrm{St}(\mathscr{H}_1, \mathscr{H}_2)$ and it is clear that $V_1 \mathscr{D}_V^{[\perp]} = \mathscr{R}_V^{[\perp]}$. The equality $\nu_+(V) = \nu_-(V) = 0$ follows from Lemma 2.11 since this equality holds for the dual pairs $(\mathscr{H}_1^+, \mathscr{H}_1^-)$ and $(\tilde{V}\mathscr{H}_1^+, \tilde{V}\mathscr{H}_1^-)$.

The converse assertion will be proved in several stages.

a) Suppose first that $\nu_+(V) = \nu_-(V) = 0$ and that $\mathscr{D}_V^{[\perp]}$ and $\mathscr{R}_V^{[\perp]}$ contain no infinite-dimensional uniformly definite subspaces. Repeating the arguments used in the proof of Theorem 2.2 we obtain that the operator V admits (J_1, J_2)-unitary extensions.

b) Let $\mathscr{D}_V^{[\perp]}$ contain infinite-dimensional uniformly definite subspaces of both signs, let $V_1 \in \mathrm{St}(\mathscr{H}_1, \mathscr{H}_2)$, $V_1 \mathscr{D}_V^{[\perp]} = \mathscr{R}_V^{[\perp]}$, and let $V' (\in \mathrm{St}(\mathscr{H}_1, \mathscr{H}_2)$ be a maximal (J_1, J_2)-isometric extension of the operator V. We go into the case when V' is a (J_1, J_2)-semi-unitary (J_1, J_2)-bi-non-contractive operator and we show that it can be 'touched up' into a (J_1, J_2)-unitary extension $\tilde{V}$ of the operator V. The remaining cases are verified (taking Corollary 2.3 into account) by the successive application of a similar procedure. Since $\mathscr{D}_V^{[\perp]}$ contains infinite-dimensional uniformly negative subspaces, $\mathscr{R}_V^{[\perp]}$ $(= V_1 \mathscr{D}_V^{[\perp]})$ also has this same property. Let $\mathscr{L}_2$ be such a subspace from $\mathscr{R}_V^{[\perp]}$ containing $\mathscr{R}_{V'}^{[\perp]}$ and let $\mathscr{L}_1 = V_1^{-1}\mathscr{L}_2$. Since $\dim \mathscr{L}_1 = \dim(V'\mathscr{L}_1 [\dot{+}] \mathscr{R}_{V'}^{[\perp]})$, and $\mathscr{L}_1$ and $V'\mathscr{L}_1 [\dot{+}] \mathscr{R}_{V'}^{[\perp]}$ are uniformly negative subspaces, there is an operator $V_1' \in \mathrm{St}(\mathscr{H}_1, \mathscr{H}_2)$ which maps $\mathscr{L}_1$ on to $V'\mathscr{L}_1 [+] \mathscr{R}_{V'}^{[\perp]}$. It only remains to put

$$\tilde{V}x = \begin{cases} V'x, & x \in \mathscr{L}_1^{[1]} \\ V_1'x, & x \in \mathscr{L}_1 \end{cases}$$

c) Now suppose that $\mathscr{D}_V^{[\perp]}$ contains infinite-dimensional uniformly definite subspaces of one sign and only finitely-dimensional ones of the other sign, and $\nu_+(V) = \nu_-(V) = 0$. In this case a combination of the arguments of stages a) and b) is applied. ■

Remark 2.13: In stage a) in the proof of the converse assertion in Theorem 2.12 the existence of the operator V_1 was not used, and in stages b) and c) in this proof the operator V_1 was used only in the search in $\mathscr{D}_V^{[\perp]}$ for an infinite-dimensional uniformly definite subspace of the same dimension as the corresponding subspace chosen in $\mathscr{R}_V^{[\perp]}$.

Corollary 2.14: *If under the conditions of Theorem 2.12* $\mathscr{D}_V^{[\perp]} = \mathscr{D}_0 [\dotplus] \mathscr{D}^{[\perp]'}$, *where* $\mathscr{D}_0 = \mathscr{D}_V \cap \mathscr{D}_V^{[\perp]}$, *and* $\mathscr{D}^{[\perp]'}$ *is a projectionally complete subspace in* $\mathscr{H}_1$, *then the existence of a* (J_1, J_2)*-unitary extension* $\tilde{V}$ *is equivalent to the equality* In $\mathscr{D}_V^{[\perp]} =$ In $\mathscr{R}_V^{[\perp]}$.

□ It follows from the conditions on $\mathscr{D}_V^{[\perp]}$ that $\mathscr{D}_V = \mathscr{D}_0 [\dotplus] \mathscr{D}'$, where $\mathscr{D}'$ is a projectionally complete subspace (see Exercise 7 on 1.§7), and therefore $\mathscr{R}_V$ and $\mathscr{R}_V^{[\perp]}$ $(= \mathscr{R}_0 [\dotplus] \mathscr{R}^{[\perp]'})$ have a similar property. Let $\mathscr{D}^{[\perp]'} = (\mathscr{D}^{[\perp]})_+ [\dotplus] (\mathscr{D}^{[\perp]})_-$ and $\mathscr{R}^{[\perp]'} = (\mathscr{R}^{[\perp]})_+ [\dotplus] (\mathscr{R}^{[\perp]})_-$. By hypothesis $\dim \mathscr{D}_0 = \dim \mathscr{R}_0$, $\dim(\mathscr{D}^{[\perp]})_\pm = \dim(\mathscr{R}^{[\perp]})_\mp$, and $(\mathscr{D}^{[\perp]})_\pm$, $(\mathscr{R}^{[\perp]})_\pm$ are uniformly definite subspaces, and so there is an operator $V_1 \in \mathrm{St}(\mathscr{H}_1, \mathscr{H}_2)$ mapping $\mathscr{D}_V^{[\perp]}$ on to $\mathscr{R}_V^{[\perp]}$. Taking as corresponding maximal dual pairs $(\mathscr{L}_i^+, \mathscr{L}_i^-)$, $i = 1, 2$ the pairs of spaces $\mathscr{L}_1^\mp \supset (\mathscr{D}^{[\perp]})_\pm$, $\mathscr{L}_2^\pm \supset (\mathscr{R}^{[\perp]})_\pm$ we obtain that $\nu_+(V) = \nu_-(V) = 0$. It only remains to apply Theorem 2.12.

The converse assertion is obvious. ■

3 We introduce the concept of $\varkappa$-regular extensions of operators $V \in \mathrm{St}(\mathscr{H}_1, \mathscr{H}_2)$, and we give a criterion for their existence. But first we prove the following simple proposition.

2.15 Suppose that $\tilde{V}$ *is a* (J_1, J_2)*-non-expansive extension or a* (J_1, J_2)*-non-contractive extension of an operator* $V \in \mathrm{St}(\mathscr{H}_1, \mathscr{H}_2)$ *on to the whole of* $\mathscr{H}_1$, *that* $\tilde{V}$ *admits a* (J_1, J_2)*-polar decomposition* $\tilde{V} = WR$, *where the operator* W *is a* (J_1, J_2)*-semi-unitary operator or is conjugate to a* (J_1, J_2)*-semi-unitary operator, and the* J_1*-selfadjoint operator* R *satisfies the conditions* $R^2 = \tilde{V}^c \tilde{V}$ *and* $\sigma(R) \subset [0, \infty)$. *Then* W *is an extension of the operator* V.

□ Let $x \in \mathscr{D}_V$. Then $R^2 x = \tilde{V}^c \tilde{V} x = x$, and since $-1 \in \rho(R)$ we also have $Rx = x$. Consequently $W \supset V$. ■

Remark 2.16: If under the conditions of Proposition 2.15 the operator W is (J_1, J_2)-semi-unitary, then it is a (J_1, J_2)-bi-extension of an operator V (see Proposition 1.10). We note also that if $\tilde{V}$ is a $\tilde{J}_1, J_2)$-bi-non-expansive operator, then by virtue of 4.Theorem 1.12 it admits the decomposition indicated in Proposition 2.15, and moreover W is a (J_1, J_2)-bi-non-expansive operator and therefore (see Proposition 1.10) W is a (J_1, J_2)-bi-extension of the operator V.

Definition 2.17: We shall say that an operator $V \in \mathrm{St}(\mathscr{H}_1, \mathscr{H}_2)$ admits *$\varkappa$-regular extensions* (and when $\varkappa = 0$, *regular extensions*) if there is a Pontryagin $\pi_\varkappa$-space $\Pi_\varkappa$ with $\varkappa$ $(< \infty)$ negative squares such that the operator V admits $(\tilde{J}_1, \tilde{J}_2)$-unitary extensions

$$\tilde{V}: \mathscr{H}_1 \oplus \Pi_\varkappa \to \mathscr{H}_2 \oplus \Pi_\varkappa, \qquad \tilde{J}_1 = J_1 \oplus \pi_\varkappa, \qquad \tilde{J}_2 = J_2 \oplus \pi_\varkappa$$

(here $\pi_\varkappa \equiv J$ in the space $\Pi_\varkappa$).

Theorem 2.18: *Let $V \cap \mathrm{St}(\mathscr{H}_1, \mathscr{H}_2)$. Then there is a J-space $\mathscr{H}$ such that the operator V admits a $(\tilde{J}_1, \tilde{J}_2)$-unitary extension*

$$V : \mathscr{H}_1 \oplus \mathscr{H} \to \mathscr{H}_2 \oplus \mathscr{H}_1, \qquad \tilde{J}_i = J_i \oplus J, \qquad i = 1, 2.$$

Moreover the following conditions are equivalent: the operator V admits a) *regular extensions;* b) *$\varkappa$-regular extensions*; c) *(J_1, J_2)-bi-non-expansive extensions.*

□ We shall suppose that $V \in \mathrm{St}(\mathscr{H}_1, \mathscr{H}_2)$ is a maximal (J_1, J_2)-isometric operator. From Corollary 2.3 its deficiency subspaces $\mathscr{D}_V^{[\perp]}$ and $\mathscr{R}_V^{[\perp]}$ are projectionally complete. To prove the first assertion it is sufficient to choose as $\mathscr{H}$ the J-space $\mathscr{H}^+ \oplus \mathscr{H}^-$ with infinite-dimensional components $\mathscr{H}^\pm$, $\dim \mathscr{H}^\pm \geqslant \max\{\dim \mathscr{D}_V^{[\perp]}, \dim \mathscr{R}_V^{[\perp]}\}$, and to use 2.14.

We pass on to the proof of the equivalence of the conditions a), b), c).

a) ⇒ b) This is trivial.

b)⇒c) Let $\tilde{V} : \mathscr{H}_1 \oplus \Pi_\varkappa \to \mathscr{H}_2 \oplus \Pi_\varkappa$ be a $\varkappa$-regular extension of the operator V, let $\Pi_\varkappa = \Pi_+ \oplus \Pi_-$, $\dim \Pi_- = \varkappa < \infty$, and let P_i' be the $\tilde{J}_i$-orthoprojectors from $\tilde{\mathscr{H}}_i$ on to the J_i'-space $\mathscr{H}_i' = \mathscr{H}_i \oplus \Pi_-$, $J_i' = J_i \oplus - I_-$, $i = 1, 2$. It follows from Exercise 34 on 2.§4 that $P_2' \tilde{V} P_1' \mid \mathscr{H}_1'$ is a (F_1', J_2')-bi-non-expansive operator, and since $\mathscr{D}_V \subset \mathscr{H}_1'$, $\mathscr{R}_V \subset \mathscr{H}_2'$, it is an extension of the operator V. We make use of Remark 2.16 and suppose that, for example, the operator V^{-1} has a (J_2', J_1')-semi-unitary (J_2', J_1')-bi-non-expansive extension $\tilde{V}'$ (if V has a $\tilde{J}_1'$, J_2')-semi-unitary (J_1', J_2')-bi-non-expansive extension, then the argument is similar). Let $\tilde{V}' = \| \tilde{V}_{ij}' \|_{i,j=1}^2$ be its matrix representation relative to the decomposition $\mathscr{H}_2' = \mathscr{H}_2 \oplus \Pi_-$, $\mathscr{H}_1' = \mathscr{H}_1 \oplus \Pi_-$, and let $\varkappa_+ = \dim \mathscr{R}_{V'}^{[\perp]}$. Then the operator $V' = \mid \mathrm{Ker}\, \tilde{V}_{21}'$ is a (J_1, J_2)-isometric extension of the operator V^{-1}, and since $\dim \mathscr{H}_2' \mid \mathrm{Ker}\, \tilde{V}_{21} < \infty$, so $\mathscr{D}_{V'}$ and $\mathscr{R}_{V'}$ decompose into the direct sum of isotropic and projectionally complete subspaces. Let $\mathrm{In}\, \mathscr{D}_{V'}^{[\perp]} = (\varkappa_0, \varkappa_+', \varkappa_-')$. Then

$$\mathrm{In}\, \mathscr{R}_{V'}^{[\perp]} = (\varkappa_0, \varkappa_-' + \varkappa_+, \varkappa_-').$$

Since $\Pi_- \subset \mathscr{D}_{V'}^{[\perp]}$ and $\Pi_- \subset \mathscr{R}_{V'}^{[\perp]}$, we have

$$\mathrm{In}(\mathscr{D}_{V'}^{[\perp]} \cap \mathscr{H}_2) = (\varkappa_0, \varkappa_+', \varkappa_-' - \varkappa)$$

and

$$\mathrm{In}(\mathscr{R}_{V'}^{[\perp]} \cap \mathscr{H}_1) = (\varkappa_0, \varkappa_+' + \varkappa_+, \varkappa_-' - \varkappa),$$

and moreover

$$\dim(\mathscr{D}_{V'}^{[\perp]} \cap \mathscr{H}_2) = \varkappa_0 + \varkappa_+' + \varkappa_-' - \varkappa < \infty.$$

Now, carrying out an argument similar to that used in the proof of Corollary 2.14 we obtain that V^{-1} has a (J_2, J_1)-semi-unitary (J_1, J_1)-bi-non-expansive extension, the (J_2, J_1)-conjugate to which will indeed by the required extension of the operator V.

c) ⇒ a) We again make use of Remark 2.16 and we shall suppose that, for example, the operator V^{-1} has a (J_2, J_1)-bi-non-expansive (J_2, J_1)-semi-unitary extension W. Since (see 2.Remark 4.29 and 2.Proposition 4.14) $\mathscr{R}_W^{[\perp]}$ is a uniformly positive subspace, so by virtue of Corollary 2.14 the operator W has a $(\tilde{J}_1, \tilde{J}_2)$-unitary extension $\widetilde{W}: \mathscr{H}_2 \oplus \mathscr{H} \to \mathscr{H}_1 \oplus \mathscr{H}$, $\tilde{J}_i = J_i \oplus I$, $i = 1.2$, where $\mathscr{H}$ is an infinite-dimensional space with $\dim \mathscr{H} \geqslant \dim \mathscr{R}_W^{[\perp]}$. Consequently $\widetilde{W}^{-1}$ is a regular extension of the operator V.

4 In conclusion we introduce some definitions concerning J-Hermitian operators, and we formulate in Exercises 6–13 the corresponding results obtained by applying as above the Cayley–Neyman transformation.

Definition 2.19: We call an operator B a *J-bi-extension* of a J-Hermitian operator A if $A \subset B$ and the graph Γ_A is isotropic in Γ_B relative to the form

$$[\langle x_1, x_2\rangle, \langle y_1, y_2\rangle]_\Gamma = i([x_1, y_2] - [x_2, y_1])$$

(2.Formula (1.4)).

The set of closed J-Hermitian operators A with points of regular type $(\lambda_0, \bar{\lambda}_0\}$ $(\lambda_0 \neq \bar{\lambda}_0)$ will be called λ_0-*standard* and will be denoted by the symbol $\mathrm{St}(\mathscr{H}; \lambda_0)$ $(= \mathrm{St}(\mathscr{H}; \bar{\lambda}_0))$.

Definition 2.20: We shall say that an operator $A \in \mathrm{St}(\mathscr{H}; \lambda_0)$ admits a *ϰ-regular* $(0 \leqslant \varkappa < \infty)$ λ_0-*standard extension* if there is a Pontryagin $\pi_\varkappa$-space $\Pi_\varkappa$ with $\varkappa$ negative squares such that in the J-space $\tilde{\mathscr{H}} = \mathscr{H} \oplus \Pi_\varkappa$ $(\tilde{J} = J \oplus \Pi_\varkappa)$ a $\tilde{J}$-Hermitian operator A has a $\tilde{J}$-self-adjoint extension $\tilde{A} \in \mathrm{St}(\tilde{\mathscr{H}}; \lambda_0)$. When $\varkappa = 0$ such an extension will be called a *regular extension*.

Exercises and problems

1 Let $V \in \mathrm{St}(\mathscr{H}_1, \mathscr{H}_2)$. Prove that $\nu_\pm(V^{-1}) = -\nu_\pm(V)$ (Azizov).
Hint: Use Formula (2.4) and Lemma 2.11.

2 Suppose that $V \in \mathrm{St}(\mathscr{H}_1, \mathscr{H}_2)$, that V has at least one (J_1, J_2)-unitary extension, and that $\mathscr{D}_V^{[\perp]}$ and $\mathscr{R}_V^{[\perp]}$ contain no infinite-dimensional uniformly definite subspaces. Prove that then every maximal (J_1, J_2)-isometric extension $V'(\in \mathrm{St}(\mathscr{H}_1, \mathscr{H}_2))$ of the operator V is (J_1, J_2)-unitary (Azizov).
Hint: Prove that $\varkappa_\pm(\tilde{V}) = 0$ for any extension $\tilde{V} \in \mathrm{St}(\mathscr{H}_1, \mathscr{H}_2)$ of the operator V.

3 Let $V \in \mathrm{St}(\Pi_{\varkappa_1}, \Pi_{\varkappa_2})$. Prove that V admits a (J_1, J_2)-unitary extension if and only if $\varkappa_1 = \varkappa_2$ and $\dim \mathscr{D}_V^{[\perp]} = \dim \mathscr{R}_V^{[\perp]}$ (*cf.* [XIV]).
Hint: Use *I*. Theorem 9.11 and Corollary 2.14.

4 Prove that under the conditions of Corollary 2.14 the projectional completeness of $\mathscr{D}^{[\perp]'}$ is essential.
Hint: Use Example 2.8.

5 Let A be a J-Hermitian operator, B an extension of it, and $\bar{\mathscr{D}}_B = \mathscr{H}$. Prove that B will be a J-bi-extension of the operator A if and only if $A \subset B^c$.
Hint: Use 2.Proposition1.3.

6 Let A be a J-Hermitian operator, B an extension of it, $\lambda \neq \bar{\lambda}$, and $\lambda \notin \sigma_p(B)$. Prove that B is a J-bi-extension of the operator A if and only if its Cayley–Neyman transform $\mathbf{K}_\lambda(B)$ is a J-bi-extension of the J-isometric operator $\mathbf{K}_\lambda(A)$.
Hint: Use the definitions of J-bi-extensions of J-symmetric and J-isometric operators, and also 2.Formulae (6.29) and (6.31).

7 Prove that $A \in \mathrm{St}(\mathscr{H}; \lambda_0)$ if and only if $\lambda_0 \notin \sigma_p(A)$ and $\mathbf{K}_{\lambda_0}(A) \in \mathrm{St}\,\mathscr{H}$.
Hint: Use 2.Proposition 6.11.

8 Let $A \in \mathrm{St}(\mathscr{H}; \lambda_0)$. Prove that among its maximal J-symmetric extensions there are operators $\tilde{A} \in \mathrm{St}(\mathscr{H}; \lambda_0)$ (Azizov).
Hint: Combine the results of Exercises 6, 7, Theorem 2.2 and Corollary 2.3.

9 Prove that *a* J-symmetric operator $A \in \mathrm{St}(\mathscr{H}; \lambda_0)$ will be a maximal J-symmetric operator if and only if at least one of the following conditions holds: a) $\lambda_0 \in \rho(A)$; b) $\bar{\lambda}_0 \in \rho(A)$; $\mathscr{R}^{[\perp]}_{A-\lambda_0 I}$ and $\mathscr{R}^{[\perp]}_{A-\bar{\lambda}_0 I}$ are uniformly definite subspaces of different signs (Azizov [14]).
Hint: Use Corollary 2.3, having first verified that a J-symmetric operator $A \in \mathrm{St}(\mathscr{H}; \lambda_0)$ is maximal if and only if the J-isometric operator $\mathbf{K}_{\lambda_0}(A) \in \mathrm{St}\,\mathscr{H}$ is maximal.

10 Prove that every J-symmetric operator $A \in \mathrm{St}(\mathscr{H}; \lambda_0)$ admits J-bi-extensions A_0 such that one of the points λ_0 or $\bar{\lambda}_0$ is regular for them, and the other is a point of regular type and $\mathscr{R}_{A-\bar{\lambda}_0 I}$, $\mathscr{R}_{A-\lambda_0 I}$ are projectionally complete subspaces (Azizov).
Hint: Use the result of Exercise 7 and Corollary 2.4.

11 Suppose that $A \in \mathrm{St}(\mathscr{H}; \lambda_0)$ and that it has a maximal J-dissipative extension A_0 with $\lambda_0 \in \rho(A_0)$. Prove that the operator A has maximal J-dissipative extensions $\tilde{A} \in \mathrm{St}(\mathscr{H}; \lambda_0)$ with $\lambda_0 \in \rho(\tilde{A})$ (Azizov [14]).
Hint: Use the result of Exercise 6, Remark 2.16, Proposition 2.15, and 2.Proposition 6.11. If necessary consider 2.Remark 4.29.

12 Let A and A_0 be the same as in Exercise 10, $\lambda_0 \in \rho(A_0)$, $\mathrm{Im}\ \lambda_0 < 0$, and let $U_0 = \mathbf{K}_{\lambda_0}(A_0)$. Prove that the values of the resolvents $R_{\lambda_0}(\tilde{A})$ of all J-bi-extensions $\tilde{A}$ of the operator A with $\lambda_0 \in \rho(\tilde{A})$ are described at the point λ_0 by the formula

$$R_{\lambda_0}(\tilde{A}) = (\tilde{I}_{\lambda_0} U_0 (U_0^c U_0)^{-1} - I)/2i\ \mathrm{Im}\ \lambda_0,$$

where $\tilde{I}_{\lambda_0}$ traverses the set of all J-bi-extensions of the operator $I \mid \mathscr{R}_{A-\bar{\lambda}_0 I}$; if, in addition, $A_0 \in \mathrm{St}(\mathscr{H}; \lambda_0)$, then

$$R_{\lambda_0}(\tilde{A}) = (\tilde{I}_{\lambda_0} U_0 - I)/2i\ \mathrm{Im}\ \lambda_0,$$

and when $A_0 = A_0^c$ the operator $\tilde{A}$ will be J-symmetric (respectively, maximal J-dissipative, J-self-adjoint) if and only if $\tilde{I}_{\lambda_0}$ is a J-semi-unitary (respectively, J-bi-non-expansive, J-unitary) operator (Azizov [14]).
Hint: Use the results of Exercise 16 on §1, Exercise 6 and 2.Formula (6.29).

13 Let $A \in \mathrm{St}(\mathscr{H}; \lambda_0)$ and $\mathrm{Im}\ \lambda_0 < 0$. Prove that A admits extensions $\tilde{A} = \tilde{A}^c \in \mathrm{St}(\tilde{\mathscr{H}}; \lambda_0)$ with exit into some J-space $\tilde{\mathscr{H}} \supset \mathscr{H}$. Moreover the following conditions are equivlent:
a) A admits $\varkappa$-regular λ_0-standard extensions $\tilde{A} = \tilde{A}^c$;
b) A admits regular λ_0-standard extensions $\tilde{A} = \tilde{A}^c$;
c) A admits maximal J-dissipative extensions $\tilde{A} \in \mathrm{St}(\mathscr{H}; \lambda_0)$ with $\lambda_0 \in \rho(\tilde{A})$ (Azizov [14]).
Hint: Use Theorem 2.18 and the result of Exercise 6.

14 Let $\mathscr{H}$ be a J-space, $\mathscr{H} \oplus \mathscr{H}$ a $\tilde{J}$-space, where $\tilde{J} = J \oplus I$, and let A be a J-symmetric operator. Prove that A admits $\tilde{J}$-self-adjoint extensions (*cf.* Exercise 13).
Hint: Use the scheme $A \to JA \to JA \oplus -JA \to \widetilde{JA \oplus -JA} = \tilde{J}\,\widetilde{JA \oplus -JA}$.

15 Let $V \in \mathrm{St}(\mathscr{H}_1, \mathscr{H}_2)$, and let the $(\mathscr{H}_1^+, \mathscr{H}_1^-)$-decomposition $\mathscr{D}_V = \mathscr{D}_0 [\dot{+}] \mathscr{D}_1 [\dot{+}] \mathscr{D}_2$ and the $(\mathscr{H}_2^+, \mathscr{H}_2^-)$-decomposition $\mathscr{R}_V = \mathscr{R}_0 [\dot{+}] \mathscr{R}_1 [\dot{+}] \mathscr{R}_2$ have the property $V\mathscr{D}_i = \mathscr{R}_i$, $i = 1, 2$. Then the following conditions are equivalent:

a) the operator A has at least one (J_1, J_2)-unitary extension;
b) there is a (J_1, J_2)-unitary operator W such that $W\mathscr{D}_V = \mathscr{R}_V$ and $W\mathscr{D}_1 = \mathscr{R}_1$;
c) there is a (J_1, J_2)-unitary operator U such that $U\mathscr{D}_V = \mathscr{R}_V$ and $U\mathscr{D}_2 = \mathscr{R}_2$;
d) there is a (J_1, J_2)-isometric operator V_1 mapping $\mathscr{D}_V^{[\perp]}$ on to $\mathscr{R}_V^{[\perp]}$ and in the Krein space $\overline{V\mathscr{D}_2 [\dot{+}] V_1(\mathscr{D}^{[\perp]})_2}$ there is a dual pair $(\mathscr{L}^+, \mathscr{L}^-)$ maximal in that space and such that $\mathscr{L}^{\pm} \cap V\mathscr{D}_2 = \{\theta\}$ and $\mathscr{L}^{\pm} \cap V_1(\mathscr{D}^{[\perp]})_2 = \{\theta\}$ (Azizov [14]).

§3 Generalized resolvents of symmetric operators

1 Before describing generalized resolvents of J-symmetric operators we introduce and prove a number of auxiliary propositions which are, incidentally, of some independent interest. One of the concepts generalizing the concept of an 'extension' of operators is that of a 'dilatation' of operators. An extension is a process which can take place both within the limits of the given spaces or with emergence from them; but a dilatation takes place necessarily with emergence from the original spaces.

Definition 3.1: Let $\mathscr{H}_i$ $(i = 1, 2, 3)$ be Hilbert spaces, and let $T: \mathscr{H}_1 \to \mathscr{H}_2$ be a bounded operator with $\mathscr{D}_T = \mathscr{H}_1$. We call a bounded operator $\tilde{T}: \mathscr{H}_1 \oplus \mathscr{H}_3 \to \mathscr{H}_2 \oplus \mathscr{H}_3$ a *dilatation of the operator* T if $\tilde{T} = \| T_{ij} \|_{i,j=1}^2$, where $T_{11} = T$ and T_{21}, T_{12}, T_{22} are operators such that $T_{12} T_{22}^{(n)} T_{21} = 0$ when $n = 0, 1, 2, \ldots$. If also the $\mathscr{H}_i$ are J_i-spaces, and if $\tilde{T}$ is a $(\tilde{J}_1, \tilde{J}_2)$-unitary operator with $\tilde{J}_1 = J_1 \oplus J_3$, $\tilde{J}_2 = J_2 \oplus J_3$, then we call $\tilde{T}$ a $(\tilde{J}_1, \tilde{J}_2)$-*unitary dilatation of the operator* T. We say that such a dilatation is $\varkappa$-regular if $\mathscr{H}_3 = \Pi_\varkappa$ with $\varkappa$ negative squares; we shall call 0-regular dilatations *regular*.

Remark 3.2: If under the conditions of Definition 3.1 $\tilde{T}$ is an extension of the operator T (this is equivalent to the equality $T_{21} = 0$), then, clearly, $\tilde{T}$ is a dilatation of the operator T.

However, not every dilatation is an extension. We shall convince ourselves of this later, for example, in Theorem 3.4.

Lemma 3.3: *Let* $T: \mathscr{H}_1 \to \mathscr{H}_2$ *be a bounded operator with* $\mathscr{D}_T = \mathscr{H}_1$, *and let* $\tilde{T}: \mathscr{H}_1 \oplus \mathscr{H}_3 \to \mathscr{H}_2 \oplus \mathscr{H}_3$ *be a dilatation of it, and let* $\tilde{\tilde{T}}: \mathscr{H}_1 \oplus \mathscr{H}_3 \oplus \mathscr{H}_4 \to \mathscr{H}_2 \oplus \mathscr{H}_3 \oplus \mathscr{H}_4$ *be a dilatation of the operator* $\tilde{T}$. Then $\tilde{\tilde{T}}$ *is a dilation of the operator* T.

By hypothesis the operators $\tilde{T}$ and $\tilde{\tilde{T}}$ admit the follwing matrix representations relative to the corresponding decompositions:

$$\tilde{T} = \| T_{ij} \|_{i,j=1}^{2}, \qquad \tilde{\tilde{T}} = \| T_{ij} \|_{i,j=1}^{3} \quad \text{with } T_{11} = T,\ T_{12} T_{22}^{n} T_{21} = 0,$$

and

$$\| T_{i3} \|_{i=1}^{2} T_{33}^{n} \| T_{3j} \|_{j=0}^{2} = 0 \ (n = 0, 1, 2, \ldots).$$

Hence

$$\| T_{ij} \|_{j=2}^{3} (\| T_{ij} \|_{i,j=2}^{3})^{n} \| T_{j1} \|_{j=2}^{3}$$

$$= T_{12} T_{22}^{n} T_{21} + T_{13} T_{33}^{n} T_{31} + T_{13} \sum_{k=1}^{n-1} T_{33}^{k} T_{32} T_{22}^{n-1-k} T_{21}$$

$$= 0 \text{ when } n = 0, 1, 2, \ldots,$$

and it only remains to use Definition 3.1. ■

Theorem 3.4: *Let $T: \mathscr{H}_1 \to \mathscr{H}_2$ be a bounded operator acting from a J_1-space $\mathscr{H}_1$ into a J_2-space $\mathscr{H}_2$ with $\mathscr{D}_T = \mathscr{H}_1$. Then there is a J_3-space $\mathscr{H}_3$ and a $(J_1 \oplus J_3, J_2 \oplus J_3)$-unitary operator $\tilde{T}: \mathscr{H}_1 \oplus \mathscr{H}_3 \to \mathscr{H}_2 \oplus \mathscr{H}_3$ which is a dilatation of the operator T.*

□ Let $J_1 - T^*J_2T = \tilde{J}_0 \,|\, J_1 - T^*J_2T \,|$ be the polar decomposition of the selfadjoint operator $J_1 - T^*J_2T$, let $\mathscr{H}_0 = \bar{\mathscr{R}}_{J_1 - T^*J_2T}$ and let $J_0 = \tilde{J}_0 \,|\, \mathscr{H}_0$. Then $J_0 = J_0^* = J_0^{-1}$. We bring into consideration the J_0'-space $\mathscr{H}_0' = \oplus_{k=0}^{\infty} \mathscr{H}_0^k$, where

$$\mathscr{H}_0^k = \mathscr{H}_0 \text{ and } J_0' = \bigoplus_{k=0}^{\infty} J_0^{(k)}, \ J_0^{(k)} = J_0 \ (k = 0, 1, 2, \ldots),$$

and we define an operator $T_1: \mathscr{H}_1 \oplus \mathscr{H}_0' \to \mathscr{H}_2 \oplus \mathscr{H}_0'$ by the formula

$$T_1 \langle x; x_0, x_1, \ldots \rangle = \langle Tx; \,| J_1 - T^*J_2T |^{1/2} x, x_0, x_1, \ldots \rangle. \tag{3.1}$$

By construction T_1 is a continuous operator with $\mathscr{D}_{T_1} = \mathscr{H}_1 \oplus \mathscr{H}_0'$ and it is $(J_1 \oplus J_0', J_2 \oplus J_0')$-isometric, i.e., T_1 is a $(J_1 \oplus J_0', J_2 \oplus J_0')$-semi-unitary operator. The subspace $\mathscr{H}_0'$ is invariant relative to T_1, and therefore T_1 is a dilatation of the operator T. By virtue of Theorem 2.18 there is a J-space $\mathscr{H}$ such that the operator T_1 admits a $(J_1 \oplus J_3, J_2 \oplus J_3)$-unitary extension $T: \mathscr{H}_1 \oplus \mathscr{H}_3 \to \mathscr{H}_2 \oplus \mathscr{H}_3$, where $\mathscr{H}_3 = \mathscr{H}_0' \oplus \mathscr{H}$ and $J_3 = J_0' \oplus J$. It only remains to use Lemma 3.3. ■

Corollary 3.5: *For an operator T under the conditions of Theorem* 3.4 *to admit a regular dilatation it is necessary and sufficient that it be a (J_1, J_2)-bi-non-expansive operator.*

□ The necessity follows from Exercise 34 on 2.§4. The proof of sufficiency can be carried out using the same scheme as in the proof of Theorem 3.4 taking into account that $J_0 = I \,|\, \bar{\mathscr{R}}_{J_1 - T^*J_2T}$ and that the operator T_1 is

$(J_1 \oplus J_0', J_2 \oplus J_0')$-bi-non-expansive. Consequently, by virtue of Theorem 2.18 we can put $J = I$. ■

2 Here we shall generalize the concept of a dilatation (see Exercise 4 below) and investigate special classes of holomorphic operator-functions.

Definition 3.6: Let $T(\mu)$ be a function holomorphic in the neighbourhood of 0 with values in a set of continuous operators acting from a space $\mathscr{H}_1$ into a space $\mathscr{H}_2$ with $\mathscr{D}_{T(\mu)} = \mathscr{H}_1$, and let $T = \| T_{ij} \|_{i,j=1}^2 : \mathscr{H}_1 \oplus \mathscr{H}_3 \to \mathscr{H}_2 \oplus \mathscr{H}_3$ be a bounded operator with $\mathscr{D}_T = \mathscr{H}_1 \oplus \mathscr{H}_3$. We shall say that *the function $T(\mu)$ is generated by the operator T* (or that *the operator T generates the function $T(\mu)$*) if in some neighbourhood of zero

$$T(\mu) = T_{11} + \mu T_{12}(I_3 - \mu T_{22})^{-1} T_{21}.$$

Definition 3.7: In Definition 3.6 suppose the $\mathscr{H}_i$ are J_i-spaces ($i = 1, 2, 3$) and that T is a $(J_1 \oplus J_3, J_2 \oplus J_3)$-unitary operator. Then we shall say that *the function $T(\mu)$ belongs to the class $\Pi(\mathscr{H}_1, \mathscr{H}_2)$*. If, in addition, $\mathscr{H}_3$ is a Pontryagin space with $\varkappa$ negative squares, then we shall say that *$T(\mu)$ belongs to the class $\Pi^{\varkappa}(\mathscr{H}_1, \mathscr{H}_2)$*. (In cases where it will not cause confusion, the symbols $\mathscr{H}_1, \mathscr{H}_2$ will be omitted from the designations of classes).

Lemma 3.8: *Let*

$$T = \| T_{ij} \|_{i,j=1}^2 : \mathscr{H}_1 \oplus \mathscr{H}_3 \to \mathscr{H}_2 \oplus \mathscr{H}_3$$

and

$$\tilde{T} = \| T_{ij} \|_{i,j=1}^3 : \mathscr{H}_1 \oplus \mathscr{H}_3 \oplus \mathscr{H}_4 \to \mathscr{H}_2 \oplus \mathscr{H}_3 \oplus \mathscr{H}_4$$

be bounded operators with $\mathscr{D}_T = \mathscr{H}_1 \oplus \mathscr{H}_3$, $\mathscr{D}_{\tilde{T}} = \mathscr{H}_1 \oplus \mathscr{H}_3 \oplus \mathscr{H}_4$, and let $\tilde{T}$ be a dilatation of the operator T. Then T and $\tilde{T}$ generate the same operator-function $T(\mu) : \mathscr{H}_1 \to \mathscr{H}_2$.

□ In accordance with Definition 3.6 we have to verify that

$$T_{11} + \mu T_{12}(I_3 - \mu T_{22})^{-1} T_{21} = T_{11} + \mu \| T_{1j} \|_{j=2}^3 (I_{3 \oplus 4} - \mu \| T_{ij} \|_{i,j=2}^3)^{-1} \| T_{i1} \|_{i=2}^3. \quad (3.2)$$

The condition '$\tilde{T}$ is a dilatation of T' is equivalent in the present case to the system of equalities $T_{i3} T_{33}^k T_{3j} = 0$ ($i, j = 2, 3$), and therefore

$$\| T_{1j} \|_{j=2}^3 (\| T_{ij} \|_{i,j=2}^3)^k \| T_{i1} \|_{i=2}^3 = T_{12} T_{22}^k T_{21} \qquad (k = 0, 1, 2).$$

It only remains to note that in a sufficiently small neighbourhood of zero the equality (3.2) is equivalent to the coincidence of the series

$$\sum_{k=0}^{\infty} \mu^k T_{12} T_{22}^k T_{21} \quad \text{and} \quad \sum_{k=0}^{\infty} \mu^k \| T_{1j} \|_{j=2}^3 (\| T_{ij} \|_{i,j=2}^3)^k \| T_{i1} \|_{i=2}^3.$$ ■

Lemma 3.9: *Let $T(\mu): \mathscr{H}_1 \to \mathscr{H}_2$ be an operator-function holomorphic in the neighbourhood of zero. Then there is a space $\mathscr{H}_3$ and an operator $T = \| T_{ij} \|_{i,j=1}^{2} : \mathscr{H}_1 \oplus \mathscr{H}_3 \to \mathscr{H}_2 \oplus \mathscr{H}_3$ such that $T(\mu)$ generates the operator T.*

□ We consider the function, holomorphic in the neighbourhood of zero,

$$K(\mu) = (T(\mu) - T(0))/\mu \quad \text{with } \mu \neq 0 \quad \text{and} \quad K(0) = T'(0).$$

By Cauchy's formula we have for it, when $r > 0$ is sufficiently small,

$$K(\mu) = \frac{r}{2\pi i} \oint_{|\zeta| = 1} \frac{K(r\zeta)}{r\zeta - \mu} \, d\zeta = \frac{1}{2\pi} \int_{-\pi}^{\pi} \frac{K(re^{i\varphi})}{1 - r^{-1} e^{-i\varphi} \mu} \, d\varphi \qquad (|\mu| < r).$$

Let $\mathscr{H}_3$ be the Hilbert space of weakly measurable functions $x_3(e^{i\varphi})$ on $[-\pi, n]$ with values in $\mathscr{H}_2$ and with the square-summable norm

$$\| x_3 \|^2 = \frac{1}{2\pi} \int_{-\pi}^{\pi} \| x_3(e^{i\varphi}) \|^2 \, d\varphi < \infty,$$

let T_{22} be the operator of multiplication in $\mathscr{H}_3$ by $r^{-1} e^{-i\varphi}$, let

$$T_{21} x_1 = K(re^{i\varphi}) x_1 \text{ when } x_1 \in \mathscr{H}_1,$$

let

$$T_{12} x_3 = \frac{1}{2\pi} \int_{-\pi}^{\pi} x_3(e^{i\varphi}) \, d\varphi \text{ when } x_3 \in \mathscr{H}_3, \text{ and let } T_{11} = T(0).$$

Then the operator $T = \| T_{ij} \|_{i,j=1}^{2}$ defined on $\mathscr{H}_1 \oplus \mathscr{H}_3$ is bounded, acts into $\mathscr{H}_2 \oplus \mathscr{H}_3$, and $T(\mu) = T(0) + \mu K(\mu) = T_{11} + \mu T_{12}(I_3 - \mu T_{22})^{-1} T_{21}$. ■

We summarize the results set out.

Theorem 3.10: *If $T(\mu): \mathscr{H}_1 \to \mathscr{H}_2$ is an operator-function holomorphic in the neighbourhood of zero, then $T(\mu) \in \Pi$.*

□ It suffices to compare Lemma 3.9, Theorem 3.4, and Lemma 3.8. ■

3 In this paragraph we introduce a number of results and constructions on the basis of which we shall obtain a criterion for $T(\mu)$ to belong to the class $\Pi^{\varkappa}$.

Definition 3.11: A function $K(\lambda, \mu)$ of two variables, defined on $\Lambda \times \Lambda$, where $\Lambda = \Lambda^* \subset \mathbb{C}$, and with values in a set of continuous operators acting in a Hilbert space $\mathscr{H}$ and defined on it, is called a *Hermitian kernel* if $K(\lambda, \mu) = K^*(\mu, \lambda)$; a *Hermitian kernel $K(\lambda, \mu)$ is said to have $\varkappa$ negative squares* if for arbitrary finite sets $\{\lambda_i\}_1^{\tau} \subset \Lambda$, $\{\xi_i\}_1^{\tau} \subset \mathbb{C}$, and $\{x_i\}_1^{\tau} \subset \mathscr{H}$ the quadratic form

$$\sum_{i,j=1}^{\tau} (K(\lambda_i, \lambda_j) x_i, x_j) \xi_i \bar{\xi}_j \tag{3.3}$$

has not more than $\varkappa$ negative squares, and for at least one set has exactly $\varkappa$ negative squares. In particular, if $\varkappa = 0$, then $K(\lambda, \mu)$ is called a *non-negative Hermitian kernel.*

□ Let $\mathscr{H}$ be a J-space, let $F(\lambda): \mathscr{H} \to \mathscr{H}$ be a function holomorphic in the neighbourhood $\mathscr{U}_F$ of zero with values in a set of continuous operators with $\mathscr{D}_{F(\lambda)} = \mathscr{H}$, and let $K(\lambda, \mu = J[F(\lambda) + F^c(\mu)]/(1 - \lambda\bar{\mu})$ be a Hermitian kernel having $\varkappa$ negative squares. We shall set up by means of the kernel a correspondence between the function $F(\lambda)$ and a certain Pontryagin space $\Pi_\varkappa(F)$, we introduce a continuous operator $\Gamma: \mathscr{H} \to \Pi_\varkappa(F)$, and a certain π-semi-unitary operator in $\Pi_\varkappa(F)$, and we express the function $F(\lambda)$ by means of them.

Let $\lambda \in \mathscr{U}_F$, let ε_λ be a symbol, and let x_λ be a vector in $\mathscr{H}$. We form a linear set $\mathscr{L}$ consisting of formal sums $f = \Sigma_{\lambda \in \mathscr{U}_F} \varepsilon_\lambda x_\lambda$ in which only a finite number of the $x_\lambda \neq \theta$. For elements f and $g = \Sigma_{\lambda \in \mathscr{U}_F} \varepsilon_\lambda g_\lambda$ we define a form $[f, g]_0 = \Sigma_{\lambda, \mu \in \mathscr{U}_F} (K(\lambda, \mu)x_\lambda, y_\mu)$. By hypothesis this form has $\varkappa$ negative squares, and if $\mathscr{L}$ is degenerate and $\mathscr{L}_0$ is it isotropic part, then the completion of the factor-lineal $\mathscr{L}/\mathscr{L}_0$ will be a Pontryagin space with $\varkappa$ negative squares (*cf.* Exercise 6 on 1.§9)—and we denote it by $\Pi_\varkappa(F)$. From now on we shall identify elements from $\mathscr{L}$ with the corresponding elements from $\mathscr{L}/\mathscr{L}_0$. On the set

$$\mathscr{D}_V = \left\{ f \mid f = \sum_{\lambda \in \mathscr{U}_F} \varepsilon_\lambda x_\lambda \in \mathscr{L}, \sum_{\lambda \in \mathscr{U}_F} x_\lambda = \theta \right\}$$

we define an operator V: $Vf = \Sigma_{\lambda \in \mathscr{U}_F} \lambda \varepsilon_\lambda x_\lambda$. Since

$$\begin{aligned}
[Vf, Vf]_0 &= \sum_{\lambda, \mu \in \mathscr{U}_F} (K(\lambda, \mu)\lambda x_\lambda, \mu x_\mu) = \sum_{\lambda, \mu \in \mathscr{U}_F} \lambda\bar{\mu}\left(\left(J \frac{F(\lambda) + F^c(\mu)}{1 - \bar{\lambda}\mu}\right) x_\lambda, x_\mu\right) \\
&= \sum_{\lambda, \mu \in \mathscr{U}_F} \frac{\lambda\bar{\mu}}{1 - \lambda\bar{\mu}} ([F(\lambda)x_\lambda, x_\mu] + [x_\lambda, F(\mu)x_\mu]) \\
&= \sum_{\lambda \in \mathscr{U}_F} \left[F(\lambda)x_\lambda, \sum_{\mu \in \mathscr{U}_F} \frac{\bar{\lambda}\mu}{1 - \bar{\lambda}\mu} x_\mu \right] + \sum_{\mu \in \mathscr{U}_F} \left[\sum_{\lambda \in \mathscr{U}_F} \frac{\lambda\bar{\mu}}{1 - \lambda\bar{\mu}} x_\lambda, F(\mu)x_\mu \right] \\
&= \sum_{\lambda \in \mathscr{U}_F} \left[F(\lambda)x_\lambda, \sum_{\mu \in \mathscr{U}_F} \frac{1}{1 - \bar{\lambda}\mu} x_\mu \right] + \sum_{\mu \in \mathscr{U}_F} \left[\sum_{\lambda \in \mathscr{U}_F} \frac{1}{1 - \lambda\bar{\mu}} x_\lambda, F(\mu)x_\mu \right] \\
&= [f, f]_0,
\end{aligned}$$

V is a π-isometric operator in $\Pi_\varkappa(F)$. Its range of values $\mathscr{R}_V$ consists of vectors $g = \Sigma_{\lambda \in \mathscr{U}_F} \varepsilon_\lambda y_\lambda \in \mathscr{L}$ such that $y_0 = \theta$ and $\Sigma_{0 \neq \lambda \in \mathscr{U}_F} y_\lambda/\lambda = 0$. This set is dense in $\Pi_\varkappa(F)$. For, if $\zeta \to 0$ and $x \in \mathscr{H}$, then, using Exercise 13 on 1.§9, we obtain $\varepsilon_\zeta x \to \varepsilon_0 x$, $\zeta\varepsilon_\lambda x \to \theta$, and therefore any element $\Sigma_{\lambda \in \mathscr{U}_F} \varepsilon_\lambda x_\lambda \in \mathscr{L}$ $(x_0 = \theta)$ is approximated by the elements $\Sigma_{\lambda \in \mathscr{U}_F} \varepsilon_\lambda x_\lambda - \zeta e_\zeta \Sigma_{0 \neq \lambda \in \mathscr{U}_F} x_\lambda/\lambda \in \mathscr{R}_V$, and since $\bar{\mathscr{L}} = \Pi_\varkappa(F)$, it follows that $\bar{\mathscr{R}}_V = \Pi_\varkappa(F)$ also. It follows from the last relation that $\mathscr{R}_V$ is non-degenerate, and this implies that V is invertible and hence (taking 2.Corollary 4.8 into account) that the operators V and V^{-1} are continuous. We keep the same notation for the closures of these operators, the second of which will be π-semi-unitary.

We bring into consideration an operator $\Gamma: \mathscr{H} \to \mathscr{L}$ by putting $\Gamma x = (1/\sqrt{2})\varepsilon_0 x$, $x \in \mathscr{H}$. For any $x, y \in \mathscr{H}$ and $\lambda \in \mathscr{U}_F$ we obtain $[\Gamma x, \Gamma x]_0 = [(F(0))_R x, y]$, $[\Gamma x, \varepsilon_\lambda y]_0 = (1/\sqrt{2})[F(0) + F^c(\lambda))x, y]$, and therefore Γ is a continuous operator (see Exercise 6 on 1§9). Consequently there is the operator $\Gamma^c: \Pi_\varkappa(F) \to \mathscr{H}$ and

$$\Gamma^c\Gamma = (F(0))_R, \qquad \Gamma^c\varepsilon_\lambda x = (1/\sqrt{2})(F^c(0) + F(\lambda))x. \tag{3.4}$$

Since (see 2.Definition 6.6) the spectrum of the π-semi-unitary operator V^{-1}, with the exception of not more than $\varkappa$ eigenvalues, is situated in the disc $\{\xi \mid |\xi| \leqslant 1\}$, the resolvent $(V - \lambda I)^{-1}$ exists everywhere in the disc $\{\lambda \mid |\lambda| < 1\}$ with the exception, possibly, of not more than $\varkappa$ points. Since $(V - \lambda I)(\varepsilon_\lambda x - \varepsilon_0 x) = \lambda\varepsilon_0 x$, we have $\Gamma^c\varepsilon_\lambda x - \Gamma^c\varepsilon_0 x = \lambda\Gamma^c(V - \lambda I)^{-1}\varepsilon_0 x$, and so we obtain from (3.4)

$$F(\lambda) = i(F(0))_I + \Gamma^c(V + \lambda I)(V - \lambda I)^{-1}\Gamma. \tag{3.5}$$

We sum up the foregoing argument:

Theorem 3.12: *Let $\mathscr{H}$ be a J-space, let $F(\lambda)$ be an operator-function holomorphic in the neighbourhood of zero and taking values in the set of continuous operators defined on $\mathscr{H}$, and let the kernel $K(\lambda, \mu) = J(F(\lambda) + F^c(\mu))/(1 - \lambda\bar{\mu})$ have $\varkappa$ negative squares. Then:*

1) *$F(\lambda)$ extends, preserving these properties, on to the disc $\{\lambda \mid |\lambda| < 1\}$ with the possible exception of $\varkappa$ points;*

2) *there is a space $\Pi_\varkappa$, a bounded operator $\Gamma: \mathscr{H} \to \Pi_\varkappa$, and a π-semi-unitary operator $V^{-1}: \Pi_\varkappa \to \Pi_\varkappa$ such that the equality* (3.5) *holds for all $\lambda \notin \sigma(V)$, $|\lambda| < 1$.* ■

□ *Remark 3.13:* In Theorem 3.12 the operator V can be regarded as π-unitary since, by virtue of Theorem 2.18, it admits regular extensions $\tilde{V}$ and $(V - \lambda\tilde{I})^{-1} \mid \mathscr{R}_\Gamma = (V - \lambda I)^{-1} \mid \mathscr{R}_\Gamma$. ■

The proof of the main Theorem 3.16 in this section depends on the following two lemmas, the first of which is a direct consequence of Theorem 3.12.

Corollary 3.14: *Let $T(\lambda): \mathscr{H} \to \mathscr{H}$ be an operator-function holomorphic in the neighbourhood of zero, let the kernel $L(\lambda, \mu) = J - T^*(\mu)JT(\lambda)/(1 - \lambda\bar{\mu})$ have $\varkappa$ negative squares, and let $1 \in \rho(T(0))$. Then $T(\lambda) \in \Pi^k$.*

Since $1 \in \rho(T(0))$, there is a neighbourhood $\mathscr{U}_0$ of zero such that $1 \in \rho(T(\lambda))$ when $\lambda \in \mathscr{U}_0$. We put $F(\lambda) = (\bar{\alpha}I + \alpha T(\lambda))(I - T\lambda)^{-1}$ when $\lambda \in \mathscr{U}_0$. The function $F(\lambda)$ is holomorphic on $\mathscr{U}_0$, and since

$$J\frac{F(\lambda) + F^c(\mu)}{1 - \bar{\mu}\lambda} = 2\,\mathrm{Re}\,\alpha(I - T^*(\mu))^{-1}\frac{J - T^*(\mu)JT(\lambda)}{1 - \bar{\mu}\lambda}(I - T(\lambda))^{-1},$$

so when Re $\alpha > 0$ the function $F(\lambda)$ satisfies the conditions of Theorem 3.12. Let $\Pi_\varkappa$, Γ and V be the same as in Theorem 3.12, and suppose (see Remark 3.13) that V is a $\Pi_\varkappa$-unitary operator. Since $T(\lambda)$ is expressed in terms of $F(\lambda)$ by the formula

$$T(\lambda) = (F(\lambda) - \bar{\alpha}I)(F(\lambda) = \alpha I)^{-1} = I - 2\ \mathrm{Re}\ \alpha(F(\lambda) + \alpha I)^{-1},$$

so, using (3.5) we obtain

$$\begin{aligned}T(\lambda) &= (F(0) - \bar{\alpha}I)(F(0) + \alpha I)^{-1}\\ &\quad + \lambda(2\sqrt{\mathrm{Re}\ \alpha}(F(0) + \alpha I)^{-1}\Gamma^c V^{-}(I - \lambda(V^{-1} - 2\Gamma(F(0) + \alpha I)^{-1}\Gamma^c V^{-1}))^{-1}\\ &\quad \times (2\sqrt{\mathrm{Re}\ \alpha}\,\Gamma(F(0)) + \alpha I)^{-1}).\end{aligned}$$

We define in the space $\mathscr{H} \oplus \Pi_\varkappa$ an operator $T = \| T_{ij} \|_{i,j=1}^2$, where

$$T_{11} = T(0) = (F(0) - \bar{\alpha}I)(F(0) + \alpha I)^{-1}, \quad T_{12} = 2\sqrt{\mathrm{Re}\ \alpha}(F(0) + \alpha I)^{-1}\Gamma^c V^{-1},$$
$$T_{21} = 2\sqrt{\mathrm{Re}\ \alpha}(F(0) + \alpha I)^{-1}, \qquad T_{22} = V^{-1} - 2\Gamma(F(0) + \alpha I)^{-1}\Gamma^c V^{-1}.$$

This operator generates the function $T(\lambda)$, and it is verified immediately (taking the equality $\Gamma^c\Gamma = (F(0))_R$ into account) that $T^c T = T^c T = I$, i.e., T is a $(J \oplus \Pi_\varkappa)$-unitary operator. ■

Lemma 3.15: *Let $T_0 : \mathscr{H}_1 \to \mathscr{H}_2$ be a continuous operator, with $\mathscr{D}_{T_0} = \mathscr{H}_1$. There will be a $\varkappa$-dimensional space Π_- and a (J_1', J_2')-bi-non-expansive operator $T = \| T_{ij} \|_{i,j=1}^2$ acting from the J_1'-space $\mathscr{H}_1' = \mathscr{H}_1 \oplus \Pi_-$ into the J_2'-space $\mathscr{H}_2' = \mathscr{H}_2 \oplus \Pi_-$, with $J_i' = J_i \oplus -I_-$ $(i = 1, 2)$, and such that $T_{11} = T_0$, if and only if the operators $J_1 - T_0^* J_2 T_0$ and $J_2 - T_0 J_1 T^*$ have the same number $\varkappa_1 \leqslant \varkappa$ of negative eigenvalues (taking multiplicity into account).*

Under these conditions the operator T_0 is expressible either in the form $T_0 = RW$, where W is a (J_1, J_2)-bi-non-expansive (J_1, J_2)-semi-unitary operator, and $R : \mathscr{H}_2 \to \mathscr{H}_2$ is a bounded operator with Ker $R \supset$ Ker W^c *and the set $[\mathbb{C}\backslash(\mathbb{R} \cup i\mathbb{R})] \cap \sigma(R)$ is no more than countable, in the form $T_0 = W'R'$, where W' is an operator (J_2, J_1)-conjugate to a (J_2, J_1)-bi-non-expansive (J_2, J_1)-semi-unitary operator, and $R' : \mathscr{H}_1 \to \mathscr{H}_1$ is a bounded operator with* Ker $R'^c \supset$ Ker W' *and the set $\sigma(R) \in [\mathbb{C}\backslash(\mathbb{R} \cup i\mathbb{R})]$ is no more than countable.*

□ Suppose the subspace Π_- and the operator T indicated in the Lemma exist. In accordance with 4.Theorem 1.12 the operator T^c admits a (J_2, J_1)-polar decomposition $T^c = V^c S_c$, where without loss of generality (otherwise we would consider the operator T_0^c instead of T_0) we can suppose that V is a (J_1, J_2)-semi-unitary (J_1, J_2)-bi-non-expansive operator, and S^c has the properties of a J_2-module indicated in 4.Theorem 1.12. Hence $T = SV$. Let P_i be an orthoprojector from $\mathscr{H}_i'$ on to $\mathscr{H}_i$, $i = 1, 2$. Then

$$T_0 = P_2 SVP_1 \mid \mathscr{H}_1.$$

If

$$S = \| S_{ij} \|_{i,j=1}^{2} : \mathscr{H}_2 \oplus \Pi_- \to \mathscr{H}_2 \oplus \Pi_-$$

and

$$V = \| V_{ij} \|_{i,j=1}^{2} : \mathscr{H}_1 \oplus \Pi_- \to \mathscr{H}_2 \oplus \Pi_-$$

are the matrix representations of the operators S and V, then $T_0 = S_{11} V_{11} + S_{12} V_{21}$. Just as in the proof of the implication b) ⇒ c) in Theorem 2.18, we prove that the operator $V_{11} | \operatorname{Ker} V_{21}$ admits a (J_1, J_2)-semi-unitary (J_1, J_2)-bi-non-expansive extension W, and we let P be a J_2-orthoprojector from $\mathscr{H}_2$ on to $\mathscr{R}_W$. Then $T_0 = (S_{11} P + S_{11}(V_{11} W^c - P) + S_{12} V_{21} W^c) W \equiv RW$. By contruction W has the properties indicated in the Lemma, and $\operatorname{Ker} R \supset \operatorname{Ker} W^c (= \mathscr{R}_W^{[\perp]})$. We verify that $\sigma(R) \cap [\mathbb{C} \backslash (\mathbb{R} \cup i\mathbb{R})]$ consists of not more than a countable set of points. Since the operators $S_{11}(V_{11} W^c - P)$ and $S_{12} V_{21} W^c$ are finite-dimensional, it is sufficient by virtue of 2.Theorem 2.11 to establish that the set $\sigma(S_{11} P) \cap [\mathbb{C} \backslash (\mathbb{R} \cup i\mathbb{R})]$ has the property mentioned. In accordance with Corollary 3.5 the J_2'-selfadjoint operator S, which is J_2'-bi-non-expansive, admits a regular dilatation $\tilde{S}$ which is a J_2''-unitary operator acting in the J_2''-space $\mathscr{H}_2'' = \mathscr{H}_2 \oplus \Pi_- \oplus \Pi_+$, where $J_2'' = J_2 \oplus -I_- \oplus I_+$. Let $\mathscr{H}_2'' = \mathscr{R}_W [\dot{+}] \mathscr{R}_W^{[\perp]}$, and let $\tilde{S} = \| \tilde{S}_{ij} \|_{i,j=1}^{2}$ be the matrix representation of the operator $\tilde{S}$ relative to this decomposition of $\mathscr{H}_2''$. Since $\sigma(S_{11} P) - \{0\} = \sigma(PS_{11} P) - \{0\}$, and $PSP | \mathscr{R}_W = \tilde{S}_{11} (= \tilde{S}_{11}^c)$, it is sufficient to verify that the set $\sigma(\tilde{S}_{11}) \cap [\mathbb{C} \backslash (\mathbb{R} \cup iR)]$ is no more that countable. Since the operator $\tilde{S}$ is J_2''-unitary, it follows that $\tilde{I}_1 - \tilde{S}_{11}^2 = \tilde{S}_{21}^c \tilde{S}_{21}$. Let λ_0 be the boundary point of the spectrum of the operator $\tilde{I}_1 - \tilde{S}_{11}^2$. Then there is a sequence $\{x_n\} \in \mathscr{R}_W$ such that $0 < \inf \| x_n \| \leqslant \sup \| x_n \| < \infty$ and $(\tilde{I}_1 - \tilde{S}_{11}^2) x_n - \lambda_0 x_n \to \theta$ as $n \to \infty$, or, what is the same thing, $\tilde{S}_{21}^c \tilde{S}_{21} x_n - \lambda_0 x_n \to \theta$. Hence it follows that $0 < \inf \| \tilde{S}_{21} x_n \| \leqslant \sup \| \tilde{S}_{21} x_n \| < \infty$ and $(\tilde{S}_{21} \tilde{S}_{21}^c - \lambda_0 \tilde{I}_1) \tilde{S}_{21} x_n \to \theta$, and this implies that the complex number λ_0 belongs to the spectrum of the operator $S_{21} \tilde{S}_{21}^c$ which is a $(I-P)^* J_2''(I-P)$-selfadjoint operator. Since W is a (J_1, J_2)-bi-non-expansive operator, the J_2''-orthogonal complement $\mathscr{R}_W^{[\perp]}$ to $\mathscr{R}_W$ is a Pontryagin space with $\varkappa$ negative squares. Therefore (see 2.Corollary 3.15) $\sigma(\tilde{I}_1 - \tilde{S}_{11}^2) \cap (\mathbb{C} \backslash \mathbb{R})$ consists of not more than $2\varkappa$ eigenvalues (taking multiplicity into account), and therefore $\sigma(\tilde{S}_{11}) \cap [\mathbb{C} \backslash (\mathbb{R} \cup i\mathbb{R})]$ consists of a finite number of eigenvalues of finite multiplicity. Thus $T_0 = RW_1$ where R and W have the properties indicated in the Lemma.

Since T is a (J_1', J_2')-bi-non-expansive operator, there is, by virtue of Corollary 3.5, a Hilbert space Π_+' and a $(\tilde{J}_1, \tilde{J}_2)$-unitary operator $T = \| \tilde{T}_{ij} \|_{i,j=1}^{2}$ acting from the $\tilde{J}$-space $\tilde{\mathscr{H}}_1 = \mathscr{H}_1 \oplus (\Pi_- \oplus \Pi_+')$ into the $\tilde{J}_2$-space $\tilde{\mathscr{H}}_2 = \mathscr{H}_2 \oplus (\Pi_- \oplus \Pi_+')$, where $\tilde{J}_i = J_i \oplus \pi_\varkappa$ and $\pi_\varkappa = -I_- + I_+'$, such that $\tilde{T}_{11} = T_0$. The equalities

$$J_1 - T_0^* J_2 T_0 = T_{21}^* \pi_\varkappa T_{21} \quad \text{and} \quad J_2 - T_0 J_1 T_0^* = T_{12} \pi_\varkappa T_{12}^*$$

can be verified immediately. Consequently (see Exercise 14 on 1§9) the operators $J_1 - T_0^* J_2 T_0$ and $J_2 - T_0 J_1 T_0^*$ have respectively $\varkappa_1 \leqslant \varkappa$ and $\varkappa_2 \leqslant \varkappa$ negative eigenvalues (taking multiplicity into account). We recall that $\varkappa_1$ and $\varkappa_2$ coincide with the number of negative squares of the forms $[(I_1 - T_0^c T_0)\cdot, \cdot]_1$ and $[(I_2 - T_0 T_0^c)\cdot, \cdot]_2$ respectively. Since $I_1 - T_0^c T_0 = I_1 - W^c R^c R W = W^c(I_2 - R^c R)W$, and $\operatorname{Ker} R \supset \operatorname{Ker} W^c$, and $\operatorname{Ker} W^c$ is a uniformly positive subspace, it follows that $\varkappa_1$ coincides with the number of negative squares of the form $[(I_2 - R^c R)\cdot, \cdot]_2$. We note also that from the condition $\operatorname{Ker} R \supset \operatorname{Ker} W^c$ the equality $I_2 - T_0 T^c = I_2 - RR^c$ follows.

Let X be a linear operator; we shall denote by the symbol X_φ the operator $e^{i\varphi}X$ ($\varphi \in \mathbb{R}$). It is easy to see that

$$I_2 - (R_\varphi)^c R_\varphi = I_2 - R^c R \quad \text{and} \quad I_2 - R_\varphi (R_\varphi)^c = I_2 - RR^c.$$

Since the set $\sigma(R) \cap [\mathbb{C}\backslash(\mathbb{R} \cup i\mathbb{R})]$ is no more than countable, there is a number φ_0 such that $1 \in \rho(R_{\varphi_0})$. Consequently

$$I_2 - R^c R = (I_2 - R_{\varphi_0})^c [(I_2 - R_{\varphi_0})^{-1} + (I_2 - (R_{\varphi_0})^c)^{-1} - I_2](I_2 - R_{\varphi_0}),$$
$$I_2 - RR^c = (I_2 - R_{\varphi_0})[(I_2 - R_{\varphi_0})^{-1} + (I_2 - (R_{\varphi_0})^c)^{-1} - I_2](I_2 - (R_{\varphi_0})^c),$$

and therefore

$$I_2 - R^c R = (I_2 - (R_{\varphi_0})^c)(I_2 - R_{\varphi_0}))^{-1}(I_2 - RR^c)(I_2 - (R_{\varphi_0})^c)^{-1}(I_2 - R_{\varphi_0}),$$

which implies that the numbers $\varkappa_1$ and $\varkappa_2$ are indeed the same.

Conversely, let the operator T_0 be such that each of the operators $J_1 - T_0^* J_2 T_0$ and $J_2 - T_0 J_1 T_0^*$ has exactly $\varkappa_1 \leqslant \varkappa$ negative eigenvectors (taking multiplicity into account). We shall first suppose that $\dim \mathscr{H}_1 = \dim \mathscr{H}_2$. We bring into consideration the operators

$$T_0^{(n)} = T_0 F_n, \quad \text{where } F_n = \sqrt{\frac{n}{n+1}}\, P_1^+ + \sqrt{\frac{n+1}{n}}\, P_1^- \qquad (n = 1, 2, \ldots)$$

For sufficiently large n the negative spectrum of each of the operators $J_1 - T_0^{(n)^*} J_2 T_0^{(n)}$ and $J_2 - T_0^{(n)} J_1 T_0^{(n)^*}$ consists of exactly $\varkappa_1$ (taking multiplicity into account) eigenvalues and these operators are boundedly invertible. We denote by $J_1^{(n)}$ and $J_2^{(n)}$ respectively the unitary parts of the polar decompositions of these operators, we put $P_{i,n}^{\pm} = \frac{1}{2}(I_i \pm J_i^{(n)})$, $i = 1, 2$, and we bring into consideration the $\tilde{J}_i^{(n)}$-spaces $\tilde{\mathscr{H}}_i^{(n)} = \mathscr{H}_i \oplus \mathscr{H}_1$, $\tilde{J}_i^{(n)} = J_i \oplus J_1^{(n)}$ and the operators

$$T^{(n)} = \| T_{ij}^{(n)} \|_{i,j=1}^2 : \tilde{\mathscr{H}}_1^{(n)} \to \tilde{\mathscr{H}}_2^{(n)},$$

where

$$T_{11}^{(n)} = T_0^{(n)} \qquad T_{12}^{(n)} = |\, J_2 - T_0^{(n)} J_1 T_0^{(n)*} |^{1/2} V_n,$$
$$T_{21}^{(n)} = |\, J_1 - T_0^{(n)*} J_2 T_0^{(n)} |^{1/2},$$
$$T_{22}^{(n)} = -J_1^{(n)} |\, J_1 - T_0^{(n)*} J_2 T_0^{(n)} |^{-1/2} T_0^{(n)*} J_2 |\, J_2 - T_0^{(n)} J_1 T_0^{(n)*} |^{1/2} V_n,$$

where V_n is a unitary operator mapping $\mathscr{H}_1$ and to $\mathscr{H}_2$ (its existence follows from the fact that dim $\mathscr{H}_1 = \dim \mathscr{H}_2$) and moreover mapping $P_{1,n}^{-}\mathscr{H}_1$ on to $P_{2,n}^{-}\mathscr{H}_2$. It can be verified immediately that $T^{(n)}$ is a $(\tilde{J}_1^{(n)}, \tilde{J}_2^{(n)})$-unitary operator. Consequently (see Exercise 34 on 2.§4) the operator

$$\tilde{T}^{(n)} = \| \tilde{T}_{ij}^{(n)} \|_{i,j=1}^{2} : \mathscr{H}_1 \oplus P_{1,n}^{-}\mathscr{H}_1 \to \mathscr{H}_2 \oplus P^{-}{}_{1,n}\mathscr{H}_1,$$

where

$$\tilde{T}_{11}^{(n)} = T_{11}^{(n)}, \tilde{T}_{21}^{(n)} = P_{1,n}^{-}T_{21}^{(n)}, \tilde{T}_{12}^{(n)} = T_{12}^{(n)}P_{1,n}^{-}, \text{ and } \tilde{T}_{22}^{(n)} = P_{1,n}^{-}T_{22}^{(n)}P_{1,n}^{-},$$

is a $(J_1 \oplus -I_{1,n}^{-}, J_2 \oplus -I_{1,n}^{-})$-bi-non-expansive operator. The operators $P_{i,n}^{-}$ converge in norm to $P_{i,0}^{-}$ when $n \to \infty$, where $P_{i,0}^{-} = \frac{1}{2}(I - P_{i,0} - J_{i,0})$ $(i = 1, 2)$, $P_{1,0}$ is the orthoprojector on to $\mathrm{Ker}(J_1 - T_0^* J_2 T_0)$, $P_{2,0}$ is the orthoprojector on to $\mathrm{Ker}(J_2 - T_0^* J_1 T_0)$, and $J_{1,0}$ and $J_{2,0}$ are the partially isometric parts of the polar decompositions of the operators $J_1 - T_0^* J_2 T_0$ and $J - T_0 J_1 T_0^*$ respectively. We choose V_n so that the operators $V_n P_{1,n}^{-}$ converge in norm to $VP_{1,0}^{-}$, where $V: P_{1,0}^{-}\mathscr{H}_1 \to P_{2,0}^{-}\mathscr{H}_2$ is an isometry. Then the operator

$$T = \| T_{ij} \|_{i,j=}^{2} : \mathscr{H}_1 \oplus P_{1,0}^{-}\mathscr{H}_1 \to \mathscr{H}_2 \oplus P_{1,0}^{-}\mathscr{H}_1$$

with

$$T = T_0, \qquad T_{12} = (P_{2,0}^{-}(T_0 J_1 T_0^* - J_2)P_{2,0}^{-})^{1/2} VP_{1,0}^{-},$$
$$T_{21} = (P_{1,0}^{-}(T_0^* J_2 T_0 - J_1)P_{1,0}^{-})^{1/2}, \ T_{22} = T^{-1} P_{1,0}^{-} T_0^* J_2 T_{12}$$

will indeed be the required (J_1', J_2')-bi-non-expansive operator

$$(J_i' = J_i \oplus -I_{1,0}', \ i = 1,2).$$

We now suppose that dim $\mathscr{H}_1 \neq \dim \mathscr{H}_2$, and that $\mathscr{H}$ is an infinite-dimensional Hilbert space with dim $\mathscr{H} \geqslant \max\{\dim \mathscr{H}_1, \dim \mathscr{H}_2\}$. We consider an operator $T_0^{(1)}$ acting from the $(J_1 \oplus I)$-space $\mathscr{H}_1 \oplus \mathscr{H}$ into the $(\tilde{J}_2 \oplus I)$-space $\mathscr{H}_2 \oplus \mathscr{H}$ and which is an extension of the operator T, and moveover $T_0^{(1)} \mid \mathscr{H} = I$. The operator $T_0^{(1)}$ satisfies the same conditions as the operator T_0, and $\dim(\mathscr{H}_1 \oplus \mathscr{H}) = \dim(\mathscr{H}_2 \oplus \mathscr{H})$, and therefore the proposition proved above holds for it. It only remains to use the result of Exercise 34 on 2.§4.

■

Theorem 3.16: *Let $T(\lambda) = \Sigma_{i=0}^{\infty} \lambda^i T_i$ be an operator function holomorphic in the neighbourhood of zero. Then the following assertions are equivalent:*

a) $T(\lambda) \in \Pi^{\varkappa}$;

b) *the negative parts of the spectra of each of the operators $J_1^{(n)} - T^{(n)*} J_2^{(n)} T^{(n)}$ consist of not more than $\varkappa$ negative eigenvalues, and the negative parts of the spectra of the operators $J_1^{(0)} - T^{(0)*} J_2^{(0)} T^{(0)}$ and $J_2^{(0)} - T^{(0)} J_1 T^{(0)*}$ consist of an equal number $\varkappa_1 \leqslant \varkappa$ of negative eigenvalues (taking multiplicity into account), where $J_k^0 = J_k$, $J_k^{(n)} = \Sigma_{k=0}^{n} \oplus J_k$, and $T^{(n)} = \| T_{ij}' \|_{i,j=0}^{n}$ are Toeplitz matrices acting from the $J_1^{(n)}$-space $\mathscr{H}_1^{(n)}$ into* *the* $J_2^{(n)}$*-space* $\mathscr{H}_2^{(n)}$, *and* $T'_{ij} = 0$ *when* $i > j$, $T'_{ij} = T_{j-1}$ *when* $i \leqslant j$,

$$\mathscr{H}_k^{(0)} = \mathscr{H}_k, \mathscr{H}_k^{(n)} = \underbrace{\mathscr{H}_k \oplus \dots \oplus \mathscr{H}_k}_{n+1}, k = 1, 2;\ i, j = 0, 1, 2, \dots, n;$$

$n = 0, ,1, 2, \dots;$

c) *the kernel* $[J_1 - T^*(\mu) J_2 T(\lambda)]/(1 - \bar{\mu}\lambda)$ *has not more than* $\varkappa$ *negative squares, and the negative parts of the spectra of each of the operators* $J_1 - T_0^* J_2 T_0$ *and* $J_2 - T_0 J_1 T_0^*$ *consist of* $\varkappa_1 \leqslant \varkappa_2$ *eigenvalues (taking multiplicity into account).*

□ a) ⇒ b). Let $T(\lambda) \in \Pi^\varkappa$, i.e., there is a Pontryagin $\pi_\varkappa$-space $\Pi_\varkappa = \Pi_- \oplus \Pi_+$ with $\dim \Pi_- = \varkappa < \infty$ and a $(J_1 \oplus \pi_\varkappa, J_2 \oplus \pi_\varkappa)$-unitary operator $T = \| T_{ij} \|_{i,j=1}^2 : \mathscr{H}_1 \oplus \Pi_\varkappa \to \mathscr{H}_{22} \oplus \Pi_\varkappa$ such that in the neighbourhood of zero $T(\lambda = T_{11} + \lambda T_{12}(I_\varkappa - \lambda T_{22})^{-1} T_{21}$. Hence

$$T_0 = T_{11}, T_i = T_{12} T_{22}^{i-1} T_{21}, i = 1, 2, \dots,$$

and

$$J_1^{(n)} - T^{(n)*} J_2^{(n)} T^{(n)} = \hat{T}^{(n)*} \hat{J}^{(n)} \hat{T}^{(n)},$$

where

$$\hat{T}^{(n)} = \| \hat{T}_{ij} \|_{i,j=0}^n : \underbrace{\mathscr{H}_1 \oplus \dots \oplus \mathscr{H}_1}_{n+1} \to \underbrace{\Pi_\varkappa \oplus \dots \oplus \Pi_\varkappa}_{n+1}$$

is a diagonal matrix with $\hat{T}_{ii} = T_{22}^i T_{21}$ $(i = 0, 1, 2, \dots)$, and

$$\hat{J}^{(n)} = \| \hat{J}_{ij} \|_{i,j=0}^n : \underbrace{\Pi_\varkappa \oplus \dots \oplus \Pi_\varkappa}_{n+1} \to \underbrace{\Pi_\varkappa \oplus \dots \oplus \Pi_\varkappa}_{n+1}, \hat{J}_{ij} = \pi_\varkappa,$$

$$i, j = 0, 1, 2, \dots, n;\ n = 0, 1, 2, \dots.$$

Since the selfadjoint operators $\hat{J}^{(n)}$ have not more than $\varkappa$ negative eigenvalues, it follows from Exercise 14 on 1.§9 that the operators $J_1^{(n)} - T^{(n)*} J_2^{(n)} T^{(n)}$ also have not more than $\varkappa$ negative eigenvalues $(n = 0, 1, 2, \dots)$.

The validity of the assertion that the operators $J_1^{(0)} - T^{(0)*} J_2 T^{(0)}$ and $J_2^{(0)} - T^{(0)} J_1^{(0)} T^{(0)*}$ have the same number of negative eigenvalues follows from Lemma 3.15 and Exercise 34 on 2.§4.

b) ⇒ c). Since $\sigma(J_1^{(n)} - T^{(n)*} J_2^{(n)} T^{(n)}) \cap (-\infty, 0)$ consists of not more than $\varkappa$ negative eigenvalues (taking multiplicity into account), and this is equivalent to the fact that the forms

$$((J_1^{(n)} - T^{(n)*} J_2^{(n)} T^{(n)}) x^{(n)}, x^{(n)}) \text{ have in } \underbrace{\mathscr{H}_1 \oplus \dots \oplus \mathscr{H}_1}_{n+1}$$

not more than $\varkappa$ negative squares, it follows that for arbitrary finite sets of vectors $\{x_i\}_i^m \subset \mathscr{H}_1$ and arbitrary complex numbers $\{\lambda_j\}_1^r$ the form

$$\sum_{r,s,p,q} ((J_1^{(n)} - T^{(n)*} J_2^{(n)} T^{(n)}) x_{r,s}^{(n)}, x_{p,q}^{(n)}),$$

where

$$x_{r,s}^{(n)} = \langle x_r, \lambda_s x_r, \ldots, \lambda_s^n x_r \rangle \in \underbrace{\mathscr{H}_1 \oplus \ldots \oplus \mathscr{H}_1}_{(n+1)\text{ times}},$$

$$s = 1, 2, \ldots, t;\ r = 1, 2, \ldots, m,$$

has no more than $\varkappa$ negative squares. This form can be rewritten as

$$\sum_{r,s,p,q} \frac{1}{1-\lambda_s\bar{\lambda}_q}\left((1-\lambda_s^n\bar{\lambda}_q^n)(J_1x_r, x_p)_1 + \sum_{i,j=0}^{n}(1-\lambda_s^{n-i}\bar{\lambda}_q^{n-j})\lambda_s^j\bar{\lambda}_q^j(T_j^*J_2T_ix_r, x_r)_1\right)$$

$$= \sum_{r,s,p,q} \frac{\left(\left(J_1 - \left(\sum_{j=0}^{n}\bar{\lambda}_q^jT_j^*\right)J_2\left(\sum_{i=0}^{n}\lambda_s^iT_i\right)\right)x_r, x_p\right)_1}{1-\lambda_s\bar{\lambda}_q}$$

$$- \sum_{r,s,p,q}\left(\frac{\lambda_s^n\bar{\lambda}_q^n(J_1x_r, x_p)_1}{1-\lambda_s\bar{\lambda}_q} + \sum_{i,j=0}^{n}\left(\frac{\lambda_s^{n-1}\bar{\lambda}_q^{n-1}}{1-\lambda_s\bar{\lambda}_q}(\bar{\lambda}_q^jT_j^*)J_2(\lambda_s^iT_i)x_r, x_p\right)_1\right).$$

To prove our assertion it suffices to verify that the second sum over r, s, p, q tends to zero as $n \to \infty$ if we take the $\{\lambda_i\}_1^t$ in a sufficiently small neighbourhood of zero. Without loss of generality we shall suppose that $\|x_r\| = 1$ $(r = 1, 2, \ldots, m)$, and $|\lambda_s| < \varepsilon$ $(s = 1, 2, \ldots, t)$ and $\varepsilon^{1/2}\|T_j\| \leqslant 1$ $(j = 1, 2, \ldots)$ (the latter is possible because $T(\lambda)$ is holomorphic in the neighbourhood of zero). Since there is only a finite number of terms in r, s, p, q, we shall verify that each of them tends to zero as $n \to \infty$; indeed,

$$\left|\frac{\lambda_s^n\bar{\lambda}_q^n(J_1x_r, x_p)_1 + \sum_{i,j=0}^{n}(\lambda_s^{n-i}\bar{\lambda}_q^{n-i}(\bar{\lambda}_q^jT_j^*)J_2(\lambda_s^iT_i)x_r, x_p)_1}{1-\lambda_s\bar{\lambda}_q}\right|$$

$$\leqslant \frac{\varepsilon^{2n} + \sum_{i,j=0}^{n}\varepsilon^{2n+(j-3i)/2}}{1-\varepsilon^2} \to 0 \quad \text{as } n \to \infty.$$

c) ⇒ a). In accordance with Lemma 3.15 we can suppose without loss of generality that $T_0 = RW$, where R and W have the properties indicated in that lemma. Let $\varphi \in [0, 2\pi)$ be a number such that $1 \in \rho(R_\varphi)$. We consider the function $R_\varphi(\lambda) = T(\lambda)(W_{-\varphi})^c$, which is holomorphic in the neighbourhood of zero. This function satisfies the conditions of Lemma 3.14 and therefore $R_\varphi(\lambda) \in \Pi^\varkappa(\mathscr{H}_2)$, i.e., there is a $\pi_\varkappa$-space $\Pi_\varkappa$ and a $(J_2 \oplus \pi_k)$-unitary operator

$$\tilde{R}_\varphi = \|R_{ij,\varphi}\|_{i,j=1}^2 : \mathscr{H}_2 \oplus \Pi_\varkappa \to \mathscr{H}_2 \oplus \Pi_\varkappa$$

such that in the neighbourhood of zero

$$R_\varphi(\lambda) = R_{11,\varphi} + \lambda R_{12,\varphi}(I_\varkappa - \lambda R_{22,\varphi})^{-1}R_{21,\varphi}.$$

Hence,

$$T(\lambda) = R_\varphi(\lambda) W_{-\varphi} = R_{11,\varphi} W_{-\varphi} + \lambda R_{12,\varphi}(I_\varkappa - \lambda R_{22,\varphi})^{-1} R_{21,\varphi} W_{-\varphi}$$

and therefore $T(\lambda)$ is generated by the matrix

$$T' = \| T'_{ij} \|_{i,j=1}^{2} : \mathcal{H}_1 \oplus \Pi_\varkappa \to \mathcal{H}_2 \oplus \Pi_\varkappa,$$

where

$$T'_{11} = R_{11,\varphi} W_{-\varphi},\ T'_{12} = R_{12,\varphi},\ T'_{21} = R_{21,\varphi} W_{-\varphi},\ T'_{22} = R_{22,\varphi}.$$

By construction T' is a $(J_1 \oplus \pi_\varkappa, J_2 \oplus \pi_\varkappa)$-semi-unitary $(J_1 \oplus \pi_\varkappa, J_2 \oplus \pi_\varkappa)$-bi-non-expansive operator. By virtue of Corollary 3.5 it admits regular dilatations, and it follows from Lemma 3.8 that each of these dilatations generates the function $T(\lambda)$, and therefore $T(\lambda) \in \Pi^\varkappa$. ■

Now let $\mathcal{H}_1 = \mathcal{H}_2 = \mathcal{H}$ be a J-space, let $T(\mu): \mathcal{H} \to \mathcal{H}$ be an operator-function holomorphic in the neighbourhood of zero, let $\mathcal{H}_0$ be a J_0-space, and let $T = \| T_{ij} \|_{i,j=1}^{2} : \mathcal{H} \oplus \mathcal{H}_0 \to \mathcal{H} \oplus \mathcal{H}_0$ be a $\tilde{J}$-unitary operator generating the function $T(\mu)$, where $\tilde{J} = J + J_0$.

Definition 3.17: The $\tilde{J}$-unitary operator $T = \| T_{ij} \|_{i,j=1}^{2}$ will be called *simple* if the operators T and T_{22} have no common completely invariant subspaces.

Theorem 3.18: *Every operator-function $T(\mu): \mathcal{H} \to \mathcal{H}$ holomorphic at zero is generated by a certain simple $\tilde{J}$-unitary operator*

$$T = \| T_{ij} \|_{i,j=1}^{2} : \mathcal{H} \oplus \mathcal{H}_0 \to \mathcal{H} \oplus \mathcal{H}_0.$$

Moreover, if $T(\mu) \in \Pi^\varkappa$, then a Pontryagin space with not more than $\varkappa$ negative squares can be taken as $\mathcal{H}_0$.

□ Let $\mathcal{H}_3$ be the J_3-space which exists by virtue of Lemma 3.9, and let $T' = \| T'_{ij} \|_{i,j=1}^{2} : \mathcal{H} \oplus \mathcal{H}_3 \to \mathcal{H} \oplus \mathcal{H}_3$ be the $(J \oplus J_3)$-unitary operator generating the function $T(\mu)$ (if $T(\mu) \in \Pi^\varkappa$, then we take as $\mathcal{H}_3$ a Pontryagin space with not more than $\varkappa$ negative squares). The subspace $\mathcal{H}'' = C\,\mathrm{Lin}\{ T'^{\,n} \mathcal{H} \}_{-\infty}^{\infty}$ is completely invariant relative to the operator T', it contains $\mathcal{H}$, and the operator T' is a dilatation of the operator $T'' = T' \mid \mathcal{H}''$, and therefore (see Lemma 3.8) the operator T'' generates the same function $T(\mu)$, and moreover T'' is an isometric operator in the indefinite metric induced by the $(J \oplus J_3)$-metric in $\mathcal{H}''$. Let P be the orthoprojector from $\mathcal{H}''$ on to its isotropic part. Then the space $\mathcal{H}'_0 = (I - P)\mathcal{H}''$ contains $\mathcal{H}$ and is a non-degenerate $(\tilde{J} \oplus G)$-space: $\mathcal{H}'_0 = \mathcal{H} \oplus \mathcal{H}'$, and the operator T'' is a $(\tilde{J} \oplus G)$-unitary dilatation of the operator $T'_0 = (I - P) T'' \mid \mathcal{H}'_0$ and therefore, again by virtue of Lemma 3.8, T'_0 generates the same function $T(\mu)$. As the space $\mathcal{H}_0$ we take the completion of $\mathcal{H}'$ relative to the norm $(|G| x, x)^{1/2}$ (*cf.* 1.Proposition 6.14); as J_0 we take the extension of the isometric part of the polar decomposition of the operator G on to $\mathcal{H}_0$; and as $T = \| T_{ij} \|_{i,j=1}^{2}$ we take the extension by continuity of T'_0 on to $\mathcal{H} \oplus \mathcal{H}_0$. It is clear that the operator T

generates the function $T(\mu)$ and is $\tilde{J}$-unitary, where $\tilde{J} = J \oplus J_0$, and $\mathscr{H} \oplus \mathscr{H}_0 = C\,\mathrm{Lin}\{T^n\mathscr{H}\}_{-\infty}^{\infty}$. We now verify that T is simple. Let $\mathscr{L}$ be the common completely invariant subspace of the operators T and T_{22}. Then $\mathscr{L}^{[\perp]}$ is completely invariant relative to T and $\mathscr{H} \subset \mathscr{L}^{[\perp]}$, and therefore $C\,\mathrm{Lin}\{T^n\mathscr{H}\}_{-\infty}^{\infty} \subset \mathscr{L}^{[\perp]}$, i.e., $\mathscr{L}^{[\perp]} = \mathscr{H} \oplus \mathscr{H}_0$ and therefore $\mathscr{L} = \{\theta\}$. We note that if $\mathscr{H}_3$ is a Pontryagin space with not more than $\varkappa$ negative squares, then $\mathscr{H}_0$ also has this property. ∎

4 *Definition 3.19:* An operator-function $\mathbf{R}_\lambda$ holomorphic in a neighbourhood of the point λ is called a *generalized resolvent of a J-symmetric operator A acting in a J-space* $\mathscr{H}$ if there is a J_1-space $\mathscr{H}_1$ such that the $\tilde{J}$-Hermitian operator A in the $\tilde{J}$-space $\tilde{\mathscr{H}} = \mathscr{H} \oplus \mathscr{H}_1$ (where $\tilde{J} = J \oplus J_1$) admits a $\tilde{J}$-selfadjoint extension $\tilde{A}$ with $\lambda_0 \in \rho(\tilde{A})$ and $\mathbf{R}_\lambda = P(\tilde{A} - \lambda\tilde{I})^{-1} \mid \mathscr{H}$, where P is the orthoprojector from $\tilde{\mathscr{H}}$ on to $\mathscr{H}$. We call such a function $\mathbf{R}_\lambda$ a *$\varkappa$-regular generalized resolvent* if some Pontryagin space with $\varkappa < \infty$ negative squares can be taken as $\mathscr{H}_1$, and a *regular generalized resolvent* if $\varkappa = 0$.

Definition 3.20: We denote by the symbol $\Pi_{\lambda_0}(A)$ the set of all operator-functions $F(\lambda): \mathscr{H} \to \mathscr{H}$, holomorphic in a neighbourhood of the point λ_0, which can be expressed in the form

$$F(\lambda) = F_{11} + \frac{\bar{\lambda}_0}{\lambda_0}\frac{\lambda - \lambda_0}{\lambda - \bar{\lambda}_0} F_{12}\left(I_1 - \frac{\bar{\lambda}_0}{\lambda_0}\frac{\lambda - \lambda_0}{\lambda - \bar{\lambda}_0} F_{22}\right)^{-1} F_{21},$$

where $F(\lambda) = \| F_{ij} \|_{i,j=1}^{2}$ is a $(J \oplus J_1)$-unitary extension of the operator $I \mid \mathscr{R}_{A - \bar{\lambda}_0 I} \equiv I_{A,\lambda_0}$ in the $(J \oplus J_1)$-space $\mathscr{H} \oplus \mathscr{H}_1$; $\mathscr{H}_1$ is a certain J_1-space; if $\mathscr{H}_1$ is a Pontryagin space with $\varkappa$ negative squares, then we denote this set by $\Pi_{\lambda_0}^{\varkappa}(A)$.

□ It follows immediately from the definition of the set $\Pi_{\lambda_0}(A)$ that for every λ from a certain neighbourhood of the point λ_0 the function $F(\lambda)$ ($\in \Pi_{\lambda_0}(A)$) is a J-bi-extension of the operator I_{A,λ_0} and the operator-function

$$G(\mu) \equiv F\left(\frac{|\lambda_0|^2(1-\mu)}{\bar{\lambda}_0 - \lambda_0\mu}\right) \in \Pi$$

is holomorphic in a neighbourhood of $\mu = 0$.

Conversely, suppose that for every λ from a certain neighbourhood of the point λ_0 the operator-function $F(\lambda)$, holomorphic in this neighbourhood, is a J-bi-extension of the operator I_{A,λ_0}. Then it belongs to the set $\Pi_{\lambda_0}(A)$. For, in accordance with Theorem 3.10 $G(\mu) \in \Pi$ and therefore

$$F(\lambda) = F_{11} + \frac{\bar{\lambda}_0\lambda - \lambda_0}{\lambda_0\lambda - \bar{\lambda}_0} F_{12}\left(I_1 - \frac{\bar{\lambda}_0}{\lambda_0}\frac{\lambda - \lambda_0}{\lambda - \bar{\lambda}_0} F_{22}\right)^{-1} F_{21},$$

and moreover by virtue of Theorem 3.18 we can regard the $(J \oplus J_1)$-unitary operator $F = \| F_{ij} \|_{i,j=1}^{2} : \mathscr{H} \oplus \mathscr{H}_1 \to \mathscr{H} \oplus \mathscr{H}_1$ as simple. (We note that we can take $\Pi_\varkappa$ as $\mathscr{H}_1$ if and only if $G(\mu) \in \Pi^{\varkappa}$.) We verify that F is an extension of the operator I_{A,λ_0} or, what is the same thing, that $F_{21}x = \theta$ when $x \in \mathscr{R}_{A-\bar{\lambda}_0 I}$. Suppose this is not so. Then there is a vector $x_0 \in \mathscr{R}_{A-\bar{\lambda}_0 I}$ such that $F_{21}x_0 \neq \theta$, and therefore $\mathscr{L} \equiv C\ \mathrm{Lin}\{F^n_{x_0}\}_{-\infty}^{\infty} \cap \mathscr{H}_1 \neq \{\theta\}$ and it is a completely invariant subspace relative to F and F_{22}—and we have obtained a contradiction of the fact that the operator F is simple.

Thus we have proved

Theorem 3.21: *Let $F(\lambda)$ be a function holomorphic in a neighbourhood of the point λ_0. Then $F(\lambda) \in \Pi_{\lambda_0}(A)$ if and only if, for every λ in a certain neighbourhood of the point λ_0, $F(\lambda)$ is a J-bi-extension of the operator I_{A,λ_0}. Moreover the conditions $\{F(\lambda) \in \Pi^{\varkappa}_{\lambda_0}(A)\}$ and*

$$\left\{G(\mu) \equiv F\left(\frac{|\lambda_0|^2(1-\mu)}{\bar{\lambda}_0 - \lambda_0\mu}\right) \in \Pi^{\varkappa}\right\}$$

are equivalent. The operator $F = \| F_{ij} \|_{i,j=1}^2$ generating the function $F(\lambda)$ can be chosen to be simple. ■

5 Let a J-symmetric operator $A \in \mathrm{St}(\mathscr{H}; \lambda_0)$ with $\mathrm{Im}\ \lambda_0 < 0$, let $A_0 \in \mathrm{St}(\mathscr{H}; \lambda_0)$ be a J-bi-extension of the operator A such that $\lambda_0 \in \rho(A_0)$ (on the existence of such an A_0 see Exercise 10 on §2), and let $U_0 = \mathbf{K}_{\lambda_0}(A)$.

Theorem 3.22: *The relations $\mathbf{R}_\lambda = (T_\lambda - \lambda I)^{-1}$ and*

$$T_\lambda = (\lambda_0 - \bar{\lambda}_0)[F(\lambda)U_0(U_0^c U_0)^{-1} - I]^{-1} + \lambda_0 I$$

describe all the generalized resolvents of the J-symmetric operator A in the neighbourhood of the point λ_0 when $F(\lambda)$ traverses the set $\Pi_{\lambda_0}(A)$. In the neighbourhood of the point $\bar{\lambda}_0$ we have

$$\mathbf{R}_\mu = [(T_{\bar{\mu}} - \bar{\mu}I)^{-1}]^c.$$

□ Let $\tilde{A}$ be a $\tilde{J}$-selfadjoint extension of the operator A acting in the $\tilde{J}$-space $\tilde{\mathscr{H}} = \mathscr{H} \oplus \mathscr{H}_1$, and let $\lambda_0 \in \rho(\tilde{A})$. It follows from Exercise 12 on §2 that

$$(\tilde{A} - \lambda_0 \tilde{I})^{-1} = \frac{1}{\lambda - \bar{\lambda}_0}\ [\tilde{I}_{\lambda_0}\tilde{U}_0(\tilde{U}_0^c\tilde{U}_0)^{-1} - I],$$

where $\tilde{U}_0$ is a $\tilde{J}$-bi-extension of the operator $U = \mathbf{K}_{\lambda_0}(A)$ coinciding on $\mathscr{H}$ with U_0, and $\tilde{U}_0 \mid \mathscr{H}_1 = (\bar{\lambda}_0/\lambda_0)I_1$, and $\tilde{I}_{\lambda_0}$ is the corresponding $\tilde{J}$-bi-extension of the operator I_{A,λ_0}. In accordance with Hilbert's theorem for the resolvent

$$\mathbf{R}_\lambda = \frac{1}{\lambda - \bar{\lambda}_0}\ P\left[\tilde{I} - \frac{\lambda - \lambda_0}{\lambda - \bar{\lambda}_0}\ (\tilde{I}_{\lambda_0}\tilde{U}_0(\tilde{U}_0^c\tilde{U}_0)^{-1} - \tilde{I}\right]^{-1} [\tilde{I}_{\lambda_0}\tilde{U}_0(\tilde{U}_0^c\tilde{U}_0)^{-1}\tilde{I}] \mid \mathscr{H}.$$

Let $\tilde{I}_{\lambda_0} = \| F_{ij} \|_{i,j=1}^2$ be the matrix representation of the operator $\tilde{I}_{\lambda_0}$. Then the operator $\tilde{U} \equiv \tilde{I}_{\lambda_0}\tilde{U}_0(\tilde{U}_0^c\tilde{U}_0)^{-1}$ has the matrix representation

$$\tilde{U} = \| \tilde{U}_{ij} \|_{i,j=1}^2, \text{ where } \tilde{U}_{i1} = F_{i1}U_0(U_0^cU_0)^{-1}, \text{ and } \tilde{U}_{i2} = \frac{\bar{\lambda}_0}{\lambda_0}\ F_{i2}, \qquad i = 1, 2.$$

Therefore in a neighbourhood of the point λ_0 Hilbert's identity given above can be rewritten in the form

$$\mathbf{R}_\lambda = \left[I - (\lambda - \lambda_0)\left(\frac{F(\lambda)U_0(U_0^c U_0)^{-1} - I}{\lambda_0 - \bar{\lambda}_0}\right)\right]^{-1} \frac{F(\lambda)U_0(U_0^c U_0)^{-1} - I}{\lambda_0 - \bar{\lambda}_0},$$

where

$$F(\lambda) = F_{11} + \frac{\bar{\lambda}_0}{\lambda_0}\frac{\lambda - \lambda_0}{\lambda - \bar{\lambda}_0} F_{12}\left(I - \frac{\bar{\lambda}_0}{\lambda_0}\frac{\lambda - \lambda_0}{\lambda - \bar{\lambda}_0} F_{22}\right)^{-1} F_{21}$$

is an operator-function holomorphic in a neighbourhood of the point λ_0 which is, for every λ from this neighbourhood, a J-bi-extension of the operator I_{A,λ_0}, i.e., $F(\lambda) \in \Pi_{\lambda_0}(A)$ by virtue of Theorem 3.21. In this same neighbourhood of the point λ_0 the operators $F(\lambda)U_0(U_0^c U_0)^{-1}$ are J-bi-extensions of the operator $U = \mathbf{K}_{\lambda_0}(A)$, and therefore it follows from $1 \in \sigma_c(U)$ that $1 \in \sigma_c(F(\lambda)U_0(U_0^c U_0)^{-1})$. Hence, $\mathbf{R}_\lambda = (T_\lambda - \lambda I)^{-1}$, where $T_\lambda = (\lambda_0 - \bar{\lambda}_0)[F(\lambda)U_0(U_0^c U_0)^{-1} - I]^{-1} + \lambda_0 I$.

Conversely, let $F(\lambda) \in \Pi_{\lambda_0}(A)$. Then $F(\lambda)U_0(U_0^c U_0)^{-1}$ is an operator-function holomorphic in a neighbourhood of the point λ_0 which, by virtue of Theorem 3.21 is a J-bi-extension of the operator $U = \mathbf{K}_{\lambda_0}(A)$ for all λ from this neighbourhood. Therefore

$$T(\mu) = F\left(\frac{|\lambda_0|^2(1-\mu)}{\bar{\lambda}_0 - \lambda_0 \mu}\right) U_0(U_0^c U_0)^{-1}$$

is a function holomorphic in a neighbourhood of the point $\mu = 0$, and therefore there is a J_1-space $\mathscr{H}_1$ and a simple $\tilde{J}$-unitary operator $\tilde{U} = \| U_{ij}\|_{i,j=1}^2 : \mathscr{H} \oplus \mathscr{H}_1 \to \mathscr{H} \oplus \mathscr{H}_1$, with $\tilde{J} = J \oplus J_1$, such that $\tilde{U}$ generates the function $T(\mu)$. It follows from the result given later in Exercise 11 that we can take as $\tilde{U}$ an extension of the operator U. Therefore by virtue of Exercise 16 on §1 $\tilde{U} = \tilde{I}_{\lambda_0}\tilde{U}_0(\tilde{U}_0^c\tilde{U}_0)^{-1}$, where $\tilde{U}_0$ is a $\tilde{J}$-bi-extension of the operator U coinciding on $\mathscr{H}$ with U_0, and $\tilde{U}_0 \mid \mathscr{H}_1 = (\bar{\lambda}_0/\lambda_0)I_1$, and $\tilde{I}_{\lambda_0}$ is the corresponding $\tilde{J}$-bi-extension of the operator I_{A,λ_0}. Since $U \subset \tilde{U}$ and $1 \in \sigma_c(U)$, it follows, because the operator $\tilde{U}$ is simple, that $1 \in \sigma_c(\tilde{U})$, i.e., there is a $\tilde{J}$-selfadjoint operator $\tilde{A}$ such that $\tilde{U} = \mathbf{K}_{\lambda_0}(\tilde{A})$. It follows from Exercise 15 on II§6 that $A \subset \tilde{A}$, and from the proof of the first theorem that $\mathbf{R}_\lambda = (T_\lambda - \lambda I)^{-1}$, where

$$\begin{aligned} T_\lambda &= (\lambda_0 - \bar{\lambda}_0)\left[T\left(\frac{\bar{\lambda}_0}{\lambda_0} \cdot \frac{\lambda - \lambda_0}{\lambda - \bar{\lambda}_0}\right) - I\right]^{-1} + \lambda_0 I \\ &= (\lambda_0 - \bar{\lambda}_0)[F(\lambda)U_0(U_0^c U_0)^{-1} - I]^{-1} + \lambda_0 I. \end{aligned}$$

It then follows from the definition of the generalized resolvent that $\mathbf{R}_\mu = (\mathbf{R}_{\bar{\mu}})^c = [(T_{\bar{\mu}} - \bar{\mu} I)^{-1}]^c$ in a neighbourhood of the point $\bar{\lambda}_0$. ■

Corollary 3.23: *Under the conditions of Theorem 3.22 let A_0 be a maximal J-dissipative J-symmetric operator. Then the function $F(\lambda) \in \Pi_{\lambda_0}^\varkappa(A)$ generates a $\varkappa$-regular generalized resolvent, and conversely, every $\varkappa$-regular generalized resolvent is generated by some function from $\Pi_{\lambda_0}^\varkappa(A)$.*

□ Let the function $F(\lambda) \in \Pi^{\varkappa}_{\lambda_0}(A)$. Then

$$T(\mu) = F\left(\frac{|\lambda_0|^2(1-\mu)}{\bar{\lambda}_0 - \lambda_0\mu}\right) U_0 \in \Pi^{\varkappa},$$

and the operator $\tilde{U}$ appearing in the second part of the proof of Theorem 3.22 can be chosen to be a $\varkappa$-regular extension of an operator $U = \mathbf{K}_{\lambda_0}(A)$. Therefore the generalized resolvent constructed there will be $\varkappa$-regular.

Conversely, let $\mathbf{R}_\lambda$ be a $\varkappa$-regular generalized resolvent generated by a $\varkappa$-regular extension $\tilde{A}: \mathscr{H} \oplus \Pi_\varkappa \to \mathscr{H} \oplus \Pi_\varkappa$ of the operator A. Then $\tilde{U} = \mathbf{K}_{\lambda_0}(\tilde{A}) = \mathbf{I}_{\lambda_0}\tilde{U}_0$, where U_0 is a $\tilde{J}$-bi-non-expansive $\tilde{J}$-semi-unitary extension of the operator U coinciding with U_0 on $\mathscr{H}$, and $\tilde{U}_0 x = (\bar{\lambda}_0/\lambda_0)x$ when $x \in \Pi_\varkappa$. Consequently $\tilde{I}_{\lambda_0}\tilde{U}_0\tilde{U}_0^c$ is a $\tilde{J}$-bi-non-expansive extension of the operator I_{A,λ_0} and therefore, by virtue of Corollary 3.5 and Lemma 3.8, the function generated by this operator belongs to $\Pi^{\varkappa}_{\lambda_0}(A)$. ■

Corollary 3.24: *Under the conditions of Corollary 3.23 let $J = I$, $\varkappa = 0$. Then the function $(T_\lambda - \lambda I)^{-1}$ admits a holomorphic extension from the neighbourhood of λ_0 on to $\mathbb{C}^-$. On $\mathbb{C}^+$ we put $(T_{\bar{\lambda}} - \bar{\lambda} I)^{-1} = [(T_\lambda - \lambda I)^{-1}]^c$. This extension coincides in $\mathbb{C}^+ \cup \mathbb{C}^-$ with the corresponding regular generalized resolvent of the operator A.*

□ By virtue of the result of Exercise 12 on §2

$$T_\lambda = (\lambda_0 - \bar{\lambda}_0)(F(\lambda)U_0 - I)^{-1} + \lambda_0 I,$$

and by virtue of Corollary 3.23 we can suppose that $F(\lambda) \in \Pi^0_{\lambda_0}(A)$, i.e., there is a Hilbert space $\mathscr{H}_1$ and a unitary extension $F = \|F_{ij}\|^2_{i,j=1}: \mathscr{H} \oplus \mathscr{H}_1 \to \mathscr{H} \oplus \mathscr{H}_1$ of the operator I_{A,λ_0} such that

$$F(\lambda) = F_{11} + \frac{\bar{\lambda}_0}{\lambda_0} \cdot \frac{\lambda - \lambda_0}{\lambda - \bar{\lambda}_0} F_{12}\left(I_1 - \frac{\bar{\lambda}_0}{\lambda_0} \cdot \frac{\lambda - \lambda_0}{\lambda - \bar{\lambda}_0} F_{22}\right)^{-1} F_{21}.$$

Since

$$\| F_{22} \| \leqslant \text{ and } \left| \frac{\bar{\lambda}_0}{\lambda_0} \cdot \frac{\lambda - \lambda_0}{\lambda - \bar{\lambda}_0} \right| < 1,$$

it follows that $F(\lambda)$ admits a holomorphic extension from the neighbourhood of the point λ_0 on to $\mathbb{C}^-$. Moreover the extenson of the function

$$G(\mu) = F\left(\frac{|\lambda_0|^2(1-\mu)}{\bar{\lambda}_0 - \lambda_0\mu}\right) \in \Pi^0(|\mu| < 1, \text{ and therefore } \|F(\lambda)\| \leqslant 1),$$

which, by virtue of 2.Remark 6.13, implies that the operators T_λ are maximally dissipative for all $\lambda \in \mathbb{C}^-$. Therefore $\mathbb{C}^- \subset \rho(T_\lambda)$ (see 2.Lemma 2.8), and therefore $(T_\lambda - \lambda I)^{-1}$ admits holomorphic extension from the vicinity of the point λ_0 on to $\mathbb{C}^-$. On $\mathbb{C}^+$ we put $(T_{\bar{\lambda}} - \bar{\lambda} I)^{-1} = [(T_\lambda - \lambda I)^{-1}]^c$.

On the other hand, the regular generalized resolvent $\mathbf{R}_\lambda$ of the operator A

defined on $\mathbb{C}^+ \cup \mathbb{C}^-$ does coincide in a neighbourhood of the point λ_0 with the function $(T_\lambda - \lambda I)^{-1}$, and therefore $\mathbf{R}_\lambda = (T_\lambda - \lambda I)^{-1}$ when $\lambda \in \mathbb{C}^-$ and $\mathbf{R}_\mu = [(T_{\bar\mu} - \bar\mu I)^{-1}]^c$ when $\mu \in \mathbb{C}^+$. ■

Exercises and problems

1 Let $\mathcal{H}_1$ and $\mathcal{H}_2$ be Hilbert spaces, and let $T: \mathcal{H}_1 \to \mathcal{H}_2$ be a bounded operator with $\mathcal{D}_T = \mathcal{H}_1$. In [XXIII] an operator $\tilde{T}: \mathcal{H}_1 \oplus \mathcal{H}_3 \to \mathcal{H}_1 \oplus \mathcal{H}_3$ with $\mathcal{D}_{\tilde{T}} = \mathcal{H}_1 \oplus \mathcal{H}_3$ is called a dilatation of the operator T if $P_1\tilde{T}^n | \mathcal{H}_1 = T^n$ $(n = 0, 1, 2, 3, \ldots)$, where P_1 is the orthoprojector from $\mathcal{H}_1 \oplus \mathcal{H}_3$ on to $\mathcal{H}_1$. Prove that this definition and Definition 3.1 are equivalent when $\mathcal{H}_1 = \mathcal{H}_2$.

2 Let T and $\tilde{T}$ be the same operators as in Exercise 1. Prove that the operator $\tilde{T}$ will be a dilatation of the operator T if and only if $P_1(\tilde{T} - \lambda\tilde{I})^{-n} | \mathcal{H}_1 = (T - \lambda I)^{-n}$, $n = 1, 2, \ldots$, $\lambda \in \rho(T) \cap \rho(\tilde{T})$ (see [XXIII]).

3 Suppose that $F(\lambda)$ is a function holomorphic in a neighbourhood of zero with values in a set of continuous operators acting in a J-space $\mathcal{H}$, and that there are: a space $\Pi_\varkappa$ with $\varkappa$ negative squares; a continuous operator $\Gamma: \mathcal{H} \to \Pi_\varkappa$; a π-semi-unitary operator $V^{-1}: \Pi_\varkappa \to \Pi_\varkappa$; and a J-selfadjoint operator S with $\mathcal{D}_S = \mathcal{H}$; such that $F(\lambda) = iS + \Gamma^c(V + \lambda I)(V - \lambda I)^{-1}\Gamma$, $|\lambda| < 1$, $\lambda \notin \sigma(V)$. Prove that the kernel $K(\lambda, \mu) = J[F(\lambda) + F^c(\mu)]/(1 - \bar\lambda\mu)$ has not more than $\varkappa$ negative squares (*cf.* M. Krein and Langer [4]).

4 Let $T(\lambda) \equiv T_0: \mathcal{H}_1 \to \mathcal{H}_2$ be an operator-function holomorphic in a neighbourhood of zero. Prove that an operator $T: \mathcal{H}_1 \oplus \mathcal{H}_0 \to \mathcal{H}_2 + \mathcal{H}_0$ will generate the function $T(\lambda)$ if and only if it is a dilatation of the operator T_0 (Azizov).

5 Prove that an operator T_0 admits a $\varkappa$-regular dilatation if and only if T_0 is a (J_1, J_2)-bi-non-expansive operator (Azizov [14]).
Hint: Use the equivalence of assertions a) and b) in Theorem 3.16 and the result of Exercise 4.

6 Give an example of an operator-function $T(\lambda): \mathcal{H}_1 \to \mathcal{H}_2$, holomorphic in a neighbourhood of zero, such that the kernel $[J_1 - T^*(\mu) J_2 T(\lambda)]/(1 - \bar\mu\lambda)$ has $\varkappa$ negative squares but nevertheless $T(\lambda) \notin \Pi^\varkappa$.
Hint: Put $T(\lambda) \equiv T_0$, where T_0 is a (J_1, J_2)-semi-unitary operator which is not (J_1, J_2)-bi-non-expansive, and use either the results of Exercises 4 and 5, or the equivalence of assertions a) and b) in Theorem 3.16.

7 Prove that under the conditions of Theorem 3.16 the function $T^c(\lambda) \in \Pi^\varkappa$.
Hint: Verify that if the operator T generates the function $T(\lambda)$, then T^c generates the function $T^c(\lambda)$.

8 Let $T(\mu): \mathcal{H}_1 \to \mathcal{H}_2$ be an operator-function holomorphic in a neighbourhood of the point $\mu = 0$ with values in a set of continuous operators ($\mathcal{D}_{T(\mu)} = \mathcal{H}_1$) acting from the J_1-space $\mathcal{H}_1$ into the J_2-space $\mathcal{H}_2$. Prove that the following assertions are equivalent:
a) the kernels

$$\frac{J_1 - T^*(\mu) J_2 T(\lambda)}{1 - \bar\mu\lambda} \quad \text{and} \quad \frac{J_2 - T(\lambda) J_1 T^*(\mu)}{1 - \bar\mu\lambda}$$

non-negative;

b) in a neighbourhood of zero the $T(\lambda)$ are (J_1, J_2)-bi-non-expansive and the function $\omega^+(T(\lambda)) \equiv T_1(\lambda)$ admits holomorphic extension on to the disc $\{\lambda \mid |\lambda| < 1\}$ and $\| T_1(\lambda)\| \leqslant 1$;
c) the kernel $[I_1 - T_1^*(\mu)T_1(\lambda)]/1 - \bar{\mu}\lambda$ is non-negative (*cf.* Arov [1]).
Hint: Verify that if $T = \| T_{ij}\|_{i,j=2}^2 : \mathscr{H}_1 \oplus \mathscr{H} \to \mathscr{H}_2 \oplus \mathscr{H}$ is a $(\tilde{J}_1, \tilde{J}_2)$-unitary operator ($\tilde{J}_i = J_i + I, i = 1, 2$) generating the function $T(\lambda)$, then $\omega^+(T)$ generates $T_1(\lambda)$, and conversely.

9 Suppose that under the conditions of Theorem 3.16 $J_i = I_i, 1 = 1, 2,$ and that $T(\mu)$ satisfies at least one of the assertions a)–c) of this theorem. Prove that then there is a $\pi_\varkappa$-space $\Pi_\varkappa$ and a $(I_1 \oplus \bar{\pi}_\varkappa, I_2 \oplus \bar{\pi}_\varkappa)$-unitary operator $T = \| T_{ij}\|_{i,j=1}^2 : \mathscr{H}_1 \oplus \Pi_\varkappa \to \mathscr{H}_2 \oplus \Pi_\varkappa$ generating $T(\mu)$ such that T_{22} has no neutral eigenvalues (Azizov).
Hint: Use Theorem 3.18.

10 Under the conditions of Problem 9 suppose $J_1 = I_1$, $J_2 = I_2$, and $T(\mu) \in \Pi_\varkappa$. Prove that then $T(\mu)$ admits holomorphic extension on to the disc $\{\lambda \mid |\lambda| < 1\}$ with the exception, possibly of not more than $\varkappa$ points which will be the poles of this extension (M. Krein and Langer [6]).
Hint: Use the result of Problem 9 and prove that the function $T(\mu)$ can be expressed in the form of a product $T(\mu) = T_1(\mu)T_2(\mu)$, where $T_1(\mu) \in \Pi^0$, and $T_1(\mu) \in \Pi^\varkappa$ and is generated by a negative $\pi_\varkappa$-space $\Pi_\varkappa$ with $\dim \Pi_\varkappa = \varkappa$.

11 Let $T(\mu)$ be an operator-function, holomorphic in a neighbourhood of the point $\mu = 0$, which is a J-bi-extension of a J-isometric operator U for every μ from a certain neighbourhood of the point $\mu = 0$. Prove that among the simple $\tilde{J}$-unitary operators generating the function $T(\mu)$ there is at least one which is an extension of the operator U (Azizov).
Hint: Use the methods of proof of Lemma 3.9, Theorem 3.4 and Lemma 3.8.

Remarks and bibliographic indications on chapter V

§1 The definition of the Potapov–Ginzburg transformation in so general a form was introduced essentially by Shmul'yan [5] (*cf.* Arov [1]). The traditional *PG*-transformation when $\mathscr{H}_1 = \mathscr{H}_2$, $J_1 = J_2$ was first introduced and studied by Potapov [1] and later Ginzburg [1] continued the investigation. The works of I. Iokhvidov [14], [18], E. Iokhvidov and I. Iokhvidov [1], Shmul'yan [4], [5], Ritzner [4] are devoted to the *PG*-transformation. The exposition in the text, sometimes new in method, is due to Azizov, although almost all the propositions can be found in parts in the works of the authors just mentioned. We mention, in particular, that Theorem 1.14 was obtained independently and simultaneously by Azizov, E. Iokhvidov and I. Iokvidov [1] and Ritsner [4].

§2 The problems of J-isometric extensions of J-isometric operators in the finite-dimensional case were taken up by Potapov [1], and in the case of $\Pi_\varkappa$ by I. Iokhvidov [15], [16] (see [XIV]; later M. Krein and Shmul'yan [3], E. Iokhvidov [2], [4], [5], Azizov [12], [14] studied these problems for operators acting in Krein spaces. In the work of M. Krein and Shmul'yan [3] $\mathscr{D}_V$ and $\mathscr{R}_V$ are regular subspaces, and in the articles of E. Iokhvidov [2], [4], [5] $\dim \mathscr{D}_V^{[\perp]} < \infty$ and $\dim \mathscr{R}_V^{[\perp]} < \infty$. The latter built his investigations on systematic use of *PG*-transformations (see, e.g., Exercises 5, 6 on §1). With due regard to the above remarks the remaining results in this section and the method used are due to Azizov.

§3.1 The concept of a dilatation on operators acting from one space into another was first, it appears, carried over and applied by Azizov [14] (Definition 3.1). The remaining results of this paragraph in the case when the operator acts in a single space are well-known (see Davis [1], Davis and Foras [1], Saktinovich [1], Kuzhel' [8], [9]. In our approach these results were proved by Azizov [14], and Corollary 3.5 we find, essentially in Arov [1].

§3.2 We find the functions $T(\mu)$ in Definition 3.6 in Arov [1], for example, where they are called 'transfer functions'. Lemma 3.8 and Theorem 3.10 were proved by Azizov [14], and Lemma 3.5 we find, essentially, in Arov [1].

§3.3 The investigations in this paragraph, carried out by Azizov [14], were stimulated by a series of papers by M. Krein and Langer [3]–[6]. In such a general formulation for Krein spaces neither Theorem 3.12 nor Lemma 3.14 are given by these authors, but our exposition follows their basic ideas, and often repeats their arguments word for word. Lemma 3.15 is due to Azizov. Theorem 3.16 is also due to Azizov [14]. Partially and in particular cases it is encountered in other authors. We single out the series of papers mentioned above of Krein and Langer, and Arov's article, which initiated the appearance of this result. Theorem 3.18 is due to Azizov [14].

§3.4 The set $\Pi_{\lambda_0}(A)$ was introduced into the investigation by Azizov. Theorem 3.21 is also due to him (see Azizov [14]).

§3.5 Theorem 3.22 is due to Azizov [14]. The form of writing the generalized resolvent $\mathbf{R}_\lambda = (T_\lambda - \lambda I)^{-1}$ was first introduced by *A* Strauss [1]. Moreover we have used M. Krein's idea (see [I]) of all the generalized resolvents by means of a fixed extension of the operator. For more detail about other approaches to the description of generalized resolvents and historical information see [IV] and also Etkin [1], [2]. Corollary 3.23 is due to Azizov [14], and Corollary 3.24 to A. Shtraus [1]; we point out that the proof in the text, due to Azizov, is not based on A. Shtraus's result.

REFERENCES

1: Monographs, Textbooks, Lectures, Survey Articles

[I] Akhiezer, N. I., and Glazman, I. M., *Theory of linear operators in Hilbert space*, 3rd Ed., *Vishcha Shkola*, Khar'kov, 1978. (English translation, Pitman. London, 1981.)

[II] Ando, T., *Linear Operators on Krein Spaces*, Sapporo, Japan, 1979.

[III] Azizov, T. Ya., and Iokhvidov, I. *Linear operators in Hilbert spaces with a G-metric*. (Russ.), *Uspekhi Mat. Nauk*, 1971, **26**(4).

[IV] Azizov, T. Ya., and Iokhvidov, I. *Linear operators in spaces with an indefinite metric* (Russ.). In the book *Mathematical Analysis*, **17** (Itogi nauka i tekhniki), 113–205, Moscow, 1979.

[V] Bognar, J. *Indefinite Inner-product Spaces*, Springer, Berlin, 1974.

[VI] Daletskiy, Yu. L., and Krein, M. G., *Stable Solutions of Differential Equations in a Banach Space* (Russ.), Nauka, Moscow, 1970.

[VII] Dunford, N., and Schwarz, J. T. *Linear Operators. General Theory*, Wiley-Interscience, New York, 1958.

[VIII] Ginzburg, Yu. P., and Iokhvidov, I. S., *Investigation by geometry of infinite-dimensional spaces with a bilinear metric* (Russ.), *Uspekhi Mat. Nauk*, 1962, **17**(4), 3–56.

[IX] Glicksberg, I. L., *A further generalization of the Kakutani fixed-point theorem, with application to Nash equilibrium points*, *Proc. Amer. Math. Soc.*, 1952, **3**.

[X] Gokhberg, I. Ts., and Krein, M. G., *Basic propositions on defect numbers, root numbers, and indices of linear operators* (Russ.), *Uspekhi Mat. Nauk*, 1957, **12**(2), 43–118. English translation. *Amer. Math. Soc. Translations* (2), **13**, 1960. 185–266.

[XI] Gokhberg, I. Ts., and Krein, M. G., *Introduction to the theory of linear non-selfadjoint operators in Hilbert space* (Russ.), Nauka, Moscow, 1965.

[XII] Greenleaf, F. P., *Invariant Means on Topological Groups and their Applications*, Van Nostrand-Reinhold Co., New York, 1970.

[XIII] Hille, E., and Phillips, R. S., *Functional analysis and semi-groups. American Math. Soc. Coll. Publications*, **31**, 1957.

[XIV] Iokhvidov, I. S., and Krein, M. G., *Spectral theory of operators in spaces with an indefinite metric, I* (Russ.), *Trudy Moskov, Mat. Obshch.*, 1965, **5**, 367–432; *Amer. Math. Soc. Translations* (2), 1960, 105–75.

[XV] Iokhvidov, I. S., and Krein, M. G., *Spectral theory of operators in spaces with an indefinite metric, II* (Russ.), *Trudy Moskov, Mat. Obshch*, 1969, **8**, 413–96; *Amer. Math. Soc. Translations* (2), **34**, 283–373.

[XVI] Iokhvidov, I. S., Krein, M. G., and Langer, H., *Introduction to the Spectral Theory of Operators in Spaces with an Indefinite Metic*, Berlin, Akademie-Verlag, 1982.

[XVII] Krein, M. G., *Introduction to the geometry of indefinite J-spaces and the theory of operators in those spaces* (Russ.). In the book *Second Summer Mathematical School*, I, Kiev, 1965, 15–92.

[XVIII] Krein, S. G., *Linear Differential Equations in a Banach Space* (Russ.), Nauka, Moscow, 1968.

[XIX] Krein, S. G., *Linear Differential Equations in a Banach Space* (Russ.), Nauka, Moscow, 1971.

[XX] Langer, H., *Spectral functions of definitizable operators in Krein spaces*, Lecture Notes in Mathematics, 1982, No. 948, 1–46.

[XXI] Nagy, K. L., *State Vector Spaces with Indefinite Metric in Quantum Field Theory*, Noordhoff, Groningen, 1966.

[XXII] Riesz, F. and Szekefaldi-Nagy, B., *Functional Analysis*, Ungar, New York, 1955.

[XXII] Szekefaldi-Nagy, B., and Foias, C., *Harmonic Analysis of Operators in a Hilbert Space* Akad. Kaido, Budapest, 1967.

2 Articles in Journals

Akopyan, R. V.

[1] On the formula for traces in the theory of perturbations for J-non-negative operators (Russ.), *Dokladi Akad. Nauk, Arm. SSR*, 1973, **57**(4), 193–9.

[2] On the theory of perturbations of a *J*-positive operator (Russ.), *Functional analysis and its applications*, 1975, **9**(2), 61–2.

[3] On the theory of the spectral function of a *J*-non-negative operator, *Izvestiaya Akad. Nauk, Arm. SSR. Series phys.-mat. nauk*. 1978, **13**(2).

[4] On the formula for traces of *J*-non-negative operators with nuclear perturbations. *Doklady Akad. Nauk, Arm. SSR*, 1983, **77** (5).

Aronszajn, N.

[1] Quadratic forms on vector spaces, *Proceedings Internat. Sympos. Linear Spaces*, Jerusalem, 1960, Acad. Press, Oxford, 29–87.

Arov, D. E.

[1] Passive linear stationary dynamic systems (Russ.), *Siber. Mat. Zhurnal*, 1979, **20**(2), 211–28.

Askerov, N. K., Krein, S. G., and Langer, G. I.

[1] A problem on oscillations of a viscous liquid and the operator equations connected with it (Russ.), *Functional analysis and its applications*, 1968, **2**(1), 21–31.

Azizov, T. Ya.

[1] Criterion for non-degeneracy of a closed linear envelope of root vectors of *J*-disipative completely continuous operators in a Pontryagin space Π_κ, Collection of Candidates' papers, Math. Faculty Voronesh Univ, 1971, ed. **1**, 1–5. (Russ.).

[2] On the spectra of certain classes of operators in Hilbert space. (Russ.), *Mat. Zarnetki*, 1971, **9**(3), 303–310.

[3] On the decomposition of a normed space into the direct or approximate sum of special subspaces (Russ.), *Trudy N. I. I. Mat, Voronezh Univ*, 1971, ed. **3**, 1–3.

[4] Invariant subspaces and criteria for completeness of a system of root vectors of *J*-dissipative operators in a Pontryagin space $\Pi_\varkappa$ (Russ.), *Doklady Akad. Nauk, USSR*, 1971, **200**(5), 1015–17. (English translation, *Soviet Math. Dokl.*, **12**(7), 1513–14.)

[5] Dissipative operators in a space with an indefinite metric (Russ.), *Candidate's dissert., Voronezh*, 1972.

[6] On invariant subspaces of a family of commutating operators in a Pontryagin space $\Pi_\varkappa$ (Russ.), *Functional analysis and its applications*, 1972, **6**(3), 59–60.

[7] On completely continuous operators selfadjoint relative to an indefinite metric. (Russ.), *Mat. Issledovaniya*, **)** (4), Kishinev, *Akad. Nauk MSSR*., 237–40.

[8] Dissipative operators in a Hilbert space with an indefinite metric (Russ.), *Izvestiaya Akad. Nauk. USSR, ser. mat*., 1973, **37**(3). (English translation, *Math. USSR Icv.*, **7**, 639–60.)

[9] On invariant subspaces of commutative families of operators in a space with an indefinite metric (Russ.), *Ukrainsk. mat. zhurnal*, 1976, **278**(3), 293–9.

[10] On uniformly *J*-expansive operators and focusing plus-operators (Russ.). In the book: *Shkola po teorii operatorov v funktsional'nykh prostranstvakh*, Minsk, 1978.

[11] Parametric representation of operators acting in a space with an indefinite metric (Russ.), *Dep. VINITI*, 1979, 1138–79.

[12] On extensions of *J*-isometric operators (Russ.). In the book: XVI *Voronezh Summer School in Mathematics*, Dep. VINITI, 1981, 5691–81, 21.

[13] On the completeness and basicity of a system of eigenvectors and associated vectors of *J*-selfadjoint operators of the class **J**(**H**) (Russ.), *Doklady Akad. Nauk, USSR*, 1980, **253** (5), 1033–36.

[14] On the theory of extensions of isometric and symmetric operators in spaces with an indefinite metric (Russ.), Dep, VINITI, 1982, 3420–82, 29.

[15] On a class of analytic operator-functions (Russ.). In the book: VIII *Shkola po teorii operatorov v funktsional'ngkh prostranstvakh*, Riga, 1983, 7–8.

Azizov, T. Ya., Gordienko, N. A., and Iokhvidov, I. S.

[1] On J-unitary groups generated by *J*-selfadjoint operators of the classes (**H**) and **K**(**H**) (Russ.), Dep. VINITI, 1980, 4213–80.

Azizov, T. Ya and Dragileva, M. D.

[1] On invariant subspaces of operator-commutative families of operators in a space with an indefinite metric (Russ.), *Trudy ALLI. mat. Voronezh Univ*., 1975, ed. **17**, 3–7.

Azizov, T. Ya, and Iokhvidov, E. J.

[1] On invariant subspaces of maximal *J*-dissipative operators (Russ.), *Mat. Zametki*, 1972, **12** (6), 747–754. (English translation, *Math. Notes*. **12** 886–9.)

Azizov, T. Ya, Iokhvidov, E. I., and Iokhvidov, I. S.

[1] On the connection between the Cayley–Neyman and the Potapov–Ginzburg transformations (Russ.). In the book: *Functional Analysis, Linear Spaces*, Ul'yanovsk, 1983, 3–8.

Azizov, T. Ya., and Iokhvidov, I. S.

[1] A criterion for completeness and basicity of root vectors of a completely continuous *J*-selfadjoint operator in a Pontryagin space Π_κ (Russ.), *Mat. Issledovaniya*, **6**, Kishinev, Akad. Nauk Mold, SSR, 1971.

Azizov, T. Ga, Iokhvidov, I. S., and Shtraus, V. A.

[1] On normally soluble extensions of a closed symmetric operator (Russ.), *Mat. Issledovaniya*, **7**(1), Kishinev, Akad. Nauk Mold. SSR, 1972.

Azizov, T. Ya., and Kondras, T. V.

[1] Criteria for the existence of maximal definite invariant subspaces for operators in a space with an indefinite metric. (Russ.). Dep. VINITI, 1977, No. 4207–77.5.

Azizov, T. Ya., and Kuznetsova. N. N.

[1] On *p*-bases in a Pontryagin space (Russ.). *Uspekhi Mat. Nauk*. 1983, **38**, (1).

Azizov, T. Ya., and Usvyatsova, E. B.

[1] Existence of special non-negative subspace of *J*-dissipative operators. (Russ.), *Siber. Mat. Zhurnal*, 1977, **18**(1).

[2] On completeness and basicity of eigenvectors and associated vectors of the class **K(H)**, *Ukrainsk. Mat. Zhurnal*, 1978, **30**(1), 86–8.

[3] On the basicity of a system of root vectors of a *J*-selfadjoint operator generated by a pencil $L(\lambda) = I - \lambda G - \lambda^{-1}H$ (Russ.), Dep. VINITI, 1980, No. 962–80, 6.

Azizov, T. Ya., and Khoroshavin, S. A.

[1] On invariant subspaces of operators acting in a space with an indefinite metric (Russ.), *Funktion analiz i ego pril.*, 1980, **14**(4), 1–7.

Azizov, T. Ya., and Shlyakman, M. Ya.

[1] On the Kühne–Iokhvidov–Shtraus effect (Russ.), *Ukrainsk. mat zhurnal*, **35** (4), 484–5.

Azizov, T. Ya., and Shtraus, V. A.

[1] On the completeness of a system of root vectors of a definitizable operator in a Banach space with an Hermitian form (Russ.), *Trudy N.I.I. mat.*, Voronezh Univ., 1972, ed. **5**, 1–6.

Bayasgalan, Ts.

[1] On the existence of a positive square-root of an operator positive relative to an indefinite metric (Russ.), *Stadia Sci. Math. Hungary*, 1980, **15**, 367–79.

Bennevitz, Ch.

[1] Symmetric relations on a Hilbert space, *Lect. Notes Math*, 1972, No. 280. 212–18.

Bognár, J.

[1] On the existence of a square-root of an operator self-adjoint relative to an indefinite metric (Russ.; English summary), *Magyar Tud. Acad. Kutató Int. Köze*, 1961. **6**(3), 351–63.

[2] On an occurrence of discontinuity of a scalar product in spaces with an indefinite metric (Russ.), *Uspekhi Mat. Nauk*, 1962, **17**, 157–9.

[3] A remark on dually strict plus-operator (Russ.), *Mat. Issledovaniya*, 1973, **8**(1) (27), 217–19, Kishinev.

[4] Non-degenerate inner-product spaces spanned by two neutral subspaces, *Studia Sci. Math. Hungary*, 1978, **13**, 463–8.

[5] A proof of the spectral theorem for *J*-positive operators, *Acta Sci. Math.*, 1983, **15**(1–2), 75–80.

Brodskiy, M. L.

[1] On the properties of an operator mapping into itself the non-negative part of a space with an indefinite metric (Russ.), *Uspekhi Mat. Nauk*, 1959, **14**(1), 147–52.

Davis, Ch.

[1] *J*-unitary dilation of general operators, *Acta Sci. Math.*, 1971, **32**, 127–99.

Davis, Ch. and Foiaş, C.

[1] Operators with bounded characterstic function and their *J*-unitary dilation, *Acta. Sci. Math.*, 1971, **32**, 127–9.

Derguzov, V. I.

[1] On the stability of solutions of Hamiltonian equations with unbounded periodic operator coefficients (Russ.), *Mat. Sbornik*, 1964, **63**, 591–619.

[2] Necessary conditions for strong stability of Hamiltonian equations with unbounded periodic operator coefficients (Russ.), Vestnik Leningr. State Univ., 1964 (**19**), 18–30.

[3] Sufficient conditions for stability of Hamiltonian equations with bounded operator coefficients (Russ.), *Mat. Sbornik*, 1964, **64**(3), 419–35.

[4] Domain of stability of linear Hamiltonian equations with periodic operator coefficients (Russ.), *Vestnik Leningrad. State University*, 1969, **13**, 20–30.

Ektov, Yu. S.

[1] *J*-non-negative completely continuous operators (Russ.), *Trudy N.I.I. Voronezh Univ.*, 1975, ed. **17**, 76–86.

[2] Completely continuous *J*-non-negative operators in a *J*-space with two norms, *Trudy, N.I.I. Mat. Voronezh Univ.*, 1975, ed. 20, 57–64.

Etkin, A. E.

[1] Generalized resolvents of a symmetric operator in a $\Pi_\varkappa$-space (Russ.), *Funktional'nyy analiz. Ul'yanovsk*, 1981, ed. 16.

[2] *Generalized resolvents of a symmetric operator in a $\Pi_\varkappa$-space* (Russ.), In the book: *Funktional'ny analiz. Lineynye prostranstva*, Ul'yanovsk, 1983, 149–57.

Ginzburg, Yu. P.

[1] On *J*-non-expansive operator-functions (Russ.), *Doklady Akad. Nauk USSR*, 1957, **117**, (2), 171–3.

[2] On *J*-non-expansive operators in a Hilbert space (Russ.), *Nauch. zap. fiz-mat. fac. Odessa Gosud. pedag. inst.*, 1958, **22**(1).

[3] On projection in a Hilbert space with a bilinear metric, *Doklad. Akad. Nauk USSR*, 1961, **139**(4), 775–8.

[4] On subspaces of a Hilbert space with an indefinite metric, *Nauch. zap. kafedr. mat., fiz. i estestvozH. Odessa Gos. ped. inst.*, 1961, **25**(2), 3–9.

Helton, J. W.

[1] Invariant subspaces of certain commuting families of operators on linear spaces with an indefinite inner product, *Doct. diss. Stanford Univ.*, 1968. (*Dissert. Abstrs.*, 1969, **29**(11), 42–67.)
[2] Unitary operators on a space with an indefinite inner-product *J. Funct. Anal.*, 1970, **6**(3), 412–40.
[3] Operators unitary in an indefinite metric and linear-fractional transformations, *Acta Sci. Math.*, 1971, **32**(3), 261–6.

Hestenes, M. R.

[1] Applications of the theory of quadratic forms in Hilbert space to the calculus of variations, *Pacific J. Math.*, 1951, **1**, 525–81.

Iokhvidov, E. I.

[1] On maximal *J*-dissipative operator (Russ.), *Sbornik studentch. nauch. rabot*, ed. S (*estestvennye nauki*), Voronezh, 1972, 74–9.
[2] On extensions of isometric operators in a Hilbert space with an indefinite metric (Russ.), *Trudy N.I.I. Mat. Voronezh Univ.*, 1974, ed.4, 21–31.
[3] On the properties of one linear-fractional operator transformation (Russ.), *Uspekhi Mat. Nauk*, 1975, **30**(4), 243–4.
[4] On *J*-unitary extensions of *J*-isometric operators with equal and finite deficiency numbers (Russ.), *Trudy N.I.I. Mat. Voronezh Univ.*, 1975, **17**, 34–41.
[5] Description of *J*-unitary extensions of *J*-isometric operators with equal and finite deficiency numbers (Russ.), *Doklady Akad. Nauk USSR*, 1976, **228**(4), 783–6.
[6] On closures and boundedness of linear operators of certain classes (Russ.), *Uspekhi Mat. Nauk*, 1979, **34**(5), 217–18.
[7] A criterion for the closability and boundedness of linear operators of certain classes (Russ.). In the book: *Funktional'nyy analiz*, Ul'yanousk, 1980, ed. **15**, 90–96.
[8] On linear operators which are *J*-non-expansive simultaneously with their adjoints. (Russ.). Dep. VINITI, 1983, No. 3284–83.
[9] On maximal *J*-isometric operators (Russ.). In the book: *VIII shkola po teopii operatorov v funktsional'nykh prostranstbakh, tezicy dokladov*, Riga, 1983, 104–05.

Iokhvidov, E. I., and Iokhvidov, I. S.

[1] On global inversion of a linear-fractional operator transformation (Russ.), *Trudy N.I.I. Mat. Voronezh Univ.*, 1973, ed. II, 64–70.

Iokhdvidov, I. S.

[1] Unitary operators in a space with an indefinite metric (Russ), *Zap, N.I.I. Mat. i Mekh. Khar'kov Gas. Univ. Mat. Obsch*, 1949, No. 21, 79–86.
[2] On the spectra of Hermitian and unitary operators in a space with an indefinite metric (Russ.), *Doklady Akad. Nauk USSR*, 1950, **71**(2).
[3] Unitary and selfadjoint operators in a space with an indefinite metric (Russ.), Dissertation, Odessa, 1950.
[4] Linear operators in spaces with an indefinite metric (Russ.), *Trudy 3rd Vsesayuzn. Mat. Sezda*, 1958, **3**, 254–61.
[5] On the boundedness of *J*-isometric operators (Russ.), *Uspekhi Mat. Nauk*, 1961, **16**(4), 167–70.
[6] On some classes of operators in a space with a general indefinite metric (Russ.). In

the book: *Funktional'nyy analiz i ego primenenie, Baku, Akad. Nauk AzSSR*, 1961, 90–5.

[7] Singular lineals in spaces with an arbitrary Hermitian-bilinear metric (Russ.), *Uspekhi Mat. Nauk*, 1962, **17** (4), 127–33.

[8] On operators with completely continuous iterations (Russ.), *Akad. Nauk USSR*, 1963, **153**(2), 258–61.

[9] A new characterization of $\Pi_\varkappa$-spaces, *Tezisy nauch. konf. Odessa inzh.-stroit Instit.*, Odessa, 1964, p. 173 (Russ.).

[10] On a lemma of K. Fan generalizing A. N. Tikhonov's principle of fixed points (Russ.), *Doklady Akad. Nauk USSR*, 1964, **159**, 501–04.

[11] On singular lineals in $\Pi_\varkappa$-spaces (Russ.; English summary), *Ukrainsk. Mat. Zhurnal*, 1964, **16**(3), 300–08.

[12] G-isometric and *J*-semi-unitary operators in a Hilbert space, (Russ.), *Uspekhi Mat. Nauk*, 1965, **20** (3), 175–81.

[13] On maximal definite lineals in a Hilbert space with a *G*-metric (Russ.), *Ukrainsk. Mat. Zhurnal*, 1965, **17** (4).

[14] Linear-fractional transformations of *J*-non-expansive operators (Russ.), *Doklady Akad. Nauk, Arm. SSR*, 1966, **42**(1), 3–8.

[15] Unitary extensions of isometric operators in the Pontryagin space Π_1 and extensions in the class $\mathscr{P}_1$ of finite sequences of the class $\mathscr{P}_{1,n}$ (Russ.), *Doklady Akad. Nauk USSR*, 1967, **173**(4).

[16] On the spectral trajectories generated by unitary extensions of isometric operators of a translation in the finite-dimensional space Π_1 (Russ.), *Doklady Akad. Nauk USSR*, 1967, **173**(5), 1002–5.

[17] On Banach spaces with a *J*-metric and certain classes of linear operators in these spaces (Russ.), *Izvestija Akad. Nauk MSSR*, series fiz. tekh u met. nauk, 1968, No. **1**, 60–80.

[18] On a class of linear-fractional operator transformations (Russ.). In the book: *Sbornik statey po funktsional'nym prostranstvam i operatornym uravnenigam*, Voronezh, 1970, 18–44.

[19] On spaces of finite sequences with a non-degenerate Hermitian form (Russ.), *Trudy N.I.I. Mat. Voronezh Univ.*, 1975, ed. 20, 30–8.

[20] On certain new ideas and results in the theory of linear operators in $\Pi_\varkappa$-spaces (Russ.). In the book: *Teoriya operatornykh urovneniy*, Voronezh: Izdat. Veronezh Univ., 1979, 33–44.

[21] Regular and projectionally complete lineals in spaces with a general Hermitian-bilinear metric (Russ.), *Doklady Akad. Nauk USSR*, 1961, **139**(4), 791–4.

Iokhvidov, I. S., and Ektov, Yu. S.

[1] Integral *J*-non-negative operators and weighted integral equations (Russ.), *Doklady Akad. Nauk Ukr. SSR*. 1981. No. 6, 15–19.

[2] Equivalence of three definitions of Hermitian-non-negative kernels and a functional realization of the Hilbert spaces generated by them (Russ.), Dep. VINITI, 1983, No. 2834–83.

Jonas, P.

[1] A condition for the existence of an eigen-spectral function for certain automorphisms of locally convex spaces (Ger.), *Math. Nachr.*, 1970, **45**, 143–60.

[2] On the maintenance of the stability of *J*-positive operators under *J*-positive and negative perturbations (Ger.), *Math. Nachr*, 1975, **65**.

[3] On the existence of eigen-spectral functions for J-positive operators, I (Ger.), *Math. Nachr.*, 1978, **82**, 241–54.

[4] On the functional calculus and the spectral function for definitizable operators in Krein space, *Belträge Anal.*, 1981, **16**, 121–35.
[5] Compact perturbations of definitiable operators, II, *J. Operator Theory*, 1882, **8**, 3–18.
[6] On spectral functions of definitizable operators, Banach Centre Publications, 1982, **8**, 301–11.
[7] On a class of *J*-unitary operators in Krein space, Preprint Akad. der Wiss. der DDR, Inst. für Mathematik, Berlin, 1983.

Jonas, P. and Langer, H.

[1] Compact perturbations of definitizable operators, *J. Operator Theory*, 1979, **2**, 63–77.
[2] Some questions on the perturbation theory of *J*-non-negative operators in Krein spaces, *Math. Nachr.*, 1983, **114**, 205–26.

Khastskevich, V. A.

[1] On some geometrical and topological properties of normed spaces with an indefinite metric (Russ.), *Izvestiya Vuzov, Matematika*, 1973.
[2] Remarks on dually plus-operators in a Hilbert *J*-space (Russ.). In the book: *Funktsional'nyy analiz i ego prilozheniya*, ed. **I**, Voronezh, 1973, 27–37.
[3] On plus-operators in a Hilbert *J*-space (Russ.), *Mat. Issledovaniya*, **9**, (2(32)), Krishinev, Akad. Nauk MSSR, 1974, 182–203.
[4] On *J*-semi-unitary operators in a Hilbert *J*-space (Russ.), *Mat. Zametki*, 1975, **17**(4), 639–47.
[5] On fixed points of generalized linear-fractional transformations (Russ.), *Izvestiya Akad. Nauk, USSR*, ser. Mat., 1975, **39**(5), 1130–41.
[6] On an application of the principle of compact mappings in the theory of operators in a space with an indefinite metric (Russ.), *Funktsion. analiz i ego pril.*, 1978, **12**(1), 88–9.
[7] Focusing and strictly plus-operators in Hilbert space (Russ.), Dep. VINITI, 1980, No. 1343–80.
[8] On the symmetry of properties of a plus-operator and its adjoint (Russ.), *Funktsion. analiz*, ed. **14**, Ul'yanovsk, 1980, 177–186.
[9] On strict plus-operators (Russ.), Dep. VINITI, 1980, No. 2440–80.
[10] Uniformly definite invariant dual pairs of plus-operators with a quasi-focusing power (Russ.), Dep. VINITI, 1981, No. 1917–81.
[11] On invariant subspaces of focusing plus-operators (Russ.), *Mat. Zametki*, 1981, **30**(5), 695–702.
[12] Invariant subspaces of plus-operators with a focusing power (Russ.), Dep. VINITI, XV Voronezh winter Mat. School.
[13] A generalized Poincaré metric on an operator ball (Russ.), *Funktsion. analiz i ego pril.*, 1983, **17** (4), 93–94.
[14] On a generalization of the Poincaré metric for a unit operator ball (Russ.), Dep. VINITI, 1983, No. 6913–83.
[15] Invariant subspaces and properties of the spectrum of plus-operators with a quasi-focusing power (Russ.), *Funktsian analiz i ego pril.*, 1984, **18**(1).

Kholevo, A. S.

[1] A generalization of Neyman's theory about the operator T^*T on spaces with an indefinite metric (Russ.), *Uchen. zapis Azerb. Univ. Ser. fiz-mat. nauk*, 1965, No. 2, 45–8.

Kopachevskiy, N. D.

[1] Theory of small osicallations of a liquid taking surface tension and rotation into account, Dissertation, Khar'kov, 1980 (Russ.).
[2] On the properties of basicity of a system of eigenvectors and associated vectors of a selfadjoint operator bundle $I - \lambda A - \lambda^{-1}B$, (Russ.), *Funktsion. analiz i ego pril.*, 1981, **15**(2).
[3] On the *p*-basicity of a system of root vectors of a selfadjoint operator bundle $I - \lambda A - \lambda^{-1}B$ (Russ.). In the book: *Funktsion. analiz i prikladnaya matematika*, Kiev, 1982, 55–70.

Kostguchenko, A. G. and Orazov, M. B.

[1] On some properties of the roots of a selfadjoint quadratic bundle (Russ.), *Funktsion. analiz i ego pril.*, 1975, **9**(4), 28–40.

Krasnosel'skiy, M. A. and Sobolev. A. V.

[1] On cones of finite rank (Russ.), *Doklady Akad. Nauk USSR*, 1975, **225**(6).

Krien, M. G.

[1] On weighted integral equations the distribution functions of which are not monotonic (Russ.), In the memorial volume to D. A. Grave, Moscow, 1940, 88–103.
[2] On linear completely continuous operators in functional spaces with two norms (Ukr.), *Zhurnal Akad. Nauk Ukr. RSR*, 1947, **9**, 104–29.
[3] Helical curves in an infinite-dimensional Lobachevskiy space and Lorents transformation (Russ.), *Uspekhi Mat. Nauk*, 1948, **3** (3). (*Amer. Math. Soc. Transl.* (2), **1**, 1955, 27–35.)
[4] On an application of the fixed-point principle in the theory of linear transformations with an indefinite metric (Russ.), *Uspekhi Mat. Nauk*. 1950. **5**(2) 180–90.
[5] A new application of the fixed-point principle in the theory of linear operators in a space with an indefinite metric (Russ.), *Doklady Akad. Nauk. USSR*, 1964, **154**(5), 1026–26. English translation: *Soviet Math. Doklady*, 5 (1964), 224–7.
[6] On the theory of weighted integral equations (Russ.), *Izvestiya Akad. Nauk. MSSR*, 1965 (7), 40–6.

Krein, M. G. and Langer, G. K.

[1] On the spectral function of a self-adjoint operator in a space with an indefinite metric (Russ.), *Doklady Akad. Nauk USSR*, 1963, **152**, 39–42. English translation: *Soviet Math. Doklady* 4, 1963, 1236–9.
[2] On some mathematical principles of the linear theory of damped oscillations of continua (Russ.). In the book: *Trudy mezhdunarod. simposiuma po prilozheniyam meopii funktsiy u mechanike splashnoy sredy*, vol. 2. Moscow, Nauka, 1966, 283–322.
[3] On definite subspaces and gneralized resolvents of an Hermitian operator in a $\Pi_\varkappa$-space (Russ.), *Funktsion. analiz i ego pril.*, 1971, **5** (2), 59–71; 1971, **5**(3), 54–69.
[4] On the generalized resolvents and the characteristic function of an isometric operator in $\Pi_\varkappa$-spaces (Ger.), *Colloquia Math. Soc. Janos Bolyai*, vol. 5; Hilbert space operators and operator algebras, Amsterdam, London, North-Holland, 1972, 353–99.

[5] On the Q-function of a π-hermitian operator in Π_κ-spaces (Ger.), *Acta Sci. Math.*, 1973, **34**, 191–230.
[6] On some extension problems closely connected with the theory of Hermitian operators in Π_κ-spaces, I: Some classes of functions and their representations (Ger.), *Math. Nachr.*, 1977, **77**, 187–236; II. Generalized resolvents, u-resolvents and entire operators (Ger.), *J. Funct. Analysis*, 1978, **30**, 3, 290–447.
[7] On some extension problems closely connected with the theory of Hermitian operators in a Π_κ-space, III: Indefinite analogue of the Hamburger and Strieltjes moment problems, *Beiträge zur Analysis*, 1979, **14**, 25–40; 1981, **15**, 27–45.

Krein, M. G. and Rutman, M. A.

[1] Linear operators leaving invariant a cone in a Banach space (Russ.), *Uspekhi Mat. Nauk*, 1948, **3** (1), 3–93; *Amer. Math. Soc. Transl.* 1950, 26.

Krein, M. G. and Shmul'yan, Yu. L.

[1] On a class of operators in a space with an indifinite metric (Russ.), *Doklady Akad. Nauk USSR*, 1966, **170**(1), 34–37; *Soviet Math. Dokl.* 1966, **7**, 1137–1141.
[2] On plus-operators in a space with an indefinite metric (Russ.), *Mat. Issledovaniya*, 1966, **1** (1), 131–61 Kishinev; *Amer. Math. Soc. Transl.*, **2**(8), 1969, 93–113.
[3] J-polar representation of plus-operators (Russ.), *Mat. Issledovan*, 1966. **1** (2), 172–210, Kishinev; *Amer. Math. Soc. Transl.* **2**(85), 1969, 115–43
[4] On linear-fractional transformations with operator coefficients (Russ.), *Mat. Issledovaniya*, 1967, **2**(3) 64–96, Kishinev.
[5] On stable plus-operators in J-spaces (Russ.). In the book: *Lineynye operatory*, Kishinev, Shtiintsa, 1980, 67–83.

Krein, S. G.

[1] Oscillations of a viscous fluid in a container (Russ.), *Doklady Akad. Nauk, USSR*, 1964, **159**(2), 262–5.

Kühne, R.

[1] On a class of J-self-adjoint operators (Ger.), *Math. Annal.*, (1964), **154**(1), 56–69.

Kuzhel, A. V.

[1] Spectral analysis of quasi-unitary operators of first rank in a space with an indefinite metric (Ukr.), *Dopovidi Akad. Nauk Ukr. RSR*, 1962, **5**, 572–4.
[2] Spectral decomposition of quasi-unitary operators of arbitrary rank in a space with an indefinite metric (Ukr.), *Dopovidi Akad. Nauk Ukr. RSR*, 1963, **4**, 430–3. (Russ. and English summaries)
[3] Spectral analysis of bounded non-selfadjoint operators in a space with an indefinite metric (Russ.), *Doklady Akad. Nauk USSR*, 1963, **151**, 772–4.
[4] On one case of the existence of invariant subspaces of quasi-unitary operators in a space with an indefinite metric (Ukr.), *Dopovidi Akad. Nauk Ukr RSR*, 1966, **5**, 583–585.
[5] Spectral analysis of quasi-unitary operators in a space with an indefinite metric (Russ.). In the book: *Teoriga funktsiy, funktsronal'nyy analiz i ikh prilozheniya*, 4, Khar'kov, 1967.
[6] Spectral analysis of unbounded non-selfadjoint operators in a space with an indefinite metric (Russ.), *Doklady Akad. Nauk USSR*, 1968, **178**, 31–3.
[7] Regular extensions of Hermitian operators in a space with an indefinite metric (Russ.), *Doklady Akad. Nauk USSR*, 1982, **265** (5).

[8] Selfadjoint and J-selfadjoint dilatations of linear operators (Russ.). In the book: *Teoriya funktsiy, funktsional'nyy analiz i ikh pril.*, 37, Khar'kov, 1982, 54–62.

[9] *J*-selfadjoint and *J*-unitary dilatations of linear operators (Russ.), *Funktsion. analiz i ego pril.*, 1983, **17** (1), 75–6.

Langer, H.

[1] On *J*-Hermitian operators (Russ.), *Doklady Akad. Nauk USSR*, **134**(2), 263–6, 1960.

[2] On the spectral theory of *J*-selfadjoint operators (Ger.), *Math. Ann.* 1962, **146**(1), 60–85.

[3] A generalization of a theorem of L. S. Pontryagin (Ger.), *Math. Ann.* 1963, **152**(5), 434–6.

[4] Spectral theory of linear operators in *J*-spaces and some applications to the bundle $L(\lambda) = \lambda^2 I + \lambda B + C$, *Habilitationschrift*, Dresden, 1965 (Ger.).

[5] Spectral functions of a class of *J*-selfadjoint operators (Ger.), *Math. Nachr.* 1967, **33**(1–2), 107–20.

[6] On strongly damped bundles in Hilbert space (Ger.), *J. Math. Mech*, 1968, **17**, N7, 685–705.

[7] On maximal dual pairs of invariant subspaces of *J*-selfadjoint operators (Russ.), *Mat. Zametki*, 1970, **7** (4), 443–7.

[8] Generalized resolvents of a *J*-non-negative operator with finite deficiency (Ger.), *J. Funct. Analysis*, 1971, **8**, 287–320.

[9] Invariant subspaces of definitizable *J*-selfadjoint operators (Ger.), *Ann. Acad. Sci. Fenn.*, 1971, Ser. A1, No. 475.

[10] On a class of polynomial bundles of self-adjoint operators in Hilbert space (Ger.), *J. Funct. Analysis*, 1973, **12**(1), 13–29; 1974, **16**, 921–34.

[11] Invariant subspaces for a class of operators in spaces with indefinite metric, *J. Funct. Analysis*, 1975, **19**, (3), 232–41.

[12] On the spectral theory of polynomial bundles of self-adjoint operators, (Ger.), *Math. Nachr.* 1975. **65**, 301–19.

[13] On invariant subspaces of linear operators acting in a space with an indefinite metric (Russ.), *Doklady Akad Nauk, USSR*, 1966, **169**, 12–15.

Langer, H. and Nayman, B.

[1] Perturbation theory for definitizable operators in Krein spaces. *J. Operator Theory*, 1983, **9**, 297–317.

Langer, H. and Sorjonen, P.

[1] Generalized resolvents of Hermitian and isometric operators in a Pontryagin sapce, *Ann. Acad. Sci. Fenn.*, 1974, Ser. AI. No. 561 (Ger.).

Larionov, V. A.

[1] Extensions of dual subspaces (Russ.), *Doklady Akad. Nauk, USSR*, 1967, **176**(3), 515–17.

[2] On a commutative family of operators in a space with an indefinite metric (Russ.), *Mar. Zametki*, 1967, **1**(5), 589–94.

[3] Extensions of dual subspaces invariant relative to an algebra (Russ.), *Mat. Zametki*, 1968, **3**(3), 253–60.

[4] On the number of fixed points of a linear-fractional transformation of an operator ball on to itself (Russ.), *Mat. Sbornik*, 1969, **78**(2), 202–13.

[5] On self-adjoint quadratic bundles (Russ.), *Izvestiya Akad. Nauk, USSR, Ser. Matem.*, 1969, **33**(1), 138–54.

[6] On bases composed of eigenvectors of an operator bundle (Russ.), *Doklady Akad. Nauk USSR*, 1972, **206**(2), 283–86.
[7] Localization of the spectrum and dual completeness of the normal motions in the problem of motion of a viscous liquid subject to surface-tension forces (Russ.), *Doklady Akad. Nauk USSR*, 1974, **217**(3), 522–25.
[8] On the principles of localization of frequencies and the principle of completeness and dual completeness of the root functions in problems of the linear theory of vibrations in continuous media (Russ.), *Trudy TsNII building construction*, 1974, **4**, 11–20.
[9] On the theory of linear operators acting in a space with an indefinite metric (Russ.). In the book: *Trudy letney shkoly po spectral'noy teorii operatorov i predstavleniy grupp*, 1968. Baku: Elm, 1975, 140–48.

Louhivaara, I. S.

[1] Remark on the theory of Nevanlinna spaces (Ger.), *Ann. Acad. Sci. Fenn.*, 1958, Ser. AI, No. 232.
[2] On the theory of subspaces in linear spaces with indefinite metric, (Ger.), *Ann. Acad. Sci. Fenn.*, 1958, Ser. AI, No. 252.
[3] On various metrics in linear spaces (Ger.), *Ann. Acad. Sci. Fenn*, 1960, Ser. AI, No. 282.

Markin, S. P.

[1] On the separability of linear spaces with a sesquilinear form (Russ.), *Trudy Mat. Fak. Voronezh Univ.*, 1973, **10**, 107–12.

Masuda, K.

[1] On the existence of invariant subspaces in spaces with indefinite metric. *Proc. Amer. Math. Soc.*, 1972, **32**(2) 440–4.

McEnnis, B. W.

[1] Fundamental reducibility of selfadjoint operators on Krein space, *J. Operator Theory*, 1982, **8**, (2), 219–25.

Naymark, M. A.

[1] On commutative unitary operators in $\Pi_\varkappa$-space (Russ.), *Doklady Acad. Nauk USSR*, 1963, **149**(6), 1261–3.
[2] On commuting unitary operators in a space with an indefinite metric *Acta Sci. Math.*, 1963, **24**(3–4), 177–89.
[3] An analogue of Stone's theorem in a space with an indefinite metric (Russ.), *Doklady Nkad. Nauk USSR*, 1966, **170**(6), 1259–61; *Soviet Math. Doklady*, 1966, 7, 1366–8.

Nevanlinna, R.

[1] Extensions of the theory of Hilbert space (Ger.), *Comm. Sen. Math. Univ. Lund. tome supplémentaire*, 1952, 160–8.
[2] On metric linear spaces, II: bilinear forms and continuity (Ger.), *Ann. Acad. Sci. Fenn.*, 1952, Ser. AI, No. 113.
[3] On metric linear spaces, III; theory of orthogonal systems (Ger.), *Ann. Acad. Sci. Fenn.*, 1952, Ser. AI, No. 115.
[4] On metric linear spaces, IV: on the theory of subspaces (Ger.), *Ann. Acad. Sci. Fenn.*, 1954, Ser. AI, No. 163.

[5] On metric linear spaces, V: relations between different metrics (Ger.), *Ann. Acad. Sci. Fenn.*, 1956, Ser. AI, No 22.

Noel, G.

[1] Strongly *J*-normal operators in a *J*-space (Fr.), *Bull. Sci. Acad. Roy. Belg.*, 1965, **51**(5), 570–85.

Ovchinnikov, V. I.

[1] On the decomposition of spaces with indefinite metric (Russ.), *Mat. Issled.*, 1968, **3**, 4, 175–7, Kishinev.

Pesonen, E.

[1] On the spectral representation of quadratic forms in linear spaces with indefinite metric (Ger.), *Ann. Acad. Sci. Fenn.*, 1956, Ser. AI, No. 227.

Phillips R.

[1] Dissipative operators and hyperbolic systems of partial differential equations, *Trans. Amer. Math. Soc.*, 1959, **90**(2), 193–254.
[2] Dissipative operators and parabolic partial differential equations, *Communs. Pure and Applied Math.*, 1959, **12**(2), 249–76.
[3] The extension of dual subspaces invariant under an algebra, *Proc. Internat. Sympos. Linear Spares*, Jerusalem, 1960. Pergamon Press, 1961, pp. 366–98.

Pontryagin, L. S.

[1] Hermitian operator in spaces with indefinite metric (Russ.), *Izvestiya Akad. Nauk USSR, Ser. Matem*, 1944, **8**, 243–80.

Potapov, V. D.

[1] The multiplicative structure of *J*-contractive matrix-functions (Russ.). *Trudy Moskov. Mat. Obshch.*, 1955, **4**, 125–236; *Amer. Math. Soc.* Transl. (2), **15**, 1960, 131–243.

Prigorskiy, V. A.

[1] On some classes of bases of a Hilbert space (Russ.), *Uspekhi Mat. Nauk*, 1965, **20**(5), 231–6.

Ritsner, V. S.

[1] Extension of *J*-Hermitian operators in *J*-spaces (Russ.). In the book: *Issledovanie roboty kleenykh derevyannykh konstruktsiy*. Khabarovsk. Kh. P. I., 1975, 186–95.
[2] Generalized resolvents of a non-densely defined accretive operator in a Hilbert space with indefinite metric, (Russ.). In the book: *Materialy nauchotekhnicheskoy konferentsii molodykh uchenykh*, Khabarovsk, 1977, 128–9.
[3] Extension of dual pairs of subspaces (Russ.), *Funktsion, analiz, Ul'yanovsk*, 1980, **15**, 135–9.
[4] Theory of linear relations (Russ.), Dep. VINITI, 1982, No. 846–82.

Sakhnovich, L. A.

[1] On a *J*-unitary dilatation of a bounded operator (Russ.), *Funktsion. analiz i ego pril.*, 1974, **8**(8), 83–4.

Savage, L. J.

[1] The application of vectorial methods to metric geometry, *Duke Math. J.*, 1946, **13**, 521–28.

Scheibe, E.

[1] On Hermitian forms in topological vector-spaces, (Ger.), I. *Ann. Acad. Sci. Fenn.*, 1960, Ser. AI, No. 294.

Shmul'yan, Yu. L.

[1] On division in the class of *J*-non-expansive operators, (Russ.), *Mat. Sbornik*, 1967, **74**, 516–25.
[2] On linear-fractional transformations with operator coefficients and operator balls (Russ.), *Mat. Sbornik*, 1968, **77** (3).
[3] On *J*-non-expansive operators operators in *J*-spaces (Russ.), *Ukrainsk. Mat. Zhurn.*, 1968, **20** (3), 352–62.
[4] Theory of extensions of operators and space with an indefinite metric (Russ.), *Izvestiya Akad. Nauk USSR*, Ser. Mat., 1974, **38**(4).
[5] Theory of linear relations and space with an indefinite metric (Russ.), *Funktsion. analiz i ego pril.*, 1976, **10**(1).
[6] *Generalized linear-fractional transformations of operator balls* (Russ.), *Siber. Mat. Zhurn*, 1978, **19**(2) 418–25.
[7] On transformers of linear relations in *J*-spaces (Russ.), *Funktsion. analiz i ego pril.*, 1980, **14**(2) 39–44.
[8] Generalized linear-fractional tranformations of operators balls (Russ.), *Siber. Mat. Zhurn.*, 1980, **21**(5), 114–31.

Shtraus, A. V.

[1] Extensions and generalized resoluents of a non-densely defined symmetric operator (Russ.), *Izvestiya Akad. Nauk USSR*, Ser. Mat., 1970, **34**(1).

Shtrauss, V. A.

[1] The Gram operator and *G*-orthonormalized systems in a Hilbert space, *Sbornik trad. aspirantov, Mat. Fac. Voronezh Univ.*, 1971, **1**, 85–90.
[2] *G*-orthonormalized systems and bases in a Hilbert space (Russ.), *Izvestiya Vuzov. Matematika*, 1973, No. 9. 108–17.
[3] On the theory of selfadjoint operators in Banach spaces with an Hermitian form (Russ.), *Siber. Mat. Zhurnal*, 1978, **19**(3).
[4] On the integral representation of a *J*-selfadjoint operator in $\Pi_\varkappa$ (Russ.), *Sbornik nauch. trudov No. 252: prikladnaya matematika. Chelyabinsk*, 1980, 114–18.

Shtraus, V. A. and Ektov, Yu. S.

[1] An analogue of Dini's theorem and *J*-non-negative operators (Russ.), *Funktsion. analiz Ul'yanovsk*, 1980, **15**, 188–91.

Shul'man, V. S.

[1] On fixed points of linear-fractional transfomations (Russ.), *Funktsion. analiz i ego pril.*, 1980, **14** (2), 93–4.

Sobolev, A. V. and Khatskevich, V. A.

[1] Invariant subspaces and properties of the spectrum of a focusing plus-operator (Russ.), Rukopis dep. VINITI, 1980, No. 1344–80.

[2] On definite invariant subspaces and the structure of the spectrum of a focusing plus-operator (Russ.), *Funktsion. analiz i ego pril.*, 1981, **15**(1), 84–5.

Sobolev, S. L.

[1] Motion of a symmetric top with a cavity filled with liquid (Russ.), *ZhPMTF.* 1960, **3**, 20–55.

Spitkovskiy, I. M.

[1] On the block structure of *J*-unitary operators (Russ.). In the book: *Teoriya funktsiy, funktsional'nyy analiz i ikh prilozheniya*, Kharkov, 1978, 129–38.

Usvyatsova, E. B.

[1] On the spectral theory of *G*-isometric operators (Russ.), *Mat. Fak. Voronezh Univ.*, 1972, 66–9.

[2] On the structure of root subspaces of J-dissipative operators (Russ.). In the book: *Melody resheniya operatornykh uravneniy*, Voronezh, 1978, 158–60.

Wittstock, G.

[1] On invariant subspaces of positive transformations in spaces with indefinite metric (Ger.), *Math. Ann*, **172**(3), 167–75.

INDEX